CAMBRIDGE STUDIES IN
ADVANCED MATHEMATICS 24

Stochastic flows and stochastic differential equations

Already published

1 W. M. L. Holcombe *Algebraic automata theory*
2 K. Petersen *Ergodic theory*
3 P. T. Johnstone *Stone spaces*
4 W. H. Schikhof *Ultrametric calculus*
5 J-P. Kahane *Some random series of functions*, 2nd edn
6 H. Cohn *Introduction to the construction of class fields*
7 J. Lambek & P. J. Scott *Introduction to higher order categorical logic*
8 H. Matsumura *Commutative ring theory*
9 C. B. Thomas *Characteristic classes and the cohomology of finite groups*
10 M. Aschbacher *Finite group theory*
11 J. L. Alperin *Local representation theory*
12 Paul Koosis *The logarithmic integral* I
13 A. Pietsch *Eigenvalues and s-numbers*
14 S. J. Patterson *An introduction to the theory of the Riemann zeta-function*
15 H. J. Baues *Algebraic homotopy*
16 V. S. Varadarajan *Introduction to harmonic analysis on semisimple Lie groups*
17 W. Dicks & M. J. Dunwoody *Groups acting on graphs*
18 L. J. Corwin & F. P. Greenleaf *Representations of nilpotent Lie groups and their applications*
19 R. Fritsch & R. Piccinini *Cellular structures in topology*
20 H. Klingen *Introductory lectures on Siegel modular form*
21 P. Koosis *The logarithmic integral* II
22 M. J. Collins *Representations and characters of finite groups*

Stochastic flows and stochastic differential equations

HIROSHI KUNITA

Professor, Department of Applied Science, Kyushu University

CAMBRIDGE
UNIVERSITY PRESS

PUBLISHED BY THE PRESS SYNDICATE OF THE UNIVERSITY OF CAMBRIDGE
The Pitt Building, Trumpington Street, Cambridge CB2 1RP, United Kingdom

CAMBRIDGE UNIVERSITY PRESS
The Edinburgh Building, Cambridge CB2 2RU, United Kingdom
40 West 20th Street, New York, NY 10011–4211, USA
10 Stamford Road, Oakleigh, Melbourne 3166, Australia

First published 1990
First paperback edition 1997

A catalogue record for this book is available from the British Library

Library of Congress Cataloguing in Publication data

Kunita, H.
Stochastic flows and stochastic differential equations / Hiroshi Kunita.
p. cm. – (Cambridge studies in advanced mathematics: 24)
Includes bibliographical references.
ISBN 0 521 35050 6
1. Stochastic analysis. 2. Flows (Differentiable dynamical systems)
3. Stochastic differential equations. I. Title. II. Series.
QA274.2.K86 1990
519.2–dc20 89-70813 CIP

ISBN 0 521 35050 6 hardback
ISBN 0 521 59925 3 paperback

Transferred to digital printing 2002

To my parents

Contents

Preface

The main purpose of this book is to develop the relationship between stochastic flows of diffeomorphisms and stochastic differential equations and through the latter, to study the properties of stochastic flows. The theory of stochastic differential equations was initiated by K. Itô in 1942. Since then the theory has been developed in various directions. An important one is the application to the study of diffusion processes associated with certain second order partial differential operators. Another direction is to study the equation itself, regarding it as the dynamical system disturbed by a noise. In this book we will mainly adopt the latter point of view, though we will sometimes return to the former problem.

Itô's stochastic differential equation can be written as

$$d\varphi_t = f_0(\varphi_t, t)\, dt + \sum_{k=1}^{m} f_k(\varphi_t, t)\, dB_t^k, \tag{1}$$

where $(B_t^1, \ldots, B_t^m)$ is a standard Brownian motion. The last term can be regarded as a noise or a random disturbance adjoined to the ordinary differential equation $d\varphi_t = f_0(\varphi_t, t)\, dt$. Here the sample paths B_t^k are not functions of bounded variations with respect to t a.s., so that dB_t^k cannot be defined as the Stieltjes integral. Nevertheless the last integrals are well defined for almost all samples if φ_t is adapted, i.e. for any t, φ_t is independent of the future Brownian motion $B_u - B_t$, $u \geq t$. The integral is called the Itô integral. Itô showed that if $f_i(x, t)$, $i = 0, \ldots, m$ are Lipschitz continuous with respect to x then for any initial state (s, x), the equation has a unique (except for measure 0) solution starting from x at time s, which we denote by $\varphi_{s,t}(x)$. Then, like the case of solutions of an ordinary differential equation, it is reasonable to ask whether the solutions $\varphi_{s,t}(x)$ define a *Brownian flow of diffeomorphisms*. That is to ask whether

(a) $\varphi_{s,t}(x)$ is continuous in s, t, x, a.s.
(b) The map $\varphi_{s,t}: \mathbb{R}^d \to \mathbb{R}^d$ is a diffeomorphism for any $s < t$ a.s.
(c) $\varphi_{s,u} = \varphi_{t,u}(\varphi_{s,t})$ for any $s < t < u$ a.s.
(d) $\varphi_{t_i, t_{i+1}}$, $i = 0, \ldots, n-1$ are independent for any $0 \leq t_0 < \cdots < t_n$.

The proof of this simple fact is not as easy as in the case of ordinary differential equations. It requires a lot of careful arguments about null sets, since the solution $\varphi_{s,t}$ is defined only outside a null set depending on the initial state (s, x).

Early work on this problem was done by Gihman [38, 39] and Blagoveščenskii–Freidlin [12]. They studied the regularity of the solutions with respect to the initial data. It was in the late 1970s that extensive attention was given to the diffeomorphic property of the solution. Contributions were made by Elworthy [29], Bismut [10], Ikeda–Watanabe [49], Kunita [74] and others.

On the other hand, independently of these works, Harris [44] studied the probabilistic aspect of Brownian flows. He showed that the infinitesimal mean and covariance of a Brownian flow determine its law in some special class of Brownian flows. Then, Baxendale [5] and Le Jan [86] considered the problem of whether a Brownian flow is represented as a solution of an Itô's stochastic differential equation. The answer was no. It turned out that an infinite number of Brownian motions, or equivalently a Brownian motion with values in vector fields, is necessary for this problem. In other words, introducing a Brownian motion $F(x, t)$ with values in the space of vector fields (infinite dimensional Brownian motion), the flow is represented as the solution of a stochastic differential equation based on F:

$$\varphi_{s,t}(x) = x + \int_s^t F(\varphi_{s,r}(x), dr). \tag{2}$$

Itô's stochastic differential equation is a special case of the above by setting

$$F(x, t) = \int_0^t f_0(x, r)\,dr + \sum_{k=1}^m \int_0^t f_k(x, r)\,dB_r^k. \tag{3}$$

Thus in this book we study the stochastic differential equation of the form (2) rather than Itô's classical equation (1).

Chapter 1 contains some preliminary materials that are needed for later discussions. Brownian motions, martingales and Markov processes are introduced briefly. Then we discuss with details the ergodic property and the recurrent potential theory of a Markov process. In the last section, we discuss Kolmogorov's two criteria for the sample continuity of a random field and the tightness of a sequence of random fields.

Chapter 2 deals with stochastic integrals by continuous semimartingales and establishes Itô's formula. Our approach is based on the quadratic variations of continuous semimartingales. Thus this chapter is a review of Itô's stochastic analysis.

In Chapter 3 we discuss continuous semimartingales with spatial parameters and develop a stochastic calculus based on them. The stochastic calculus of Chapter 2 is efficient for the analysis of the time parameter. However we also need the stochastic calculus of spatial parameter simultaneously for the study of the stochastic flows. Sections 3.1–3.3 are devoted to the stochastic analysis which is applicable both to spatial and time variables. Then in Section 3.4 we study the stochastic differential equation of the form (2).

Chapter 4 is the central part of this book: we discuss a variety of topics on stochastic flows. In the first part (Sections 4.1–4.4) we show that a given Brownian flow or a semimartingale flow is governed by a certain stochastic differential equation of the form (2), and as an application we discuss the asymptotic properties of image measures by a Brownian flow. In the second part (Sections 4.5–4.8), we show that solutions of a given stochastic differential equation define stochastic flows of diffeomorphisms or local diffeomorphisms, treating the case of Euclidean spaces and the case of manifolds separately. In the third part (Section 4.9), we establish some Itô formulas for stochastic flows and apply them to various problems.

Chapter 5 is devoted to limit theorems concerning stochastic flows. We intend to treat the following three types of limit theorems in a unified method.

(a) Approximation theorems of stochastic differential equations and stochastic flows, due to Bismut [10], Ikeda–Watanabe [49], Malliavin [92] and Dowell [27].
(b) Limit theorems for driven processes due to Papanicolaou–Stroock–Varadhan [109].
(c) Limit theorems for stochastic ordinary differential equations due to Khasminskii [68], Papanicolaou–Kohler [108] and others.

In Chapter 6, we study stochastic partial differential equations by applying the theory of stochastic flows. A first order parabolic stochastic partial differential equation can be solved using the stochastic characteristic curve, like the case of the first order deterministic partial differential equations. Then we show that a certain linear second order stochastic partial differential equation can be solved by taking the partial expectation (i.e. conditional expectation) of the solution of an associated first order linear stochastic partial differential equation. The application to the filtering theory will be discussed.

This book has grown out of two series of lectures given by the author. The first [77] was given at the summer school at St Flour in 1982 and the second

[79] at the Tata Institute of Fundamental Research in Bangalore in 1985. Series of talks given by the author in Tokyo in 1984 (Seminar on probability, Japan) and at Warwick in 1985 (Symposium on stochastic dynamical systems) were helpful in preparing the book. The author is grateful for having had these opportunities.

Hiroshi Kunita
Fukuoka, Japan

1

Stochastic processes and random fields

1.1 Preliminaries

Probability spaces and random variables

Let Ω be a set. A collection $\mathscr{F}$ of subsets of Ω is called a *σ-field* if it contains an empty set and is closed under the operations of countable unions and complements. The pair $(\Omega, \mathscr{F})$ is called a *measurable space*. Elements of Ω are called *samples* and those of $\mathscr{F}$ are called *events*. Let P be a σ-additive measure on $(\Omega, \mathscr{F})$. It is called a *probability* if $P(\Omega) = 1$. The triple $(\Omega, \mathscr{F}, P)$ is called a *probability space*.

A finite collection of events $\{A_1, \ldots, A_n\}$ is called *independent* if $P(\bigcap_{l=1}^k A_{i_l}) = \prod_{l=1}^k P(A_{i_l})$ holds for any subset $\{A_{i_1}, \ldots, A_{i_k}\}$ of $\{A_1, \ldots, A_n\}$. A collection of infinite events $\{A_\lambda : \lambda \in \Lambda\}$ is called independent if any finite subcollection of $\{A_\lambda : \lambda \in \Lambda\}$ is independent. Let $\mathscr{G}$ be a subset of $\mathscr{F}$. If $\mathscr{G}$ is a σ-field, it is called a *sub σ-field* of $\mathscr{F}$. Suppose now we are given a finite collection $\{\mathscr{F}_1, \ldots, \mathscr{F}_n\}$ of sub σ-fields of $\mathscr{F}$. It is called independent if for every choice $A_i \in \mathscr{F}_i$. $i = 1, \ldots, n$, the collection $\{A_1, \ldots, A_n\}$ is independent. An infinite collection $\{\mathscr{F}_\lambda : \lambda \in \Lambda\}$ of sub σ-fields of $\mathscr{F}$ is called independent if its arbitrary finite subcollection is independent.

A real valued measurable function $X(\omega)$ defined on $(\Omega, \mathscr{F})$ is called a *real random variable* or simply a *random variable*. The random variable $X(\omega)$ may take values $\pm\infty$, but we assume that $X(\omega)$ takes finite values for almost all ω unless otherwise mentioned. We often suppress the sample ω and write it as X. If we can define the integral of X by the measure P, we denote it by $E[X]$ and call it the *expectation* of X, i.e.,

$$E[X] = \int_\Omega X(\omega)\, \mathrm{d}P(\omega). \tag{1}$$

In later discussions we will often use the following notation

$$E[X : A] = \int_A X(\omega)\, \mathrm{d}P(\omega). \tag{2}$$

Let S be a complete separable metric space and $\mathscr{B}(S)$ be its topological Borel field. A measurable map X from $(\Omega, \mathscr{F})$ into $(S, \mathscr{B}(S))$ is called a *random variable with values in S* or *S-valued random variable*. If X is a random variable with values in S and B is an element of $\mathscr{B}(S)$, then the set $\{\omega : X(\omega) \in B\}$ belongs to $\mathscr{F}$. For simplicity this set is denoted by $\{X \in B\}$ or $X \in B$. Now the collection of the sets $\{X \in B : B \in \mathscr{B}(S)\}$ is a sub σ-field of $\mathscr{F}$. It is called the *σ-field generated by the random variable* X and is denoted by $\sigma(X)$. Let $\{X_\lambda : \lambda \in \Lambda\}$ be a collection of random variables. The smallest sub σ-field of $\mathscr{F}$ containing $\bigcup_{\lambda \in \Lambda} \sigma(X_\lambda)$ is denoted by $\sigma(X_\lambda : \lambda \in \Lambda)$ and is called the σ-field generated by $\{X_\lambda : \lambda \in \Lambda\}$.

Two families of random variables $\{X_\lambda : \lambda \in \Lambda\}$ and $\{X_\gamma : \gamma \in \Gamma\}$ are called independent if $\sigma(X_\lambda : \lambda \in \Lambda)$ and $\sigma(X_\gamma : \gamma \in \Gamma)$ are independent. The independence of infinite families of random variables is defined similarly.

An event A is called a null event or a null set if $P(A) = 0$ holds. If a proposition holds except for ω belonging to a certain null set, it is said to hold *almost everywhere* (abbreviated as a.e.) or *almost surely* (abbreviated as a.s.). As an example for two random variables X and Y, '$X = Y$ a.s.' means that $\{\omega : X(\omega) \neq Y(\omega)\}$ is a null set. We often do not distinguish these X and Y and write simply $X = Y$.

For a sequence $X_1, X_2, \ldots, X$ of real random variables, we introduce three types of convergence.

(a) $\{X_n\}$ is said to *converge to X almost everywhere* or *almost surely* if for almost all ω, $\{X_n(\omega)\}$ converges to $X(\omega)$.

(b) Let $p \geq 1$. Denote by L^p the totality of random variables Y such that $E[|Y|^p] < \infty$ and define the L^p-norm by $\|Y\|_p = E[|Y|^p]^{1/p}$. If $X_1, X_2, \ldots, X$ are in L^p and $\|X_n - X\|_p \to 0$ is satisfied, $\{X_n\}$ is said to *converge to X in L^p*.

(c) $\{X_n\}$ is said to *converge to X in probability* if for any $\varepsilon > 0$ $P(|X_n - X| > \varepsilon)$ converges to 0.

We give the well known relations on these three convergences without proofs.

Theorem 1.1.1 *The almost everywhere convergence implies the convergence in probability. The L^p-convergence implies the convergence in probability.* □

A collection of real random variables $\{X_\lambda\}$ is called *uniformly integrable* if

$$\sup_\lambda \int_{|X_\lambda| > c} |X_\lambda| \, dP \xrightarrow[c \to \infty]{} 0 \tag{3}$$

is satisfied. For a sequence of uniformly integrable random variables, the convergence in probability implies the convergence in L^1.

Theorem 1.1.2 *Let $\{X_n\}$ be a sequence of uniformly integrable random variables. If $\{X_n\}$ converges to X in probability, then it converges in L^1.* □

Let $P_n : n = 1, 2, \ldots, P$ be a sequence of probabilities on $(S, \mathscr{B}(S))$. The sequence $\{P_n\}$ is said to *converge weakly* to P if for any bounded continuous function f on S, $\{\int f \, dP_n\}$ converges to $\int f \, dP$. Now for an S-valued random variable X, we define its *law* by the probability P_X on $(S, \mathscr{B}(S))$ such that

$$P_X(B) = P(X \in B), \qquad \text{for all } B \in \mathscr{B}(S). \tag{4}$$

A sequence $\{X_n\}$ of S-valued random variables is then said to *converge weakly* if the corresponding sequence of the laws converges weakly.

We shall quote some basic properties of the weak convergence. Proofs of the following theorems (1.1.3–1.1.5) can be found in Billingsley [8] and Ikeda–Watanabe [49].

Theorem 1.1.3 *Let $\{P_n : n = 1, 2, \ldots, P\}$ be a sequence of probabilities on $(S, \mathscr{B}(S))$. The following statements are equivalent.*

(a) *$\{P_n\}$ converges to P weakly.*
(b) *$\overline{\lim}_{n\to\infty} P_n(F) \le P(F)$ holds for any closed subset F of S.*
(c) *$\underline{\lim}_{n\to\infty} P_n(G) \ge P(G)$ holds for any open subset G of S.* □

A family $\{P_\lambda : \lambda \in \Lambda\}$ of probabilities over $(S, \mathscr{B}(S))$ is called *relatively compact* if any subset of $\{P_\lambda : \lambda \in \Lambda\}$ contains a subsequence converging weakly. A useful criterion for the relative compactness of the measures is the tightness: a family $\{P_\lambda : \lambda \in \Lambda\}$ of probabilities is called *tight* (or *uniformly tight*) if for any $\varepsilon > 0$ there exists a compact subset K_ε of S such that $P_\lambda(K_\varepsilon) > 1 - \varepsilon$ holds for all $\lambda \in \Lambda$.

Theorem 1.1.4 *The family of probabilities $\{P_\lambda : \lambda \in \Lambda\}$ on $(S, \mathscr{B}(S))$ is relatively compact if and only if it is tight.* □

The following theorem, due to Skorohod, shows that the weak convergence and the strong convergence are equivalent if the corresponding random variables are defined on a suitable probability space.

Theorem 1.1.5 *Let $\{P_n\}$ be a sequence of probabilities on $(S, \mathscr{B}(S))$ converging weakly to P. Then on a suitable probability space $(\tilde{\Omega}, \tilde{\mathscr{F}}, \tilde{P})$ we can construct S-valued random variables $\tilde{X}_n$, $n = 1, 2, \ldots$ and $\tilde{X}$ satisfying the following properties.*

(a) *The laws of* $\tilde{X}_n$, $n = 1, 2, \ldots$, *and* $\tilde{X}$ coincide with P_n, $n = 1, 2, \ldots$ *and* P, *respectively.*
(b) $\{\tilde{X}_n\}$ *converges to* $\tilde{X}$ *almost everywhere.* □

Conditional expectations
Let $\mathscr{G}$ be a sub σ-field of $\mathscr{F}$ and let X be an integrable random variable. An integrable $\mathscr{G}$-measurable random variable $\hat{X}$ is called the *conditional expectation of X with respect to* $\mathscr{G}$ if $\hat{X}$ satisfies

$$\int_A X \, dP = \int_A \hat{X} \, dP, \qquad \text{for all } A \in \mathscr{G}. \tag{5}$$

The conditional expectation exists uniquely. We denote it by $E[X|\mathscr{G}]$.

Theorem 1.1.6 *Let* $X, Y, X_n, n = 1, 2, \ldots$ *be integrable random variables and* $\mathscr{G}, \mathscr{H}$ *be sub* σ*-fields of* $\mathscr{F}$.

(c.1) *If* a, b *are constants, then* $E[aX + bY|\mathscr{G}] = aE[X|\mathscr{G}] + bE[Y|\mathscr{G}]$ *a.s.*
(c.2) *If* X *is* $\mathscr{G}$*-measurable and* XY *is integrable, then* $E[XY|\mathscr{G}] = XE[Y|\mathscr{G}]$ *a.s.*
(c.3) *If* $\mathscr{H} \subset \mathscr{G}$, *then* $E[E[X|\mathscr{G}]|\mathscr{H}] = E[E[X|\mathscr{H}]|\mathscr{G}] = E[X|\mathscr{H}]$ *a.s.*
(c.4) (Jensen's inequality) *Let* $f(x)$ *be a convex function. If* $f(X)$ *is integrable, then* $f(E[X|\mathscr{G}]) \le E[f(X)|\mathscr{G}]$ *a.s. In particular, if* $|X|^p$ *is integrable for* $p \ge 1$, *then* $|E[X|\mathscr{G}]|^p \le E[|X|^p|\mathscr{G}]$ *a.s.*
(c.5) *Let* $p \ge 1$. *If* $\{X_n\}$ *converges to* X *in* L^p, *then* $\{E[X_n|\mathscr{G}]\}$ *converges to* $E[X|\mathscr{G}]$ *in* L^p.
(c.6) *If* $\{X_n\}$ *converges to* X *with respect to the weak topology of* L^p, *then* $\{E[X_n|\mathscr{G}]\}$ *converges to* $E[X|\mathscr{G}]$ *with respect to the weak topology of* L^p. □

For the proof, see Neveu [103].

The *conditional probability of the event A given the* σ*-field* $\mathscr{G}$ is defined by

$$P(A|\mathscr{G}) = E[\chi(A)|\mathscr{G}], \tag{6}$$

where $\chi(A)$ is the indicator function of the set A. Then it has these three properties:

(a) $0 \le P(A|\mathscr{G}) \le 1$ a.s.
(b) $P(\Omega|\mathscr{G}) = 1$, $P(\phi|\mathscr{G}) = 0$ a.s.
(c) for any pairwise disjoint sets $A_1, A_2, \ldots$, $P(\bigcup_{n=1}^{\infty} A_n|\mathscr{G}) = \sum_{n=1}^{\infty} P(A_n|\mathscr{G})$ a.s.

These are easily verified from the definition of the conditional probability.

Exercise 1.1.7 Let X and Y be independent random variables with values in complete separable metric spaces S and S', respectively. Let $g(x, y)$ be a real measurable function on $S \times S'$ such that $g(X, Y)$ is integrable. Show that

$$E[g(X, Y)|\sigma(Y)] = \int g(x, Y)P_X(\mathrm{d}x) \qquad \text{a.s.}$$

where P_X is the law of X.

1.2 Stochastic processes

Brownian motions

A collection of random variables X_t, $t \in \mathbb{T}$ with values in a complete separable metric space S where $\mathbb{T}$ is a time set is called a *stochastic process with state space S*. If $\mathbb{T}$ is an interval, it is called a *stochastic process with continuous parameter*. If $\mathbb{T}$ is a discrete subset of $\mathbb{R}$, it is called a *stochastic process with discrete parameter*. When a sample ω is fixed, $X_t(\omega)$, $t \in \mathbb{T}$ can be regarded as a function of t. It is called a *sample path* (or *sample function*) of the stochastic process. In this book we shall mainly consider stochastic processes with continuous parameter. In most cases the time set $\mathbb{T}$ will be the finite interval $[0, T]$, but the infinite interval $\mathbb{T} = [0, \infty)$ or $\mathbb{T} = (-\infty, 0]$ will be dealt with in some cases.

In this section we define three basic stochastic processes called Brownian motions, martingales and Markov processes, which are the central topics in this book. We first introduce some general notions on stochastic processes.

Let X_t, $t \in \mathbb{T}$ be a stochastic process with continuous parameter. It is called *measurable* if $X : \mathbb{T} \times \Omega \to S$ is measurable with respect to the product σ-field $\mathscr{B}(\mathbb{T}) \otimes \mathscr{F}$. The continuity of a stochastic process is defined similarly as the convergence of random variables. Let X_t, $t \in \mathbb{T}$ be a real valued stochastic process. It is called *continuous in probability* if for any $t \in \mathbb{T}$, X_{t+h} converges to X_t in probability as h tends to 0. If X_t is in L^p and $\lim_{h\to 0} \|X_{t+h} - X_t\|_p = 0$ holds for any t, it is called *continuous in L^p*. Obviously a stochastic process continuous in L^p is continuous in probability.

If the sample function $X_t(\omega)$, $t \in \mathbb{T}$ is a continuous function of t for almost all ω, X_t is called a *continuous stochastic process*. If the sample function $X_t(\omega)$, $t \in \mathbb{T}$ is a right continuous function of t for almost all ω, X_t is called a *right continuous stochastic process*.

A stochastic process $\tilde{X}_t$, $t \in \mathbb{T}$ is called a *modification* of X_t, $t \in \mathbb{T}$ if $P(X_t = \tilde{X}_t) = 1$ holds for all t of $\mathbb{T}$. In most cases we do not distinguish between a stochastic process and its modification. However the properties of sample functions can depend on the choice of modification, so it is sometimes necessary to take a good modification of a given stochastic process. In Section 1.4 we shall give a criterion for a given stochastic process to have a modification of a continuous stochastic process.

Now let $X_t = (X_t^1, \ldots, X_t^d)$ be a continuous process with values in $\mathbb{R}^d$ having the mean vector $m(t) = E[X_t]$ and covariance matrix $V(s, t) = E[(X_s - m(s))(X_t - m(t))^t]$ where $(\ )^t$ stands for the transpose of the vector $(\ )$. It is called a *Brownian motion* if it has independent increments, i.e. for any $0 \le t_0 < t_1 < \cdots < t_n$ of $\mathbb{T}$, $X_{t_0}, X_{t_{i+1}} - X_{t_i}: i = 0, \ldots, n-1$ are independent random variables. Now if X_t is a Brownian motion, $X_t' = X_t - m(t)$ is also a Brownian motion. Further increments X_{t_0}', $X_{t_{i+1}}' - X_{t_i}': i = 0, \ldots, n-1$ are orthogonal to each other, i.e.

$$E[X_{t_0}'(X_{t_{j+1}}' - X_{t_j}')^t] = 0, \qquad E[(X_{t_{i+1}}' - X_{t_i}')(X_{t_{j+1}}' - X_{t_j}')^t] = 0$$

holds for any $i \ne j$. (See Exercise 1.1.7.) Then the covariance $V(s, t)$ satisfies the following:

(i) $V(s, t) = V(r, r)$ where $r \equiv \min\{s, t\}$,
(ii) $V(t) \equiv V(t, t)$ increases with t.

A Brownian motion is called *standard* if $m(t) = 0$ and $V(t, t) = tE$ where E is the identity matrix.

Martingales

Let $\{\mathscr{F}_t : t \in \mathbb{T}\}$ be a family of sub σ-fields of $\mathscr{F}$. It is called a *filtration* of sub σ-fields of $\mathscr{F}$ if it satisfies the following three properties:

(i) $\mathscr{F}_s \subset \mathscr{F}_t$ if $s < t$,
(ii) $\bigcap_{\varepsilon > 0} \mathscr{F}_{t+\varepsilon} = \mathscr{F}_t$,
(iii) each $\mathscr{F}_t$ contains all null sets of $\mathscr{F}$.

A stochastic process X_t, $t \in \mathbb{T}$ is called $(\mathscr{F}_t)$-*adapted* if for each t, X_t is $\mathscr{F}_t$-measurable.

Suppose that we are given a stochastic process X_t, $t \in \mathbb{T}$, first. Let $\mathscr{F}_t$ be the smallest σ-field including $\bigcap_{\varepsilon > 0} \sigma(X_s : s \le t + \varepsilon)$ and all null sets of $\mathscr{F}$. Then $\{\mathscr{F}_t\}$ is a filtration and X_t is an $(\mathscr{F}_t)$-adapted process. It is called the *filtration generated by the process* X_t.

In the sequel we assume that a filtration $\{\mathscr{F}_t\}$ is given and is fixed unless otherwise mentioned. Let X_t be a real valued $(\mathscr{F}_t)$-adapted process such that for each t, X_t is integrable. It is called a *martingale* if it satisfies

$$E[X_t | \mathscr{F}_s] = X_s \qquad \text{a.s. for any } t > s. \qquad (1)$$

It is called a *submartingale* if it satisfies

$$E[X_t|\mathscr{F}_s] \geq X_s \qquad \text{a.s. for any } t > s. \tag{2}$$

Further if the converse inequalities $E[X_t|\mathscr{F}_s] \leq X_s$ hold a.s., it is called a *supermartingale.*

Let X_t be a (sub)martingale and let $f: \mathbb{R} \to \mathbb{R}$ be an (increasing) convex function. If $f(X_t)$ is integrable for any t, it is a submartingale because of Jensen's inequality in Theorem 1.1.6. In particular $X_t^+ \equiv \max\{X_t, 0\}$ is a submartingale if X_t is a submartingale. Next let X_t be a martingale and let $p \geq 1$. If $E[|X_t|^p] < \infty$ holds for any t, it is called an *L^p-martingale.* In this case $|X_t|^p$ is a submartingale.

One of the most important examples of martingales is a Brownian motion. We will give Lévy's characterization of a Brownian motion through certain martingale properties.

Theorem 1.2.1 *Let $X_t = (X_t^1, \ldots, X_t^d)$, $t \in [0, T]$ be a continuous stochastic process with $X_0 = 0$ having the mean vector 0 and the covariance matrix $V(s, t)$. The following statements are equivalent.*

(i) *X_t is a Brownian motion.*
(ii) *Both X_t^i and $X_t^i X_t^j - V^{ij}(t, t)$ $i, j = 1, \ldots, d$ are martingales with respect to the filtration generated by X_t.*
(iii) *X_t is a Gaussian process, i.e. $(X_{t_0}, \ldots, X_{t_n})$ is subject to a Gaussian distribution for any $0 \leq t_0 < \cdots < t_n \leq T$. Further its covariance $V(s, t)$ coincides with $V(r, r)$, where $r \equiv \min\{s, t\}$.* □

The proof will be given in Section 2.3, see Theorem 2.3.13.

We next quote some theorems on martingales due to Doob, without giving proofs (Theorems 1.2.2, 1.2.3, 1.2.5–1.2.7 below). The details are found in Doob [26], Meyer [99] and other books dealing with the martingale theory.

Theorem 1.2.2 *Let X_t be a submartingale such that $E[X_t]$ is right continuous with respect to t. Then it has a modification $\tilde{X}_t$ such that its sample paths are right continuous with left hand limits a.s.* □

Theorem 1.2.3

(i) *Let X_t, $t \in [0, \infty)$ be a right continuous submartingale. Suppose* $\sup_t E[X_t^+] < \infty$. *Then $X_\infty = \lim_{t\uparrow\infty} X_t$ exists a.s. and X_∞ is integrable. Furthermore if $\{X_t^+ : t \in [0, \infty)\}$ is uniformly integrable, then X_t,*

$t \in [0, \infty]$ is a submartingale. In particular if X_t, $t \in [0, \infty)$ is a uniformly integrable martingale, then X_t, $t \in [0, \infty]$ is a martingale.

(ii) *Let X_t, $t \in (-\infty, 0]$ be a right continuous submartingale. Then $X_{-\infty} = \lim_{t\to-\infty} X_t$ exists a.s.* □

Stopping times play important roles in the theory of submartingales. Let $\mathbb{T}$ be $[0, \infty)$ or $[0, T]$. A random variable τ with values in $\bar{\mathbb{T}}$ (closure of $\mathbb{T}$ in $[0, \infty]$) is called a *stopping time* if $\{\tau \le t\} \in \mathscr{F}_t$ holds for any t. For a given stopping time τ, we set

$$\mathscr{F}_\tau = \{A \in \mathscr{F} : A \cap \{\tau \le t\} \in \mathscr{F}_t \text{ holds for all } t \in \mathbb{T}\}. \quad (3)$$

It is easily verified that $\mathscr{F}_\tau$ is a sub σ-field of $\mathscr{F}$.

Now let X_t be an $(\mathscr{F}_t)$-adapted process, right continuous with left hand limits. Then X_τ is $\mathscr{F}_\tau$-measurable. Indeed, if X_t is an $(\mathscr{F}_t)$-adapted *simple process*, i.e. there exists a finite partition $\{0 = t_0 < t_1 < \cdots < t_l\}$ of $\mathbb{T}$ such that $X_t = X_{t_k}$ holds for any $t \in [t_k, t_{k+1})$, then X_τ is written as $\sum_k X_{t_k}\chi(t_k \le \tau < t_{k+1})$ where $\chi(A)$ is the indicator function of the set A. It is clearly $\mathscr{F}_\tau$-measurable. Since any process which is right continuous with left hand limits is approximated uniformly by a sequence of the above simple processes a.s., X_τ is $\mathscr{F}_\tau$-measurable.

Let τ and σ be stopping times. Then the following properties hold.

(i) The sets $\{\tau < \sigma\}$, $\{\tau = \sigma\}$ and $\{\tau \le \sigma\}$ belong to both $\mathscr{F}_\tau$ and $\mathscr{F}_\sigma$.

(ii) If $\tau \le \sigma$, then $\mathscr{F}_\tau \subset \mathscr{F}_\sigma$.

Indeed, we have $\{\tau < \sigma\} \cap \{\tau \le t\} = \bigcup \{\tau < s, s < \sigma\}$, where the union is taken for all rationals s less than or equal to t. Then the set belongs to $\mathscr{F}_t$. This proves that the set $\{\tau < \sigma\}$ belongs to $\mathscr{F}_\tau$. The other assertions in (i) can be shown similarly. Next suppose $\tau \le \sigma$. If $B \in \mathscr{F}_\tau$, then $B \cap \{\sigma \le t\} = B \cap \{\tau \le t\} \cap \{\sigma \le t\} \in \mathscr{F}_t$. Therefore we have $B \in \mathscr{F}_\sigma$. This proves the second assertion.

We give an example of a stopping time, which will be used later in order to localize martingales.

Example 1.2.4 Let X_t be a right continuous $(\mathscr{F}_t)$-adapted process. Let G be an open subset of $\mathbb{R}$. The *hitting time* of X_t to the set G or the *first time such that $X_t \in G$* is defined by

$$\tau_G = \inf\{t \in \mathbb{T} : X_t \in G\} \qquad (= \sup\{t \in \mathbb{T}\} \text{ if } \{\cdots\} = \emptyset). \quad (4)$$

Then τ_G is a stopping time. In fact, we have $\{\tau_G \ge t\} = \bigcap (X_r \in G^c\}$, where the intersection is taken for all rationals r less than t. Therefore $\{\tau_G \geq t\} \in \mathscr{F}_t$ holds for any t.

Any stopping time can be approximated from the above by a decreasing sequence of stopping times with discrete values. Indeed, given a stopping time τ, we define the sequence τ_n, $n = 1, 2, \dots$ by

$$\tau_n = \begin{cases} 0, & \text{if } \tau = 0, \\ \min\left\{\frac{k+1}{2^n}, \sup\{t \in \mathbb{T}\}\right\}, & \text{if } \frac{k}{2^n} < \tau \leq \frac{k+1}{2^n}, \quad k = 0, 1, 2, \dots \end{cases}$$

We can easily verify that each τ_n is a stopping time and the sequence $\{\tau_n\}$ decreases to τ.

We will now quote Doob's *optional stopping time theorem* and its two consequences called *Doob's inequalities*. In the first theorem, we consider submartingales with finite time interval.

Theorem 1.2.5 *Let X_t, $t \in [0, T]$ be a continuous submartingale and let τ, σ be any two stopping times. Then X_τ is integrable and satisfies*

$$E[X_\tau | \mathscr{F}_\sigma] \geq X_{\min\{\tau, \sigma\}}. \tag{5}$$

In particular if X_t is a martingale, the equality holds in (5). □

Theorem 1.2.6 *Let X_t be a submartingale. Then*

$$cP\left(\sup_{s \leq t} X_s > c\right) \leq \int_{\sup_{s \leq t} X_s > c} X_t \, dP \tag{6}$$

holds for any $c > 0$ and $t \in \mathbb{T}$. □

Theorem 1.2.7 *Let X_t be a positive submartingale. Then for any $p > 1$ we have*

$$E\left[\sup_{s \leq t} X_s^p\right] \leq q^p E[X_t^p] \qquad \textit{for all } t \in \mathbb{T}, \tag{7}$$

where q is a positive number such that $p^{-1} + q^{-1} = 1$. □

Markov processes

Let S be a locally compact, complete separable metric space and let $\mathscr{B}(S)$ be the set of all Borel subsets of S. By a *Borel measure* on S we mean a regular measure μ on $\mathscr{B}(S)$ such that $\mu(K) < \infty$ holds for any compact subset K of S. In particular if $\mu(S) = 1$ it is called a probability. A family of Borel measures $\{K(x, \cdot): x \in S\}$ on S is called a *kernel* if $K(x, E)$ is $\mathscr{B}(S)$-measurable with respect to x for each E of $\mathscr{B}(S)$. In the following a $\mathscr{B}(S)$-measurable function is called simply measurable. Let f be a real valued measurable function on S. We use the notation

$$Kf(x) = \int K(x, \mathrm{d}y)f(y) \tag{8}$$

if the integral is well defined for any x. The function $Kf(x)$ is measurable obviously.

Let $\{P_{s,t}(x, \cdot)\}$ be a family of kernels consisting of probability distributions on S, where $s < t$ are elements of $\mathbb{T}$. It is called a *transition probability* if it satisfies the Chapmann–Kolmogorov equation:

$$P_{s,u}(x, E) = \int_S P_{t,u}(y, E)P_{s,t}(x, \mathrm{d}y), \tag{9}$$

for every $s < t < u$, $x \in S$ and $E \in \mathscr{B}(S)$.

Suppose that a filtration $\{\mathscr{F}_t : t \in \mathbb{T}\}$ of sub σ-fields of $\mathscr{F}$ is given. Let X_t, $t \in \mathbb{T}$ be a stochastic process with state space S adapted to $(\mathscr{F}_t)$. The process X_t is called a *Markov process with transition probability* $\{P_{s,t}(x, \cdot)\}$ if it has the *Markov property* with respect to $\{\mathscr{F}_t\}$:

$$P(X_t \in E|\mathscr{F}_s) = P_{s,t}(X_s, E) \qquad \text{for every } s < t \text{ and } E \in \mathscr{B}(S). \tag{10}$$

A Markov process X_t is called *temporally homogeneous* if the transition probability $\{P_{s,t}(x, A)\}$ depends only on $t - s$.

In the following we will consider a temporally homogeneous Markov process with continuous time parameter $\mathbb{T} = [0, \infty)$. The transition probability $P_{0,t}(x, \cdot)$ is often denoted by $P_t(x, \cdot)$. For a bounded measurable function f on S, we denote $P_t f$ by $T_t f$, namely

$$T_t f(x) = \int_S P_t(x, \mathrm{d}y)f(y). \tag{11}$$

It is again a bounded measurable function of x. The family of operators $\{T_t : t \in \mathbb{T}\}$ satisfies the semigroup property $T_{t+s}f = T_t T_s f$ by the Chapmann–Kolmogorov equation.

Let $C(S)$ be the set of all real valued continuous functions on S. If S is a compact space, it is a separable Banach space with the supremum norm. If S is a noncompact space, denote by $C_\infty(S)$ the subset of $C(S)$ such that $\lim_{x\to\infty} f(x)$ exists and equals 0, where ∞ is the infinity adjoined to S as a one point compactification. Then $C_\infty(S)$ is also a Banach space with the supremum norm. Now suppose that S is compact (or noncompact). If T_t defined by (11) maps $C(S)$ (or $C_\infty(S)$) into itself and is strongly continuous i.e. for every t, $T_{t+h}f$ converges to $T_t f$ strongly as $h \to 0$ for any f, the semigroup $\{T_t\}$ of linear operators on $C(S)$ (or $C_\infty(S)$) is called a *Feller semigroup.*

Theorem 1.2.8 *Let X_t be a Markov process with a Feller semigroup $\{T_t\}$. Then X_t has a modification such that its sample functions are right continuous with left hand limits.*

Proof We prove the theorem in the case where the state space S is compact. The case for noncompact S is left to the reader (see Exercises 1.2.11 and 1.2.12). By the Markov property (10), we have for $s < t < u$

$$E[T_{u-t}f(X_t)|\mathscr{F}_s] = T_{t-s}T_{u-t}f(X_s) = T_{u-s}f(X_s) \quad \text{a.s.}$$

Then if $\alpha > 0$ and f is non-negative we obtain

$$E\left[\int_t^\infty e^{-\alpha u}T_{u-t}f(X_t)\,du\,\middle|\,\mathscr{F}_s\right] \le \int_s^\infty e^{-\alpha u}T_{u-s}f(X_s)\,du.$$

Setting

$$U_\alpha f(x) = \int_0^\infty e^{-\alpha u}T_u f(x)\,du, \tag{12}$$

the above inequality is written as $e^{-\alpha t}E[U_\alpha f(X_t)|\mathscr{F}_s] \le e^{-\alpha s}U_\alpha f(X_s)$. Therefore $e^{-\alpha t}U_\alpha f(X_t)$ is a bounded supermartingale. Then by Theorem 1.2.2, $e^{-\alpha t}U_\alpha f(X_t)$ has a modification such that its sample functions are right continuous with left hand limits. The latter property is valid for any f of $C(S)$.

Now let $\{f_n\}$ be a countable dense subset of $C(S)$. Then the set of functions $g_n \equiv U_\alpha f_n$, $n = 1, 2, \ldots$ where $\alpha > 0$ is fixed, is again a dense subset of $C(S)$. Indeed, in view of the resolvent equation $U_\alpha f - U_\beta f + (\alpha - \beta)U_\alpha U_\beta f = 0$ (*cf.* Lemma 1.3.1) the range of $C(S)$ by the map U_α is independent of α, and further it is a dense subset of $C(S)$ since $\alpha U_\alpha f$ converges to f strongly as α tends to infinity. Now let $g_n(X_t)^\sim$ be a modification of $g_n(X_t)$ such that its sample functions are right continuous and have left hand limits. Let $\tilde{\Omega}$ be the set of all samples ω such that $g_n(X_t)^\sim(\omega)$, $n = 1, 2, \ldots$ are all right continuous with left hand limits. Take any ω from $\tilde{\Omega}$. Then for every t there exists a unique point $\tilde{X}_t(\omega)$ in S such that $g_n(X_t)^\sim(\omega) = g_n(\tilde{X}_t(\omega))$ holds for any n. Thus $g_n(\tilde{X}_t(\omega))$ is right continuous with left hand limits with respect to t for all n. This implies that $\tilde{X}_t(\omega)$ itself is right continuous with left hand limits in the space S since $\{g_n\}$ is dense in $C(S)$. Since $P(\tilde{\Omega}) = 1$, we have $g_n(X_t) = g_n(\tilde{X}_t)$ for all n a.s. for each t. Therefore $\tilde{X}_t$ is a modification of X_t. □

Suppose we are given a temporally homogeneous transition probability $\{P_t(x,\cdot)\}$ such that it defines a Feller semigroup. We shall construct a Markov process X_t associated with $\{P_t(x,\cdot)\}$ such that its sample paths are right continuous with left hand limits. Let $\overline{W}$ be the set of all maps $\bar{w}$ from $\mathbb{T}$ into S. We denote the elements of $\overline{W}$ by $\bar{w}$ and their values at $t \in \mathbb{T}$ by $\bar{w}(t)$ or $\bar{w}_t$. A subset A of $\overline{W}$ represented by

$$A = \{\bar{w} : (\bar{w}(t_1), \ldots, \bar{w}(t_n)) \in E_n\}, \tag{13}$$

where $0 \le t_1 < \cdots < t_n$ and E_n is a Borel set in S^n, is called a *cylinder set of* $\overline{W}$. Let $\mathscr{B}_{t_1,\ldots,t_n}$ be the collection of all cylinder sets represented above where $t_1, \ldots, t_n$ are fixed and E_n are running over all Borel sets in S^n. Then it is a σ-field of $\overline{W}$. For each x of S we define a probability measure $P_x^{(t_1,\ldots,t_n)}$ on $\mathscr{B}_{t_1,\ldots,t_n}$ by

$$P_x^{(t_1,\ldots,t_n)}(A)$$

$$= \int \cdots \int_{E_n} P_{t_1}(x, dx_1) P_{t_2-t_1}(x_1, dx_2) \ldots P_{t_n-t_{n-1}}(x_{n-1}, dx_n), \tag{14}$$

where A is an element of $\mathscr{B}_{t_1,\ldots,t_n}$ defined by (13). Then the family of measures $\{P_x^{(t_1,\ldots,t_n)}\}$ is consistent for each x, i.e. if $(t'_1, \ldots, t'_m)$ is a subset of $(t_1, \ldots, t_n)$ and A is an element of $\mathscr{B}_{t_1,\ldots,t_n}$ belonging to $\mathscr{B}_{t'_1,\ldots,t'_m}$, then $P_x^{(t_1,\ldots,t_n)}(A) = P_x^{(t'_1,\ldots,t'_m)}(A)$ holds. Now let $\mathscr{A}(\overline{W})$ be the algebra $\bigcup \mathscr{B}_{t_1,\ldots,t_n}$ where the union runs over all $t_1, \ldots, t_n$ of $\mathbb{T}$ and $n = 1, 2, \ldots$ Then there exists a unique measure $\hat{P}_x$ on the algebra $\mathscr{A}(\overline{W})$ such that its restriction to $\mathscr{B}_{t_1,\ldots,t_n}$ coincides with $P_x^{(t_1,\ldots,t_n)}$. Let $\mathscr{B}(\overline{W})$ be the smallest σ-field containing $\mathscr{A}(\overline{W})$. Then the measure $\hat{P}_x$ can be extended uniquely to a measure $\bar{P}_x$ on $\mathscr{B}(\overline{W})$ by Kolmogorov–Hopf's theorem. Denote by $\mathscr{F}(\overline{W})$ the completion of $\mathscr{B}(\overline{W})$ with respect to $\bar{P}_x$ (x being fixed).

Let $\{\mathscr{F}_t\}$ be the filtration generated by the stochastic process $\bar{w}(t)$. We show that $\bar{w}(t)$ has the Markov property with respect to $\{\mathscr{F}_t\}$ for each measure $\bar{P}_x$. Let $s < t$ and A be an element of $\mathscr{A}(\overline{W})$ represented as in (13) where $t_n \le s$. Then we have by (14)

$$\bar{P}_x(A \cap \{\bar{w}(t) \in E\}) = \int \cdots \int_{E_n \times S} P_{t_1}(x, dx_1) \ldots P_{s-t_n}(x_n, dy) P_{t-s}(y, E).$$

Denote by $\bar{E}_x$ the expectation with respect to the measure $\bar{P}_x$. Then $\bar{E}_x[P_{t-s}(\bar{w}(s), E) : A]$ is also written as the right hand side of the above. This implies

$$\bar{E}_x[f(\bar{w}(t)) : A] = \bar{E}_x[T_{t-s} f(\bar{w}(s)) : A] \tag{15}$$

for any f of $C_\infty(S)$ (or of $C(S)$ if S is compact) and A of $\sigma(\bar{w}(r): r \le s)$. Further $f(\bar{w}(t))$ is right continuous in L^2. In fact we have by (15)

$$\bar{E}_x[(f(\bar{w}(s+\varepsilon)) - f(\bar{w}(s)))^2]$$
$$= \bar{E}_x[T_\varepsilon(f^2)(\bar{w}(s)) - 2T_\varepsilon f(\bar{w}(s))f(\bar{w}(s)) + f(\bar{w}(s))^2]$$

which converges to 0 as $\varepsilon \to 0$. Then the above equality (15) is valid for any A of $\mathscr{F}_s$. Therefore $\bar{w}(t)$ is a Markov process with the Feller semigroup $\{T_t\}$. Then $\bar{w}(t)$ has a modification $X_t(\bar{w})$ which is right continuous with left hand limits by the previous theorem.

Now let W be the set of all $w \in \bar{W}$ such that $w(t)$ is right continuous with left hand limits. Then for almost all $\bar{w}$, the right continuous modification $X(\bar{w}) \equiv \{X_t(\bar{w}): t \in \mathbb{T}\}$ can be regarded as an element of W. Let $\mathscr{B}(W)$ be the smallest σ-field of W containing all cylinder sets of W. Then the set $\{\bar{w}: X(\bar{w}) \in B\}$ belongs to $\mathscr{F}(\bar{W})$ for any B of $\mathscr{B}(W)$. We define the law of $(X_t, \bar{P}_x)$ on the space $(W, \mathscr{B}(W))$ by

$$P_x(B) = \bar{P}_x(\{\bar{w}: X(\bar{w}) \in B\}). \tag{16}$$

The expectation by the measure P_x is denoted by E_x. Note that $P_x(B)$ is measurable with respect to x for any B of $\mathscr{B}(W)$. The triple $(W, \mathscr{B}(W), P_x: x \in S)$ is called a *right continuous Markov process with the Feller semigroup* $\{T_t\}$ or simply a *Feller process*.

Now for $s \in [0, \infty)$, let θ_s be a map from W into itself such that $(\theta_s w)_t = w(s+t)$ holds for all t. It is a measurable map from $(W, \mathscr{B}(W))$ into itself. The family $\{\theta_s\}$ satisfies the semigroup property $\theta_s\theta_t = \theta_{s+t}$.

We shall extend the Markov property of the Feller process.

Theorem 1.2.9 *Let $(W, \mathscr{B}(W), P_x: x \in S)$ be a Feller process. Then each P_x satisfies*

$$P_x(\theta_s^{-1}B|\mathscr{F}_s) = P_{w(s)}(B) \qquad \text{for every } B \in \mathscr{B}(W), \tag{17}$$

where $\{\mathscr{F}_t\}$ is the filtration generated by $w(t)$.

Proof Let A and B be the cylinder sets $\{w: (w(t_1), \dots, w(t_n)) \in E_n\}$ and $\{w: (w(u_1), \dots, w(u_m)) \in E_m\}$ respectively, where $t_n \le s$ and E_n and E_m are Borel sets of S^n and S^m respectively. Using the Markov property, we can show that both of $P_x(\theta_s^{-1}B \cap A)$ and $E_x[P_{w(s)}(B): A]$ are represented by the same quantity below:

$$\int \cdots \int_{E_n \times E_m} P_{t_1}(x, dx_1) \cdots P_{u_1+s-t_n}(x_n, dy_1) \cdots P_{u_m-u_{m-1}}(y_{m-1}, dy_m).$$

Therefore we have $P_x(\theta_s^{-1}B \cap A) = E_x[P_{w(s)}(B) : A]$ for the above cylinder sets A and B. The equality can be extended to any B of $\mathscr{B}(W)$ and A of $\mathscr{F}_s$. Thus the theorem is established. □

Exercise 1.2.10 Suppose that a filtration $\{\mathscr{F}_t : t \in [0, \infty)\}$ (or $\{\mathscr{F}_t : t \in (-\infty, 0]\}$) of sub σ-fields of $\mathscr{F}$ is given. Let X be a real random variable such that $E[|X|^p] < \infty$ for some $p > 1$. Show that $X_t = E[X|\mathscr{F}_t]$ is a martingale. Let $\tilde{X}_t$ be a right continuous modification of X_t. Show that $X_\infty = \lim_{t\to\infty} \tilde{X}_t$ (or $X_{-\infty} = \lim_{t\to-\infty} \tilde{X}_t$ respectively) exists and equals $E[X|\mathscr{F}_\infty]$ (or $E[X|\mathscr{F}_{-\infty}]$ respectively) where $\mathscr{F}_\infty$ is the least σ-field including $\bigcup_t \mathscr{F}_t$ (or $\mathscr{F}_{-\infty} = \bigcap_t \mathscr{F}_t$, respectively). (*Hint*: Show that $\{\tilde{X}_t : t \in [0, \infty)\}$ is uniformly integrable and then show $E[X_\infty : A] = E[X : A]$ holds for any A of $\mathscr{F}_\infty$.)

Exercise 1.2.11 (*Strong Markov property*) Let X_t be a right continuous Markov process associated with the Feller semigroup $\{T_t\}$. Let $\{\mathscr{F}_t\}$ be the filtration generated by X_t. Let τ be a stopping time and $\mathscr{F}_\tau$ be the σ-field defined by (3). Show that

$$E[f(X_{t+\tau}) : A] = E[T_t f(X_\tau) : A] \qquad \text{for every } A \in \mathscr{F}_\tau$$

holds for any $f \in C(S)$ (or $C_\infty(S)$ if S is noncompact). (*Hint*: show the above first in the case where τ is a stopping time with discrete values. Then for general τ, approximate it from the above by a decreasing sequence of stopping times with discrete values.)

Exercise 1.2.12 (*Proof of Theorem* 1.2.8 *for noncompact case*) Suppose that S is noncompact in Theorem 1.2.8. Let $\hat{S} = S \cup \{\infty\}$ be the one point compactification of S.

(i) Show that X_t has a modification $\tilde{X}_t$ with values in $\hat{S}$ such that its sample paths are right continuous with left hand limits with respect to the topology of $\hat{S}$.

(ii) Let $\{G_n\}$ be a decreasing sequence of open neighborhoods of ∞ such that $\bigcap_n G_n = \{\infty\}$. Let σ_n be the first time that $\tilde{X}_t$ hits the set G_n and let $\sigma_\infty = \lim_n \sigma_n$. Let $\alpha > 0$. Show

$$E\left[\int_0^\infty e^{-\alpha t} f(\tilde{X}_t)\,dt\right] = E\left[\int_0^{\sigma_\infty} e^{-\alpha u} f(\tilde{X}_u)\,du\right] + E[e^{-\alpha\sigma_\infty} U_\alpha f(\tilde{X}_{\sigma_\infty -})],$$

where $\tilde{X}_{\sigma_\infty -} = \lim_n \tilde{X}_{\sigma_n}$ and $U_\alpha f$ is defined by (12). Deduce from this that $\sigma_\infty = \infty$ a.s. (*Hint*: use the strong Markov property of $\tilde{X}_t$).

1.3 Ergodic properties of Markov processes

Recurrent and transient processes

Let $(W, \mathscr{B}(W), P_x : x \in S)$ be a Feller process constructed in the previous section associated with the Feller semigroup $\{T_t\}$, where S is a locally compact complete separable metric space. We assume that the transition probability has a positive, continuous density function.

Condition (A) *There exists a Borel measure μ on S supported by S, and a strictly positive function $p_t(x, y)$ continuous in $(t, x, y) \in (0, \infty) \times S^2$ such that the transition probability $P_t(x, \mathrm{d}y)$ equals $p_t(x, y)\mu(\mathrm{d}y)$.* □

Let $B(S)$ be the set of all bounded measurable functions on S and let $BC(S)$ be the set of all bounded continuous functions on S. Each T_t maps $B(S)$ into $BC(S)$. Indeed if f of $B(S)$ satisfies $0 \le f \le 1$, both of $T_t f$ and $T_t(1 - f)$ are lower semicontinuous and their sum is constant 1. Then $T_t f$ has to be continuous. Obviously the last property is valid for any f of $B(S)$. The semigroup with this property is called a *strong Feller semigroup*.

The Feller process is called *recurrent in the sense of Harris* if

$$\int_0^\infty \chi(A)(w(t))\,\mathrm{d}t = \infty \qquad \text{a.s. } P_x \tag{1}$$

is satisfied for every $x \in S$ whenever $\mu(A) > 0$. Further, the Feller process is called *transient* if

$$\sup_{x \in S} E_x\left[\int_0^\infty \chi(K)(w(t))\,\mathrm{d}t\right] < \infty \tag{2}$$

holds for any compact subset K of S.

We will show that any Feller process satisfying Condition (A) is either transient or recurrent in the sense of Harris. Our discussion of this problem is similar to Revuz [112]. As an intermediate step we introduce a suitable Markov process with discrete time parameter and show first a similar transient–recurrent dichotomy for this process (Lemma 1.3.3 below). Then it will be applied to our Feller process.

Let h be a bounded non-negative measurable function on S. Set

$$U_h f(x) = E_x\left[\int_0^\infty \exp\left\{-\int_0^t h(w(s))\,\mathrm{d}s\right\} f(w(t))\,\mathrm{d}t\right]. \tag{3}$$

It is well defined; at least f is a non-negative measurable function. Setting $U_h(x, E) \equiv U_h\chi(E)(x)$, it defines a kernel if $U_h(x, K) < \infty$ holds for any compact K.

Lemma 1.3.1

(i) *The following relation holds:*

$$U_h h(x) = 1 - E_x\left[\exp\left\{-\int_0^\infty h(w(s))\,\mathrm{d}s\right\}\right]. \tag{4}$$

(ii) *Let $h \geq k \geq 0$ and $f \geq 0$. Then U_h and U_k satisfy the resolvent equation:*

$$U_k f - U_h f = U_h(h-k)U_k f = U_k(h-k)U_h f. \tag{5}$$

(iii) *If $U_h f$ is a bounded function, it is continuous.*

Proof Note the equality

$$\int_0^\infty \exp\left\{-\int_0^t h(w(s))\,\mathrm{d}s\right\} h(w(t))\,\mathrm{d}t = 1 - \exp\left\{-\int_0^\infty h(w(s))\,\mathrm{d}s\right\}.$$

Then equality (4) is immediate. We have by Theorem 1.2.9 (Markov property),

$$U_k f(w_t) = E_x\left[\int_t^\infty \exp\left\{-\int_t^v k(w_u)\,\mathrm{d}u\right\} f(w_v)\,\mathrm{d}v \,\middle|\, \mathscr{F}_t\right].$$

Therefore,

$$\begin{aligned}
&U_h(h-k)U_k f(x)\\
&= E_x\left[\int_0^\infty \exp\left\{-\int_0^t h(w_s)\,\mathrm{d}s\right\}(h(w_t)-k(w_t))\right.\\
&\quad \left.\times\left(\int_t^\infty \exp\left\{-\int_t^v k(w_u)\,\mathrm{d}u\right\} f(w_v)\,\mathrm{d}v\right)\mathrm{d}t\right]\\
&= E_x\left[\int_0^\infty f(w_v)\left\{\int_0^v \exp\left\{-\int_0^t (h(w_s)-k(w_s))\,\mathrm{d}s\right\}\right.\right.\\
&\quad \left.\left.\times (h(w_t)-k(w_t))\,\mathrm{d}t\right\}\exp\left\{-\int_0^v k(w_u)\,\mathrm{d}u\right\}\mathrm{d}v\right]\\
&= E_x\left[\int_0^\infty f(w_v)\left[1-\exp\left\{-\int_0^v (h(w_s)-k(w_s))\,\mathrm{d}s\right\}\right]\right.\\
&\quad \left.\times \exp\left\{-\int_0^v k(w_u)\,\mathrm{d}u\right\}\mathrm{d}v\right]\\
&= U_k f(x) - U_h f(x).
\end{aligned} \tag{6}$$

This proves the resolvent equation (5).

Now if α is a positive constant, $U_\alpha f$ is represented by (12) of the previous section. It is bounded continuous if f is bounded measurable, since the semigroup has the strong Feller property. Then in view of the resolvent equation $U_h f = U_\alpha f + U_\alpha(\alpha - h)U_h f$, $U_h f(x)$ is also bounded continuous if $U_h f$ is bounded measurable. □

In the following, the Greek letters α, β stand for positive constants and Roman letters a, b, h, k stand for bounded non-negative functions. Set

$$u_\alpha(x, y) = \int_0^\infty e^{-\alpha t} p_t(x, y)\, dt. \tag{7}$$

It is a strictly positive lower semicontinuous function. We have

$$U_\alpha f(x) = \int u_\alpha(x, y) f(y)\, d\mu(y). \tag{8}$$

The family $\{U_\alpha : \alpha > 0\}$ is called the *resolvent* of the semigroup $\{T_t\}$.

Lemma 1.3.2 *Let $\alpha_0 > 0$ be a positive constant.*

(i) *There exist bounded strictly positive continuous functions $a(x)$ and $b(y)$ such that*

$$u_{\alpha_0}(x, y) \geq a(x)b(y) \qquad \text{for every } x, y. \tag{9}$$

(ii) *Choose the above function a so that $\alpha_0 > 2a(x)$ holds for any x. Then*

$$U_a(x, dy) \geq \frac{\alpha_0}{2} U_a a(x) b(y)\, d\mu(y) \qquad \text{for every } x. \tag{10}$$

Proof From the resolvent equation for constants α_0, $\alpha_0/2$, we have

$$u_{\alpha_0}(x, y) \geq \frac{\alpha_0}{2} \int u_{\alpha_0}(x, z) u_{\alpha_0/2}(z, y)\, d\mu(z).$$

Let K be a compact subset of S such that $\mu(K) > 0$ and let β be a positive constant. Set

$$a(x) = \frac{\alpha_0}{2} \int_K u_{\alpha_0}(x, z)\, d\mu(z), \qquad b(y) = \min\left\{ \inf_{z \in K} u_{\alpha_0/2}(z, y), \beta \right\}.$$

Then both are bounded strictly positive continuous functions and satisfy (9).

Next using the resolvent equation for the function a and constant α_0, and inequality (9), we obtain

$$U_a(x, dy) \geq U_a(\alpha_0 - a)U_{\alpha_0}(x, dy) \geq \frac{\alpha_0}{2} U_a U_{\alpha_0}(x, dy)$$
$$\geq \frac{\alpha_0}{2}\left(\int U_a(x, dz)a(z)\right)b(y)\, d\mu(y) \geq \frac{\alpha_0}{2} U_a a(x) b(y)\, d\mu(y).$$

Therefore (10) is established. □

Now we will fix the function a of the above lemma and define a kernel Q by

$$Q(x, E) = \int_E U_a(x, dy)a(y). \tag{11}$$

Then $Q(x, S) \leq 1$ holds for any x by Lemma 1.3.1 (i). It maps $B(S)$ into $BC(S)$ by Lemma 1.3.1 (iii). We can regard Q as a transition probability of a Markov process with discrete parameter. The n-step transition probability Q^n is defined by induction as $Q^{n-1}Q$. A Borel measure π on S is called *Q-invariant* if $\pi Q \equiv \int \pi(dx)Q(x, \cdot)$ coincides with π.

Lemma 1.3.3

(i) *Assume that $Q(x, S) < 1$ holds for some x. Then there exists a positive constant c less then or equal to $\frac{1}{2}$ such that*

$$Q^n(x, S) \leq (1 - c)^{n-1} \tag{12}$$

holds for every x and for every $n \geq 2$.

(ii) *Assume $Q(x, S) = 1$ holds for every x. Then there exists a Q-invariant probability π satisfying*

$$\|Q^n(x, \cdot) - \pi(\cdot)\| \to 0, \tag{13}$$

for every x as $n \to \infty$, where $\|\lambda\|$ denotes the total variation of the signed measure λ. Further, any Q-invariant measure is a constant multiple of π.

Proof Consider case (i). Set $c = (\alpha_0/2)\int (1 - U_a a)ab\, d\mu$. Since $U_a a(x)$ is a continuous function of x, the integrand is positive in a certain non-void open set. Therefore c is positive. Also it is less than or equal to $\frac{1}{2}$ since $(\alpha_0/2)\int ab\, d\mu \leq (\alpha_0/2)\int u_{\alpha_0}(x, y)\, d\mu(y) = 1/2$ holds by (9). Then using inequality (10) we obtain

$$Q^2 1 = U_a a - U_a(a(1 - U_a a)) \leq U_a a - cU_a a = (1 - c)Q1.$$

Repeating this inductively, we obtain $Q^n 1 \leq (1 - c)^{n-1} Q1$. Hence assertion (i) is established.

Next consider case (ii). Let ν be a probability on S. Set $\nu Q(E) = \int \nu(dx)Q(x, E)$ and $\mu'(E) = (\alpha_0/2)\int_E a(y)b(y)\,d\mu(y)$. We can regard the latter as a kernel $\mu'(x, E)$ not depending on the variable x. Then, we have $(\nu - \nu Q^m)Q = (\nu - \nu Q^m)(Q - \mu')$ for any positive integer m, since $(\nu - \nu Q^m)(S) = 0$ holds. By induction we get $(\nu - \nu Q^m)Q^n = (\nu - \nu Q^m)(Q - \mu')^n$ for any positive integer n. Since $\sup_x \|Q(x, \cdot) - \mu'\| = 1 - c < 1$, we obtain $\|\nu Q^n - \nu Q^{m+n}\| \to 0$ as $n, m + n \to \infty$. Consequently $\{\nu Q^n\}$ converges to a probability π. Clearly π is Q-invariant.

Now let ν' be another probability measure. Then

$$\|\nu'Q^n - \pi\| = \|(\nu' - \pi)Q^n\| = \|(\nu' - \pi)(Q - \mu')^n\| \xrightarrow[n\to\infty]{} 0.$$

Therefore $\{\nu'Q^n\}$ converges to π. This proves that any Q-invariant probability is unique. Moreover, taking $\nu' = \delta_x$, the delta measure concentrated at x, we get the convergence (13).

Finally let $\tilde{\pi}$ be an arbitrary Q-invariant measure. Then for any Borel set A we have

$$\tilde{\pi}(A) = \tilde{\pi}Q^n(A) \geq \int_K \tilde{\pi}(dx)Q^n(x, A) \xrightarrow[n\to\infty]{} \tilde{\pi}(K)\pi(A).$$

where K is a compact set in S. Therefore we have $\tilde{\pi}(A) \geq \tilde{\pi}(S)\pi(A)$, proving that $\tilde{\pi}$ is a bounded measure. Then the normalized measure $\tilde{\pi}(S)^{-1}\tilde{\pi}$ coincides with π. □

Theorem 1.3.4 *Assume that a Feller process satisfies Condition* (A). *Then it is either transient or recurrent in the sense of Harris.*

Proof We will apply Lemma 1.3.3. We first consider case (i) of the lemma. The resolvent equation implies $U_0 a = U_a a + QU_0 a$ where $Qf(x) = U_a(af)(x)$ and $U_0 a(x) = E_x[\int_0^\infty a(w(s))\,ds]$. Therefore $U_0 a = \sum_{n\geq 0} Q^n U_a a \leq 2 + \sum_{n\geq 0}(1 - c)^n$. Consequently $U_0 a$ is a bounded function. For any compact set K there exists a positive constant c' such that $a/c' \geq \chi(K)$, since a is a strictly positive continuous function. Then $U_0\chi(K) \leq c'^{-1} U_0 a$ holds and hence $U_0\chi(K)(x)$ is bounded. This proves that the Feller process is transient.

We next consider case (ii) of Lemma 1.3.3. Let h be any bounded nonnegative measurable function such that $h \leq a$ and $\int h\,d\mu(x) > 0$. Set $g = 1 - U_h h$. It is a continuous function with values in $[0, 1]$. Since $U_h h = U_a h + U_a(a - h)U_h h$ and $1 = U_a h + U_a(a - h)$, the function g satisfies $g = U_a((a - h)g)$. Therefore $g \leq Qg \leq \cdots \leq Q^n g \to \int g\,d\pi$. This implies $g \leq U_a(a - h)\int g\,d\pi$ and we obtain

$$\int g\,d\pi \le \left(\int U_a(a-h)\,d\pi\right)\int g\,d\pi \le \left(1-\frac{\alpha_0}{2}\int bh\,d\mu\right)\int g\,d\pi.$$

Since $\int bh\,d\mu > 0$ we have $\int g\,d\pi = 0$ and hence $U_h h \equiv 1$. Thus $\int_0^\infty h(w(s))\,ds = \infty$ a.s. P_x for any x by equality (4). Consequently the process is recurrent in the sense of Harris. □

Let Λ be a Borel measure on S. It is called an *invariant measure* of the Feller semigroup $\{T_t\}$ or $\{T_t\}$*-invariant measure* if

$$\int_S T_t f(x)\Lambda(dx) = \int_S f(x)\Lambda(dx), \tag{14}$$

holds for every t for any non-negative function f of compact support. Taking the Laplace transform of each term of the above and exchanging the order of integrations by $\Lambda(dx)$ and dt, we obtain

$$\int U_\alpha f(x)\Lambda(dx) = \alpha^{-1}\int f(x)\Lambda(dx), \qquad \text{for every } \alpha > 0. \tag{15}$$

Conversely if the above is valid, then (14) is satisfied by taking the inverse Laplace transform. Therefore Λ is (T_t)-invariant if and only if (15) is satisfied for any non-negative function f of compact support.

Theorem 1.3.5 *Assume that a Feller process satisfies Condition* (A). *If it is recurrent in the sense of Harris, it has a* (T_t)*-invariant measure* Λ. *Further any* (T_t)*-invariant measure is a constant multiple of* Λ. *The measure* Λ *is mutually absolutely continuous with respect to the Borel measure* μ.

Proof Let Q be the kernel defined by (11) and let π be the invariant measure of the kernel Q. Let a, b be functions of Lemma 1.3.2. Define the measure Λ by $\Lambda(dx) \equiv a(x)^{-1}\pi(dx)$. It is a Borel measure. We show that it is a (T_t)-invariant measure. By the resolvent equation $U_a = U_\alpha + U_a(\alpha - a)U_\alpha$, we get

$$\begin{aligned}\int U_\alpha f\,d\Lambda &= \int a^{-1}U_\alpha f\,d\pi = \int Q(a^{-1}U_\alpha f)\,d\pi = \int U_a U_\alpha f\,d\pi\\ &= \alpha^{-1}\int (U_a f - U_\alpha f + U_a a U_\alpha f)\,d\pi = \alpha^{-1}\int U_a f\,d\pi\\ &= \alpha^{-1}\int a^{-1} f\,d\pi = \alpha^{-1}\int f\,d\Lambda. \end{aligned}\tag{16}$$

Therefore Λ is (T_t)-invariant. Now, let Λ' be another (T_t)-invariant measure.

Set $\pi' = a\Lambda'$. Then by the resolvent equation $U_a = U_\alpha + U_\alpha(\alpha - a)U_a$, we get

$$\int Qf\, d\pi' = \int aU_a(af)\, d\Lambda' = \alpha \int U_\alpha a U_a(af)\, d\Lambda'$$

$$= \alpha \int \Big\{ U_a(af) - U_\alpha(af) + U_\alpha \alpha U_a(af) \Big\}\, d\Lambda'$$

$$= \alpha \int U_a(af)\, d\Lambda' = \int af\, d\Lambda' = \int f\, d\pi'.$$

Therefore π' is a constant multiple of π and hence Λ' is also a constant multiple of Λ.

Now since $\pi = \pi Q$ and the kernel $Q(x, \cdot)$ is absolutely continuous with respect to the measure μ for any x, the measure Λ is also absolutely continuous with respect to μ. Further since $\pi(E) = \int \pi(dx)Q(x, E) \geq (\alpha_0/2)\int_E ab\, d\mu$ holds by Lemma 1.3.2, we have $\Lambda(E) \geq (\alpha_0/2)\int_E b\, d\mu$. Consequently Λ is mutually absolutely continuous with respect to μ. □

Let us quickly look at the case where the state space of the Feller process is compact. The process is certainly not transient. Hence it is recurrent in the sense of Harris and has an invariant measure, which has to be a bounded Borel measure. Therefore we have the following.

Theorem 1.3.6 *Assume that a Feller process with compact state space satisfies Condition* (A). *Then it is recurrent in the sense of Harris and has a unique invariant probability.* □

Ergodic properties

Let $(W, \mathscr{B}(W), P_x : x \in S)$ be a Feller process. An element B of $\mathscr{B}(W)$ is called (θ_t)-invariant if $\theta_t^{-1}B = B$ holds for any t. We denote by $\mathscr{I}$ the set of all (θ_t)-invariant sets. It is a sub σ-field of $\mathscr{B}(W)$, called the *invariant σ-field.* The invariant σ-field $\mathscr{I}$ is called *trivial a.s.* if for any B of $\mathscr{I}$, $P_x(B) = 0$ holds for all x or $P_x(B) = 1$ holds for all x. The Feller process is called *ergodic* if its invariant σ-field is trivial a.s. Now let $Z(w)$ be a $\mathscr{B}(W)$-measurable random variable. It is $\mathscr{I}$-measurable if and only if $Z(\theta_t w) = Z(w)$ is satisfied for any t and w. It is called *(θ_t)-invariant.* Then the invariant σ-field is trivial if and only if any bounded (θ_t)-invariant function equals a constant c a.s. for any P_x.

Now let h be a bounded measurable function on S. It is called (T_t)-*invariant* if $T_t h(x) = h(x)$ holds for any t and x. We will characterize the ergodic property by means of (T_t)-invariant functions.

***Theorem 1.3.7** A Feller process is ergodic if every bounded measurable (T_t)-invariant function is a constant function. Conversely if the Feller process is ergodic, every bounded continuous (T_t)-invariant function is a constant function.*

Proof Assume that any bounded (T_t)-invariant function is constant. Let B be any (θ_t)-invariant set. Set $f(x) = P_x(B)$. It is (T_t)-invariant since by Theorem 1.2.9 we have $T_t f(x) = P_x(\theta_t^{-1} B) = P_x(B) = f(x)$. Therefore f is a constant function. Further $f(w(t))$ is a martingale for any P_x, because we have by the Markov property,

$$E[f(w(t))|\mathscr{F}_s] = T_{t-s}f(w(s)) = f(w(s))$$

for every $s \leq t$. Therefore $\lim_{t\to\infty} f(w(t)) = \chi(B)$ holds a.s. P_x (see Exercise 1.2.10). Then we have $f \equiv 1$ or $f \equiv 0$, proving that the process is ergodic.

Assume next that the process is ergodic. Let h be a bounded continuous invariant function. Then $h(w(t))$ is a martingale for any P_x as we have seen above. Set $Z(w) = \overline{\lim}_{t\to\infty} h(w(t))$. It is (θ_t)-invariant. By the convergence theorem of martingales (Theorem 1.2.3), we have $h(x) = E_x[Z]$ for any x. Therefore $E_x[Z]$ is a constant function. □

A sub σ-field $\mathscr{T}$ of $\mathscr{F}$ defined below is called the *tail σ-field*.

$$\mathscr{T} = \bigcap_{t\geq 0} \theta_t^{-1}\mathscr{B}(W) = \bigcap_{t>0} \sigma(w(u) : u \geq t). \tag{17}$$

Notice that a set A belongs to $\mathscr{T}$ if and only if for every t there exists a set A_t of $\mathscr{B}(W)$ such that $A = \theta_t^{-1} A_t$. Therefore $\mathscr{I}$ is a sub σ-field of $\mathscr{T}$. Consequently if the tail σ-field is trivial a.s., the Feller process is ergodic. We wish to show that any Feller process satisfying Condition (A) has this property.

Before we proceed to this problem, we will define a space–time invariant function. Let $h(x, t)$, $(x, t) \in \mathbb{R}^d \times \mathbb{T}$ be a bounded measurable function. It is called *space–time invariant* if

$$h(x, t) = \int_S h(y, t + s) P_s(x, \mathrm{d}y) \tag{18}$$

is satisfied for any x, t, s. Note that if $h(x, t)$ is space–time invariant, then for every t_0, $h(w(t), t + t_0)$ is a martingale with respect to P_x for any x, since we have by the Markov property,

$$E_x[h(w(t), t + t_0)|\mathscr{F}_s] = E_{w(s)}[h(w(t - s), t + t_0)] = h(w(s), s + t_0).$$

Theorem 1.3.8 *For a Feller process the following three statements are equivalent:*

(i) *the tail σ-field $\mathscr{T}$ is trivial a.s.*
(ii) *any bounded space–time invariant function is a constant function.*
(iii) *for every pair of probabilities π_1 and π_2, we have*

$$\lim_{t \uparrow \infty} \|(\pi_1 - \pi_2)P_t\| = 0, \tag{19}$$

where $\pi_1 P_t(\cdot) = \int \pi_1(\mathrm{d}x)P_t(x, \cdot)$ and $\| \ \|$ is the norm of the total variation.

Proof We first prove that (ii) implies (i). Let $A \in \mathscr{T}$. Let A_t be an element of $\mathscr{B}(W)$ such that $\theta_t^{-1}A_t = A$. We set $h(x, t) = P_x(A_t)$. Then, noting the relation $\theta_s^{-1}A_{t+s} = A_t$ and using the Markov property (Theorem 1.2.9), it can be seen immediately that $h(x, t)$ is a bounded space–time invariant function. Therefore it is a constant by assumption (ii). The constant should be 0 or 1 since $\lim_{t \uparrow \infty} h(w(t), t) = \chi(A)$ a.s. holds. Therefore $\mathscr{T}$ is trivial a.s.

We next suppose (i) holds and prove (iii). Set $\mathscr{T}_t \equiv \theta_t^{-1}\mathscr{B}(W)$. Then for any probability ν,

$$\lim_{t \uparrow \infty} \sup_{A \in \mathscr{T}_t} |P_\nu(A \cap B) - P_\nu(A)P_\nu(B)| = 0. \tag{20}$$

Indeed, we have

$$\begin{aligned}\sup_{A \in \mathscr{T}_t} |P_\nu(A \cap B) - P_\nu(A)P_\nu(B)| &= \sup_{A \in \mathscr{T}_t} \left| \int_A (P_\nu(B|\mathscr{T}_t) - P_\nu(B))\, \mathrm{d}P_\nu \right| \\ &\le \int_\Omega |P_\nu(B|\mathscr{T}_t) - P_\nu(B)|\, \mathrm{d}P_\nu.\end{aligned}$$

Define $\hat{\mathscr{F}}_t$ by $\mathscr{T}_{-t}$ for $t \in (-\infty, 0]$. Then $\{\hat{\mathscr{F}}_t : t \in (-\infty, 0]\}$ is a filtration such that $\bigcap_t \hat{\mathscr{F}}_t = \mathscr{T}$. Then $P_\nu(B|\hat{\mathscr{F}}_t)$ converges to $P_\nu(B|\mathscr{T}) = P_\nu(B)$. Therefore the last member of the above converges to 0. Now let $\pi_1 - \pi_2 = \pi^+ - \pi^-$ be the Jordan–Hahn decomposition of the signed measure $\pi_1 - \pi_2$ and let S^+, S^- be the associated sets of the decomposition. Put $\pi = \pi^+ + \pi^-$. Then we have by (20)

$$\lim_{t \to \infty} \sup_{\Gamma \in \mathscr{B}(S)} |P_{\pi^+}(w(t) \in \Gamma) - P_\pi(w(0) \in S^+)P_\pi(w(t) \in \Gamma)| = 0,$$

$$\lim_{t \to \infty} \sup_{\Gamma \in \mathscr{B}(S)} |P_{\pi^-}(w(t) \in \Gamma) - P_\pi(w(0) \in S^-)P_\pi(w(t) \in \Gamma)| = 0.$$

These two imply

$$\lim_{t\to\infty} \sup_{\Gamma\in\mathscr{B}(S)} |P_{\pi^+}(w(t)\in\Gamma) - P_{\pi^-}(w(t)\in\Gamma)| = 0,$$

which is equivalent to (19).

Finally we will show that (iii) implies (ii). Let h be a bounded space–time invariant function. Then

$$|h(x,t) - h(y,t)| = \left|\int h(z, t+s)(P_s(x, dz) - P_s(y, dz))\right|$$

$$\leq \|h\|\,\|P_s(x,\cdot) - P_s(y,\cdot)\| \xrightarrow[s\to\infty]{} 0,$$

where $\|h\|$ is the supremum norm of the function h. Therefore $h(x,t) = h(y,t)$ for any x, y, t. This implies immediately that h is a constant function. □

We can now prove the trivialness of the tail σ-field.

Theorem 1.3.9 *Assume that a Feller process satisfies condition* (A) *and is recurrent in the sense of Harris. Then the tail σ-field $\mathscr{T}$ is trivial a.s.*

Proof Let h be a bounded space–time invariant function. We will first show that h does not depend on t. Suppose on the contrary that h does depend on t. Then there exists $(x_0, t_0) \in \mathbb{R}^d \times \mathbb{T}$ such that $h(x_0, t_0) \neq h(x_0, 0)$. Let $Z_0(t)$ and $Z_{t_0}(t)$ be the right continuous modifications of the martingales $h(w(t), t)$ and $h(w(t), t + t_0)$ with respect to P_{x_0}, respectively. Set $Z_0 = \lim_{t\to\infty} Z_0(t)$ and $Z_{t_0} = \lim_{t\to\infty} Z_{t_0}(t)$. Then $P_{x_0}(Z_0 \neq Z_{t_0}) > 0$ holds. We may assume $P_{x_0}(Z_0 < Z_{t_0}) > 0$. Then there exist real numbers a, b such that $a < b$ and $P_{x_0}(Z_0 < a < b < Z_{t_0}) = \delta > 0$ holds. Set

$$g(x,s) = P_x(Z_s(t) \leq a,\ Z_{s+t_0}(t) \geq b \text{ for all } t \in \mathbb{T}).$$

Then by the Markov property we have

$$g(w(s), s) = P_x(Z_0(t) \leq a \text{ and } Z_{t_0}(t) \geq b \text{ for all } t \geq s \,|\, \mathscr{F}_s).$$

Letting $s \to \infty$, $g(w(s), s)$ converges to 1 on the set $\{Z_0 < a < b < Z_{t_0}\}$. On the other hand, by Condition (A) for any bounded set C of $\mathbb{R}^d$ with $\mu(C) > 0$ there exists $\varepsilon > 0$ such that

$$\inf\{p_{t_0}(x,y) : (x,y) \in C \times C\} \geq \varepsilon, \qquad \inf\{p_{2t_0}(x,y) : (x,y) \in C \times C\} \geq \varepsilon.$$

Since the process is recurrent in the sense of Harris, for any s there exists $t \geq s$ such that $w(t) \in C$ a.s. Thus there exists $(x_1, t_1) \in \mathbb{R}^d \times \mathbb{T}$ such that $x_1 \in C$ and $1 - g(x_1, t_1) \leq \frac{\varepsilon}{4}\mu(C)$. Set

$$C' = C \cap \{x : h(x, t_1 + t_0) \le a \text{ and } h(x, t_1 + 2t_0) \ge b\}^c.$$

Then we have

$$\begin{aligned} \varepsilon\mu(C') &\le P_{x_1}(w(t_0) \in C') \\ &\le P_{x_1}(h(w(t_0), t_1 + t_0) > a \text{ or } h(w(t_0), t_1 + 2t_0) < b) \\ &\le 1 - g(x_1, t_1) \le \frac{\varepsilon}{4}\mu(C). \end{aligned}$$

Set further $C'' = C \cap \{x : h(x, t_1 + 2t_0) \le a \text{ and } h(x, t_1 + 3t_0) \ge b\}^c$. In the same way we can show $\mu(C'') \le \frac{1}{4}\mu(C)$. Therefore $C \cap (C' \cup C'')^c$ is non-empty. Choose a point x from this set. Then we have $h(x, t_1 + 2t_0) \ge b$ and $h(x, t_1 + 2t_0) \le a$. This is a contradiction. We have thus proved that h does not depend on t.

We will next show that the function $h(x) = h(x, t)$ is a constant function. If h is not a constant function, then there exist two real numbers $a < b$ such that both $\mu(\{x : h(x) < a\})$ and $\mu(\{x : h(x) > b\})$ are positive. Then $\varliminf_{t\to\infty} h(w(t)) \le a$ and $\varlimsup_{t\to\infty} h(w(t)) \ge b$ holds with positive probability with respect to P_x since the process is recurrent in the sense of Harris. This contradicts the existence of Z_0. Therefore h is a constant function, proving that the tail σ-field is trivial. □

Ergodic theorems

We first establish an ergodic theorem for a Feller semigroup.

Theorem 1.3.10 *Assume that a Feller process satisfies Condition* (A).

(i) *If it is recurrent in the sense of Harris and has an invariant probability* Λ, *then for any bounded measurable function* f,

$$T_t f(x) \xrightarrow[t\to\infty]{} \int f \, d\Lambda \tag{21}$$

holds for every $x \in S$.

(ii) *If it is recurrent in the sense of Harris and has an infinite invariant measure* Λ, *then for any bounded measurable function* f *such that* $\int |f| \, d\Lambda < \infty$,

$$T_t f(x) \xrightarrow[t\to\infty]{} 0 \tag{22}$$

holds for every $x \in S$.

(iii) *If it is transient, for any bounded measurable function* f *with compact support, the convergence* (22) *holds for every* x.

Proof The proof of (i) is immediate from Theorems 1.3.8 and 1.3.9. Indeed, since the tail σ-field is trivial, we have (19). Set $\pi_1 = \delta_x$ (the delta measure concentrated at the point x) and $\pi_2 = \Lambda$ in (19). Then we obtain

$$\overline{\lim_{t\to\infty}} \|P_t(x, \cdot) - \Lambda\| = 0. \tag{23}$$

This implies (21) for any bounded measurable function.

We next consider case (ii). It is sufficient to prove (22) for a bounded non-negative measurable function f such that $\int f \, d\Lambda < \infty$. For any given x_0 of S and $\varepsilon > 0$, set $B_t = \{y : T_t f(y) > T_t f(x_0) - \varepsilon\}$. Since $\|P_t(x_0, \cdot) - P_t(y, \cdot)\|$ converges to 0 for any y by Theorem 1.3.8, $\underline{\lim}_{t\to\infty} B_t = S$ holds. Further we have

$$\int f \, d\Lambda = \int T_t f(y) \, d\Lambda(y) \geq \{T_t f(x_0) - \varepsilon\} \Lambda(B_t).$$

Since $\Lambda(B_t)$ tends to infinity as t tends to infinity, we obtain $\overline{\lim}_{t\to\infty} T_t f(x_0) \leq \varepsilon$. Therefore $T_t f(x_0)$ converges to 0 as t tends to infinity.

We shall next prove (iii). Let f be a bounded non-negative function with compact support. Set $U_0 f(x) = \int_0^\infty T_u f(x) \, du$. It is a bounded function since the semigroup $\{T_t\}$ is transient. It is continuous by Lemma 1.3.1. Further $U_0 f(w(t))$ is a supermartingale because

$$E_x[U_0 f(w(t)) | \mathscr{F}_s] = T_{t-s} U_0 f(w(s)) = \int_{t-s}^\infty T_u f(w(s)) \, du \leq U_0 f(w(s)). \tag{24}$$

Then $\lim_{t\to\infty} U_0 f(w(t)) \equiv Y$ exists by Theorem 1.2.3. Further we have by (24)

$$E_x[U_0 f(w(t))] = \int_t^\infty T_u f(x) \, du \xrightarrow[t\to\infty]{} 0.$$

Therefore $Y = 0$ a.s. P_x for any x. Now since $U_0 f(x) > 0$ holds for any x of S by Condition (A), the above fact implies that the paths $w(t)$ tend to infinity as $t \to \infty$ a.s. P_x for any x. This proves that $f(w(t))$ also converges to 0 a.s. P_x for any x if f is a bounded function of compact support. Then $T_t f(x)$ converges to 0 for any x. □

Suppose now that the semigroup $\{T_t\}$ has an invariant probability Λ. Define the measure P_Λ on $(W, \mathscr{B}(W))$ by

$$P_\Lambda(B) = \int P_x(B) \Lambda(dx) \quad \text{for all } B \in \mathscr{B}(W), \tag{25}$$

and denote by E_Λ the expectation. Then for each t, θ_t is a P_Λ-preserving transformation i.e. $P_\Lambda(\theta_t^{-1}B) = P_\Lambda(B)$ holds for any B of $\mathscr{B}(W)$. Indeed, setting $f(x) \equiv P_x(B)$ we have by the Markov property

$$P_\Lambda(\theta_t^{-1}B) = \int E_x[P_{w(t)}(B)]\, d\Lambda(x) = \int T_t f(x)\, d\Lambda(x) = \int f(x)\, d\Lambda(x)$$
$$= P_\Lambda(B).$$

We give here a pointwise ergodic theorem without proof.

Theorem 1.3.11 *Let* Φ *be any element of* $L^p(W, \mathscr{B}(W), P_\Lambda)$ *where* $1 \le p < \infty$. *Then*

$$\tilde{\Phi}(w) = \lim_{t\to\infty} \frac{1}{t}\int_0^t \Phi(\theta_s w)\, ds \tag{26}$$

exists for a.e. w. *The limit function* $\tilde{\Phi}$ *is* (θ_t)*-invariant a.s.* P_Λ. *Further the above convergence takes place in* L^p. □

For the proof see e.g. Doob [26, Chapter XI].

The measure P_Λ is called *ergodic* if $P_\Lambda(B) = 0$ or 1 holds for any invariant set B. Then P_Λ is ergodic if and only if the limit function $\tilde{\Phi}$ of (26) is a constant function a.s. for any Φ. The constant is $E[\Phi]$.

We will summarize pathwise ergodic theorems of Feller processes.

Theorem 1.3.12 *Assume that a Feller process satisfies Condition* (A).

(i) *If it is recurrent in the sense of Harris and has an invariant probability* Λ, *then for any bounded measurable function* f,

$$\frac{1}{t}\int_0^t f(w(s))\, ds \xrightarrow[t\to\infty]{} \int f\, d\Lambda \qquad \text{a.s. } P_x \tag{27}$$

holds for every $x \in S$.

(ii) *If it is recurrent in the sense of Harris and has an infinite invariant measure* Λ, *then for any bounded measurable function such that* $\int |f|\, d\Lambda < \infty$,

$$f(w(t)) \xrightarrow[t\to\infty]{} 0 \qquad \text{in } L^1(P_x) \tag{28}$$

holds for every $x \in S$.

(iii) *If it is transient, then for any bounded measurable function* f *with compact support,* (28) *holds a.s.* P_x *for every* $x \in S$.

Proof We first consider case (i). Since the measure P_Λ is ergodic we have by the pointwise ergodic theorem (Theorem 1.3.11),

$$\lim_{t\to\infty} \frac{1}{t}\int_0^t f(w(s))\,\mathrm{d}s = \int f\,\mathrm{d}\Lambda$$

a.s. P_Λ. Now let B be the set of all w such that the above limit exists and is equal to the right hand side. The set B is (θ_t)-invariant and satisfies $P_\Lambda(B) = 1$. Therefore $P_x(B) = 1$ holds for all x by Theorem 1.3.9, proving (27). The assertions (ii) and (iii) are immediate from the corresponding assertions (ii) and (iii) of Theorem 1.3.10. □

Potential kernels

Let $\{U_\alpha\}$ be the resolvent of a Feller semigroup $\{T_t\}$ satisfying Condition (A). A positive kernel U is called a *potential kernel* for the resolvent $\{U_\alpha\}$ or the semigroup $\{T_t\}$ if it satisfies

$$(I - \alpha U_\alpha)Uf = U_\alpha f, \tag{29}$$

for every positive number α and every function f such that Uf is bounded and continuous. The function Uf is called a *potential* of f. Note that if Uf belongs to the domain of the infinitesimal generator A of the semigroup $\{T_t\}$, the potential satisfies $AUf = (\alpha - U_\alpha^{-1})Uf = -f$. Therefore Uf is a solution of the Poisson equation $Au = -f$ satisfying condition $\lim_{x\to\infty} u(x) = 0$. Now if the process is transient,

$$U(x, E) = \int_0^\infty P_t(x, E)\,\mathrm{d}t \tag{30}$$

is a potential kernel. Indeed if f is a bounded measurable function with compact support, Uf is a bounded continuous function and satisfies $Uf = \lim_{\beta\to 0} U_\beta f$. Therefore it satisfies (29) since $\{U_\alpha\}$ satisfies the resolvent equation $U_\beta f - (\alpha - \beta)U_\alpha U_\beta f = U_\alpha f$.

Now assume that the Feller process is recurrent in the sense of Harris with the invariant measure Λ. Then (30) diverges if $\Lambda(E) > 0$. However as is shown below, we can construct another kernel $W(x, \cdot)$ such that Wf is a bounded continuous function and satisfies (29) if f is a bounded measurable function with compact support satisfying $\int f\,\mathrm{d}\Lambda = 0$. Such a kernel W is called a *recurrent potential kernel* and Wf, a *recurrent potential of f*. The function f with $\int f\,\mathrm{d}\Lambda = 0$ is said to be of *null charge*.

For the construction of such a kernel W, we need two lemmas.

Lemma 1.3.13 *There exists a bounded strictly positive continuous function h such that $U_h h(x) = 1$ and $U_h(x, \cdot) \geq \Lambda(\cdot)$ holds for all x.*

Proof Define the measure μ' by $\mathrm{d}\mu' = (\alpha_0/2)b\,\mathrm{d}\mu$ where α_0 and b are those introduced in Lemma 1.3.2. Then the invariant measure Λ is mutually absolutely continuous with respect to μ'. We can take the Radon–Nikodym density $d\Lambda/d\mu'$ as a strictly positive continuous function. Indeed, $g(y) = (2\int p_{t_0}(x, y)\, d\Lambda(x))(\alpha_0 b(y))^{-1}$ is a density function. Set $h = a(1+g)^{-1}$ where a is the function introduced in Lemma 1.3.2. Then $a - h = gh$. By the resolvent equation and Lemma 1.3.2 we have for any $f \geq 0$,

$$U_h f \geq U_a(a-h)U_h f \geq \int (a-h)U_h f\, d\mu' = \int ghU_h f\, d\mu' = \int hU_h f\, d\Lambda. \tag{31}$$

Now integrate the resolvent equation $U_\alpha f - U_h f = U_\alpha(h-\alpha)U_h f$ by the measure Λ and apply the relation $\alpha \int U_\alpha f\, d\Lambda = \int f\, d\Lambda$. Then we obtain $\int f\, d\Lambda = \int hU_h f\, d\Lambda$. Therefore the last member of (31) is $\int f\, d\Lambda$, proving $U_h(x, \cdot) \geq \Lambda(\cdot)$ for all x. Finally since $h \leq a$ we have $U_h h \equiv 1$ (see the proof of Theorem 1.3.4). □

Now define kernels V and Vh by $V(x, dy) = U_h(x, dy) - \Lambda(dy)$ and $(Vh)(x, dy) = V(x, dy)h(y)$, and then define a kernel W by

$$W = \sum_{n \geq 0} (Vh)^n V. \tag{32}$$

We denote by $B_K(S)$ the set of all bounded measurable functions with compact supports.

Lemma 1.3.14 *The kernel W maps $B_K(S)$ into $BC(S)$. Further, it satisfies*

$$W = U_\alpha + \alpha U_\alpha W - \left(\int h\, d\Lambda\right)^{-1} U_\alpha h \otimes \Lambda, \tag{33}$$

where the last term represents the kernel $K(x, dy) = (\int h\, d\Lambda)^{-1} U_\alpha h(x)\Lambda(dy)$.

Proof Since $U_h h \equiv 1$ and h is a strictly positive continuous function, $U_h(x, K)$ is a bounded function of x if K is a compact subset of S. (See the proof of Theorem 1.3.4.) Then V maps $B_K(S)$ into $BC(S)$. Further the operator Vh maps $B(S)$ into $BC(S)$. Its supremum norm is estimated as $\|Vh\| = \sup|V(h1)| = 1 - \int h\, d\Lambda$. It is less than 1. Then the infinite sum $\sum_{n \geq 0} (Vh)^n Vh$ is the uniform convergence. Therefore Wf is in $BC(S)$ if f is in $B_K(S)$.

By the resolvent equation and the invariant property of Λ, we have $V = U_\alpha + U_\alpha(\alpha - h)V - U_\alpha h \otimes \Lambda$. Since $W = V + VhW$ holds, we get

$$\begin{aligned} W = U_\alpha + U_\alpha(\alpha - h)V - U_\alpha h \otimes \Lambda + U_\alpha hW + U_\alpha(\alpha - h)VhW \\ - U_\alpha h \otimes (h\Lambda)W. \end{aligned} \tag{34}$$

Now $\int hU_h f\, d\Lambda = \int f\, d\Lambda$ for any f, i.e. $(h\Lambda)U_h = \Lambda$. Therefore we have

$(h\Lambda)Vh = (h\Lambda)(U_h - \Lambda)h = (1 - \int h\,d\Lambda)(h\Lambda)$ and by induction $(h\Lambda)(Vh)^n = (1 - \int h\,d\Lambda)^n(h\Lambda)$. Therefore,

$$(h\Lambda)W = \sum_{n\geq 0}(h\Lambda)(Vh)^n V = \sum_{n\geq 0}\left(1 - \int h\,d\Lambda\right)^{n+1}\Lambda.$$

Consequently, (34) yields

$$\begin{aligned} W &= U_\alpha + U_\alpha hW + U_\alpha(\alpha - h)(V + VhW) - U_\alpha h \otimes (1+q)\Lambda \\ &= U_\alpha + \alpha U_\alpha W - (1+q)U_\alpha h \otimes \Lambda, \end{aligned}$$

where $q = \sum_{n\geq 0}(1 - \int h\,d\Lambda)^{n+1} = (\int h\,d\Lambda)^{-1}(1 - \int h\,d\Lambda)$. This establishes the lemma. □

Theorem 1.3.15 *Assume that a Feller process satisfies Condition* (A) *and is recurrent in the sense of Harris with an invariant measure* Λ. *Then the kernel* W *defined by* (32) *is a recurrent potential. Assume further that the process has an invariant probability* Λ. *If* f *is a function of* $B_K(S)$ *with null charge, then* $\int_0^t T_s f(x)\,ds$ *is bounded in* (t, x) *and satisfies*

$$\int_0^t T_s f(x)\,ds \xrightarrow[t\to\infty]{} Wf(x) - \int Wf\,d\Lambda \tag{35}$$

for every x.

Proof By the previous lemma Wf satisfies $Wf = U_\alpha f + \alpha U_\alpha Wf$ if $\int f\,d\Lambda = 0$. Therefore Wf is a recurrent potential. The above relation is written as

$$\int_0^\infty e^{-\alpha t}\{T_t Wf - Wf\}\,dt = -\int_0^\infty e^{-\alpha t}\left\{\int_0^t T_s f\,ds\right\}dt.$$

Since the above is valid for any $\alpha > 0$ we have by taking the inverse Laplace transform,

$$T_t Wf - Wf = -\int_0^t T_s f\,ds.$$

If the invariant measure Λ is a probability, $T_t Wf$ converges to $\int Wf\,d\Lambda$ boundedly by Theorem 1.3.10. Therefore the convergence (35) is verified. □

Corollary 1.3.16 *Assume that the state space* S *is compact in Theorem* 1.3.15. *Then* W *is a bounded kernel. It satisfies*

$$W(x, E) - \int W(x, E)\,d\Lambda(x) = \lim_{t\to\infty}\int_0^t \{P_s(x, E) - \Lambda(E)\}\,ds \tag{36}$$

for every $E \in \mathscr{B}(S)$. □

Exercise 1.3.17 Show that any Feller semigroup on compact space has an invariant probability. (*Hint*: set $\Lambda_t(\cdot) = t^{-1} \int_0^t P_u(x_0, \cdot)\, du$, $t \in [0, \infty)$, where x_0 is a fixed element of S. Then any weak limit of its subsequence $\{\Lambda_{t_n}\}$ as $t_n \to \infty$ is an invariant probability.)

Exercise 1.3.18 Suppose that a Feller semigroup on non-compact space has no invariant probability. Let f be a bounded continuous function on S such that $\lim_{x \to \infty} f(x) \equiv f(\infty)$ exists. Show the following:

(i) the family of probabilities $\{\Lambda_t\}$ of Exercise 1.3.17 converges vaguely to 0 as $t \to \infty$, i.e. $\int f\, d\Lambda_t$ converges to 0 for any continuous function f with compact support.

(ii) show further that

$$\frac{1}{t}\int_0^t T_s f(x)\, ds \to f(\infty), \qquad \frac{1}{t}\int_0^t f(w(s))\, ds \to f(\infty) \quad \text{in } L^1(P_x)$$

for every x of S.

1.4 Random fields

Kolmogorov's continuity criterion of random fields

Let $\mathbb{D}$ be a domain in $\mathbb{R}^d$. A collection of S-valued random variables $X(x)$, $x \in \mathbb{D}$ is called a *random field with parameter* $\mathbb{D}$. If $\mathbb{D}$ is a time set $\mathbb{T}$, the random field is a stochastic process. Thus similarly to the case of stochastic processes, *modifications of random fields, measurable random fields* and *continuous random fields* are defined.

We shall introduce Kolmogorov's criterion for a given random field to have a modification of a continuous random field.

Theorem 1.4.1 *Let $X(x)$, $x \in \mathbb{D}$ be a random field with values in a Banach space B where $\mathbb{D}$ is a domain in $\mathbb{R}^d$. Assume that there exist positive constants γ, C and α_i, $i = 1, \ldots, d$ with $\sum_{i=1}^d \alpha_i^{-1} < 1$ satisfying*

$$E[\|X(x) - X(y)\|^\gamma] \le C\left(\sum_{i=1}^d |x^i - y^i|^{\alpha_i}\right), \qquad \text{for every } x, y \in \mathbb{D}, \quad (1)$$

where $\|\ \|$ stands for the norm of the Banach space and $x = (x^1, \ldots, x^d)$, $y = (y^1, \ldots, y^d)$. Then $X(x)$ has a continuous modification $\tilde{X}(x)$.

Let β_i, $i = 1, \ldots, d$ be arbitrary positive numbers less than $\alpha_i(\alpha_0 - d)/\alpha_0\gamma$, $i = 1, \ldots, d$ respectively, where α_0 is defined by $\alpha_0^{-1} d = \sum_{i=1}^d \alpha_i^{-1}$. Then for any hypercube $\mathbb{I}$ included in $\mathbb{D}$, there exists a positive random variable $K(\omega)$ with $E[K^\gamma] < \infty$ such that

$$\|\tilde{X}(x, \omega) - \tilde{X}(y, \omega)\| \le K(\omega)\left(\sum_{i=1}^{d} |x^i - y^i|^{\beta_i}\right), \qquad \text{for every } x, y \in \mathbb{I} \quad (2)$$

holds for almost all ω. *Further if* $E[\|X(x_0)\|^\gamma] < \infty$ *holds for some* x_0, *then*

$$E\left[\sup_{x \in K} \|\tilde{X}(x)\|^\gamma\right] < \infty \tag{3}$$

holds for any compact subset K *of* $\mathbb{D}$. □

We shall call the above $\tilde{X}(x)$ a $(\beta_1, \ldots, \beta_d)$-*Hölder continuous random field.*

Before the proof we need an inequality concerning the modulus of continuity of a Banach space valued function defined on the hypercube $\mathbb{I} = [0, 1]^d$.

Some terminology is needed. Let q be a positive integer greater than 1. Then any real number x of $[0, 1]$ is represented as $\sum_{k=1}^{\infty} a_k q^{-k}$ where a_k, $k = 1, 2, \ldots$ are non-negative integers less than q. It is called a *q-adic expansion of* x. The finite partial sum $\sum_{k=1}^{N} a_k q^{-k}$ of the q-adic expansion of x is denoted by x_N. If $x = x_N$ holds for some N, x is called a *q-adic rational.* Let $\mathbf{q} = (q_1, \ldots, q_d)$ be a vector of positive integers greater than 1. A point $x = (x^1, \ldots, x^d)$ of $\mathbb{I}$ is called $\mathbf{q}$-*adic rational* if each x^i is q_i-adic rational. The set of all $\mathbf{q}$-adic rationals x of length N, i.e. $x^i = x_N^i$ for any $i = 1, \ldots, d$ is denoted by Δ_N. The number of elements of Δ_N is at most $\prod_{i=1}^{d} q_i^N$. Set $\Delta = \bigcup_N \Delta_N$. It is a countable dense subset of $\mathbb{I}$. Now let $\boldsymbol{\beta} = (\beta_1, \ldots, \beta_d)$ be a vector of positive numbers such that $\beta_i/\beta_j = \log q_j/\log q_i$ holds for any $i, j = 1, \ldots, d$. Then $q_1^{\beta_1} = \cdots = q_d^{\beta_d}$ holds. Write the common value as δ.

Given a map $f\colon \Delta \to B$ we define for each positive integer N the *modulus of continuity* $\Delta_N(f)$ and the *modulus of* $\boldsymbol{\beta} = (\beta_1, \ldots, \beta_d)$-*Hölder continuity* $\Delta_N^{\boldsymbol{\beta}}(f)$ by

$$\Delta_N(f) = \max_{x, y \in \Delta_N, |x^i - y^i| \le q_i^{-N}, i = 1, \ldots, d} \|f(x) - f(y)\|, \tag{4}$$

$$\Delta_N^{\boldsymbol{\beta}}(f) = \delta^N \Delta_N(f) = \Delta_N(f)/(q_i^{-N})^{\beta_i}. \tag{5}$$

Lemma 1.4.2 The inequality

$$\|f(x) - f(y)\| \le 4|\mathbf{q}|\left(\sum_{N=1}^{\infty} \Delta_N^{\boldsymbol{\beta}}(f)\right)\left(\max_i |x^i - y^i|^{\beta_i}\right),$$

$$\text{for every } x, y \in \Delta \tag{6}$$

holds for any map $f\colon \Delta \to B$, *where* $|\mathbf{q}| = \sum_{i=1}^{d} q_i$.

Proof It is enough to prove the above for the case where $\sum_{N=1}^{\infty} \Delta_N^{\boldsymbol{\beta}}(f) < \infty$. For every $x = (x^1, \ldots, x^d)$ of $\mathbb{I}$ denote the associated $\mathbf{q}$-adic rational of length N by $x_N = (x_N^1, \ldots, x_N^d)$. We define a simple function g_N on $\mathbb{I}$ by $g_N(x) = f(x_N)$. Then we have clearly

$$\|g_{N+1}(x) - g_N(x)\| \le |\mathbf{q}|\Delta_{N+1}(f)$$

for every $x \in \mathbb{I}$. Therefore,

$$\sum_{N=1}^{\infty} \|g_{N+1}(x) - g_N(x)\| \le |\mathbf{q}| \sum_{N=1}^{\infty} \Delta_{N+1}(f) \le |\mathbf{q}| \sum_{N=1}^{\infty} \Delta_{N+1}^{\boldsymbol{\beta}}(f) < \infty.$$

Then $\{g_N\}$ converges uniformly on $\mathbb{I}$. Let $g(x)$ be the limit function. Then $g(x) = f(x)$ is satisfied for any x of Δ. In the sequel we shall prove the lemma for the function g instead of f.

Let x, y be any two points in Δ. There exists a non-negative integer k such that $\delta^{-(k+1)} \le max_i |x^i - y^i|^{\beta_i} \le \delta^{-k}$ since $\delta > 1$. Then,

$$\begin{aligned}\|g(x) - g_{k+1}(x)\| &\le |\mathbf{q}| \sum_{N=k+1}^{\infty} \Delta_{N+1}(f) \le |\mathbf{q}|\delta^{-(k+1)} \sum_{N=k+1}^{\infty} \Delta_N^{\boldsymbol{\beta}}(f) \\ &\le |\mathbf{q}| \left(\sum_{N=1}^{\infty} \Delta_N^{\boldsymbol{\beta}}(f)\right)\left(\max_i |x^i - y^i|^{\beta_i}\right).\end{aligned}$$

Further since $|x^i - y^i| \le \delta^{-k/\beta_i} = q_i^{-k}$ holds for any $i = 1, \ldots, d$, we have

$$\begin{aligned}\|g_{k+1}(x) - g_{k+1}(y)\| &\le 2|\mathbf{q}|\Delta_{k+1}(f) \le 2|\mathbf{q}|\delta^{-(k+1)}\Delta_{k+1}^{\boldsymbol{\beta}}(f) \\ &\le 2|\mathbf{q}| \left(\sum_{N=1}^{\infty} \Delta_N^{\boldsymbol{\beta}}(f)\right)\left(\max_i |x^i - y^i|^{\beta_i}\right).\end{aligned}$$

Therefore we have

$$\begin{aligned}\|g(x) - g(y)\| &\le \|g(x) - g_{k+1}(x)\| + \|g_{k+1}(x) - g_{k+1}(y)\| \\ &\quad + \|g_{k+1}(y) - g(y)\| \\ &\le 4|\mathbf{q}| \left(\sum_{N=1}^{\infty} \Delta_N^{\boldsymbol{\beta}}(f)\right)\left(\max_i |x^i - y^i|^{\beta_i}\right).\end{aligned}$$

The proof is complete. □

By the above lemma the map $f: \Delta \to B$ satisfying $\sum_{N=1}^{\infty} \Delta_N^{\boldsymbol{\beta}}(f) < \infty$ has a continuous extension $\tilde{f}: \mathbb{I} \to B$ which satisfies the inequality (6) for all x, $y \in \mathbb{I}$. We shall call $\tilde{f}$ the *$\boldsymbol{\beta}$-Hölder continuous extension of f.*

We shall apply the above lemma to the random field $X(x)$. Observe that for each ω, $X(\cdot, \omega)$ can be regarded as a map from $\mathbb{I}$ into B. Then we have

$$\|X(x, \omega) - X(y, \omega)\| \le 4|\mathbf{q}|\left(\sum_{N=1}^{\infty} \Delta_N^{\boldsymbol{\beta}}(X(\omega))\right)\left(\sum_{i=1}^{d} |x^i - y^i|^{\beta_i}\right)$$

for every $x, y \in \Delta$. We want to prove that $\sum_{N=1}^{\infty} \Delta_N^{\boldsymbol{\beta}}(X(\omega))$ is finite a.e.

Lemma 1.4.3 *Suppose that $X(x)$ satisfies (1). If $\beta_1, \ldots, \beta_d$ satisfy $\beta_j < \alpha_j(\gamma + \sum_{i=1}^d \beta_i^{-1})^{-1}$, $j = 1, \ldots, d$, then we have*

$$E\left[\left|\sum_{N=1}^{\infty} \Delta_N^{\boldsymbol{\beta}}(X)\right|^{\gamma}\right] < \infty. \tag{7}$$

Proof We first consider the case $\gamma < 1$. Since $(a + b)^{\gamma} < a^{\gamma} + b^{\gamma}$ holds for any positive numbers a, b, we have

$$E\left[\left|\sum_{N=1}^{\infty} \Delta_N^{\boldsymbol{\beta}}(X)\right|^{\gamma}\right] \le \sum_{N=1}^{\infty} E[\Delta_N^{\boldsymbol{\beta}}(X)^{\gamma}].$$

Observe that

$$\Delta_N^{\boldsymbol{\beta}}(X)^{\gamma} \le \sum \|X(x) - X(y)\|^{\gamma}\delta^{N\gamma}$$

where the summation is taken for all $x, y \in \Delta_N$ such that $|x^i - y^i| \le q_i^{-N}$, $i = 1, \ldots, d$. The number of the summation is at most $2^d(\Pi_{i=1}^d q_i^N)$. Therefore

$$E[\Delta_N^{\boldsymbol{\beta}}(X)^{\gamma}] \le 2^d(\Pi_{i=1}^d q_i^N)\max E[\|X(x) - X(y)\|^{\gamma}]\delta^{N\gamma}.$$

By (1), the above is bounded by

$$2^d(\Pi_{i=1}^d q_i^N)C\left(\sum_{j=1}^{d} q_j^{-N\alpha_j}\right)\delta^{N\gamma} \le 2^d C\left\{\sum_{j=1}^{d} \delta^{N\left(\sum_{i=1}^{d} \beta_i^{-1} - \alpha_j\beta_j^{-1} + \gamma\right)}\right\}.$$

Therefore we have

$$E\left[\left|\sum_{N=1}^{\infty} \Delta_N^{\boldsymbol{\beta}}(X)\right|^{\gamma}\right] \le C' \sum_{j=1}^{d}\sum_{N=1}^{\infty} \delta^{N\left(\sum_{i=1}^{d} \beta_i^{-1} - \alpha_j\beta_j^{-1} + \gamma\right)},$$

where $C' = 2^d C$. The infinite sum is finite if and only if $\sum_{i=1}^d \beta_i^{-1} - \alpha_j\beta_j^{-1} + \gamma < 0$.

The case $\gamma > 1$ can be shown similarly. In fact we have

$$\begin{aligned} \mathrm{E}\left[\left|\sum_{N} \Delta_N^{\boldsymbol{\beta}}(X)\right|^{\gamma}\right]^{1/\gamma} &\le \sum_{N} E[\Delta_N^{\boldsymbol{\beta}}(X)^{\gamma}]^{1/\gamma} \\ &\le (C')^{1/\gamma} \sum_{j=1}^{d}\sum_{N=1}^{\infty} (\delta^{1/\gamma})^{N(\ldots)\left(\sum_{i=1}^{d} \beta_i^{-1} - \alpha_j\beta_j^{-1} + \gamma\right)}. \end{aligned}$$

It is finite if and only if $\sum_i \beta_i^{-1} - \alpha_j\beta_j^{-1} + \gamma < 0$. □

Proof of Theorem 1.4.1 Let β_i, $i = 1, \ldots, d$ be arbitrary numbers less than $\alpha_i(\alpha_0 - d)/\alpha_0\gamma$, respectively. Take $\varepsilon > 0$ such that $\beta_i < \alpha_i(\alpha_0 - d) \times$

$\{\alpha_0\gamma(1+\alpha_0^{-1}d\varepsilon)\}^{-1}$ holds for all i. We can choose $\beta'_1, \ldots, \beta'_d$ such that $\beta'_i \geq \beta_i$, $\beta'_i/\beta'_j = \log q'_j/\log q'_i$ hold for some positive integers $q'_i, \ldots, q'_d$ greater than 1 and $\beta'_1, \ldots, \beta'_d$ satisfy

$$\frac{\alpha_i(\alpha_0 - d)}{\alpha_0\gamma(1+\varepsilon)} < \beta'_i < \frac{\alpha_i(\alpha_0 - d)}{\alpha_0\gamma(1+\alpha_0^{-1}d\varepsilon)}.$$

Then $\beta'_1, \ldots, \beta'_d$ satisfy the condition of Lemma 1.4.3. In fact,

$$\frac{\alpha_j}{\gamma + \sum_i \frac{1}{\beta'_i}} \geq \frac{\alpha_j}{\gamma + \sum_i \frac{\alpha_0\gamma(1+\varepsilon)}{\alpha_i(\alpha_0 - d)}} = \frac{\alpha_j(\alpha_0 - d)}{\gamma(\alpha_0 + d\varepsilon)} > \beta'_j.$$

Consequently $\sum_{N=1}^{\infty} \Delta_N^{\boldsymbol{\beta}'}(X) < \infty$ holds a.e. by Lemma 1.4.3. Note $\beta'_i \geq \beta_i$. Then

$$\|X(x) - X(y)\| \leq 4|\mathbf{q}'|\left(\sum_{N=1}^{\infty} \Delta_N^{\boldsymbol{\beta}'}(X)\right)\left(\sum_{i=1}^{d} |x^i - y^i|^{\beta_i}\right),$$

where $\mathbf{q}' = (q'_1, \ldots, q'_d)$. Therefore, $X(x)$, $x \in \Delta$ has a $\boldsymbol{\beta}$-Hölder continuous extension $\tilde{X}(x)$, $x \in \mathbb{I}$. Note that both $\tilde{X}(x)$, $x \in \mathbb{I}$ and $X(x)$, $x \in \mathbb{I}$ are continuous in probability and that $P(X(x) = \tilde{X}(x)) = 1$ holds for any $x \in \Delta$. Then the same property holds for all x in $\mathbb{I}$. Therefore $\tilde{X}(x)$ is a continuous modification of $X(x)$ and it satisfies the inequality (2).

Finally the domain $\mathbb{D}$ is covered by a countable number of hypercubes included in $\mathbb{D}$. Therefore $X(x)$, $x \in \mathbb{D}$ has a continuous modification. The moment property (3) is immediate from (2) since the random variable $K(\omega)$ in (2) satisfies $E[K^\gamma] < \infty$. □

Remark Theorem 1.4.1 and its proof show that any random field $X(x)$, $x \in \Delta$ where Δ is the set of all $\mathbf{q}$-adic rationals can be extended to a continuous random field $\tilde{X}(x)$, $x \in \mathbb{I}$ if the inequality (1) holds for any x, y in Δ.

We shall next discuss the regularity of a random field with double parameters. Some notation is needed. Let $f(x, y)$, $x \in \mathbb{D}_1$, $y \in \mathbb{D}_2$ be a function with values in a Banach space B. Set

$$\Delta_{(x,x')}f(y) = f(x', y) - f(x, y), \tag{8}$$

$$\Delta_{(y,y')}f(x) = f(x, y') - f(x, y),$$

$$\Delta_{(x,x'),(y,y')}f = \Delta_{(x,x')}f(y') - \Delta_{(x,x')}f(y)$$

$$= \Delta_{(y,y')}f(x') - \Delta_{(y,y')}f(x).$$

Theorem 1.4.4 *(Kolmogorov–Čencov). Let $X(x, y)$, $x \in \mathbb{D}_1$, $y \in \mathbb{D}_2$ be a random field with values in a Banach space B, where $\mathbb{D}_1$ and $\mathbb{D}_2$ are domains in $\mathbb{R}^{d_1}$ and $\mathbb{R}^{d_2}$, respectively. Assume that there exist positive constants C, γ and $\alpha_1 > d_1$, $\alpha_2 > d_2$ such that*

$$E[\|\Delta_{(x,x'),(y,y')}X\|^\gamma] \le C|x - x'|^{\alpha_1}|y - y'|^{\alpha_2}, \tag{9}$$

$$E[\|\Delta_{(x,x')}X(y)\|^\gamma] \le C|x - x'|^{\alpha_1}, \tag{10}$$

$$E[\|\Delta_{(y,y')}X(x)\|^\gamma] \le C|y - y'|^{\alpha_2} \tag{11}$$

hold for any x, $x' \in \mathbb{D}_1$ and y, $y' \in \mathbb{D}_2$. Then *$X(x, y)$ has a continuous modification $\tilde{X}(x, y)$. The modification $\tilde{X}$ is (β_1, β_2)-Hölder continuous for any $\beta_1 < (\alpha_1 - d_1)/\gamma$ and $\beta_2 < (\alpha_2 - d_2)/\gamma$. Furthermore if $\mathbb{D}_1$ and $\mathbb{D}_2$ are hypercubes, then there exists a finite valued random variable K such that*

$$\|\Delta_{(x,x'),(y,y')}\tilde{X}\| \le K|x - x'|^{\beta_1}|y - y'|^{\beta_2} \tag{12}$$

holds for every x, $x' \in \mathbb{D}_1$ and y, $y' \in \mathbb{D}_2$. □

Proof is left to the reader (see Exercises 1.4.15 and 1.4.16).

Weak convergence of continuous random fields

Let $\{X_n(x) : x \in \mathbb{I}\}$ be a sequence of continuous random fields with values in a complete separable metric space S, where $\mathbb{I}$ is a hypercube in $\mathbb{R}^d$. We shall define two types of laws and discuss their weak convergences. One is the weak convergence of the random fields and the other is the weak convergence of the finite dimensional distributions.

We will first introduce the laws of the random fields. Let $W = C(\mathbb{I} : S)$ be the set of all continuous maps from $\mathbb{I}$ into S. We denote its element by X and its projection at $x \in \mathbb{I}$ by $X(x)$. Introducing the metric

$$\mathbf{d}(X, Y) = \sup_{x \in \mathbb{I}} d(X(x), Y(x))$$

where d is a metric on S, W is again a complete separable metric space. Its topological Borel field is denoted by $\mathscr{B}(W)$. The *law of a continuous random field* $X(x, \omega)$ with values in S is then defined by

$$P(A) = P(\{\omega : X(\cdot, \omega) \in A\}), \qquad A \in \mathscr{B}(W). \tag{13}$$

We shall next define the finite dimensional distributions of the random field $X(x, \omega)$. Let $x_1, \ldots, x_k$ be elements of $\mathbb{I}$. We define the law of $(X(x_1), \ldots, X(x_k))$ by

$$P_{x_1,\ldots,x_k}(A) = P(\{\omega : (X(x_1), \ldots, X(x_k)) \in A\}), \qquad A \in \mathscr{F}(S^k). \quad (14)$$

It is called a *finite dimensional distribution of the random field* $X(x)$.

Now let $\{X_n\}$ be a sequence of continuous random fields with values in S. Let P_n be the law of X_n on the space $(W, \mathscr{B}(W))$. The sequence $\{X_n\}$ is said to converge weakly if $\{P_n\}$ converges weakly. Similarly $\{X_n\}$ is said to be tight if the corresponding $\{P_n\}$ is tight.

If $\{X_n\}$ converges weakly, then the δ finite dimensional distributions $\{P^{(n)}_{x_1,\ldots,x_k}\}$ converge weakly for any $x_1, \ldots, x_k$. However the converse is false in general. We need the tightness of measures additionally.

Theorem 1.4.5 *The sequence $\{X_n\}$ converges weakly if and only if the following two conditions are satisfied.*

(a) *The sequence $\{X_n\}$ is tight.*
(b) *The sequence of finite dimensional distributions $\{P^{(n)}_{x_1,\ldots,x_k}\}$ converges weakly for any $x_1, \ldots, x_k$ of $\mathbb{R}^d$, where k runs over the set of positive integers.* □

In order to check the tightness of measures $\{P_n\}$, we have to find for any $\varepsilon > 0$ a compact subset K of W satisfying $P_n(K) > 1 - \varepsilon$ for all n. Therefore some characterization of a compact subset of W is required. The following characterization of a relatively compact set of W is immediate from the Ascoli–Arzela theorem.

Theorem 1.4.6 *A subset K of $W = C(\mathbb{I} : S)$ is relatively compact if and only if the following two properties are satisfied.*

(a) *For each $x \in \mathbb{I}$, there exists a compact subset Γ_x of S such that $X(x) \in \Gamma_x$ holds for all X in K.*
(b) *The set K is equicontinuous, i.e. for any $\varepsilon > 0$ there exists $\delta > 0$ such that $\sup_{|x-y|<\delta} d(X(x), X(y)) < \varepsilon$ holds for all X in K.* □

Now if S is an infinite dimensional Frechet space, property (a) is not easily verified in practice since any compact subset of S with respect to the strong topology is very restrictive. In application it is convenient to consider the weak topology of S instead of the strong one.

In the following we will assume that S is a separable semi-reflexive Frechet space with seminorms $\| \ \|_N$, $N = 1, 2, \ldots$ Then a subset Γ of S is relatively compact with respect to the weak topology if and only if the set Γ is bounded, i.e. $\sup_{f \in \Gamma} \|f\|_N < \infty$ holds for any N.

On the space W we shall define seminorms $\| \ \|_\xi$, $\xi \in S^*$ by setting $\|X\|_\xi = \sup_{x\in\mathbb{I}} |\langle X(x), \xi\rangle|$. Then W is a Frechet space by the above seminorms. $\| \ \|_\xi$, $\xi \in S^*$ define the *semi-weak topology*, while the seminorms $\| \ \|_N$, $N = 1, 2, \ldots$ define the *strong topology*. Now a subset K of W is relatively compact with respect to the semi-weak topology if K satisfies property (b) of Theorem 1.4.6 and the following (a') instead of (a).

(a') *For each $x \in \mathbb{I}$, there exists a bounded set Γ_x of S such that $X(x) \in \Gamma_x$ holds for all X in K.*

Kolmogorov's tightness criterion
We shall now establish the Kolmogorov criterion for the tightness of a sequence of continuous random fields with values in a Frechet space.

Theorem 1.4.7 *Let $\{X_n(x) : x \in \mathbb{I}\}$ be a sequence of continuous random fields with values in a real separable semi-reflexive Frechet space S with seminorms $\| \ \|_N$, $N = 1, 2, \ldots$ Assume that for each N there exist positive constants γ, C and $\alpha_1, \ldots, \alpha_d$ with $\sum_{i=1}^d \alpha_i^{-1} < 1$ such that*

$$E[\|X_n(x) - X_n(y)\|_N^\gamma] \le C\left(\sum_{i=1}^d |x^i - y^i|^{\alpha_i}\right), \qquad \textit{for every } x, y \in \mathbb{I}, \quad (15)$$

$$E[\|X_n(x)\|_N^\gamma] \le C, \qquad \textit{for every } x \in \mathbb{I} \quad (16)$$

holds for any n. Then $\{X_n\}$ is tight with respect to the semi-weak topology of $W = C(\mathbb{I} : S)$.

Proof We shall only prove the case where S is a Banach space with norm $\| \ \|$, since the case of Frechet space can be reduced to the above case immediately.

Let $\mathbf{q} = (q_1, \ldots, q_d)$ be a vector of positive integers greater than 1 and $\boldsymbol{\beta} = (\beta_1, \ldots, \beta_d)$ be a vector of positive numbers such that $\beta_i/\beta_j = \log q_j/\log q_i$ and $\sum_k \beta_k^{-1} - \alpha_j\beta_j^{-1} + \gamma < 0$ holds for any i and j. The existence of such $\mathbf{q}$ and $\boldsymbol{\beta}$ is shown in the proof of Theorem 1.4.1. Let M be a positive integer and let $\Delta_M^{\boldsymbol{\beta}}(X_n)$ be the modulus of $\boldsymbol{\beta}$-Hölder continuity of $X_n(x)$. The proofs of Lemmas 1.4.2 and 1.4.3 tell us that

$$\|X_n(x) - X_n(y)\| \le 4|\mathbf{q}|\left(\sum_{M=1}^\infty \Delta_M^{\boldsymbol{\beta}}(X_n)\right)\left(\sum_{i=1}^d |x^i - y^i|^{\beta_i}\right), \quad (17)$$

$$\sup_n E\left[\left|\sum_{M=1}^\infty \Delta_M^{\boldsymbol{\beta}}(X_n)\right|^\gamma\right] < \infty. \quad (18)$$

Then by Chebyschev's inequality for any $\varepsilon > 0$ there exists $a > 0$ such that

$$P\left(\sum_{M=1}^{\infty} \Delta_M^{\beta}(X_n) > a\right) < \frac{\varepsilon}{2}, \tag{19}$$

$$P(\|X_n(0)\| > a) < \frac{\varepsilon}{2}. \tag{20}$$

Now let P_n, $n = 1, 2, \ldots$ be the laws of X_n, $n = 1, 2, \ldots$ defined on $W = C(\mathbb{I} : S)$, respectively. Let K be a subset of W defined by

$$K = \left\{X : \sum_{M=1}^{\infty} \Delta_M^{\beta}(X) \le a \text{ and } \|X(0)\| \le a\right\}. \tag{21}$$

Then $P_n(K) > 1 - \varepsilon$ holds for any n by (19) and (20). Furthermore the set K is relatively compact in W with respect to the semi-weak topology. Indeed, since

$$\|X(x)\| \le \|X(0)\| + \|X(x) - X(0)\| \le a + 4a|\mathbf{q}|\left(\sum_{i=1}^{d} |x^i|^{\beta_i}\right)$$

holds for any $X \in K$, property (a) of Theorem 1.4.6 is satisfied, The set K is equicontinuous since

$$\|X(x) - X(y)\| \le 4a|\mathbf{q}|\left(\sum_{i=1}^{d} |x^i - y^i|^{\beta_i}\right)$$

is satisfied for any X in K. □

Exercise 1.4.8 Let $X(x)$, $x \in \mathbb{I}$ be an $\mathbb{R}$-valued random field with mean 0 and covariance $v(x, y) = E[X(x)X(y)]$ where $\mathbb{I} = [0, 1]^d$. Show that if $v(x, y)$ is continuous in (x, y), $X(x, \omega)$ has a modification $\tilde{X}(x, \omega)$ such that it is a measurable random field. Such $\tilde{X}$ is called a *measurable modification* of X. (*Hint*: for $n = 1, 2, \ldots$ set $X_n(x) = X([2^n x]/2^n)$ where $[2^n x]$ is the integer part of $2^n x$. Choose a subsequence $n_1 < n_2 < \cdots$ such that $\sup_{x \in \mathbb{I}} \sum_{k=1}^{\infty} E[|X_{n_{k+1}}(x) - X_{n_k}(x)|^2]^{1/2} < \infty$. Then show that $\tilde{X}(x) = \overline{\lim}_{k\to\infty} X_{n_k}(x)$ is a measurable modification of $X(x)$.)

Exercise 1.4.9 (continued from 1.4.8) Suppose that the covariance function $v(x, y)$ is continuously differentiable. Show that there exist measurable random fields $X_i'(x)$, $i = 1, \ldots, d$ with mean 0 and covariance $\partial^2 v(x, y)/\partial x^i \partial y^i$ such that

$$E\left[\left|X_i'(x) - \frac{1}{h}(X(x + he_i) - X(x))\right|^2\right] \xrightarrow[h\to 0]{} 0$$

hold for any x, where $e_i = (0, \ldots, 0, 1, 0, \ldots, 0)$ (1 is at the ith component). Such X'_i is called the *mean square derivative of* X. Show further that for almost all ω, $X'_i(\cdot, \omega)$ coincides with the distributional derivative $\partial X(\cdot, \omega)/\partial x^i$.

Exercise 1.4.10 (continued from 1.4.9) Suppose that the covariance $v(x, y)$ is m-times continuous differentiable. Show that $X(x, \omega)$ has a modification such that for almost all ω, it is $m - ([d/2] + 1)$ times continuously differentiable. (*Hint*: show that $X(x, \omega)$ is m-times differentiable in the sense of the mean-square. Then apply Sobolev's imbedding theorem.)

Exercise 1.4.11 Let $X(x, \omega)$, $x \in \mathbb{D}$ be a measurable random field with values in $\mathbb{R}$ such that $E[|X(x)|] < \infty$ for any x. Let $\mathscr{G}$ be a sub σ-field of $\mathscr{F}$. Show that there exists a measurable modification $Y(x)$ of the conditional expectations $E[X(x)|\mathscr{G}]$. Suppose now $f: \Omega \to \mathbb{D}$ is a $\mathscr{G}$-measurable map such that $E[|X(f)|] < \infty$. Show that $E[X(f)|\mathscr{G}] = Y(f)$ holds a.s.

Exercise 1.4.12 Let $X(x, \omega)$, $x \in \mathbb{D}$ be a random field with values in $\mathbb{R}$ satisfying (1) with $\gamma \geq 1$. Show that $E[X(x)|\mathscr{G}]$ has a modification of a continuous random field. (*Hint*: show that $Y(x) = E[X(x)|\mathscr{G}]$ satisfies (1).)

Exercise 1.4.13 Let $X(x, t)$, $x \in \mathbb{D}$, $t \in [0, T]$ be a continuous random field with values in $\mathbb{R}$ and be continuously differentiable in x. Set $Z(x, t) = |X(x, t)| + \Sigma_i |(\partial x/\partial x^i)(x, t)|$. Suppose that $E[Z(x, t)]$ is bounded on compact sets in $\mathbb{D} \times [0, T]$. Show that $E[\int_0^T X(x, t)\, dt|\mathscr{G}]$ has a modification of a continuous random field. (*Hint*: for any compact subset K of $\mathbb{D}$, set $\tau_n = \inf\{t > 0\colon \sup_{x \in K}|Z(x, t)| \geq n\}$ ($= T$ if $\{\cdots\} = \emptyset$). Show that $P(\tau_n < T) \to 0$ as $n \to \infty$ and each $E[\int_0^{\tau_n} X(x, t)\, dt|\mathscr{G}]$ has a modification of a continuous random field.)

Exercise 1.4.14 Let X_t, $t \in [0, \infty)$ be a continuous process with $X_0 = 0$. Suppose that for any $p > 1$ there exists a constant $C > 0$ such that $E[|X_t - X_s|^p] \leq C|t - s|^{p/2}$ for any t, s. Show that for any $\varepsilon > 0$ $\lim_{t \uparrow \infty} X_t/t^{(1/2)+\varepsilon}$ exists and equals 0 a.s. (*Hint*: set $\hat{X}(t) = t^{(1/2)+\varepsilon} X_{1/t}$ if $t > 0$ and $\hat{X}(0) = 0$. Show that for any $p > 1$ there exists $C' > 0$ such that $E[|\hat{X}(t) - \hat{X}(s)|^p] \leq C'|t - s|^{p\varepsilon}$. Apply Kolmogorov's theorem to derive $\lim_{t \to 0} \hat{X}(t) = 0$.)

Exercise 1.4.15 Let β_1, β_2 be positive numbers less than or equal to 1, and d_1, d_2 be positive integers. Let $M, N = 1, 2, \ldots$ For a Banach space valued function $f(x, y)$, $x \in [0, 1]^{d_1}$, $y \in [0, 1]^{d_2}$ define the (β_1, β_2)-modulus of continuity by

$$\Delta_{M,N}^{\beta_1,\beta_2}(f) = \max \|\Delta_{(\mathbf{k}/2^M, \mathbf{k}'/2^M),(\mathbf{l}/2^N, \mathbf{l}'/2^N)}(f)\| 2^{\beta_1 M} 2^{\beta_2 N},$$

where the maximum is taken for all $\mathbf{k} = (k_1, \ldots, k_{d_1})$, $\mathbf{k}' = (k_1', \ldots, k_{d_1}')$, $\mathbf{l} = (l_1, \ldots, l_{d_2})$ and $\mathbf{l}' = (l_1', \ldots, l_{d_2}')$ which are multi-indices of non-negative integers such that k_i, $k_i' \le 2^M - 1$, l_i, $l_i' \le 2^N - 1$ and $|\mathbf{k} - \mathbf{k}'| \le 1$, $|\mathbf{l} - \mathbf{l}'| \le 1$. Prove the following inequalities

$$\|\Delta_{(x,x'),(y,y')}(f)\| \le 64 d_1 d_2 \left\{ \sum_{M,N} \Delta_{M,N}^{\beta_1,\beta_2}(f) \right\} |x - x'|^{\beta_1} |y - y'|^{\beta_2},$$

$$\|\Delta_{(x,x')} f(y)\| \le 8 d_1 \left\{ \sum_M \Delta_M^{\beta_1}(f(\cdot, 0)) + 8 d_2 \sum_{M,N} \Delta_{M,N}^{\beta_1,\beta_2}(f) |y|^{\beta_2} \right\} |x - x'|^{\beta_1}.$$

(*Hint*: apply Lemma 1.4.2.)

Exercise 1.4.16 Complete the proof of Theorem 1.4.4. (*Hint*: show

$$E\left[\left| \sum_M \Delta_M^{\beta_1}(X(\cdot, 0)) \right|^\gamma \right] < \infty, \qquad E\left[\left| \sum_{M,N} \Delta_{M,N}^{\beta_1,\beta_2}(X) \right|^\gamma \right] < \infty \text{ etc.}$$

and use Exercise 1.4.15.)

Exercise 1.4.17 Let $\{X_n(x),\ x \in \mathbb{I}\}$, $n = 1, 2, \ldots$ be a sequence of $\mathbb{R}$-valued continuous random fields with mean 0 and covariance $v_n(x, y)$, where $\mathbb{I} = [0, 1]^d$, respectively. Suppose that $v_n(x, y)$ are m-times continuously differentiable with respect to x and y respectively, and satisfy $\sup_{x,y,n} |D_x^\alpha D_y^\beta v_n(x, y)| < \infty$ for all $|\alpha|$, $|\beta| \le m$, where $\alpha = (\alpha_1, \ldots, \alpha_d)$ is a multi-index of integers, $D_x^\alpha = (\partial/\partial x^1)^{\alpha_1} \ldots (\partial/\partial x^d)^{\alpha_d}$ and $|\alpha| = \alpha_1 + \cdots + \alpha_d$. Show that if $m \ge [d/2] + 1$, the sequence $\{X_n\}$ is tight. (*Hint*: apply Sobolev's inequality. See Exercise 1.4.10.)

Exercise 1.4.18 Let S_1 and S_2 be separable Frechet spaces and let $\{X_n(x), x \in \mathbb{I}\}$ be a sequence of S_1-valued random fields satisfying (15) and (16) of Theorem 1.4.7. Suppose that $S_1 \subset S_2$ and the injection map $i: S_1 \to S_2$ is compact. Show that the sequence $\{i(X_n(x)), x \in \mathbb{I}\}$ is tight with respect to the strong topology of $W_2 = C(\mathbb{I} : S_2)$.

Exercise 1.4.19 (*Kolmogorov–Čencov's tightness critierion*) Let $\{X_n(x, y), x \in \mathbb{I}_1,\ y \in \mathbb{I}_2\}$, $n = 1, 2, \ldots$ be a sequence of continuous real random fields where $\mathbb{I}_1 = [0, 1]^{d_1}$ and $\mathbb{I}_2 = [0, 1]^{d_2}$, respectively. Suppose that there exist positive constants $C, \gamma, \alpha_1 > d_1, \alpha_2 > d_2$ such that

$$E[|\Delta_{(x,x'),(y,y')}X_n|^\gamma] \le C|x-x'|^{\alpha_1}|y-y'|^{\alpha_2},$$
$$E[|\Delta_{(x,x')}X_n(y)|^\gamma] \le C|x-x'|^{\alpha_1},$$
$$E[|\Delta_{(y,y')}X_n(x)|^\gamma] \le C|y-y'|^{\alpha_2},$$
$$E[|X_n(x,y)|^\gamma] \le C$$

holds for any $x, x' \in \mathbb{I}_1$, $y, y' \in \mathbb{I}_2$ and n. Show that the sequence of the random fields $\{X_n\}$ is tight.

2

Continuous semimartingales and stochastic integrals

2.1 Preliminaries

Localmartingales and semimartingales

In this chapter we introduce stochastic integrals based on continuous semimartingales and establish the so-called Itô's formula which gives us a rule for stochastic calculus concerning semimartingales.

In Section 1.2 we defined martingales and submartingales. Localmartingales and semimartingales which we introduce in this section are generalizations of these. We begin with defining localmartingales. Let $(\Omega, \mathscr{F}, P)$ be a complete probability space equipped with the filtration $\{\mathscr{F}_t\}$ of sub σ-fields of $\mathscr{F}$ with finite time interval $\mathbb{T} = [0, T]$. A continuous real valued $\{\mathscr{F}_t\}$-adapted process X_t is called a *localmartingale* if there exists an increasing sequence of stopping times $\{\tau_n\}$ such that $P(\tau_n < T) \to 0$ as $n \to \infty$ and each stopped process $X_t^{\tau_n} \equiv X_{t \wedge \tau_n}$ is a martingale where $t \wedge \tau_n = \min\{t, \tau_n\}$. A continuous *local-submartingale* and a continuous *local-supermartingale* are defined similarly. A martingale is a localmartingale obviously. (Set $\tau_n \equiv T$ for all n.) The following theorem gives us a criterion for a localmartingale to be a martingale.

Theorem 2.1.1 *Let X_t be a continuous localmartingale. If $E[\sup_s|X_s|] < \infty$, then X_t is a martingale.*

Proof Let $\{\tau_n\}$ be an increasing sequence of stopping times such that $P(\tau_n < T) \to 0$ as $n \to \infty$ and each $X_t^{\tau_n}$ is a martingale. Then for each t, $X_t^{\tau_n}$ converges to X_t a.s. Further since $|X_t^{\tau_n} - X_t|$ is bounded by an integrable random variable $2\sup_s|X_s|$, $E[|X_t^{\tau_n} - X_t|]$ converges to 0 as $n \to \infty$. Then the limit X_t is a martingale because we have by Theorem 1.1.6 (c.5),

$$E[X_t|\mathscr{F}_s] = \lim_{n\to\infty} E[X_t^{\tau_n}|\mathscr{F}_s] = \lim_{n\to\infty} X_s^{\tau_n} = X_s. \quad \square$$

Remark

(a) Let X_t be a localmartingale and let σ be a stopping time. Then the stopped process X^σ is also a localmartingale. The fact follows immediately from the optional stopping time theorem.

(b) Let X_t be a localmartingale. Then there exists an increasing sequence of stopping times $\{\sigma_n\}$ such that $P(\sigma_n < T) \to 0$ as $n \to \infty$ and each stopped process X^{σ_n} is a bounded martingale. In fact, let σ_n be the hitting time of X_t to the open set $[-n, n]^c$, i.e. $\sigma_n = \inf\{t > 0 : |X_t| > n\}$ $(= T$ if $\{\ldots\} = \emptyset)$. Then $P(\sigma_n < T) \to 0$ as $n \to \infty$. Further $X_t^{\sigma_n}$ is a localmartingale such that $\sup_s |X_s^{\sigma_n}| \le n$. Then it is a martingale by Theorem 2.1.1.

Let $\mathscr{L}_c$ be the linear space consisting of all real valued continuous stochastic processes. We introduce the metric ρ by

$$\rho(X - Y) = \rho(X, Y) = E\left[\frac{\sup_t |X_t - Y_t|^2}{1 + \sup_t |X_t - Y_t|^2}\right]^{1/2}. \tag{1}$$

It is equivalent to the topology of the uniform convergence in probability: a sequence $\{X_t^n\}$ of $\mathscr{L}_c$ is a Cauchy sequence if and only if for any $\varepsilon > 0$ $P(\sup_t |X_t^n - X_t^m| > \varepsilon) \to 0$ holds as $n, m \to \infty$.

We introduce the norm $\| \ \|$ by $\|X\| = E[\sup_t |X_t|^2]^{1/2}$ and denote by $\mathscr{L}_c^2$ the set of all elements of $\mathscr{L}_c$ with finite norms. We may say that the topology of $\mathscr{L}_c^2$ is the uniform convergence in L_2. Since $\rho(X) \le \|X\|$, the topology by $\| \ \|$ is stronger than that by ρ. It is easy to see that $\mathscr{L}_c^2$ is a dense subset of $\mathscr{L}_c$.

Let $\mathscr{M}_c$ be the set of all continuous square integrable martingales X_t with $X_0 = 0$. From Doob's inequality, the norm $\|X\|$ is finite for any X of $\mathscr{M}_c$. Hence $\mathscr{M}_c$ is a subset of $\mathscr{L}_c^2$. We denote by $\mathscr{M}_c^{loc}$ the set of all continuous localmartingales X_t such that $X_0 = 0$. It is a subset of $\mathscr{L}_c$.

Theorem 2.1.2 *$\mathscr{M}_c$ is a closed subspace of $\mathscr{L}_c^2$. $\mathscr{M}_c^{loc}$ is a closed subspace of $\mathscr{L}_c$. Furthermore, $\mathscr{M}_c$ is dense in $\mathscr{M}_c^{loc}$.*

Proof. Obviously $\mathscr{M}_c$ is a linear subspace of $\mathscr{L}_c^2$. Let $\{X_n\}$ be a Cauchy sequence in $\mathscr{M}_c$ and let X be its limit in $\mathscr{L}_c^2$. Then for any t, $\{X_t^n\}$ converges to X_t in L^2 and hence $\{E[X_t^n|\mathscr{F}_s]\}$ converges to $E[X_t|\mathscr{F}_s]$ in L^2 for any $s < t$ by Theorem 1.1.6. Since $E[X_t^n|\mathscr{F}_s] = X_s^n$ holds a.s. for any n, the limits satisfy $E[X_t|\mathscr{F}_s] = X_s$ a.s. Thus X_t belongs to $\mathscr{M}_c$.

Let $\{X^n\}$ be a Cauchy sequence in $\mathscr{M}_c^{loc}$ converging to X of $\mathscr{L}_c$ by the topology ρ. Choosing a subsequence if necessary, we may assume that $\{X^n\}$ converges to X uniformly a.s. Set $A_t = \sup_n \sup_{s \le t} |X_s^n|$. Then it is a continuous increasing process. Let $k = 1, 2, \ldots$ and let τ_k be the first time such that $A_t > k$. Then $\sup_t |X_t^{n,\tau_k}| \le k$ for all n. Therefore, for each k, $\{X_t^{n,\tau_k}\}$ is a sequence of martingales converging to $X_t^{\tau_k}$ boundedly. Therefore X^{τ_k} is a martingale for each k, proving that X is an element of $\mathcal{M}_c^{loc}$. The last assertion is immediate. □

We shall next define semimartingales and other related stochastic processes. Let X_t be a continuous $(\mathcal{F}_t)$-adapted process. It is called an *increasing process* if for almost all ω, $X_t(\omega)$ is an increasing function of t such that $X_0(\omega) = 0$. It is called a *process of bounded variation* if it is written as the difference of two increasing processes. It is called a *semimartingale* if it is written as the sum of a localmartingale and a process of bounded variation.

Remark A submartingale X_t is said to belong to *the class* (D) if the family of random variables $\{X_\tau : \tau \in \mathcal{G}\}$ is uniformly integrable where $\mathcal{G}$ is the set of all stopping times. P. A. Meyer has shown that any continuous submartingale of the class (D) is decomposed uniquely to the sum of a continuous martingale and a continuous increasing process. The decomposition is called the *Doob–Meyer decomposition.* See Meyer [99], Liptser–Shiryaev [91], Ikeda–Watanabe [49].

Now for any continuous submartingale X_t, let σ_n be the hitting time of X_t to the set $[-n, n]^c$. Then each stopped process $X_t^{\sigma_n}$ is a bounded submartingale by the optional stopping time theorem. Hence it belongs to the class (D) and has a Doob–Meyer decomposition. This shows that the continuous submartingale is decomposed to the sum of a continuous localmartingale and a continuous increasing process. Hence in particular any submartingale is a semimartingale. Similarly any continuous supermartingale is a semimartingale.

Local semimartingales

In the next chapter we shall define the stochastic differential equation and study its solutions. Like solutions of ordinary differential equations, the stochastic differential equation will not always have a global solution. It may explode at finite random time. Thus the solution will be defined as a *local process* of the form X_t, $t \in [0, \tau)$ where τ is a stopping time. The stopping time τ is often called the *terminal time* of the local process X_t. Further if $\lim_{t\uparrow\tau} X_t = \infty$ is satisfied, τ is called the *explosion time.*

Local processes are defined more precisely as follows. A stopping time τ with values in $[0, \infty]$ is called *accessible* if there exists an increasing sequence $\{\tau_n\}$ of stopping times such that $\tau_n < \tau$ and $\lim_{n\to\infty} \tau_n = \tau$ hold a.s. Let τ be an accessible stopping time. A family of random variables X_t with the random time parameter $t \in [0, \tau)$ is called a *local process.* Note that if X_t, $t \in [0, \tau)$ is a continuous local process and $\{\tau_n\}$ is the sequence of stopping times mentioned above, the stopped process $X_t^{\tau_n} = X_{t\wedge\tau_n}$ is a usual (global) process for every n. The local process is called $(\mathscr{F}_t)$-*adapted* if $X_t^{\tau_n}$ is $(\mathscr{F}_t)$-adapted for every n. Further it is called a *local semimartingale* if $X_t^{\tau_n}$ is a semimartingale for every n. In particular if $X_t^{\tau_n}$ is a localmartingale for every n, X_t is called a *local localmartingale.*

2.2 Quadratic variational processes

Quadratic variations of continuous semimartingales

This section is devoted to the study of the quadratic variation of a continuous stochastic process X_t, $t \in [0, T]$. Let Δ be a partition of the interval $[0, T]: \Delta = \{0 = t_0 < \cdots < t_l = T\}$ and let $|\Delta| = \max_k(t_{k+1} - t_k)$. Associated with the partition Δ, we define a continuous process $\langle X\rangle_t^\Delta$ by

$$\langle X\rangle_t^\Delta = \sum_{k=0}^{l-1} (X_{t\wedge t_{k+1}} - X_{t\wedge t_k})^2. \tag{1}$$

We call it the *quadratic variational process* or simply the *quadratic variation of* X_t *associated with the partition* Δ.

Now let $\{\Delta_n\}$ be a sequence of partitions such that $|\Delta_n| \to 0$. If for every t the limit of $\{\langle X\rangle_t^{\Delta_n}\}$ exists in probability and it is independent of the choice of sequences $\{\Delta_n\}$ a.s., it is called the *quadratic variational process* or simply the *quadratic variation* of X_t and is denoted by $\langle X\rangle_t$ or by $\langle X_t\rangle$.

The quadratic variation is not well defined to any continuous stochastic process. We will see in the sequel that a natural class of processes where quadratic variations are well defined is that of continuous semimartingales.

We begin the discussion with a process of bounded variation.

Lemma 2.2.1 *Let X_t be a continuous process of bounded variation. Then the quadratic variation exists and equals* 0 *a.s.*

Proof. Let $|X|_t(\omega)$ be the total variation of the sample function $X_s(\omega)$, $0 \le s \le t$. Then

$$\langle X\rangle_t^\Delta \le \left(\sum_{j=0}^{l-1} |X_{t\wedge t_{j+1}} - X_{t\wedge t_j}|\right)\max_k|X_{t_{k+1}} - X_{t_k}|$$

$$\le |X|_t \max_k |X_{t_{k+1}} - X_{t_k}|.$$

The right hand side converges to 0 as $|\Delta| \to 0$ a.s. □

We next consider the quadratic variation of a bounded continuous martingale.

Theorem 2.2.2 *Let M_t be a bounded continuous martingale. Then for any sequence of partitions $\{\Delta_n\}$ such that $|\Delta_n| \to 0$, $\{\langle M\rangle_t^{\Delta_n}\}$ converges uniformly to a continuous increasing process $\langle M\rangle_t$ in L^2.* □

For the proof of the theorem we introduce another process:

$$N_t^\Delta = \sum_{k=0}^{l-1} M_{t\wedge t_k}(M_{t\wedge t_{k+1}} - M_{t\wedge t_k}). \tag{2}$$

A direct computation yields the equality

$$M_t^2 - M_0^2 = \langle M\rangle_t^\Delta + 2N_t^\Delta. \tag{3}$$

Hence the convergence of $\{\langle M\rangle_t^{\Delta_n}\}$ is reduced to the convergence of $\{N_t^{\Delta_n}\}$. To show that $\{N_t^{\Delta_n}\}$ converges, we need two lemmas.

Lemma 2.2.3 *N_t^Δ is a martingale with mean 0. Further,*

$$E[(N_t^\Delta)^2] \le 2C^4, \tag{4}$$

where C is a constant such that $\sup_t |M_t| \le C$.

Proof Let $s < t$. Choose t_j such that $t_j \le s < t_{j+1}$. Then

$$N_t^\Delta - N_s^\Delta = \sum_{k\ge j+1} M_{t\wedge t_k}(M_{t\wedge t_{k+1}} - M_{t\wedge t_k}) + M_{t_j}(M_{t_{j+1}} - M_s).$$

The conditional expectation with respect to $\mathscr{F}_s$ of each term of the right hand side is 0. For example,

$$E[M_{t\wedge t_k}(M_{t\wedge t_{k+1}} - M_{t\wedge t_k})|\mathscr{F}_s] = E[M_{t\wedge t_k}E[M_{t\wedge t_{k+1}} - M_{t\wedge t_k}|\mathscr{F}_{t\wedge t_k}]|\mathscr{F}_s]$$
$$= 0$$

holds for any $k \ge j+1$. Therefore N_t^Δ is a martingale with mean 0.

Now let $\langle N^\Delta\rangle_t^\Delta$ be the quadratic variation of N_t^Δ associated with the partition Δ. Then,

$$\langle N^\Delta\rangle_t^\Delta = \sum_k M_{t\wedge t_k}^2(M_{t\wedge t_{k+1}} - M_{t\wedge t_k})^2 \le C^2\langle M\rangle_t^\Delta.$$

Since $E[\langle M\rangle_t^\Delta] = E[M_t^2 - M_0^2]$ holds by (3), we have

$$E[\langle N^\Delta\rangle_t^\Delta] \le C^2 E[M_t^2 - M_0^2] \le 2C^4.$$

Since $E[(N_t^\Delta)^2] = E[\langle N^\Delta\rangle_t^\Delta]$ holds, the inequality (4) follows. The proof is complete. □

Lemma 2.2.4 *We have* $\lim_{n,m\to\infty} E[\sup_t |N_t^{\Delta_n} - N_t^{\Delta_m}|^2] = 0$.

Proof. For the partition $\Delta = \{0 = t_0 < \cdots < t_l = T\}$ and the process M_t, we define a simple process f_t^Δ by setting $f_t^\Delta = M_{t_k}$ if $t_k \le t < t_{k+1}$. Denote the joint partition $\Delta_n \cup \Delta_m$ by $\{0 = u_0 < \cdots < u_{l'} = T\}$. Then we have

$$N_t^{\Delta_n} - N_t^{\Delta_m} = \sum_k (f_{t\wedge u_k}^{\Delta_n} - f_{t\wedge u_k}^{\Delta_m})(M_{t\wedge u_{k+1}} - M_{t\wedge u_k}).$$

Therefore we have

$$\begin{aligned}\langle N^{\Delta_n} - N^{\Delta_m}\rangle_t^{\Delta_n\cup\Delta_m} &= \sum_k (f_{t\wedge u_k}^{\Delta_n} - f_{t\wedge u_k}^{\Delta_m})^2(M_{t\wedge u_{k+1}} - M_{t\wedge u_k})^2 \\ &\le \sup_s |f_s^{\Delta_n} - f_s^{\Delta_m}|^2 \langle M\rangle_t^{\Delta_n\cup\Delta_m}.\end{aligned}$$

Then by Doob's inequality and Schwarz's inequality,

$$\begin{aligned}E\left[\sup_{s\le t} |N_s^{\Delta_n} - N_s^{\Delta_m}|^2\right] &\le 4E[|N_t^{\Delta_n} - N_t^{\Delta_m}|^2] \qquad (5)\\ &\le 4E[\langle N^{\Delta_n} - N^{\Delta_m}\rangle_t^{\Delta_n\cup\Delta_m}] \\ &\le 4E\left[\sup_s |f_s^{\Delta_n} - f_s^{\Delta_m}|^4\right]^{1/2} E[|\langle M\rangle_t^{\Delta_n\cup\Delta_m}|^2]^{1/2}.\end{aligned}$$

Further we have from (3) and the previous lemma,

$$E[|\langle M\rangle_t^{\Delta_n\cup\Delta_m}|^2] \le 2\{E[(M_t^2 - M_0^2)^2] + 4E[(N_t^{\Delta_n\cup\Delta_m})^2]\} \le 24C^4.$$

Since $\sup_s|f_s^{\Delta_n} - f_s^{\Delta_m}|$ converges to 0 in L^4 as $n, m \to \infty$, the last term of (5) converges to 0 as $n, m \to 0$. The proof is complete. □

Proof of Theorem 2.2.2 By the above lemma $\{N_t^{\Delta_n}\}$ converges uniformly in L^2. Then so does the sequence $\{\langle M\rangle_t^{\Delta_n}\}$. Choosing a subsequence if necessary, $\{\langle M\rangle_t^{\Delta_n}\}$ converges uniformly a.s. Denote the limit as $\langle M\rangle_t$. It is a continuous process. We will prove that $\langle M\rangle_t$ is increasing in t a.s. Taking joint partitions if necessary, we may and do assume that Δ_{n+1} is a refined partition of Δ_n for each n and the set $\bigcup_n \Delta_n$ is dense in $[0, T]$. Let $s < t$ be two points in $\bigcup_n \Delta_n$. There exists a positive integer n_0 such that s, $t \in \Delta_{n_0}$. Then $\langle M\rangle_t^{\Delta_n} \ge \langle M\rangle_s^{\Delta_n}$ is satisfied for all $n \ge n_0$. Therefore $\langle M\rangle_t \ge \langle M\rangle_s$ is satisfied. The inequality is then satisfied for any real numbers $s < t$, since $\langle M\rangle_t$ is continuous in t a.s. □

We next consider the quadratic variation of a localmartingale. This time the quadratic variations associated with partitions do not converge in L^2 in general, but they converge in probability. In fact, we have the following theorem.

Theorem 2.2.5 *Let M_t be a continuous localmartingale. Then there exists a continuous increasing process $\langle M\rangle_t$ such that $\langle M\rangle_t^\Delta$ converges uniformly to $\langle M\rangle_t$ in probability.*

Proof Let $\{\tau_n\}$ be an increasing sequence of stopping times such that $P(\tau_n < T) \to 0$ and each stopped process $M_t^n \equiv M_t^{\tau_n}$ is a bounded martingale (see Remark after Theorem 2.1.1). Then $\langle M^n\rangle_{t\wedge\tau_m} = \langle M^m\rangle_t$ if $m \le n$, because $\langle M^n\rangle_{t\wedge\tau_m}^\Delta = \langle M^m\rangle_t^\Delta$ is satisfied for any Δ. Hence there exists a continuous increasing process $\langle M\rangle_t$ such that $\langle M\rangle_{t\wedge\tau_n} = \langle M^n\rangle_t$ holds for $t < \tau_n$. This $\langle M\rangle_t$ satisfies

$$\rho(\langle M\rangle^\Delta - \langle M\rangle)^2 \le \rho(\langle M\rangle^{\Delta,\tau_n} - \langle M\rangle^{\tau_n})^2 + P(\tau_n < T)$$

where ρ is the metric defined by (1) of Section 2.1. For any $\varepsilon > 0$ choose n so large that $P(\tau_n < T) < \varepsilon$ and let $|\Delta|$ tend to 0. Then we have $\overline{\lim}_{|\Delta|\to 0}\rho(\langle M\rangle^\Delta - \langle M\rangle)^2 < \varepsilon$, proving that $\{\langle M\rangle_t^\Delta\}$ converges uniformly to $\langle M\rangle_t$ in probability. □

Corollary 2.2.6 *$M_t^2 - \langle M\rangle_t$ is a localmartingale if M_t is a continuous localmartingale.*

Proof Let $\{\tau_n\}$ be the sequence of stopping times as in the proof of Theorem 2.2.5. For each n, the process $N_{t\wedge\tau_n}^\Delta$ is a martingale by Lemma 2.2.3. Therefore N_t^Δ is a localmartingale. Further N_t^Δ converges to $\frac{1}{2}(M_t^2 - M_0^2 - \langle M\rangle_t)$ uniformly in t in probability as $|\Delta| \to 0$ by Theorem 2.2.5. Then $M_t^2 - \langle M\rangle_t$ is a localmartingale by Theorem 2.1.2. □

Corollary 2.2.7 *An element M of $\mathscr{M}_c^{loc}$ belongs to $\mathscr{M}_c$ if and only if $\langle M\rangle_T$ is integrable. In this case, $M_t^2 - \langle M\rangle_t$ is a martingale.*

Proof Let $M_t \in \mathscr{M}_c^{loc}$ and $\{\tau_n\}$ be an increasing sequence of stopping times as in the proof of Theorem 2.2.5. Suppose that $\langle M\rangle_T$ is integrable. Then

$$E\left[\sup_t |M_t|^2\right] = E\left[\lim_{n\to\infty}\sup_t |M_t^{\tau_n}|^2\right] \le \underline{\lim}_{n\to\infty} E\left[\sup_t |M_t^{\tau_n}|^2\right]$$
$$\le 4\,\underline{\lim}_{n\to\infty} E[|M_T^{\tau_n}|^2] = 4\,\underline{\lim}_{n\to\infty} E[\langle M\rangle_T^{\tau_n}] = 4E[\langle M\rangle_T] < \infty.$$

Therefore M_t is an L^2-martingale by Theorem 2.1.1. Furthermore, $\sup_t|M_t^2 - \langle M\rangle_t|$ is dominated by an integrable random variable $\sup_t|M_t|^2 + \langle M\rangle_T$. Then $M_t^2 - \langle M\rangle_t$ is also a martingale by the same theorem.

Conversely if M_t is an L^2-martingale, then

$$E\left[\sup_t |M_t|^2\right] \geq E\left[\sup_t |M_t^{\tau_n}|^2\right] \geq E[|M_T^{\tau_n}|^2] = E[\langle M\rangle_T^{\tau_n}].$$

Therefore $E[\langle M\rangle_T] \leq E[\sup_t |M_t|^2] < \infty$. The proof is complete. □

The following characterization of the quadratic variation is useful for finding the quadratic variation of a given localmartingale, explicitly.

Theorem 2.2.8 *Let M_t be a continuous localmartingale. A continuous increasing process A_t coincides with the quadratic variation of M_t if and only if $M_t^2 - A_t$ is a localmartingale.*

Proof 'Only if' is clear from Corollary 2.2.6. Suppose that $M_t^2 - A_t$ is a localmartingale. Then $\langle M\rangle_t - A_t$ is a continuous localmartingale of bounded variation, whose quadratic variation is 0. Therefore by Corollary 2.2.7, $\langle M\rangle_t - A_t$ is an L^2-martingale and $(\langle M\rangle_t - A_t)^2$ is an L^1-martingale. This implies $E[(\langle M\rangle_t - A_t)^2] = 0$, and we get $\langle M\rangle_t = A_t$. □

Remark Corollary 2.2.7 indicates that the submartingale M_t^2 is decomposed to the sum of the martingale $N_t = M_t^2 - \langle M\rangle_t$ and the increasing process $\langle M\rangle_t$. The decomposition is simply the Doob–Meyer decomposition of the submartingale. Note that we did not use the decomposition theorem for the proof of Theorem 2.2.2. If we apply it, then we can prove the theorem more easily (see Métivier [98]).

We will next consider the quadratic variation of a local localmartingale. Let M_t, $t \in [0, \tau)$ be a continuous local localmartingale. Then $\langle M\rangle_t^\Delta$, $t \in [0, \tau)$ converges uniformly on any compact subset of $[0, \tau)$ in probability. Therefore the quadratic variation $\langle M\rangle_t$, $t \in [0, \tau)$ is well defined. It is characterized as a continuous increasing local process such that $M_t^2 - \langle M\rangle_t$, $t \in [0, \tau)$ is a local localmartingale. Set $\langle M\rangle_{\tau-} = \lim_{t\uparrow\tau} \langle M\rangle_t$. It is a non-negative random variable possibly taking the value $+\infty$ with positive probability.

Theorem 2.2.9 *Let M_t, $t \in [0, \tau)$ be a continuous local localmartingale. Then we have*

$$P\left(\{\langle M\rangle_{\tau-} < \infty\}\Delta\left\{\lim_{t\uparrow\tau} M_t \text{ exists and is finite}\right\}\right) = 0, \tag{6}$$

$$P\left(\{\langle M\rangle_{\tau-} = \infty\}\Delta\left\{\overline{\lim_{t\uparrow\tau}} |M_t| = \infty\right\}\right) = 0, \tag{7}$$

where $A\Delta B$ denotes the symmetric difference $A \cup B - A \cap B$.

Proof We may assume $M_0 = 0$. For a positive constant c, let σ_c be the infimum of $t \in [0, \tau)$ such that $|M_t| > c$ ($=\tau$ if there is no such t). Let $\{\tau_n\}$ be an increasing sequence of stopping times such that $\tau_n < \tau$ and $\lim \tau_n = \tau$ hold a.s. Then $M^2_{t\wedge\tau_n\wedge\sigma_c} - \langle M\rangle_{t\wedge\tau_n\wedge\sigma_c}$ is a martingale with mean 0 for every n. This yields $E[\langle M\rangle_{t\wedge\tau_n\wedge\sigma_c}] = E[|M_{t\wedge\tau_n\wedge\sigma_c}|^2] \le c^2$. Let $t \uparrow \infty$ and $n \to \infty$. Then we obtain $E[\langle M\rangle_{\sigma_c-}] \le c^2$ and $\langle M\rangle_{\sigma_c-} < \infty$ a.s. Now if $\overline{\lim}_{t\uparrow\tau}|M_t(\omega)| < \infty$, then $\sigma_c(\omega) = \tau(\omega)$ is satisfied for sufficiently large c (the number c may depend on ω). Therefore we have $\langle M\rangle_{\tau-}(\omega) < \infty$. We have thus proved

$$\left\{\omega : \lim_{t\uparrow\tau} M_t(\omega) \text{ exists and is finite}\right\} \subset \{\omega : \langle M\rangle_{\tau-}(\omega) < \infty\} \quad \text{a.s.}$$

Next for $c > 0$ let ρ_c be the infimum of $t \in [0, \tau)$ such that $\langle M\rangle_t \ge c$ ($=\tau$ if there is no such t). Then $M_{t\wedge\tau_n\wedge\rho_c}$, $t \in [0, \infty)$ is an L^2-martingale such that $E[|M_{t\wedge\tau_n\wedge\rho_c}|^2] \le c$ for every n. By the optional stopping time theorem, we have $E[M_{t\wedge\tau_n\wedge\rho_c}|\mathscr{F}_{\tau_m\wedge\rho_c}] = M_{t\wedge\tau_m\wedge\rho_c}$ if $n > m$. Then for each fixed t, $\{M_{t\wedge\tau_n\wedge\rho_c}, n = 1, 2, \ldots\}$ is an L^2-bounded martingale with discrete parameter. Therefore $M^c_t \equiv \lim_{n\to\infty} M_{t\wedge\tau_n\wedge\rho_c}$ exists a.s. and in L^2. The convergence is uniform with respect to t by Doob's inequality (Theorem 1.2.7). Therefore the limit M^c_t is a continuous L^2-bounded martingale and hence $\lim_{t\uparrow\tau} M^c_t$ exists a.s. This shows that $\lim_{t\uparrow\tau} M_{t\wedge\rho_c}$ exists a.s. Now if $\langle M\rangle_{\tau-}(\omega) < \infty$, then $\rho_c(\omega) = \tau(\omega)$ holds for sufficiently large c. Therefore $\lim_{t\uparrow\tau} M_t(\omega)$ exists and is finite. We have thus proved

$$\{\omega : \langle M\rangle_{\tau-}(\omega) < \infty\} \subset \left\{\omega : \lim_{t\uparrow\tau} M_t(\omega) \text{ exists and is finite}\right\} \quad \text{a.s.}$$

Consequently (6) is established. The relation (7) can be shown similarly. □

We will finally consider the quadratic variation of a continuous semimartingale. Let X_t be a continuous semimartingale and let $X_t = M_t + A_t$ be the decomposition to a localmartingale M_t and a process of bounded variation A_t. The quadratic variation $\langle X\rangle^\Delta_t$ associated with the partition Δ satisfies

$$|\langle X\rangle^\Delta_t - \langle M\rangle^\Delta_t - \langle A\rangle^\Delta_t| \le 2\{\langle M\rangle^\Delta_t \langle A\rangle^\Delta_t\}^{1/2}.$$

Since $\langle M\rangle^\Delta_t$ converges uniformly to $\langle M\rangle_t$ in probability and $\langle A\rangle^\Delta_t$ converges uniformly to 0 a.s., $\langle X\rangle^\Delta_t$ converges uniformly to $\langle M\rangle_t$ in probability. We then have the following theorem.

Theorem 2.2.10 *Let X_t be a continuous semimartingale. Then $\langle X\rangle^\Delta_t$ converges uniformly to $\langle M\rangle_t$ in probability as $|\Delta| \to 0$, where M_t is the local-martingale part of X_t.* □

Joint quadratic variations

Let M and N be two elements of $\mathcal{M}_c^{loc}$. The *joint quadratic variation of* M, N *associated with the partition* $\Delta = \{0 = t_0 < \cdots < t_l = T\}$ is defined by

$$\langle M, N\rangle_t^\Delta = \sum_{k=0}^{l-1} (M_{t\wedge t_{k+1}} - M_{t\wedge t_k})(N_{t\wedge t_{k+1}} - N_{t\wedge t_k}). \tag{8}$$

Theorem 2.2.11 *$\langle M, N\rangle_t^\Delta$ converges uniformly to a continuous process of bounded variation in probability as $|\Delta| \to 0$.* □

Proof is immediate from $\langle M, N\rangle_t^\Delta = \frac{1}{4}\{\langle M + N\rangle_t^\Delta - \langle M - N\rangle_t^\Delta\}$ and Theorem 2.2.5. The limit is denoted by $\langle M, N\rangle_t$ or $\langle M_t, N_t\rangle$ and is called the *joint quadratic variation* of M_t and N_t. Note the relation $\langle M, M\rangle_t = \langle M\rangle_t$.

The following is immediate from Theorem 2.2.8.

Theorem 2.2.12 *Given M, N of $\mathcal{M}_c^{loc}$, a continuous process of bounded variation A_t coincides with the joint quadratic variation $\langle M, N\rangle_t$ if and only if $M_tN_t - A_t$ is a localmartingale.* □

We list some important properties of joint quadratic variations.

Theorem 2.2.13 *Let M, M^1, M^2, N be arbitrary elements of $\mathcal{M}_c^{loc}$ and let a, b be real numbers.*

(i) **Stopping time property:** *let τ be any stopping time. Then $M_t^\tau \equiv M_{\tau\wedge t}$ belongs to $\mathcal{M}_c^{loc}$ and satisfies $\langle M, N\rangle_{t\wedge\tau} = \langle M^\tau, N\rangle_{t\wedge\tau} = \langle M^\tau, N\rangle_t$ a.s.*

(ii) **Bilinear:** *$\langle aM^1 + bM^2, N\rangle = a\langle M^1, N\rangle + b\langle M^2, N\rangle$ holds a.s.*

(iii) **Symmetric:** *$\langle M, N\rangle = \langle N, M\rangle$ holds a.s.*

(iv) **Positive definite:** *$\langle M\rangle_t - \langle M\rangle_s \geq 0$ holds a.s.. for any $t \geq s$ and the equality holds a.s. if and only if $M_r = M_s$ holds for all $r \in [s, t]$ a.s. In particular $\langle M\rangle_t \equiv 0$ a.s. if and only if $M_t \equiv 0$ a.s.*

(v) **Schwarz's inequality:** *$|\langle M, N\rangle_t - \langle M, N\rangle_s| \leq (\langle M\rangle_t - \langle M\rangle_s)^{1/2} \times (\langle N\rangle_t - \langle N\rangle_s)^{1/2}$ holds for any $s < t$ a.s.*

(vi) **Extended Schwarz's inequality**: *there exists a null set N such that for any $\omega \notin N$ the inequality*

$$\left|\int_s^t f_u g_u \, \mathrm{d}\langle M, N\rangle_u\right| \leq \left(\int_s^t |f_u|^2 \, \mathrm{d}\langle M\rangle_u\right)^{1/2} \left(\int_s^t |g_u|^2 \, \mathrm{d}\langle N\rangle_u\right)^{1/2} \tag{9}$$

holds for any measurable processes f_u, g_u such that $\int_s^t |f_u|^2 \, \mathrm{d}\langle M\rangle_u < \infty$ and $\int_s^t |g_u|^2 \, \mathrm{d}\langle N\rangle_u < \infty$.

Proof The property $M^\tau \in \mathcal{M}_c^{loc}$ is immediate from the optional stopping time theorem. Then the equality of (i) is obvious from the definition of the joint quadratic variation. Properties (ii)–(iv) are obvious. Property (v) follows immediately from (ii)–(iv). We will prove (vi). Suppose that f_u and g_u are simple functions: there exists a partition $\Delta = \{s = t_0 < \cdots < t_l = t\}$ of $[s, t]$ and bounded random variables f_k and g_k such that $f_u = f_k$, $g_u = g_k$ if $t_k \le u < t_{k+1}$. Then

$$\left|\int_s^t f_u g_u \, d\langle M, N\rangle_u\right| = \left|\sum_{k=0}^{l-1} f_k g_k (\langle M, N\rangle_{t_{k+1}} - \langle M, N\rangle_{t_k})\right|$$

$$\le \sum_{k=0}^{l-1} |f_k||g_k| (\langle M\rangle_{t_{k+1}} - \langle M\rangle_{t_k})^{1/2} (\langle N\rangle_{t_{k+1}} - \langle N\rangle_{t_k})^{1/2}$$

$$\le \left\{\sum_{k=0}^{l-1} |f_k|^2 (\langle M\rangle_{t_{k+1}} - \langle M\rangle_{t_k})\right\}^{1/2} \left\{\sum_{k=0}^{l-1} |g_k|^2 (\langle N\rangle_{t_{k+1}} - \langle N\rangle_{t_k})\right\}^{1/2}$$

$$= \left\{\int_s^t |f_u|^2 \, d\langle M\rangle_u\right\}^{1/2} \left\{\int_s^t |g_u|^2 \, d\langle N\rangle_u\right\}^{1/2}.$$

The extension to general f and g will be clear. □

Finally we will mention the joint quadratic variations of continuous semimartingales. Let X_t and Y_t be continuous semimartingales. The joint quadratic variation associated with the partition Δ is defined in the same way as (8) and is written as $\langle X, Y\rangle_t^\Delta$. The following theorem is immediate.

Theorem 2.2.14 *$\langle X, Y\rangle_t^\Delta$ converges uniformly in probability to a continuous process of bounded variation $\langle X, Y\rangle_t$. If M_t and N_t are the parts of local-martingales of X_t and Y_t, respectively, then $\langle X, Y\rangle_t$ coincides with $\langle M, N\rangle_t$.* □

The above process $\langle X, Y\rangle_t$ is also written as $\langle X_t, Y_t\rangle$ and is called the *joint quadratic variation of X_t and Y_t.*

Continuity of quadratic variations

Quadratic variations are continuous in the space $\mathcal{M}_c$ and $\mathcal{M}_c^{loc}$ in their topologies.

Theorem 2.2.15

(i) *Let $\{M^n\}$ be a sequence in $\mathcal{M}_c$. It converges to M of $\mathcal{M}_c$ if and only if $\{\langle M^n - M\rangle_T\}$ converges to 0 in L^1-norm.*

(ii) *Let $\{M^n\}$ be a sequence in $\mathcal{M}_c^{loc}$. It converges to M of $\mathcal{M}_c^{loc}$ if and only if $\{\langle M^n - M\rangle_T\}$ converges to 0 in probability.*

Proof The first assertion (i) is immediate from the relation

$$\|M^n - M\|^2 \ge E[|M_T^n - M_T|^2] = E[\langle M^n - M\rangle_T] \ge \tfrac{1}{4}\|M^n - N\|^2.$$

Suppose next that $\{M^n\}$ of $\mathcal{M}_c^{loc}$ converges to M of $\mathcal{M}_c^{loc}$. If $\{\langle M^n - M\rangle_T\}$ does not converge to 0 in probability, there exist $\varepsilon > 0$ and a subsequence $\{M^{n_i}\}$ such that $\underline{\lim}_{i\to\infty} P(\langle M^{n_i} - M\rangle_T > \varepsilon) > 0$. Choosing a subsequence if necessary, we may assume that $\{M^{n_i}\}$ converges to M uniformly a.s. Then there exists an increasing sequence of stopping times $\{\tau_m\}$ with $P(\tau_m < T) \to 0$ such that the stopped processes $M^{n_i}_{t\wedge\tau_m}$ are in $\mathcal{M}_c$ and converge to $M_{t\wedge\tau_m}$ in the space $\mathcal{M}_c$ as $n_i \to \infty$ (see the proof of Theorem 2.1.2). Then $\{\langle M^{n_i} - M\rangle_{\tau_m}\}$ converges to 0 in L^1-norm. Since it is valid for any τ_m, $\{\langle M^{n_i} - M\rangle_T\}$ converges to 0 in probability. This is a contradiction. We have thus shown that $\{\langle M^n - M\rangle_T\}$ converges to 0 in probability.

Conversely suppose that $\{\langle M^n - M\rangle_T\}$ converges to 0 in probability. If $\{M^n\}$ does not converge to M in $\mathcal{M}_c^{loc}$, there exist $\varepsilon > 0$ and a subsequence $\{M^{n_i}\}$ such that $\underline{\lim}_{i\to\infty} P(\sup_{t\le T}|M_t^{n_i} - M_t| > \varepsilon) > 0$. Choosing a subsequence if necesssary we may assume that $\{\langle M^{n_i} - M\rangle_T\}$ converges to 0 a.s. Then $\{\langle M^{n_i}\rangle_t\}$ converges uniformly a.s., because

$$\sup_{t\le T} |\langle M^{n_i}\rangle_t^{1/2} - \langle M\rangle_t^{1/2}| \le \langle M^{n_i} - M\rangle_T^{1/2}.$$

Now set $B_t = \sup_i \langle M^{n_i}\rangle_t$. It is a continuous increasing process. For a positive integer m, let σ_m be the first time such that $B_t > m$. Then $\langle M^{n_i}\rangle_{t\wedge\sigma_m} \le m$, so that we have

$$\langle M^{n_i} - M\rangle_{t\wedge\sigma_m} \le 2(\langle M^{n_i}\rangle_{t\wedge\sigma_m} + \langle M\rangle_{t\wedge\sigma_m}) \le 4m$$

for all n_i, t. Therefore $\{\langle M^{n_i} - M\rangle_{t\wedge\sigma_m}\}$ converges to 0 in L^1-norm as $n_i \to \infty$. Consequently, $\{M^{n_i}_{t\wedge\sigma_m}, i = 1, 2, \ldots\}$ is in $\mathcal{M}_c$ and converges to $M_{t\wedge\sigma_m}$ in $\mathcal{M}_c$ by the assertion (i). Hence $\{M^{n_i}\}$ converges to M uniformly in probability. This is a contradiction. We have thus shown that $\{M^n\}$ converges to M in $\mathcal{M}_c^{loc}$. □

The joint quadratic variations are also continuous in $\mathcal{M}_c$ and $\mathcal{M}_c^{loc}$.

Corollary 2.2.16 *Let $\{M^n\}$ be a sequence in $\mathcal{M}_c$ (or $\mathcal{M}_c^{loc}$) converging to M of $\mathcal{M}_c$ (or $\mathcal{M}_c^{loc}$, respectively). Then $\{\langle M^n, N\rangle_t\}$ converges to $\langle M, N\rangle_t$ uniformly in t in L^2 (or in probability respectively) for any N of $\mathcal{M}_c$ (or $\mathcal{M}_c^{loc}$, respectively).* □

Theorem 2.2.17

(i) *Let $\{M^n\}$ be a sequence in $\mathcal{M}_c$ converging to M of $\mathcal{M}_c$. Then*

$$\sup_{\Delta} E\left[\sup_t \langle M^n - M\rangle_t^{\Delta}\right] \xrightarrow[n\to\infty]{} 0. \qquad (10)$$

(ii) *Let $\{M^n\}$ be a sequence in $\mathcal{M}_c^{\text{loc}}$ converging to M of $\mathcal{M}_c^{\text{loc}}$. Then for any $\varepsilon > 0$*

$$\sup_\Delta P\left(\sup_t \langle M^n - M\rangle_t^\Delta > \varepsilon\right) \xrightarrow[n\to\infty]{} 0. \tag{11}$$

Proof The first assertion is obvious since

$$E\left[\sup_t \langle M^n - M\rangle_t^\Delta\right] \le 4E[|M_T^n - M_T|^2].$$

For the proof of (ii), suppose on the contrary that the assertion is not valid. Then for some $\varepsilon > 0$ there is a sequence $n_1 < n_2 < \cdots$ and $\Delta_1, \Delta_2, \ldots$ such that $\underline{\lim}_{i\to\infty} P(\sup_{t\le T}\langle M^{n_i} - M\rangle_t^{\Delta_i} > \varepsilon) > 0$. Choose a subsequence denoted by $\{M^{n_{i'}}\}$ converging to M uniformly on $[0, T]$ a.s. There exists an increasing sequence of stopping times $\{\tau_m\}$ with $P(\tau_m < T) \to 0$ such that for each m, $\{M^{n_{i'}}_{t\wedge\tau_m}\}$ converges to $M_{t\wedge\tau_m}$ in $\mathcal{M}_c$ as $n_{i'} \to \infty$. Therefore

$$\begin{aligned}
&P\left(\sup_t \langle M^{n_{i'}} - M\rangle_t^{\Delta_{i'}} > \varepsilon\right)\\
&\le P\left(\sup_t \langle M^{n_{i'}} - M\rangle_{t\wedge\tau_m}^{\Delta_{i'}} > \varepsilon, \tau_m = T\right) + P(\tau_m < T)\\
&\le \frac{1}{\varepsilon}E\left[\sup_t \langle M^{n_{i'}} - M\rangle_{t\wedge\tau_m}^{\Delta_{i'}}\right] + P(\tau_m < T)\\
&\le \frac{4}{\varepsilon}E[|M^{n_{i'}}_{\tau_m} - M_{\tau_m}|^2] + P(\tau_m < T).
\end{aligned}$$

This converges to 0 as $i' \to \infty$ and $m \to \infty$. This is a contradiction. The proof is complete. □

For later applications, we shall extend the above result to continuous semimartingales.

Corollary 2.2.18 *Let $\{X_t^n, n = 1, 2, \ldots, X\}$ be a sequence of continuous semimartingales decomposed as $X_t^n = M_t^n + A_t^n$, $X_t = M_t + A_t$, where M_t^n, M_t are continuous local martingales and A_t^n, A_t are continuous processes of bounded variation. Suppose that $M_t^n \to M_t$ in $\mathcal{M}_c^{\text{loc}}$ and $A_t^n \to A_t$ uniformly in probability and that $\sup_n |A^n|_T < \infty$ a.e., where $|A^n|_t(\omega)$ is the total variation of the function $A_s^n(\omega)$, $0 \le s \le t$. Then for any $\varepsilon > 0$*

$$\lim_{n\to\infty}\sup_\Delta P\left(\sup_t \langle X^n - X\rangle_t^\Delta > \varepsilon\right) = 0. \tag{12}$$

Proof Note the inequality

$$\sup_\Delta \langle A^n - A\rangle_t^\Delta \le |A^n - A|_t \sup_\Delta \sup_k |(A^n_{t_{k+1}} - A_{t_{k+1}}) - (A^n_{t_k} - A_{t_k})|.$$

It converges to 0 uniformly in t in probability as $n \to \infty$. Since

$$\langle X^n - X\rangle_t^\Delta \le 2\langle M^n - M\rangle_t^\Delta + 2\langle A^n - A\rangle_t^\Delta,$$

we obtain the assertion of the corollary by the previous theorem. □

Corollary 2.2.19 *Let $\{X_t^n\}$ and $\{Y_t^n\}$ be two sequences of continuous semimartingales converging to continuous semimartingales X_t and Y_t respectively in the sense of Corollary 2.2.18. Then*

$$\lim_{n\to\infty} \sup_{\Delta} P\left(\sup_t |\langle X^n, Y^n\rangle_t^\Delta - \langle X, Y\rangle_t^\Delta| > \varepsilon \right) = 0 \tag{13}$$

holds for any $\varepsilon > 0$. □

2.3 Stochastic integrals and Itô's formula

Itô's integrals and Stratonovich's integrals

Let M_t be a continuous localmartingale and let f_t be a continuous $(\mathscr{F}_t)$-adapted process. We will define the stochastic integral of f_t by the differential dM_t. Here, the differential does not mean a signed measure, since the sample function of a continuous local martingale is not of bounded variation, except a trivial martingale $M_t \equiv$ constant a.s. Nevertheless, the integral is well defined if the integrand f_t is $(\mathscr{F}_t)$-adapted. Our discussion will be based on the properties of martingales, especially those of quadratic variations.

The stochastic integral was first defined by K. Itô [51]. It was based on the standard Brownian motion. Later it was extended to arbitrary localmartingales and semimartingales by the work of Doob [26], Motoo–Watanabe [102], Kunita–Watanabe [81], Meyer [100], among others. In this book we shall restrict the definition of integrals to continuous localmartingales and continuous semimartingales, though many results can be extended to right continuous localmartingales and semimartingales.

Let $\Delta = \{0 = t_0 < \cdots < t_l = T\}$ be a partition of $[0, T]$. Define

$$L_t^\Delta \equiv \sum_{k=0}^{l-1} f_{t\wedge t_k}(M_{t\wedge t_{k+1}} - M_{t\wedge t_k}). \tag{1}$$

Lemma 2.3.1 *L_t^Δ is a continuous localmartingale. Its quadratic variation is*

$$\langle L^\Delta\rangle_t = \int_0^t |f_s^\Delta|^2 \, d\langle M\rangle_s, \tag{2}$$

where $\langle M\rangle_t$ is the quadratic variation of M_t and f_t^Δ is the simple process defined from f_t by setting $f_t^\Delta \equiv f_{t_k}$ if $t_k \le t < t_{k+1}$.

Proof If M_t belongs to $\mathscr{M}_c$ and f_t is square integrable, then L_t^Δ is a continuous martingale (see the proof of Lemma 2.2.3). For the general case, define an increasing sequence of stopping times $\{\tau_n\}$ by $\tau_n = \inf\{t \in [0, T] : |M_t| > n \text{ or } |f_t| > n\}$ ($=T$ if $\{\dots\} = \emptyset$). Then for each n, $M_{t\wedge\tau_n}$ and $f_{t\wedge\tau_n}$ have the above property. Therefore $L_{t\wedge\tau_n}^\Delta$ is a martingale. Since $P(\tau_n < T)$ tends to 0 as $n \to \infty$, L_t^Δ is a localmartingale. We shall consider the quadratic variation of L_t^Δ. The property

$$\langle L^\Delta\rangle_{t_{k+1}} - \langle L^\Delta\rangle_{t_k} = f_{t_k}^2(\langle M\rangle_{t_{k+1}} - \langle M\rangle_{t_k})$$

is obvious from the definition of L_t^Δ. Summing the above for $k = 0, 1, 2, \dots$, we obtain (2). The proof is complete. □

Let Δ' be another partition of $[0, T]$. We define $L_t^{\Delta'}$ similarly using the same f_s and M_s. Denote by $\{0 = u_0 < \cdots < u_{l'} = T\}$ the joint partition $\Delta \cup \Delta'$. Then we have

$$L_t^\Delta - L_t^{\Delta'} = \sum_{k=0}^{l'-1} (f_{t\wedge u_k}^\Delta - f_{t\wedge u_k}^{\Delta'})(M_{t\wedge u_{k+1}} - M_{t\wedge u_k}).$$

Its quadratic variation is given by

$$\langle L^\Delta - L^{\Delta'}\rangle_t = \int_0^t |f_s^\Delta - f_s^{\Delta'}|^2 \, d\langle M\rangle_s.$$

Now let $\{\Delta_n\}$ be a sequence of partitions of $[0, T]$ such that $|\Delta_n| \to 0$. Then $\langle L^{\Delta_n} - L^{\Delta_m}\rangle_T$ converges to 0 in probability as $n, m \to \infty$. Hence $\{L^{\Delta_n}\}$ is a Cauchy sequence in $\mathscr{M}_c^{loc}$ by Theorem 2.2.15. We denote the limit by L_t. This L_t is called the *Itô integral of f_t by* dM_t and is denoted by $\int_0^t f_s \, dM_s$.

The Itô integral can be defined on a more general class of stochastic processes called predictable stochastic processes. Here the *predictable σ-field* is, by definition, the least σ-field on the product space $[0, T] \times \Omega$ for which all continuous $(\mathscr{F}_t)$-adapted processes are measurable. A *predictable process* is, by definition, a process measurable with respect to the predictable σ-field. A continuous $(\mathscr{F}_t)$-adapted process is predictable, obviously.

Let A_t be a continuous increasing process and let p be a positive number greater than or equal to 1. We denote by $L^p(A)$ the set of all predictable processes f_t such that $\int_0^T |f_s|^p \, dA_s < \infty$ a.s. Then the set of continuous $(\mathscr{F}_t)$-adapted processes is dense in $L^p(A)$, i.e. for any f_t of $L^p(A)$, there exists a sequence of continuous $(\mathscr{F}_t)$-adapted processes $\{f_t^n\}$ such that

$\int_0^T |f_s^n - f_s|^p \, d\langle M \rangle_s$ converges to 0 a.s. Now let M be a continuous local-martingale and let $\langle M \rangle_s$ be the quadratic variation. For any $f \in L^2(\langle M \rangle)$, choose a sequence of continuous predictable processes $\{f^n\}$ converging to f in the above sense. Then stochastic integrals $\{\int_0^t f_s^n \, dM_s\}$ form a Cauchy sequence in $\mathcal{M}_c^{loc}$. Denote the limit as $\int_0^t f_s \, dM_s$ and call it the *Itô integral of f_t by* dM_t.

Theorem 2.3.2

(i) *Let $M \in \mathcal{M}_c^{loc}$ and $f \in L^2(\langle M \rangle)$. Then the Itô integral $\int f dM$ is an element of $\mathcal{M}_c^{loc}$ and satisfies*

$$\left\langle \int f \, dM, N \right\rangle_t = \int_0^t f_s \, d\langle M, N \rangle_s, \qquad \textit{for every } N \in \mathcal{M}_c^{loc}. \tag{3}$$

(ii) *Conversely assume that L of $\mathcal{M}_c^{loc}$ satisfies*

$$\langle L, N \rangle_t = \int_0^t f_s \, d\langle M, N \rangle_s, \qquad \textit{for every } N \in \mathcal{M}_c^{loc}. \tag{4}$$

Then L is the Itô integral of f_t by dM_t, i.e. the Itô integral is characterized as the unique element L in $\mathcal{M}_c^{loc}$ satisfying (4).

Proof If f is a simple process, the relation (3) is direct from the computation of the joint quadratic variation. If f is an arbitrary process in $L^2(\langle M \rangle)$, we may choose a sequence of simple processes $\{f^n\}$ converging to f in $L^2(\langle M \rangle)$ a.s. Then, noting the continuity of the joint quadratic variation, we see that (3) is valid for this f. Now, for the proof of the second assertion, set $L'_t = \int_0^t f_s \, dM_s$. Then $\langle L - L', N \rangle_t = 0$ is satisfied for all N of $\mathcal{M}_c^{loc}$ because of (3) and (4). Hence $\langle L - L' \rangle_t = 0$, proving $L_t = L'_t$. The proof is complete. □

Corollary 2.3.3 The quadratic variation of the stochastic integral satisfies

$$\left\langle \int f \, dM \right\rangle_t = \int_0^t f_s^2 \, d\langle M \rangle_s. \quad \square \tag{5}$$

Let X be a continuous semimartingale decomposed to a continuous local-martingale M and a continuous process of bounded variation A. Let $|A|_t$ be the total variation of $A_s : 0 \le s \le t$. It is a continuous increasing process. For an arbitrary element f of $L^2(\langle M \rangle) \cap L^1(|A|)$ we define the *Itô integral by* dX_t:

$$\int_0^t f_s \, dX_s \equiv \int_0^t f_s \, dM_s + \int_0^t f_s \, dA_s. \tag{6}$$

It is a continuous semimartingale. Its joint quadratic variation with a continuous semimartingale Y_t satisfies

$$\left\langle \int f\,\mathrm{d}X, Y \right\rangle_t = \int_0^t f_s\,\mathrm{d}\langle X, Y\rangle_s = \int_0^t f_s\,\mathrm{d}\langle M, N\rangle_s, \tag{7}$$

where N_t is the localmartingale part of Y.

We will list some properties of the Itô integrals.

Theorem 2.3.4

(i) *If f, g are in $L^2(\langle M\rangle)\cap L^1(|A|)$ and a, b are constants, then $af + bg$ is in $L^2(\langle M\rangle)\cap L^1(|A|)$ and satisfies*

$$\int_0^t (af_s + bg_s)\,\mathrm{d}X_s = a\int_0^t f_s\,\mathrm{d}X_s + b\int_0^t g_s\,\mathrm{d}X_s. \tag{8}$$

(ii) *Let $f \in L^2(\langle M\rangle)\cap L^1(|A|)$ and $N_t = \int_0^t f_s\,\mathrm{d}M_s$, $B_t = \int_0^t f_s\,\mathrm{d}A_s$ and $L_t = N_t + B_t$. Let g be a predictable process such that $fg \in L^2(\langle M\rangle)\cap L^1(|A|)$. Then $g \in L^2(\langle N\rangle)\cap L^1(|B|)$ and satisfies*

$$\int_0^t g_s\,\mathrm{d}L_s = \int_0^t f_s g_s\,\mathrm{d}X_s. \tag{9}$$

(iii) *Let τ be a stopping time. If f, g are in $L^2(\langle M\rangle)\cap L^1(|A|)$ and $f_s = g_s$ holds for $s < \tau$, then*

$$\int_0^{t\wedge\tau} f_s\,\mathrm{d}X_s = \int_0^{t\wedge\tau} g_s\,\mathrm{d}X_s = \int_0^t f_s\,\mathrm{d}X_s^\tau = \int_0^t g_s\,\mathrm{d}X_s^\tau. \tag{10}$$

(iv) *Let $\{f^n\}$ be a sequence in $L^2(\langle M\rangle)\cap L^1(|A|)$ such that $\int_0^T |f_s^n - f_s|^2\,\mathrm{d}\langle M\rangle_s \to 0$, $\int_0^T |f_s^n - f_s|\,\mathrm{d}|A|_s \to 0$ in probability as $n \to \infty$. Then $\int_0^t f_s^n\,\mathrm{d}X_s \to \int_0^t f_s\,\mathrm{d}X_s$ uniformly in probability.*

(v) *Let f_t be a continuous $(\mathscr{F}_t)$-adapted process. For a partition $\Delta = \{0 = t_0 < \cdots < t_l = T\}$, set*

$$L_t^\Delta = \sum_{k=0}^{l-1} f_{t\wedge t_k}(X_{t\wedge t_{k+1}} - X_{t\wedge t_k}).$$

Then $\{L_t^\Delta\}$ converges to $\int_0^t f_s\,\mathrm{d}X_s$ uniformly in probability as $|\Delta| \to 0$. □

The proof is left to the reader (see Exercise 2.3.15).

Now let X_t, $t \in [0, \tau)$ be a continuous local semimartingale decomposed as $X_t = M_t + A_t$ where M_t is a continuous local localmartingale and A_t is a continuous local predictable process of bounded variation. Let f_t, $t \in [0, \tau)$ be a predictable process. Set

$$\sigma = \inf\left\{t \in [0, \tau): \int_0^t |f_s|^2 \, d\langle M \rangle_s = \infty \text{ or } \int_0^t |f_s| \, d|A|_s = \infty\right\}$$

$$(= \tau \text{ if } \{\dots\} = \emptyset).$$

Then the Itô integral $\int_0^t f_s \, dX_s$ is well defined for $t < \sigma$. It is a continuous local semimartingale.

We can define the Itô integral for a more general integrand. A stochastic process g_t is called *progressively measurable* if for each t the map from $[0, t] \times \Omega$ into $\mathbb{R}$ defined by $(s, \omega) \to g_s(\omega)$ is measurable with respect to the product σ-field $\mathscr{B}([0, t]) \times \mathscr{F}_t$. Now let M_t be a continuous local martingale and g_t be a progressively measurable process such that $\int_0^T |g| \, d\langle M \rangle < \infty$. Then $A_t \equiv \int_0^t g \, d\langle M \rangle$ is a continuous $(\mathscr{F}_t)$-adapted process. Hence it is predictable. Define

$$f_t = \overline{\lim_{h \downarrow 0}} \frac{A_t - A_{t-h}}{\langle M \rangle_t - \langle M \rangle_{t-h}}.$$

Then f_t is predictable and satisfies $\int_0^T |f - g| \, d\langle M \rangle = 0$. We shall identify f_t and g_t and write $\int f \, dM$ as $\int g \, dM$.

In many of the works concerning stochastic integrals based on continuous semimartingales or Brownian motions, the integrands are taken from the class of progressively measurable processes, e.g. Stroock–Varadhan [118], Ikeda–Watanabe [49]. The definitions of stochastic integrals in these works are the same as ours by the above identification.

We will define another stochastic integral by the differential $\circ dX_t$:

$$\int_0^t f_s \circ dX_s = \lim_{|\Delta| \to 0} \sum_{k=0}^{l-1} \frac{1}{2}(f_{t \wedge t_{k+1}} + f_{t \wedge t_k})(X_{t \wedge t_{k+1}} - X_{t \wedge t_k}). \tag{11}$$

If the above limit exists in the sense of the convergence in probability, it is called the *Stratonovich integral* of f_s by $\circ dX_s$.

Theorem 2.3.5 *If f_t is a continuous semimartingale, the Stratonovich integral is well defined. Further it satisfies*

$$\int_0^t f_s \circ dX_s = \int_0^t f_s \, dX_s + \frac{1}{2}\langle f, X \rangle_t. \tag{12}$$

Proof Observe the relation

$$\sum_{k=0}^{l-1} \frac{1}{2}(f_{t \wedge t_{k+1}} + f_{t \wedge t_k})(X_{t \wedge t_{k+1}} - X_{t \wedge t_k}) = \sum_{k=0}^{l-1} f_{t \wedge t_k}(X_{t \wedge t_{k+1}} - X_{t \wedge t_k}) + \frac{1}{2}\langle f, X \rangle_t^{\Delta}.$$

Then the theorem follows from Theorem 2.3.4 and Theorem 2.2.14. □

We list some properties of the Stratonovich integral.

Theorem 2.3.6 *Let X_t, f_t, g_t be continuous semimartingales.*

(i)
$$\int_0^t (af_s + bg_s)\circ \mathrm{d}X_s = a\int_0^t f_s\circ \mathrm{d}X_s + b\int_0^t g_s\circ \mathrm{d}X_s. \tag{13}$$

(ii) *Set $L_t = \int_0^t f_s\circ \mathrm{d}X_s$. Then*
$$\int_0^t g_s\circ \mathrm{d}L_s = \int_0^t (f_s g_s)\circ \mathrm{d}X_s. \tag{14}$$

(iii) *Let τ be a stopping time. If $f_s = g_s$ holds for $s < \tau$, then*
$$\int_0^{t\wedge\tau} f_s\circ \mathrm{d}X_s = \int_0^{t\wedge\tau} g_s\circ \mathrm{d}X_s = \int_0^t f_s\circ \mathrm{d}X_s^\tau = \int_0^t g_s\circ \mathrm{d}X_s^\tau. \tag{15}$$

(iv) *Let $\{f_t^n\}$ be a sequence of continuous semimartingales converging to f_t in the sense of Corollary 2.2.18. Then $\lim_{n\to\infty}\int_0^t f_s^n\circ \mathrm{d}X_s = \int_0^t f_s\circ \mathrm{d}X_s$ holds uniformly in probability.* □

The proof is left to the reader (see Exercise 2.3.16).

Orthogonal decomposition of localmartingales

Let M, N be elements of $\mathcal{M}_c^{loc}$. If $\langle M, N\rangle = 0$, M and N are called *orthogonal* and denoted by $M \perp N$. Since $M_t N_t - \langle M, N\rangle_t$ is a localmartingale, M and N are orthogonal if and only if $M_t N_t$ is a localmartingale.

In the case where $M, N \in \mathcal{M}_c$, the property $M \perp N$ is equivalent to
$$E[(M_t - M_s)(N_t - N_s)|\mathcal{F}_s] = 0 \qquad \text{for all } t > s.$$

Hence the orthogonality means that $M_t - M_s$ and $N_t - N_s$ are orthogonal in L^2-space with respect to the conditional probability $P(\cdot|\mathcal{F}_s)$. Therefore the orthogonal property of martingales is stronger than that in L^2-space. However, the property is weaker than the independence. Indeed if $M_t - M_s$, $N_t - N_s$ and $\mathcal{F}_s$ are independent and $E[M_t - M_s] = 0$ or $E[N_t - N_s] = 0$ holds,
$$E[(M_t - M_s)(N_t - N_s)|\mathcal{F}_s] = E[M_t - M_s)(N_t - N_s)] = 0,$$

so that M_t and N_t are orthogonal. As an example, consider a d-dimensional standard Brownian motion $X_t = (X_t^1, \ldots, X_t^d)$. Let $\{\mathcal{F}_t\}$ be the filtration generated by X_t. Then $X_t^1, \dots, X_d^d$ are martingales adapted to $(\mathscr{F}_t)$ and orthogonal to each other.

Now for a given M of $\mathcal{M}_c^{loc}$, set

$$\mathscr{L}(M) = \left\{ \int f \, \mathrm{d}M : f \in L^2(\langle M \rangle) \right\}.$$

Then it is a linear subspace of $\mathcal{M}_c^{loc}$. Note that if $M \perp N$, then $\langle \int f \, \mathrm{d}M, N \rangle = \int f \, \mathrm{d}\langle M, N \rangle = 0$. Therefore N is orthogonal to all elements of $\mathscr{L}(M)$.

Given M, N of $\mathcal{M}_c^{loc}$, we shall show that N is decomposed to the sum of an element of $\mathscr{L}(M)$ and an element of $\mathcal{M}_c^{loc}$ orthogonal to $\mathscr{L}(M)$. We first prove a lemma.

Lemma 2.3.7 *Given M, N of $\mathcal{M}_c^{loc}$, there exists a unique $f \in L^2(\langle M \rangle)$ satisfying $\langle M, N \rangle_t = \int_0^t f_s \, \mathrm{d}\langle M \rangle_s$. In particular if N belongs to $\mathcal{M}_c$, then the above f satisfies $E[\int_0^T |f_s|^2 \, \mathrm{d}\langle M \rangle_s] < \infty$.*

Proof Let A be an element of the predictable σ-field $\mathscr{P}$. If $\int \chi(A) \, \mathrm{d}\langle M \rangle = 0$ a.s., we have $\int \chi(A) \, \mathrm{d}\langle M, N \rangle = 0$ a.s. by Theorem 2.2.13, (vi). Therefore for almost all ω, $\langle M, N \rangle_t(\omega)$ is absolutely continuous with respect to $\langle M \rangle_t(\omega)$. We can take a predictable process as its Radon–Nikodym derivative. Indeed since $\langle M, N \rangle_t$ and $\langle M \rangle_t$ are predictable,

$$f_s = \overline{\lim_{h \downarrow 0}} \frac{\langle M, N \rangle_s - \langle M, N \rangle_{s-h}}{\langle M \rangle_s - \langle M \rangle_{s-h}}$$

is a predictable process satisfying $\int f \, \mathrm{d}\langle M \rangle = \langle M, N \rangle$.

We shall prove $f \in L^2(\langle M \rangle)$. For a positive constant c, set $f_s^c = f_s$ if $|f_s| \le c$, $= 0$ if $|f_s| > c$. Then since $|f_s^c|^2 = f_s^c f_s$ we have by Theorem 2.2.13, (vi),

$$\int_0^t |f_s^c|^2 \, \mathrm{d}\langle M \rangle_s = \int_0^t f_s^c \, \mathrm{d}\langle M, N \rangle_s \le \left(\int_0^t |f_s^c|^2 \, \mathrm{d}\langle M \rangle_s \right)^{1/2} \langle N \rangle_t^{1/2}.$$

Therefore $\int |f^c|^2 \, \mathrm{d}\langle M \rangle \le \langle N \rangle$ is satisfied. Let c tend to infinity. Then we obtain $\int |f|^2 \, \mathrm{d}\langle M \rangle \le \langle N \rangle$. Consequently f belongs to $L^2(\langle M \rangle)$. In particular if $N \in \mathcal{M}_c$ then $E[\int_0^T |f_s|^2 \, \mathrm{d}\langle M \rangle_s] < \infty$. The uniqueness of f will be obvious. The proof is complete. □

Theorem 2.3.8 *Let M, N be of $\mathcal{M}_c^{loc}$. Then N is decomposed uniquely to the sum of an element of $\mathscr{L}(M)$ and an element of $\mathcal{M}_c^{loc}$ orthogonal to $\mathscr{L}(M)$. In particular, if N belongs to $\mathcal{M}_c$, both elements belong to $\mathcal{M}_c$.*

Proof Let f be the predictable process of Lemma 2.3.7. Set $N^{(1)} = \int f \, \mathrm{d}M$ and $N^{(2)} = N - N^{(1)}$. By Theorem 2.3.2 and Lemma 2.3.7 we have

$\langle N^{(1)}, M\rangle = \int f\,d\langle M\rangle = \langle N, M\rangle$. Therefore we have $\langle N^{(2)}, M\rangle = 0$. This shows the existence of the orthogonal decomposition. In particular if $N \in \mathcal{M}_c$, then $\int f\,dM \in \mathcal{M}_c$ by the previous lemma. Then both $N^{(1)}$ and $N^{(2)}$ are in $\mathcal{M}_c$.

We shall next prove the uniqueness of the decomposition. Suppose that N is decomposed to the sum of an element $\hat{N}^{(1)}$ of $\mathcal{L}(M)$ and an element $\hat{N}^{(2)}$ orthogonal to $\mathcal{L}(M)$. Then since $\hat{N}^{(2)} - N^{(2)} = \hat{N}^{(1)} - N^{(1)}$, $\hat{N}^{(2)} - N^{(2)}$ belongs to $\mathcal{L}(M)$. Clearly it is orthogonal to $\mathcal{L}(M)$. Therefore we have $\hat{N}^{(2)} - N^{(2)} = 0$, proving the uniqueness of the decomposition. The proof is complete. □

The element belonging to $\mathcal{L}(M)$ in the decomposition is called the *orthogonal projection of N to the space* $\mathcal{L}(M)$ and is denoted by $P_{\mathcal{L}(M)}N$.

Now suppose we are given n elements of $\mathcal{M}_c^{loc}$; $M_t^{(1)}, \ldots, M_t^{(n)}$. We shall define $N^{(1)}, \ldots, N^{(n)}$ by the method of Gram–Schmidt's orthogonalization:

$$N^{(1)} = M^{(1)},$$

$$N^{(n)} = M^{(n)} - \sum_{k=1}^{n-1} P_{\mathcal{L}(N^{(k)})}M^{(n)}, \qquad n \geq 2.$$

Then $N^{(1)}, \ldots, N^{(n)}$ are orthogonal. Further each $M^{(k)}$, $k = 1, \ldots, n$ belongs to $\mathcal{L}(N^{(1)}) \oplus \cdots \oplus \mathcal{L}(N^{(n)})$. Note that if $M^{(k)}$, $k = 1, \ldots, n$ belong to $\mathcal{M}_c$, then $N^{(i)}$, $i = 1, \ldots, n$ belong to M_c.

Let $\{M^{(n)}\}$ be an orthogonal system in $\mathcal{M}_c$. If $\langle M, M^{(n)}\rangle = 0$ for all n implies $M = 0$, the system $\{M^{(n)}\}$ is called an *orthogonal basis.*

Theorem 2.3.9 *Let $\{M^{(n)}\}$ be an orthogonal system. It is an orthogonal basis if and only if any M of $\mathcal{M}_c$ is expanded as $M = \sum_k P_{\mathcal{L}(M^{(k)})}M$.* □

The proof is similar to the case of the orthogonal expansion in a Hilbert space and is left to the reader.

Theorem 2.3.10 *$\mathcal{M}_c$ has an orthogonal basis consisting of at most countable elements provided that $(\Omega, \mathcal{F}, P)$ is separable.*

Proof Let $\mathcal{M}$ be the set of all square integrable martingales X_t with $X_0 = 0$, which may or may not be continuous in t. It is a real Hilbert space by the L^2-norm $\|X\| = E[X_T^2]^{1/2}$. Further the space is separable. Indeed, taking a countable dense subset $\{X^k\}$ in $L^2(\Omega, \mathcal{F}, P)$, set $Y_t^k = E[X^k|\mathcal{F}_t] - E[X^k|\mathcal{F}_0]$. Then $\{Y_t^k\}$ is a dense subset of $\mathcal{M}$. Now since $\mathcal{M}_c$ is a closed subset of $\mathcal{M}$, it is also separable. Let $\{M_t^k\}$ be a countable dense subset of $\mathcal{M}_c$. Define the orthogonal system $\{N_t^k\}$ from $\{M_t^k\}$ by the method of

Gram–Schmidt's orthogonalization. Then $\{N_t^k\}$ is an orthogonal basis. □

Remark The number of elements in the orthogonal basis is not necessarily infinite even though the dimension of L^2-space on $(\Omega, \mathscr{F}, P)$ is infinite. See Exercise 2.3.17 for such an example.

Itô's formula

Let $F(x^1, \ldots, x^d)$ be a C^1-function and let $X_t = (X_t^1, \ldots, X_t^d)$ be a continuous semimartingale. If each X_t^i is a process of bounded variation, it is well known that the formula for the change of variables holds:

$$F(X_t) - F(X_0) = \sum_{i=1}^{d} \int_0^t F_{x^i}(X_s)\, dX_s^i.$$

However if X_t contains a martingale part, the above formula does not hold. We have instead the celebrated *Itô's formula.*

Theorem 2.3.11 *Let $X_t = (X_t^1, \ldots, X_t^d)$ be a continuous semimartingale. If $F(x^1, \ldots, x^d)$ is a C^2-function, then $F(X_t)$ is a continuous semimartingale and satisfies the formula*

$$F(X_t) - F(X_0) = \sum_{i=1}^{d} \int_0^t F_{x^i}(X_s)\, dX_s^i + \frac{1}{2} \sum_{i,j=1}^{d} \int_0^t F_{x^i x^j}(X_s)\, d\langle X^i, X^j\rangle_s. \tag{16}$$

Furthermore if F is a C^3-function, then we have

$$F(X_t) - F(X_0) = \sum_{i=1}^{d} \int_0^t F_{x^i}(X_s) \circ dX_s^i. \tag{17}$$

Proof Let $\Delta = \{0 = t_0 < \cdots < t_l = t\}$ be a partition of $[0, t]$. We have by the mean value theorem,

$$\begin{aligned} F(X_t) - F(X_0) &= \sum_{k=0}^{l-1} \{F(X_{t_{k+1}}) - F(X_{t_k})\} \\ &= \sum_{i=1}^{d} \left\{ \sum_{k=0}^{l-1} F_{x^i}(X_{t_k})(X_{t_{k+1}}^i - X_{t_k}^i) \right\} \\ &\quad + \frac{1}{2} \sum_{i,j=1}^{d} \left\{ \sum_{k=0}^{l-1} F_{x^i x^j}(\xi_k^{ij})(X_{t_{k+1}}^i - X_{t_k}^i)(X_{t_{k+1}}^j - X_{t_k}^j) \right\}, \end{aligned}$$

where ξ_k^{ij} are $\mathscr{F}_{t_{k+1}}$-measurable random variables such that $|\xi_k^{ij} - X_{t_k}| \le |X_{t_{k+1}} - X_{t_k}|$. The first term converges to $\sum_{i=1}^{d} \int_0^t F_{x^i}(X_s)\, dX_s^i$ in probability as $|\Delta| \to 0$ by Theorem 2.3.4 (iv). The second term is written as

$$\frac{1}{2}\sum_{i,j}\left\{\sum_k F_{x^ix^j}(X_{t_k})(X^i_{t_{k+1}}-X^i_{t_k})(X^j_{t_{k+1}}-X^j_{t_k})\right\}$$
$$+\frac{1}{2}\sum_{i,j}\left\{\sum_k (F_{x^ix^j}(\xi^{ij}_k)-F_{x^ix^j}(X_{t_k}))(X^i_{t_{k+1}}-X^i_{t_k})(X^j_{t_{k+1}}-X^j_{t_k})\right\}.$$

Set

$$L^\Delta_t=\sum_{k=0}^{l-1}F_{x^ix^j}(X_{t_k})(X^i_{t_{k+1}}-X^i_{t_k}),$$
$$L_t\doteq\int_0^t F_{x^ix^j}(X_s)\,\mathrm{d}X^i_s.$$

Then L^Δ_t converges to L_t in probability. This implies that $\langle L^\Delta_t, X^j_t\rangle^\Delta$ converges to $\langle L_t, X^j_t\rangle$ in probability by Corollary 2.2.19. $\langle L_t, X^j_t\rangle$ is equal to

$$\left\langle\int_0^t F_{x^ix^j}(X_s)\,\mathrm{d}X^i_s, X^j_t\right\rangle=\int_0^t F_{x^ix^j}(X_s)\,\mathrm{d}\langle X^i, X^j\rangle_s.$$

On the other hand we have

$$\left|\sum_k (F_{x^ix^j}(\xi^{ij}_k)-F_{x^ix^j}(X_{t_k}))(X^i_{t_{k+1}}-X^i_{t_k})(X^j_{t_{k+1}}-X^j_{t_k})\right|$$
$$\le\sup_k|F_{x^ix^j}(\xi^{ij}_k)-F_{x^ix^j}(X_{t_k})|\left\{\sum_k(X^i_{t_{k+1}}-X^i_{t_k})^2\right\}^{1/2}\left\{\sum_k(X^j_{t_{k+1}}-X^j_{t_k})^2\right\}^{1/2}. \tag{18}$$

Note that $\sup_k|\ldots|$ of the above converges to 0 a.s. as $|\Delta|\to 0$ and $\sum_k(X^i_{t_{k+1}}-X^i_{t_k})^2$converges to $\langle X^i\rangle_t$ in probability. Then (18) converges to 0 in probability. We have thus proved the formula (16). The right hand side is obviously a continuous semimartingale.

Suppose next that F is a C^3-function. Then $F_{x^i}(X_t)$ is a continuous semimartingale. By formula (16), we have

$$F_{x_i}(X_t)=F_{x^i}(X_0)t\sum_{j=1}^d\int_0^t F_{x^ix^j}(X_s)\,\mathrm{d}X^j_s+\frac{1}{2}\sum_{k,l}\int_0^t F_{x^ix^kx^l}(X_s)\,\mathrm{d}\langle X^k, X^l\rangle_s.$$

The Stratonovich integral of $F_{x^i}(X_t)$ by $\circ\,\mathrm{d}X^i_t$ is then equal to

$$\int_0^t F_{x^i}(X_s)\circ\mathrm{d}X^i_s=\int_0^t F_{x^i}(X_s)\,\mathrm{d}X^i_s+\frac{1}{2}\langle F_{x^i}(X_t), X^i_t\rangle$$
$$=\int_0^t F_{x^i}(X_s)\,\mathrm{d}X^i_s+\frac{1}{2}\sum_{j=1}^d\int_0^t F_{x^ix^j}(X_s)\,\mathrm{d}\langle X^j, X^i\rangle_s.$$

Therefore the right hand side of (17) is equal to the right hand side of (16). This proves (17). The proof is complete. □

Itô's formula, given above, is obviously valid for continuous local semimartingales.

Applications of Itô's formula

As an application of Itô's formula, we shall prove the *Burkholder inequality.*

Theorem 2.3.12 *For any $p \geq 2$ there exist positive constants $C_i = C_i(p)$ $i = 1, 2$ such that*

$$C_1 E[\langle M\rangle_t^{p/2}] \leq E[|M_t|^p] \leq C_2 E[\langle M\rangle_t^{p/2}], \qquad \text{for all } t \in [0, T] \tag{19}$$

holds for any continuous martingale M_t such that $M_0 = 0$ and $E[|M_T|^p] < \infty$.

Proof We shall first prove the inequality (19) for a class of martingales M_t such that both $|M_t|$ and $\langle M\rangle_t$ are bounded by a constant. Apply Itô's formula by setting $F(x) = |x|^p$ and $X_t = M_t$. Then

$$|M_t|^p = p\int_0^t |M_s|^{p-1}\,\text{sign}(M_s)\,\mathrm{d}M_s + \frac{1}{2}p(p-1)\int_0^t |M_s|^{p-2}\,\mathrm{d}\langle M\rangle_s,$$

where $\text{sign}(M_s)$ is 1 if $M_s \geq 0$ and is -1 if $M_s < 0$. The first term of the right hand side belongs to $\mathscr{M}_c$ since its quadratic variation is bounded by a constant. Taking the expectation of each of the above, we have

$$\begin{aligned}
E[|M_t|^p] &= \frac{1}{2}p(p-1)E\left[\int_0^t |M_s|^{p-2}\,\mathrm{d}\langle M\rangle_s\right] \\
&\leq \frac{1}{2}p(p-1)E\left[\sup_{0\leq s\leq t} |M_s|^{p-2}\langle M\rangle_t\right] \\
&\leq \frac{1}{2}p(p-1)E\left[\sup_{0\leq s\leq t} |M_s|^p\right]^{(p-2)/p} E[\langle M\rangle_t^{p/2}]^{2/p} \\
&\leq \frac{1}{2}p(p-1)q^{p-2}E[|M_t|^p]^{(p-2)/p}E[\langle M\rangle_t^{p/2}]^{2/p}.
\end{aligned}$$

Here we have used Hölder's inequality and Doob's inequality of martingales (Theorem 1.2.7). This proves the second half of the inequality.

To prove the other side of the inequality, set $N_t \equiv \int_0^t \langle M\rangle_s^{(p/4)-(1/2)}\,\mathrm{d}M_s$. Then

$$\langle N\rangle_t = \int_0^t \langle M\rangle_s^{(p/2)-1}\,\mathrm{d}\langle M\rangle_s = \frac{2}{p}\langle M\rangle_t^{p/2}.$$

Since

$$M_t\langle M\rangle_t^{(p/4)-(1/2)} = \int_0^t \langle M\rangle_s^{(p/4)-(1/2)}\,\mathrm{d}M_s + \int_0^t M_s\,\mathrm{d}(\langle M\rangle_s^{(p/4)-(1/2)})$$

holds by Itô's formula, we have $|N_t| \le 2\sup_{s\le t}|M_s|\langle M\rangle_t^{(p/4)-(1/2)}$. Therefore

$$\begin{aligned} E[\langle M\rangle_t^{p/2}] &\le \frac{p}{2}E[\langle N\rangle_t] = \frac{p}{2}E[N_t^2] \\ &\le 2pE\left[\sup_{s\le t}|M_s|^2\langle M\rangle_t^{(p/2)-1}\right] \\ &\le 2pE\left[\sup_{s\le t}|M_s|^p\right]^{p/2} E[\langle M\rangle_t^{p/2}]^{1-(2/p)}. \end{aligned}$$

This proves the first half of the inequality.

Now let M_t be an arbitrary continuous martingale satisfying conditions of the theorem. For $m = 1, 2, \ldots$ let τ_m be the first time such that $|M_t| \ge m$ or $\langle M\rangle_t \ge m$ holds. Then each stopped process $M_t^{\tau_m}$ satisfies inequality (19), where constants C_1 and C_2 do not depend on m. Let m tend to infinity. Then we have (19) for M_t. The proof is complete. □

As another application of Itô's formula we shall characterize a Brownian motion by a certain martingale property. A d-dimensional continuous $(\mathscr{F}_t)$-adapted process $X_t = (X_t^1, \ldots, X_t^d)$ is called a (standard) $(\mathscr{F}_t)$-Brownian motion if it is a (standard) Brownian motion such that $X_t - X_s$ is independent of $\mathscr{F}_s$ for any $s < t$.

Theorem 2.3.13 *Let $X_t \equiv (X_t^1, \ldots, X_t^d)$ be a continuous process with mean 0 and covariance $V(s, t)$ adapted to $(\mathscr{F}_t)$. The following statements are equivalent:*

(i) *X_t is an $(\mathscr{F}_t)$-Brownian motion.*
(ii) *$X_t^1, \ldots, X_t^d$ are L^2-martingales and $\langle X^i, X^j\rangle_t = V^{ij}(t, t)$.*
(iii) *$\exp\{(\xi, X_t) - \frac{1}{2}(V(t, t)\xi, \xi)\}$ is a martingale for any $\xi \in \mathbb{R}^d$.*
(iv) *$\exp\{i(\xi, X_t) + \frac{1}{2}(V(t, t)\xi, \xi)\}$ is a martingale for any $\xi \in \mathbb{R}^d$ where $i = \sqrt{-1}$.*
(v) *X_t is a Gaussian process such that $X_t - X_s$ is independent of $\mathscr{F}_s$ for any $0 < s < t$.*

Proof Suppose first that (i) is satisfied. Let $s < t$. Then for any i, $X_t^i - X_s^i$ is independent of $\mathscr{F}_s$. Therefore we have

$$E[X_t^i - X_s^i|\mathscr{F}_s] = E[X_t^i - X_s^i] = 0.$$

(See Exercise 1.1.7.) This shows that X_t^i, $i = 1, \ldots, d$ are martingales. Similarly we have

$$\begin{aligned} E[X_t^i X_t^j - X_s^i X_s^j | \mathscr{F}_s] &= E[(X_t^i - X_s^i)(X_t^j - X_s^j) | \mathscr{F}_s] \\ &= E[(X_t^i - X_s^i)(X_t^j - X_s^j)] \\ &= V^{ij}(t, t) - V^{ij}(s, s). \end{aligned}$$

Therefore $X_t^i X_t^j - V^{ij}(t, t)$ are martingales, and (ii) is verified.

Next suppose (ii) holds. Set

$$\alpha_t(\xi) \equiv \exp\left\{(\xi, X_t) - \frac{1}{2}(V(t, t)\xi, \xi)\right\}.$$

By Ito's formula we have

$$\alpha_t(\xi) = 1 + \sum_{i=1}^d \xi_i \int_0^t \alpha_s(\xi)\, dX_s^i.$$

Therefore $\alpha_t(\xi)$ is a localmartingale and its quadratic variation $\langle \alpha_t(\xi) \rangle$ is given by $\int_0^t \alpha_s(\xi)^2\, dF(s)$, where $F(t) = (V(t, t)\xi, \xi)$. We wish to show that $\alpha_t(\xi)$ is an L^2-martingale or equivalently $\langle \alpha_t(\xi) \rangle$ is integrable. Observe the equality $\alpha_t(\xi)^2 = \alpha_t(2\xi) \exp\{F(t)\}$, where $\alpha_t(2\xi)$ is a positive localmartingale. There exists an increasing sequence of stopping times τ_n such that $P(\tau_n < T) \to 0$ as $n \to \infty$ and each stopped process $\alpha_{t \wedge \tau_n}(2\xi)$ is a martingale with mean 1. Then we have $E[\alpha_t(2\xi)] \leq 1$ by Fatou's lemma. Consequently $E[\alpha_t(\xi)^2]$ is bounded by $\exp\{F(t)\}$. We have thus shown that $\alpha_t(\xi)$ is an L^2-martingale for any ξ.

Conversely suppose (iii). Then $\partial \alpha_t(\xi)/\partial \xi^i$ and $\partial^2 \alpha_t(\xi)/\partial \xi^i \partial \xi^j$ are localmartingales by Theorem 2.1.2. Setting $\xi = 0$ in these expressions, we see the X_t^i and $X_t^i X_t^j - V^{ij}(t, t)$ are localmartingales. Then $V^{ij}(t, t)$ is the joint quadratic variation of X_t^i and X_t^j, so X_t^i should be an L^2-martingale by Corollary 2.2.7. Therefore property (ii) is satisfied.

The equivalence of (ii) and (iv) can be shown similarly. Suppose next (iv) is satisfied. Then we have

$$E[\exp\{i(\xi, X_t - X_s)\} | \mathscr{F}_s] = \exp\{-\tfrac{1}{2}((V(t, t) - V(s, s))\xi, \xi)\}$$

for any ξ. This proves that $X_t - X_s$ is independent of $\mathscr{F}_s$ and is subject to the Gaussian distribution with mean 0 and covariance $V(t, t) - V(s, s)$. Hence (v) is satisfied. Property (v) implies (i) immediately. The proof is complete. □

We shall establish *Girsanov's theorem*, applying the above theorem.

Theorem 2.3.14 *Let $X_t \equiv (X_t^1, \ldots, X_t^d)$ be a standard $(\mathscr{F}_t)$-Brownian motion and $f_t \equiv (f_t^1, \ldots, f_t^d)$ be a predictable process such that $\int_0^T |f_s|^2\, ds$ is finite* *a.s. and the expectation of*

$$\alpha_t = \exp\left\{\sum_j \int_0^t f_s^j \, dX_s^j - \frac{1}{2}\int_0^t |f_s|^2 \, ds\right\} \tag{20}$$

is 1 *for any* t. *Define the probability* Q *by* $Q(dw) = \alpha_T \, dP(w)$. *Then* $\tilde{X}_t \equiv X_t - \int_0^t f_s \, ds$ *is a standard* $(\mathscr{F}_t)$-*Brownian motion with respect to* Q.

Proof Let us first observe that α_t of (20) is a positive martingale. By Itô's formula, the functional α_t satisfies $\alpha_t = 1 + \sum_j \int_0^t \alpha_s f_s^j \, dX_s^j$. Therefore it is a positive local martingale. Let $\{\tau_n\}$ be an increasing sequence of stopping times such that $P(\tau_n < T) \to 0$ as $n \to \infty$ and each stopped process $\alpha_{t \wedge \tau_n}$ is a martingale. Then we have

$$E[\alpha_t | \mathscr{F}_s] \le \lim_{n\to\infty} E[\alpha_{t\wedge\tau_n} | \mathscr{F}_s] = \lim_{n\to\infty} \alpha_{s\wedge\tau_n} = \alpha_s.$$

Therefore α_t is a supermartingale. Since its expectation is identically 1, it is a martingale.

For any $\xi = (\xi^1, \ldots, \xi^d)$ of $\mathbb{R}^d$, set $\beta_t = \exp\{i(\xi, \tilde{X}_t) + \frac{1}{2}|\xi|^2 t\}$ where $i = \sqrt{-1}$. It is sufficient to show that β_t is a martingale with respect to the probability Q by the previous theorem. A direct computation yields

$$\alpha_t \beta_t = \exp\left\{\sum_j \int_0^t (f_s^j + i\xi^j) \, dX_s^j - \frac{1}{2}\int_0^t |f_s + i\xi|^2 \, ds\right\}.$$

By Itô's formula it satisfies the equality

$$\alpha_t \beta_t = 1 + \sum_j \int_0^t \alpha_s \beta_s (f_s^j + i\xi^j) \, dX_s^j.$$

Therefore $\alpha_t \beta_t$ is a localmartingale with respect to P. Let $\{\tau_n\}$ be an increasing sequence of stopping times such that $P(\tau_n < T) \to 0$ as $n \to \infty$ and each stopped process $\alpha_{t\wedge\tau_n}\beta_{t\wedge\tau_n}$ is a martingale. Now by the optional stopping time theorem, $Q(B) = E[\alpha_{t\wedge\tau_n} : B]$ holds for any B of $\mathscr{F}_{t\wedge\tau_n}$. Therefore we have

$$E_Q[\beta_{t\wedge\tau_n} | \mathscr{F}_{s\wedge\tau_n}] = \alpha_{s\wedge\tau_n}^{-1} E[\alpha_{t\wedge\tau_n}\beta_{t\wedge\tau_n} | \mathscr{F}_{s\wedge\tau_n}] = \beta_{s\wedge\tau_n}.$$

This shows that β_t is a localmartingale with respect to Q. Since $|\beta_t|$ is a bounded process it is a martingale with respect to Q. The proof is complete. □

Exercise 2.3.15 Prove Theorem 2.3.4. (*Hint*: for (i)–(iii), it suffices to prove the case that X_t is a localmartingale. For the proof of equalities (8) and (9), apply Theorem 2.3.2 (ii). For the proof of (10), apply the same theorem and Theorem 2.2.13 (i). For the proof of (iv) and (v), apply Theorem 2.2.15.)

Exercise 2.3.16 Prove Theorem 2.3.6 (*Hint*: reduce the proof of (13) and (15) to Exercise 2.3.15 using formula (12). For the proof of (14) show first that $f_t g_t$ is a continuous semimartingale and is represented by $\int_0^t f_s \, dg_s + \int_0^t g_s \, df_s + \langle f, g \rangle_t$ (Itô's formula). Then use formula (12).

Exercise 2.3.17 Let $X_t = (X_t^1, \ldots, X_t^d)$ be a standard Brownian motion and let $\{\mathscr{F}_t\}$ be the filtration generated by X_t. Show that $\{X_t^1, \ldots, X_t^d\}$ is an orthogonal basis of the space $\mathscr{M}_c$ of continuous L^2-martingales adapted to $(\mathscr{F}_t)$. (*Hint*: let N_t be a positive continuous $(\mathscr{F}_t)$-martingale with mean 1, orthogonal to $X_t^1, \ldots, X_t^d$. Let Q be a probability such that $Q(A) = \int_A N_T \, dP$. Show that $(X_t^1, \ldots, X_t^d)$ is a standard Brownian motion with respect to Q and deduce that $N_T \equiv 1$. *Cf.* Dellacherie [25].)

Exercise 2.3.18 Let M_t, $t \in [0, \infty)$, be a continuous martingale with the

(i) Show that $\alpha_t = \exp\{M_t - \frac{1}{2}\langle M \rangle_t\}$ is a positive continuous local-martingale and a continuous supermartingale.
(ii) Show that if $\exp\{\langle M \rangle_t\}$ is integrable, then α_t is a martingale. (*Hint*: use Itô's formula.)

Exercise 2.3.19 Let X_t, $t \in [0, \infty)$, be a continuous martingale with the quadratic variation $\langle X \rangle_t = \int_0^t a(s) \, ds$ where $a(s)$ is a bounded predictable process. Show that $\lim_{t \uparrow \infty} X_t / t^{(1/2)+\epsilon} = 0$ holds a.s. for any $\varepsilon > 0$. (*Hint*: derive the estimate of Exercise 1.4.14 using Theorem 2.3.12.)

Exercise 2.3.20 Let M_t, $t \in [0, \tau)$ be a local localmartingale.

(i) Show that $\alpha_t = \exp\{M_t - \frac{1}{2}\langle M \rangle_t\}$, $t \in [0, \tau)$ is a positive local super-martingale and $\lim_{t \uparrow \tau} \alpha_t$ exists and is finite.
(ii) Prove $P(\{\lim_{t \uparrow \tau} \alpha_t = 0\} \Delta \{\lim_{t \uparrow \tau} \langle M \rangle_t = \infty\}) = 0$.

(*Hint*: for the proof of (ii), define β_t and γ_t by

$$\beta_t = \exp\{-M_t - \tfrac{1}{2}\langle M \rangle_t\}, \qquad \gamma_t = \exp\{\tfrac{1}{2}M_t - \tfrac{1}{2}\langle \tfrac{1}{2}M \rangle_t\}$$

and note the relations $\alpha_t \beta_t = \exp\{-\langle M \rangle_t\}$ and $\alpha_t = \gamma_t^2 \exp\{-\langle M \rangle_t / 4\}$.)

3

Semimartingales with spatial parameters and stochastic integrals

3.1 Preliminaries

Martingales with spatial parameters

In the previous chapter we defined the Itô integral of the form $\int f_s \, \mathrm{d}X_s$ and the Stratonovich integral of the form $\int f_s \circ \mathrm{d}X_s$. The integrands f_s are a predictable process and a continuous semimartingale, respectively. The $\mathrm{d}X_t$ and $\circ \mathrm{d}X_t$ are the Itô and the Stratonovich differentials respectively based on a continuous semimartingale X_t. However in applications to various stochastic problems concerning stochastic differential equations and stochastic flows, it is necessary to define stochastic integrals of the form

$$\int F(f_s(\lambda), \mathrm{d}s) \qquad \text{and} \qquad \int F(f_s(\lambda), \circ \mathrm{d}s)$$

where $F(x, t)$ is a continuous semimartingale with spatial parameter $x \in \mathbb{R}^d$. Kernels $F(x, \mathrm{d}t)$ and $F(x, \circ \mathrm{d}t)$ are Itô's and Stratonovich's differential, respectively. The integrands $f_s(\lambda)$ are a predictable process and a continuous semimartingale, respectively with spatial parameter λ. A main objective of this chapter is to define these integrals rigorously and develop the differential rule for these integrals with respect to the spatial parameter as well as the time parameter. Another objective will be to discuss the existence and the uniqueness of the solution of the stochastic differential equation of the form

$$\mathrm{d}\varphi_t = F(\varphi_t, \mathrm{d}t) \qquad \text{or} \qquad \mathrm{d}\varphi_t = F(\varphi_t, \circ \mathrm{d}t).$$

Before we proceed further let us observe that these problems require careful arguments. As an example consider the Itô integral of a predictable process $f_t(\lambda)$ with parameter λ based on the continuous local martingale M_t. Suppose $\int_0^T f_s(\lambda)^2 \mathrm{d}\langle M \rangle_s < \infty$ holds for each λ. Then the Itô integral $M(\lambda, t) = \int_0^t f_s(\lambda) \, \mathrm{d}M_s$ is well defined for any t except for a null set. It is a continuous local martingale with the spatial parameter λ. We denote the null set by N_λ since it depends on the parameter λ. Then $M(\lambda, t)$ is well defined for all (λ, t) if $\omega \in (\bigcup_\lambda N_\lambda)^c$. However the exceptional set $\bigcup_\lambda N_\lambda$ may not be a null set since it is an uncountable union of null sets. To overcome this technical problem, we must take a good modification of the random field $M(\lambda, t)$ so that it is well defined for all (λ, t) a.s. and is continuous or continuously differentiable with respect to λ for all t a.s. We shall achieve it later by applying Kolmogorov's theorem (Theorem 1.4.1.).

In this section we shall discuss the continuity and differentiability of continuous semimartingales with the spatial parameter in more general settings. Stochastic integrals based on them and stochastic differential equations will be discussed in the subsequent sections.

We shall first introduce some notations. Let $\mathbb{D}$ be a domain in $\mathbb{R}^d$ and let $\mathbb{R}^e$ be another Euclidean space. Let m be a non-negative integer. Denote by $C^m(\mathbb{D}:\mathbb{R}^e)$ or C^m the set of all maps $f: \mathbb{D} \to \mathbb{R}^e$ which are m-times continuously differentiable. In case $m = 0$, it is often denoted by $C(\mathbb{D}:\mathbb{R}^e)$. For the multi-index of non-negative integers $\alpha = (\alpha_1, \ldots, \alpha_d)$, we define the differential operator D_x^α or D^α by

$$D_x^\alpha = \frac{\partial^{|\alpha|}}{(\partial x^1)^{\alpha_1} \ldots (\partial x^d)^{\alpha_d}},$$

where $|\alpha| = \sum \alpha_i$. Let K be a subset of $\mathbb{D}$. We set

$$\|f\|_{m:K} = \sup_{x \in K} \frac{|f(x)|}{(1 + |x|)} + \sum_{1 \le |\alpha| \le m} \sup_{x \in K} |D^\alpha f(x)|. \tag{1}$$

Then $C^m(\mathbb{D}:\mathbb{R}^e)$ is a Frechet space by seminorms $\{\| \ \|_{m:K} : K$ are compacts in $\mathbb{D}\}$. When $K = \mathbb{D}$, we write $\| \ \|_{m:K}$ as $\| \ \|_m$. Denote by $C_b^m(\mathbb{D}:\mathbb{R}^e)$ or C_b^m the set $\{f \in C^m : \|f\|_m < \infty\}$. It is a Banach space with the norm $\| \ \|_m$.

Now let δ be a positive number less than or equal to 1. Denote by $C^{m,\delta}(\mathbb{D}:\mathbb{R}^e)$ or simply by $C^{m,\delta}$ the set of all f of C^m such that $D^\alpha f$, $|\alpha| = m$ are δ-Hölder continuous. By the seminorms

$$\|f\|_{m+\delta:K} = \|f\|_{m:K} + \sum_{|\alpha|=m} \sup_{\substack{x,y \in K \\ x \ne y}} \frac{|D^\alpha f(x) - D^\alpha f(y)|}{|x - y|^\delta}, \tag{2}$$

it is again a Frechet space. When $K = \mathbb{D}$ we write $\| \ \|_{m+\delta:K}$ as $\| \ \|_{m+\delta}$. Denote by $C_b^{m,\delta}(\mathbb{D}, \mathbb{R}^e)$ or $C_b^{m,\delta}$ the set $\{f \in C^{m,\delta} : \|f\|_{m+\delta} < \infty\}$.

A continuous function $f(x, t)$, $x \in \mathbb{D}$, $t \in [0, T]$ is said to belong to the class $C^{m,\delta}$ if for every t, $f(t) \equiv f(\cdot, t)$ belongs to $C^{m,\delta}$ and $\|f(t)\|_{m+\delta:K}$ is integrable on $[0, T]$ with respect to t for any compact subset K. If the set K is replaced by $\mathbb{D}$, f is said to belong to the class $C_b^{m,\delta}$. Furthermore, if $\|f(t)\|_{m+\delta}$ is bounded in t, it is said to belong to the class $C_{ub}^{m,\delta}$.

Still other classes of function spaces and seminorms are needed. We denote by $\tilde{C}^m$ the set of all $\mathbb{R}^e$-valued functions $g(x, y)$, $x, y \in \mathbb{D}$ which are m-times continuously differentiable with respect to each variable x and y. For $g \in \tilde{C}^m$, we define a seminorm

$$\|g\|_{m:K}^{\sim} = \sup_{x,y\in K} \frac{|g(x, y)|}{(1 + |x|)(1 + |y|)} + \sum_{1\le|\alpha|\le m} \sup_{x,y\in K} |D_x^\alpha D_y^\alpha g(x, y)|, \quad (3)$$

and for $0 < \delta \le 1$,

$$\|g\|_{m+\delta:K}^{\sim} = \|g\|_{m:K}^{\sim} + \sum_{|\alpha|=m} \|D_x^\alpha D_y^\alpha g\|_{\delta:K}^{\sim}, \quad (4)$$

where

$$\|g\|_{\delta:K}^{\sim} = \sup_{\substack{x,y,x',y'\in K \\ x\ne x', y\ne y'}} \frac{|g(x, y) - g(x', y) - g(x, y') + g(x', y')|}{|x - x'|^\delta |y - y'|^\delta}.$$

The function g is said to belong to the space $\tilde{C}^{m,\delta}$ if $\|g\|_{m+\delta:K}^{\sim} < \infty$ for any compact set K in $\mathbb{D}$. We denote $\| \;\|_{m:\mathbb{D}}^{\sim}$ and $\| \;\|_{m+\delta:\mathbb{D}}^{\sim}$ by $\| \;\|_m^{\sim}$ and $\| \;\|_{m+\delta}^{\sim}$, respectively. We set $\tilde{C}_b^m = \{g : \|g\|_m^{\sim} < \infty\}$ and $\tilde{C}_b^{m,\delta} = \{g : \|g\|_{m+\delta}^{\sim} < \infty\}$.

A continuous function $g(x, y, t)$, $x, y \in \mathbb{D}$, $t \in [0, T]$ is said to belong to the class $\tilde{C}^{m,\delta}$ if for every t, $g(t) \equiv g(\cdot, \cdot, t)$ belongs to the space $\tilde{C}^{m,\delta}$ and $\|g(t)\|_{m+\delta:K}^{\sim}$ is integrable on $[0, T]$ with respect to t for any compact subset K. The classes $\tilde{C}_b^{m,\delta}$ and $\tilde{C}_{ub}^{m,\delta}$ are defined similarly.

Let $F(x, t)$ be a family of real valued processes with parameter $x \in \mathbb{D}$ where $\mathbb{D}$ is a domain in $\mathbb{R}^d$. We can regard it as a random field with double parameters x and t. If $F(x, t, \omega)$ is a continuous function of x for almost all ω for any t, we can regard $F(\cdot, t)$ as a *stochastic process with values in* $C = C(\mathbb{D} : \mathbb{R}^1)$ or a *C-valued process*. If $F(x, t, \omega)$ is m-times continuously differentiable with respect to x a.s. for any t, it can be regarded as a *stochastic process with values in C^m or a C^m-valued process*. In the case where $F(x, t)$ is a continuous process with values in C^m, it is called a *continuous C^m-process*. A *$C^{m,\delta}$-valued process* and *continuous $C^{m,\delta}$-process* are defined similarly.

Let $G(x, y, t)$ be a stochastic process with parameters $x, y \in \mathbb{D}$. If it is m-times continuously differentiable with respect to each x and y a.s. for any t, it is called a *stochastic process with values in $\tilde{C}^m$* or a *$\tilde{C}^m$-valued process*. The *$\tilde{C}^{m,\delta}$-valued process* and *continuous $\tilde{C}^{m,\delta}$-valued process* are defined similarly.

A continuous $C^{m,\delta}$-process $F(t)$ is called a *$C^{m,\delta}$-valued Brownian motion* if for any $0 \le t_0 < \cdots < t_l \le T$, $F(t_0)$, $F(t_{i+1}) - F(t_i)$, $i = 0, \ldots, l - 1$ are independent $C^{m,\delta}$-valued random variables. Note that if $F(t)$ is a $C^{m,\delta}$-valued Brownian motion, its finite dimensional restriction $(F(x_1, t), \ldots, F(x_n, t))$ is an nd-dimensional Brownian motion for any $(x_1, \ldots, x_n)$. Therefore if $F(0) = 0$, $F(t)$ is subject to a Gaussian distribution on the space $C^{m,\delta}$.

Conversely let $F(x, t)$, $x \in \mathbb{D}$ be a family of continuous real processes such that $(F(x_1, t), \ldots, F(x_n, t))$ is a Brownian motion for any $x_1, \ldots, x_n$. If its mean $E[F(x, t)]$ and covariance $E[F(x, t)F(y, t)^t]$ are smooth functions of

x and x, y, respectively, then $F(x, t)$ should have a modification of a $C^{m,\delta}$-valued Brownian motion for some $m \geq 0$ and $\delta > 0$. In the next subsection we shall prove this fact for martingales and semimartingales, more generally.

Regularity of martingales with respect to spatial parameters

Now we assume that on a separable probability space $(\Omega, \mathscr{F}, P)$ a filtration $\{\mathscr{F}_t : t \in [0, T]\}$ of sub σ-fields of $\mathscr{F}$ is given. Semimartingales, predictable processes etc. appearing in this chapter are understood to be those with respect to the above filtration $\{\mathscr{F}_t\}$ unless otherwise mentioned. Let $M(x, t)$, $x \in \mathbb{D}$ and $N(y, t)$, $y \in \mathbb{D}$ be two families of continuous local martingales with spatial parameters x, y and let $A^{MN}(x, y, t)$ be the joint quadratic variation of $M(x, t)$ and $N(y, t)$. We shall discuss the regularity of $M(x, t)$, $N(y, t)$ with respect to the spatial parameter in connection with the regularity of the joint quadratic variation. In the next theorem, we will derive the Hölder continuity of $M(x, t)$ with respect to x. The differentiability with respect to x will be discussed in the subsequent theorem.

Theorem 3.1.1 *Let $M(x, t)$, $x \in \mathbb{D}$ be a family of continuous localmartingales such that $M(x, 0) \equiv 0$. Assume that the joint quadratic variation $\langle M(x, t), M(y, t)\rangle$ has a modification of a continuous $\tilde{C}^{0,\delta}$-process. Then $M(x, t)$ has a modification of a continuous $C^{0,\varepsilon}$-process for any $\varepsilon < \delta$.*

Proof Denote the continuous modification of $\langle M(x, t), M(y, t)\rangle$ by $A(t) \equiv A(x, y, t)$. Let K be a compact convex subset of $\mathbb{D}$. For each positive integer n set

$$\tau_n = \inf\{t \in [0, T] : \|A(t)\|^{\sim}_{0+\delta:K} \geq n\} \quad (= T \text{ if } \{\dots\} = \emptyset).$$

Then $P(\tau_n < T) \to 0$ as $n \to \infty$. Therefore it is enough to prove that each stopped process $M^{\tau_n}(x, t) \equiv M(x, t \wedge \tau_n)$ has a modification of a continuous $C^{0,\varepsilon}$-process.

Now set $A^{\tau_n}(x, y, t) \equiv A(x, y, t \wedge \tau_n)$. Since $M^{\tau_n}(x, t)M^{\tau_n}(y, t) - A^{\tau_n}(x, y, t)$ is a localmartingale by the optional stopping time theorem, $A^{\tau_n}(x, y, t)$ is the joint quadratic variation of $M^{\tau_n}(x, t)$ and $M^{\tau_n}(y, t)$. Then the quadratic variation of the difference $M^{\tau_n}(x, t) - M^{\tau_n}(y, t)$ is estimated as

$$\begin{aligned}\langle M^{\tau_n}(x, t) - M^{\tau_n}(y, t)\rangle &= A^{\tau_n}(x, x, t) - 2A^{\tau_n}(x, y, t) + A^{\tau_n}(y, y, t) \\ &\leq n|x - y|^{2\delta}\end{aligned}$$

if $x, y \in K$. In view of Burkholder's inequality (Theorem 2.3.12), for any $p \geq 2$ there exists a positive constant $C = C(p)$ such that

$$\begin{aligned}E[|M^{\tau_n}(x, t) - M^{\tau_n}(y, t)|^{2p}] &\leq CE[\langle M^{\tau_n}(x, t) - M^{\tau_n}(y, t)\rangle^p] \\ &\leq Cn^p|x - y|^{2\delta p}.\end{aligned}$$

Then Doob's inequality (Theorem 1.2.7) yields

$$E[\|M^{\tau_n}(x,\cdot)-M^{\tau_n}(y,\cdot)\|^{2p}]\le q^pCn^p|x-y|^{2\delta p},$$

where $\|\ \|$ is the supremum norm for the space $C([0,T];\mathbb{R}^d)$ and q is the conjugate of $2p$. Take p as large as $p>d/2\delta$. Then by Kolmogorov's theorem (Theorem 1.4.1), $M^{\tau_n}(x,\cdot)$ has a modification of a continuous C-valued process. It is ε-Hölder continuous in x for any $\varepsilon<(2\delta p-d)/2p$. We can choose p as large as we wish. Therefore $M^{\tau_n}(x,t)$ has a modification of a continuous $C^{0,\varepsilon}$-process for any ε less than δ. The proof is complete. □

Theorem 3.1.2 *Let $M(x,t)$, $x\in\mathbb{D}$ be a family of continuous localmartingales such that $M(x,0)\equiv 0$. Assume that its joint quadratic variation has a modification $A(x,y,t)$ of a continuous $\tilde{C}^{m,\delta}$-process for some $m\ge 1$ and $\delta\in(0,1]$. Then $M(x,t)$ has a modification of a continuous $C^{m,\varepsilon}$-process for any $\varepsilon<\delta$. Furthermore, for each α with $|\alpha|\le m$, $D_x^\alpha M(x,t)$, $x\in\mathbb{D}$ is a family of continuous localmartingales with the joint quadratic variation $D_x^\alpha D_y^\alpha A(x,y,t)$.*

Proof We shall prove the differentiability of $M(x,t)$ with respect to x. Given $\gamma>0$, let $\mathbb{D}_\gamma$ be a subdomain of $\mathbb{D}$ defined by $\mathbb{D}_\gamma=\{x\in\mathbb{R}^d:\rho(x,\mathbb{D}^c)\le\gamma\}^c$, where ρ is the Euclidean metric. Set

$$N(x,h,t)=\frac{1}{h}\{M(x+he_i,t)-M(x,t)\} \tag{5}$$

for $(x,h)\in\mathbb{D}_\gamma\times\{(-\gamma,0)\cup(0,\gamma)\}$ where e_i is the unit vector $(0,\dots,0,1,0,\dots,0)$ (1 is at the ith component). It is a continuous random field. We wish to show that $N(x,h,t)$ converges uniformly in (x,t) on compact sets as $h\to 0$ a.s. (almost surely). If it is established, the limit $N(x,0,t)$ is a continuous localmartingale for any x by Theorem 2.1.2. This means that $M(x,t)$ is continuously differentiable with respect to x at any x a.s. and the partial derivative $(\partial M/\partial x^i)(x,t)$ is a continuous localmartingale $N(x,0,t)$.

Let us first observe that

$$A^N((x,h),(y,k),t)=\frac{1}{hk}\{A(x+he_i,y+ke_i,t)-A(x+he_i,y,t)$$
$$-A(x,y+ke_i,t)+A(x,y,t)\} \tag{6}$$

is the joint quadratic variation of $N(x,h,t)$ and $N(y,k,t)$. By the mean value theorem it can be written as

$$A^N((x,h),(y,k),t)=\int_0^1 d\tau\int_0^1 d\theta\frac{\partial^2 A}{\partial x^i\partial y^i}(x+\theta he_i,y+\tau ke_i,t). \tag{7}$$

It can be extended continuously at $h=0$, $k=0$ by setting $h=0$, $k=0$ in the right hand side of (7) a.s. It is a continuous $\tilde{C}^{0,\delta}$-process. Then by the previous theorem, $N(x, h, t)$ can be extended to a continuous random field on $\mathbb{D}_\gamma \times (-\gamma, \gamma)$. Therefore, $N(x, 0, t)$ exists and is a continuous $C^{0,\varepsilon}$-process.

We have thus proved that $M(x, t)$ is continuously differentiable at any (x, t) a.s. The partial derivative $(\partial M/\partial x^i)(x, t) = N(x, 0, t)$ has the joint quadratic variation $(\partial^2 A/\partial x^i \partial y^i)(x, y, t)$ a.s. by (7). This yields the theorem in the case $m = 1$. The case $m > 1$ can be proved inductively and is left to the reader. □

We shall call the random field $M(x, t)$ with the property of Theorem 3.1.2 a *continuous localmartingale with values in* $C^{m,\varepsilon}$ or a *continuous* $C^{m,\varepsilon}$-*localmartingale.*

We next discuss the converse problem. We will show the regularity of the joint quadratic variation of continuous $C^{m,\delta}$-localmartingales.

Theorem 3.1.3 *Let $M(x, t)$, $x \in \mathbb{D}$ and $N(y, t)$, $y \in \mathbb{D}$ be continuous localmartingales with values in $C^{m,\delta}$ where $m \geq 0$ and $\delta > 0$. Then the joint quadratic variation has a modification of a continuous $\tilde{C}^{m,\varepsilon}$-process for any $\varepsilon < \delta$. Furthermore, the modification satisfies*

$$D_x^\alpha D_y^\beta \langle M(x, t), N(y, t)\rangle = \langle D_x^\alpha M(x, t), D_y^\beta N(y, t)\rangle \quad a.s. \tag{8}$$

for any t if $m \geq 1$ and $|\alpha|, |\beta| \leq m$.

Proof Let $A(x, y, t)$ be the joint quadratic variation of $M(x, t)$ and $N(y, t)$. We first show that $A(x, y, t)$ has a modification of a continuous $C^{m,\varepsilon}$-process. We may assume that $M(x, 0) \equiv 0$ and $N(y, 0) \equiv 0$. Given a compact subset K of $\mathbb{D}$, let

$$\tau_n = \inf\{t \in [0, T] : \|M(t)\|_{0+\delta:K} + \|N(t)\|_{0+\delta:K} \geq n\} \quad (= T \text{ if } \{\ldots\} = \emptyset).$$

Then $\{\tau_n\}$ is an increasing sequence of stopping times such that $P(\tau_n < T) \to 0$ as $n \to \infty$. Using Burkholder's inequality, we have

$$E\left[\sup_t |A^{\tau_n}(x, y, t) - A^{\tau_n}(x, y', t) - A^{\tau_n}(x', y, t) + A^{\tau_n}(x', y', t)|^p\right]$$
$$\leq E[\langle M^{\tau_n}(x, T) - M^{\tau_n}(x', T)\rangle^p]^{1/2} E[\langle N^{\tau_n}(y, T) - N^{\tau_n}(y', T)\rangle^p]^{1/2}$$
$$\leq CE[|M^{\tau_n}(x, T) - M^{\tau_n}(x', T)|^{2p}]^{1/2} E[|N^{\tau_n}(y, T) - N^{\tau_n}(y', T)|^{2p}]^{1/2},$$

where C is a positive constant. The last term is not greater than $Cn^{2p}|x - x'|^{\delta p}|y - y'|^{\delta p}$ if $x, x', y, y' \in K$, since $|M^{\tau_n}(x, t) - M^{\tau_n}(x', t)| \leq n|x - x'|^\delta$ etc. hold. By a similar argument there exists a positive constant

C' such that

$$E\left[\sup_t |A^{\tau_n}(x, y, t) - A^{\tau_n}(x, y', t)|^p\right] \le C'|y - y'|^{\delta p},$$

$$E\left[\sup_t |A^{\tau_n}(x, y, t) - A^{\tau_n}(x', y, t)|^p\right] \le C'|x - x'|^{\delta p},$$

hold for all $x, x', y, y' \in K$. Then $A^{\tau_n}(x, y, \cdot)$ has a modification of a continuous $\tilde{C}^{0,\varepsilon}$-process for $\varepsilon < (\delta p - d)/p$ if $p > d/\delta$ by Kolmogorov–Chentzov's theorem (Theorem 1.4.4). Since we can choose p as large as we wish, the above holds for any $\varepsilon < \delta$.

In the following we denote the continuous modification by the same symbol $A(x, y, t)$. We next discuss the differentiability of $A(x, y, t)$ in the case $m \ge 1$. Given a compact convex subset K of $\mathbb{D}$, let σ_n be the infimum of $t \in [0, \tau_n]$ such that $\|M(t)\|_{1+\delta:K} + \|N(t)\|_{1+\delta:K}$ is greater than n. (We set $\sigma_n = \tau_n$ if there exists no such t.) Then $\{\sigma_n\}$ is an increasing sequence of stopping times such that $P(\sigma_n < T) \to 0$ as $n \to \infty$. Set

$$B(x, h, y, t) \equiv \frac{1}{h}\{A(x + he_i, y, t) - A(x, y, t)\}, \quad (x, h) \in \mathbb{D}_\gamma \times \{(-\gamma, 0) \cup (0, \gamma)\}.$$

Then we have

$$\begin{aligned} &|B^{\sigma_n}(x, h, y, t) - B^{\sigma_n}(x', h', y', t)| \\ &\quad \le \langle N^{\sigma_n}(y, t)\rangle^{1/2} \langle L^{\sigma_n}(x, h, t) - L^{\sigma_n}(x', h', t)\rangle^{1/2} \\ &\quad\quad + \langle L^{\sigma_n}(x', h', t)\rangle^{1/2} \langle N^{\sigma_n}(y, t) - N^{\sigma_n}(y', t)\rangle^{1/2}, \end{aligned}$$

where

$$L(x, h, t) = \frac{1}{h}\{M(x + he_i, t) - M(x, t)\}.$$

Therefore using Burkholder's inequality we obtain

$$\begin{aligned} &E\left[\sup_t |B^{\sigma_n}(x, h, y, t) - B^{\sigma_n}(x', h', y', t)|^p\right] \\ &\quad \le CE[|N^{\sigma_n}(y, T)|^{2p}]^{1/2} E[|L^{\sigma_n}(x, h, T) - L^{\sigma_n}(x', h', T)|^{2p}]^{1/2} \\ &\quad\quad + CE[|L^{\sigma_n}(x', h', T)|^{2p}]^{1/2} E[|N^{\sigma_n}(y, T) - N^{\sigma_n}(y', T)|^{2p}]^{1/2}, \end{aligned}$$

where C is a positive constant. The right hand side of the above is bounded by $C'\{|x - x'|^{\delta p} + |h - h'|^{\delta p} + |y - y'|^{\delta p}\}$ where C' is a positive constant. Then Kolmogorov's theorem asserts that $B^{\sigma_n}(x, h, y, t)$ can be extended continuously at $h = 0$ a.s. This proves that $A(x, y, t)$ is differentiable with respect to x and the derivative is continuous in (x, y, t) a.s. We have further

$$\frac{\partial A}{\partial x^i}(x, y, t) = \lim_{h \to 0} B(x, h, y, t) = \left\langle \frac{\partial M}{\partial x^i}(x, t), N(y, t) \right\rangle.$$

Here the latter equality follows from Corollary 2.2.16. We can show the differentiability of $A(x, y, t)$ with respect to y and the equality

$$\frac{\partial^2 A}{\partial y^j \partial x^i}(x, y, t) = \left\langle \frac{\partial M}{\partial x^i}(x, t), \frac{\partial N}{\partial y^j}(y, t) \right\rangle$$

in the same way. Repeating this argument inductively we get the assertion of the theorem. □

Corollary 3.1.4 *Under the same conditions as in Theorem* 3.1.3, *we have*

$$\frac{\partial}{\partial x^i}(\langle M(x, t), N(x, t) \rangle) = \left\langle \frac{\partial M}{\partial x^i}(x, t), N(x, t) \right\rangle + \left\langle M(x, t), \frac{\partial N}{\partial x^i}(x, t) \right\rangle$$

if $m \geq 1$.

Proof The formula is obvious from the equality

$$\frac{\partial}{\partial x^i}(A(x, x, t)) = \frac{\partial A}{\partial x^i}(x, y, t)|_{y=x} + \frac{\partial A}{\partial y^i}(x, y, t)|_{y=x}. \quad \square$$

Exercise 3.1.5 (*cf.* Theorem 3.3.3) Let M_t be a continuous localmartingale and $f(x, t)$, $x \in \mathbb{D}$ be a predictable process with values in $C^{m,\delta}$ such that $\| f(s) \|_{m+\delta:K} \in L^2(\langle M \rangle)$ holds for any compact set K. Show that $M(x, t) = \int_0^t f(x, s)\, dM_s$ has a modification of a $C^{m,\varepsilon}$-localmartingale for any $\varepsilon < \delta$. Show further that it satisfies

$$D_x^\alpha \int_0^t f(x, s)\, dM_s = \int_0^t D_x^\alpha f(x, s)\, dM_s, \qquad \text{for } |\alpha| \leq m.$$

(*Hint*: apply Theorem 3.1.2.)

Exercise 3.1.6 (Continued from Exercise 3.1.5) (*cf.* Theorem 3.3.4). Suppose that $f(x, t)$ is a continuous $C^{m,\delta}$-semimartingale. Show that the Stratonovich integral $L(x, t) = \int_0^t f(x, s) \circ dM_s$ has a modification of a continuous $C^{m,\varepsilon}$-semimartingale for any $\varepsilon < \delta$. Show further that it satisfies

$$D_x^\alpha \int_0^t f(x, s) \circ dM_s = \int_0^t D_x^\alpha f(x, s) \circ dM_s, \qquad \text{for } |\alpha| \leq m.$$

(*Hint*: use the relation $\int_0^t f(x, s) \circ dM_s = \int_0^t f(x, s)\, dM_s + \frac{1}{2}\langle f(x, t), M_t \rangle$.)

3.2 Stochastic integrals based on semimartingales with spatial parameters

Itô integrals: case of martingales

Let $F(x, t)$, $x \in \mathbb{D}$ be a continuous localmartingale with values in $C = C(\mathbb{D} : \mathbb{R})$ such that $F(x, 0) \equiv 0$ and let f_t be a predictable process with values in $\mathbb{D}$. In this section we shall define two types of stochastic integrals of f_t based on the kernel $F(x, dt)$ taking the forms $\int_0^t F(f_s, ds)$ and $\int_0^t F(f_s, \circ ds)$. These are generalizations of the Itô integral and the Stratonovich integral respectively, defined in the previous chapter. In fact, if $F(x, t) = xX_t$ where X_t is a one-dimensional continuous localmartingale and $x \in \mathbb{R}^1$, our definitions in this section coincide with the stochastic integrals $\int_0^t f_s \, dX_s$ and $\int_0^t f_s \circ dX_s$ in Chapter 2. However, the Itô integral $F_t(f) = \int_0^t F(f_s, ds)$ (and the Stratonovich integral $\mathring{F}_t(f) = \int_0^t F(f_s, \circ ds)$) that we will define in this section does not have the linear property: $F_t(af + bg) = aF_t(f) + bF_t(g)$, where a, b are constants, because $F(x, t)$ is not a linear function of x in general.

In the first step, we shall define the Itô integral in the case where the kernel is a localmartingale. We denote it by $M(x, t)$.

We begin by classifying the joint quadratic variations of $M(x, t)$ according to the regularity of their density functions. Let $A(x, y, t)$ be the joint quadratic variation of $M(x, t)$ and $M(y, t)$. Then there exists a continuous increasing process A_t such that $A(x, y, t)$ is absolutely continuous with respect to A_t a.s. for any x and y. (See Exercise 3.2.10 for the existence of such an A_t.) We define a measure μ on $([0, T] \times \Omega, \mathscr{P})$ by $\mu(B) = E[\int_0^T \chi(B)(s, w) \, dA_s]$, $B \in \mathscr{P}$, where $\mathscr{P}$ is the predictable σ-field and χ is the indicator function. Then for each x and y, there exists a predictable process $a(x, y, t)$ such that

$$A(x, y, t) = \int_0^t a(x, y, r) \, dA_r \tag{1}$$

holds for any t a.s. The density function $a(x, y, t, \omega)$ is determined up to μ-measure 0 for each x and y. The pair $(a(x, y, t), A_t)$ is called a *local characteristic* of $M(x, t)$, $x \in \mathbb{D}$. When A_t is identical to t, the random field $a(x, y, t)$ itself is often called the *local characteristic* of $M(x, t)$.

We shall classify the local characteristic according to its regularity. Let m be a non-negative integer and let δ be a non-negative number less than or equal to 1. The local characteristic $(a(x, y, t), A_t)$ is said to belong to the *class* $B^{m,\delta}$ if $a(x, y, t)$ has a modification of a predictable process with values in $\tilde{C}^{m,\delta}$ and $\|a(t)\|^{\sim}_{m+\delta:K}$ belongs to $L^1(A)$ a.s. for any compact subset K of $\mathbb{D}$. In particular if $\|a(t)\|^{\sim}_{m+\delta}$ belongs to $L^1(A)$, the pair (a, A) is said to belong to the *class* $B_b^{m,\delta}$. Furthermore, if there exists a positive constant c such that $\|a(t)\|_{m+\delta}^{\sim} \le c$ holds for any t and ω, the pair (a, A) is said to belong to the *class* $B_{ub}^{m,\delta}$.

Throughout this and the subsequent sections, we will only consider continuous C^m-localmartingales with the local characteristics belonging to classes $B^{m,\delta}$ for some $m \ge 0$ and $\delta \ge 0$. Observe that if $M(x, t)$, $x \in \mathbb{D}$ is a family of continuous localmartingales with local characteristic belonging to the class $B^{m,\delta}$ where $\delta > 0$, then M has a modification of a continuous $C^{m,\varepsilon}$-localmartingale where $\varepsilon < \delta$ by Theorem 3.1.2.

Let f_t be a predictable process with values in $\mathbb{D}$ satisfying

$$\int_0^T a(f_r, f_r, r)\, \mathrm{d}A_r < \infty \quad \text{a.s.} \tag{2}$$

Then we can define Itô's stochastic integral of f_t based on the kernel $M(x, \mathrm{d}t)$. We shall define it step by step. In the first step, we will look at the case where f_t is a simple process, i.e. there exists a partition $\Delta = \{0 = t_0 < t_1 < \cdots < t_l = T\}$ of the interval $[0, T]$ such that $f_t = f_{t_k}$ holds for any $t \in [t_k, t_{k+1})$. In the second step, we consider the general case.

Let f_t be a simple process as described above. We define the integral of f_t based on the kernel $M(x, \mathrm{d}t)$ by the finite sum of the form:

$$M_t(f) = \sum_{k=0}^{l-1} \{M(f_{t_k \wedge t}, t_{k+1} \wedge t) - M(f_{t_k \wedge t}, t_k \wedge t)\}. \tag{3}$$

It is a continuous $(\mathscr{F}_t)$-adapted process with values in $\mathbb{R}$.

Lemma 3.2.1 *$M_t(f)$ is a continuous localmartingale.*

Proof Suppose for a moment that $M(x, t)$ is a square integrable martingale for each x and the simple process f_t satisfies

$$E\left[\int_0^T a(f_r, f_r, r)\, \mathrm{d}A_r\right] < \infty. \tag{4}$$

We want to prove that $M_t(f)$ is then a square integrable martingale. Let s, t be any fixed times such that $s < t$. Adjoining s and t to the partition Δ, we may assume that $s = t_i$ and $t = t_j$ for some t_i, $t_j \in \Delta$. Since for each x $M(x, t)$ is a square integrable martingale with the quadratic variation $\int_0^t a(x, x, r)\, \mathrm{d}A_r$, we have

$$E[(M(x, t_{k+1}) - M(x, t_k))^2 | \mathscr{F}_{t_k}] = E\left[\int_{t_k}^{t_{k+1}} a(x, x, r)\, \mathrm{d}A_r \middle| \mathscr{F}_{t_k}\right] \tag{5}$$

a.s. for any x. We can choose modifications of both sides so that they are $\mathscr{F}_{t_k} \otimes \mathscr{B}(\mathbb{D})$-measurable. Further, writing the modifications in the same notation, we have the equalities

$$E[(M(x, t_{k+1}) - M(x, t_k))^2|\mathscr{F}_{t_k}]|_{x=f_{t_k}} = E[(M(f_{t_k}, t_{k+1}) - M(f_{t_k}, t_k))^2|\mathscr{F}_{t_k}],$$

$$E\left[\int_{t_k}^{t_{k+1}} a(x, x, r)\, dA_r \middle| \mathscr{F}_{t_k}\right]\Bigg|_{x=f_{t_k}} = E\left[\int_{t_k}^{t_{k+1}} a(f_{t_k}, f_{t_k}, r)\, dA_r \middle| \mathscr{F}_{t_k}\right].$$

(See Exercise 1.4.11.) Therefore the right hand sides of the two equalities above coincide with each other and their expectation is finite. This proves that $M_t(f)$ is square integrable.

We will prove that $M_t(f)$ is a martingale. Observe the equality

$$E[M_t(f) - M_s(f)|\mathscr{F}_s] = \sum_{k=i}^{j-1} E[E[M(f_{t_k}, t_{k+1}) - M(f_{t_k}, t_k)|\mathscr{F}_{t_k}]|\mathscr{F}_s]. \quad (6)$$

Since

$$E(M(f_{t_k}, t_{k+1}) - M(f_{t_k}, t_k)|\mathscr{F}_{t_k}] = E[M(x, t_{k+1}) - M(x, t_k)|\mathscr{F}_{t_k}]|_{x=f_{t_k}} = 0, \quad (7)$$

the right hand side of (6) is 0, proving that $M_t(f)$ is a martingale.

We next consider the case where $M(x, t)$, $x \in \mathbb{D}$ is an arbitrary family of continuous localmartingales. Given a compact subset K of $\mathbb{D}$ and a positive number n, we define $\tau_{K,n} = \inf\{t \in [0, T] : \int_0^t \|a(s)\|_{0:K}^{\sim} \, dA_s \geq n\}$ $(=T$ if $\{\dots\} = \emptyset)$. Then $P(\tau_{K,n} < T) \to 0$ as $n \uparrow \infty$ for each K. The stopped process $M^{\tau_{K,n}}(x, t) = M(x, t \wedge \tau_{K,n})$ is a square integrable martingale for any x of K. In fact, we have

$$\langle M^{\tau_{K,n}}(x, t)\rangle = \int_0^{t \wedge \tau_{K,n}} a(x, x, s)\, dA_s \leq n \times \sup_{x' \in K} (1 + |x'|) \qquad \text{for all } x \in K.$$

Given an arbitrary simple process f_t, we define a *truncation* of f associated with the set K by $f_t^K = f_t$ if $f_t \in K$, $f_t^K = x_0$ if $f_t \in K^c$ where x_0 is a fixed point in K. Then $M_t^{\tau_{K,n}}(f^K)$ is a square integrable martingale. Clearly $\{M_t^{\tau_{K,n}}(f^K)\}$ converges to $M_t(f)$ uniformly in t a.s. as $n \to \infty$ and $K \uparrow \mathbb{D}$. Therefore $M_t(f)$ is a continuous localmartingale by Theorem 2.1.2. The proof is complete. □

Let g_t be another simple process with values in $\mathbb{D}$. Then $M_t(g)$ is also a continuous localmartingale. We shall compute the joint quadratic variation of $M_t(f)$ and $M_t(g)$.

Lemma 3.2.2 *The joint quadratic variation of $M_t(f)$ and $M_t(g)$ is given by*

$$\langle M_t(f), M_t(g)\rangle = \int_0^t a(f_r, g_r, r)\,\mathrm{d}A_r \qquad a.s. \tag{8}$$

Proof It is enough to prove it for the case where $M(x, t)$ is a family of square integrable martingales and both f and g satisfy the integrability condition (4). The general case can be shown by using stopped processes and truncated ones similarly to the proof of Lemma 3.2.1. Furthermore, we may assume that both f and g have the common jump times, i.e. there exists $0 = t_0 < \cdots < t_l = T$ such that $f_t = f_{t_k}$, $g_t = g_{t_k}$ hold for any $t \in [t_k, t_{k+1})$. We may assume $s = t_i$ and $t = t_j$ as before. Then

$$\begin{aligned}
&E[(M_t(f) - M_s(f))(M_t(g) - M_s(g))|\mathscr{F}_s] \\
&= \sum_{i \le h < k < j} E[E[M(f_{t_k}, t_{k+1}) - M(f_{t_k}, t_k)|\mathscr{F}_{t_k}](M(g_{t_h}, t_{h+1}) - M(g_{t_h}, t_h))|\mathscr{F}_s] \\
&+ \sum_{i \le h < k < j} E[E[M(g_{t_k}, t_{k+1}) - M(g_{t_k}, t_k)|\mathscr{F}_{t_k}](M(f_{t_h}, t_{h+1}) - M(f_{t_h}, t_h))|\mathscr{F}_s] \\
&+ \sum_{k=i}^{j-1} E[E[(M(f_{t_k}, t_{k+1}) - M(f_{t_k}, t_k))(M(g_{t_k}, t_{k+1}) - M(g_{t_k}, t_k))|\mathscr{F}_{t_k}]|\mathscr{F}_s].
\end{aligned} \tag{9}$$

The first term of the right hand side is 0 by (7). The second term is also 0 for the same reason. As for the third term, similarly to the proof of Lemma 3.2.1, we have

$$\begin{aligned}
&E[(M(f_{t_k}, t_{k+1}) - M(f_{t_k}, t_k))(M(g_{t_k}, t_{k+1}) - M(g_{t_k}, t_k))|\mathscr{F}_{t_k}] \\
&\qquad = E\left[\int_{t_k}^{t_{k+1}} a(f_{t_k}, g_{t_k}, r)\,\mathrm{d}A_r \,\middle|\, \mathscr{F}_{t_k}\right].
\end{aligned}$$

Therefore the last term of (9) is equal to

$$E\left[\int_s^t a(f_r, g_r, r)\,\mathrm{d}A_r \,\middle|\, \mathscr{F}_s\right].$$

This proves that $M_t(f)M_t(g) - \int_0^t a(f_r, g_r, r)\,\mathrm{d}A_r$ is a martingale. Then the assertion follows. □

Now let f_t be a predictable process with values in a compact subset K of $\mathbb{D}$. Then there exists a sequence $\{f^n\}$ of simple $(\mathscr{F}_t)$-adapted processes with values in K such that

$$\int_0^T \{a(f_r^n, f_r^n, r) - 2a(f_r^n, f_r^m, r) + a(f_r^m, f_r^m, r)\}\,\mathrm{d}A_r \to 0 \quad \text{a.s.} \tag{10}$$

as $n, m \to \infty$. Then $\langle M(f^n) - M(f^m)\rangle_T \to 0$ holds a.s. Therefore $\{M_t(f^n)\}$ converges uniformly in probability to a continuous local martingale $M_t(f)$ by Theorem 2.2.15.

Next let f_t be an arbitrary predictable process satisfying (2). Let $\{K_n\}$ be a sequence of compact subsets of $\mathbb{D}$ such that $K_n \uparrow \mathbb{D}$. Let f^n be a truncation of f associated with the set K_n. The convergence (10) holds again. Therefore $\{M_t(f^n)\}$ converges uniformly in probability. The limit $M_t(f)$ is a continuous local martingale.

The continuous localmartingale $M_t(f)$ defined above is called the *Itô integral of f_t by the kernel* $M(x, dt)$. It is denoted by $\int_0^t M(f_r, dr)$.

Let $N(x, t)$, $x \in \mathbb{D}$ be another continuous localmartingale with values in C, with local characteristic $(a^N(x, y, t), A_t)$ belonging to the class $B^{m,\delta}$, where A_t is the same continuous increasing process as that of the local characteristic $(a^M(x, y, t), A_t)$ of $M(x, t)$. Let g_t be a predictable process satisfying $\int_0^t a^N(g_r, g_r, r)\, dA_r < \infty$. Then the stochastic integral $N_t(g) = \int_0^t N(g_r, dr)$ is well defined as a continuous local martingale. We shall study the joint quadratic variation of $M_t(f)$ and $N_t(g)$.

Let us first observe that the joint quadratic variation of $M(x, t)$ and $N(y, t)$ denoted by $A^{MN}(x, y, t)$ is absolutely continuous with respect to A_t. For each x, y, there exists a predictable process $a^{MN}(x, y, t)$ such that

$$A^{MN}(x, y, t) = \int_0^t a^{MN}(x, y, r)\, dA_r, \qquad \text{for all } t \text{ a.s.} \tag{11}$$

The pair (a^{MN}, A) is called the *joint local characteristic* of M and N. We shall prove a simple lemma for the joint local characteristic.

Lemma 3.2.3 *The joint local characteristic has a modification such that $a^{MN}(x, y, t)$ is continuous with respect to (x, y) for almost all (t, ω) with respect to the measure μ. Furthermore, it satisfies*

$$|a^{MN}(x, y, t)| \le a^M(x, x, t)^{1/2} a^N(y, y, t)^{1/2} \qquad \text{a.e. } t, \omega. \tag{12}$$

Proof Set $A^{MN}(x, y, (s, t]) = A^{MN}(x, y, t) - A^{MN}(x, y, s)$, $t > s$. We have

$$\begin{aligned} &|A^{MN}(x, y, (s, t]) - A^{MN}(x', y', (s, t])| \\ &\le A^N(y, y, (s, t])^{1/2}(A^M(x, x, (s, t]) - 2A^M(x, x', (s, t]) + A^M(x', x', (s, t]))^{1/2} \\ &+ A^M(x', x', (s, t])^{1/2}(A^N(y, y, (s, t]) - 2A^N(y, y', (s, t]) + A^N(y', y', (s, t]))^{1/2}. \end{aligned}$$

Then the density function satisfies

$$|a^{MN}(x, y, t) - a^{MN}(x', y', t)|$$
$$\leq a^N(y, y, t)^{1/2}(a^M(x, x, t) - 2a^M(x, x', t) + a^M(x', x', t))^{1/2}$$
$$+ a^M(x', x', t)^{1/2}(a^N(y, y, t) - 2a^N(y, y', t) + a^N(y', y', t))^{1/2} \quad \text{a.e. } t, \omega$$

for each x, x', y, y'. Since a^M and a^N are continuous in x, y for almost every (t, ω) with respect to μ, a^{MN} has a modification with respect to μ such that it is also continuous in x, y for almost every (t, ω). The inequality (12) will be immediate from

$$A^{MN}(x, y, (s, t]) \leq A^M(x, x, (s, t])^{1/2} A^N(y, y, (s, t])^{1/2} \quad \text{a.s.} \quad (13)$$

□

Theorem 3.2.4 *The joint quadratic variation of Itô integrals* $\int M(f_s, \mathrm{d}s)$ *and* $\int N(g_s, \mathrm{d}s)$ *satisfies*

$$\left\langle \int_0^t M(f_s, \mathrm{d}s), \int_0^t N(g_s, \mathrm{d}s) \right\rangle = \int_0^t a^{MN}(f_r, g_r, r)\, \mathrm{d}A_r. \quad (14)$$

□

The proof can be carried out similarly to the proof of Lemma 3.2.2 and is omitted.

Itô integrals: case of semimartingales

We will next define the Itô integral based on the semimartingale kernel $F(x, \mathrm{d}t)$. This is not difficult, since we have already defined it based on the martingale kernel. The stochastic integral based on the absolutely continuous kernel, for example $B(x, \mathrm{d}t) = b(x, t)\, \mathrm{d}A_t$, is immediate.

Let $F(x, t), x \in \mathbb{D}$ be a family of continuous semimartingales decomposed as $F(x, t) = M(x, t) + B(x, t)$, where $M(x, t)$ is a continuous localmartingale and $B(x, t)$ is a continuous process of bounded variation. A family of continuous semimartingales $F(x, t), x \in \mathbb{D}$ is said to *belong to the class* $C^{m,\delta}$ or simply to be a $C^{m,\delta}$*-semimartingale* if $M(x, t)$ is a continuous $C^{m,\delta}$-localmartingale and $B(x, t)$ is a continuous $C^{m,\delta}$-process such that $D_x^\alpha B(x, t), x \in \mathbb{D}, |\alpha| \leq m$ are all processes of bounded variation. Further if $\delta = 0$ it is called a *semimartingale of the class* C^m. We will assume that $M(x, t), x \in \mathbb{D}$ has a local characteristic $(a(x, y, t), A_t)$ and $B(x, t)$ is written as

$$B(x, t) = \int_0^t b(x, s)\, \mathrm{d}A_s \quad (15)$$

where $b(x, t), x \in \mathbb{D}$ is a family of predictable processes. The triple $(a(x, y, t), b(x, t), A_t)$ is called the *local characteristic of the family of semimartingales*

$F(x, t)$, $x \in \mathbb{D}$. When A_t is identical to t, the pair $(a(x, y, t), b(x, t))$ is often called the *local characteristic* of $F(x, t)$.

We shall classify the family of semimartingales $F(x, t)$, $x \in \mathbb{D}$ according to the regularity of its local characteristic. The local characteristic (b, A) is said to belong to the class $B^{m',\delta'}$ if $b(x, t)$ is a predictable process with values in $C^{m',\delta'}(\mathbb{D} : \mathbb{R})$ and for any compact subset K of $\mathbb{D}$ $\|b(t)\|_{m'+\delta':K} \in L^1(A)$. In particular if $\|b(t)\|_{m'+\delta'} \in L^1(A)$ holds, (b, A) is said to belong to the class $B_b^{m',\delta'}$. Furthermore if there exists a constant c such that $\|b(t)\|_{m'+\delta'} \le c$, it is said to belong to the class $B_{ub}^{m',\delta'}$.

The triple (a, b, A_t) is then said to belong to the class $(B^{m,\delta}, B^{m',\delta'})$ $((B_b^{m,\delta}, B_b^{m',\delta'})$ or $(B_{ub}^{m,\delta}, B_{ub}^{m',\delta'}))$ if (a, A_t) belongs to the class $B^{m,\delta}$ ($B_b^{m,\delta}$ or $B_{ub}^{m,\delta}$ respectively) and (b, A_t) belongs to the class $B^{m',\delta'}$ ($B_b^{m',\delta'}$ or $B_{ub}^{m',\delta'}$ respectively). When $m = m'$ and $\delta = \delta'$, the triple is simply said to belong to the class $B^{m,\delta}$ ($B_b^{m,\delta}$ or $B_{ub}^{m,\delta}$ respectively).

Remark If the local characteristic (a, b, A_t) belongs to the class $B_b^{m,\delta}$, there exists a local characteristic (a', b', A_t') equivalent to (a, b, A_t) belonging to the class $B_{ub}^{m,\delta}$. Indeed, set $K_t = \|b(t)\|_{m+\delta} + \|a(t)\|_{m+\delta}^{\sim}$ and

$$b'(x, t) = b(x, t)(1 + K_t)^{-1}, \qquad a'(x, y, t) = a(x, y, t)(1 + K_t)^{-1},$$
$$A_t' = \int_0^t (1 + K_s)\, dA_s. \tag{16}$$

Then (a', b', A_t') is a local characteristic belonging to the class $B_{ub}^{m,\delta}$.

Now let $F(x, t)$, $x \in \mathbb{D}$ be a continuous C-semimartingale with local characteristic (a, b, A_t) belonging to the class $B^{0,\delta}$ for some $\delta > 0$. Let f_t be a predictable process with values in $\mathbb{D}$ satisfying

$$\int_0^T a(f_s, f_s, s)\, dA_s < \infty, \qquad \int_0^T |b(f_s, s)|\, dA_s < \infty \qquad \text{a.s.} \tag{17}$$

Then Itô's stochastic integral of f_t based on the kernel $F(x, dt)$ is defined by

$$\int_0^t F(f_s, ds) = \int_0^t M(f_s, ds) + \int_0^t b(f_s, s)\, dA_s. \tag{18}$$

Sometimes it is written as $F_t(f)$.

If f_t is a continuous predictable process, condition (17) is always satisfied. The integral is approximated as

$$\int_0^t F(f_s, ds) = \lim_{|\Delta| \to 0} \sum_{k=0}^{l-1} \{F(f_{t_k \wedge t}, t_{k+1} \wedge t) - F(f_{t_k \wedge t}, t_k \wedge t)\}, \tag{19}$$

(convergence in probability) where $\Delta = \{0 = t_0 < \cdots < t_l = T\}$ (see Theorem 2.2.15).

Stratonovich integrals

We shall next define the Stratonovich integral of f_t based on the kernel $F(x, dt)$. It requires more regularity conditions both for f_t and $F(x, t)$ than the Itô integral does.

Let $F(x, t)$, $x \in \mathbb{D}$ be a continuous C-semimartingale and let f_t be a continuous process with values in $\mathbb{D}$. For a partition $\Delta = \{0 = t_0 < \cdots < t_l = T\}$, we set

$$\mathring{F}_t^{\Delta}(f) = \sum_{k=0}^{l-1} \frac{1}{2}\{F(f_{t_{k+1}\wedge t}, t_{k+1} \wedge t) + F(f_{t_k \wedge t}, t_{k+1} \wedge t)$$

$$- F(f_{t_{k+1}\wedge t}, t_k \wedge t) - F(f_{t_k\wedge t}, t_k \wedge t)\}. \quad (20)$$

If the sequence $\{\mathring{F}_t^{\Delta_m}(f)\}$ converges uniformly in t in probability for any sequence of partitions Δ_m such that $|\Delta_m| \to 0$, then the limit $\mathring{F}_t(f)$ is called the *Stratonovich integral of f_t based on the kernel* $F(x, t)$ and is denoted by $\int_0^t F(f_s, \circ ds)$.

Theorem 3.2.5 *Assume that $F(x, t)$ is a continuous C^1-semimartingale with local characteristic belonging to the class $(B^{2,\delta}, B^{1,0})$ for some $\delta > 0$ and f_t is a continuous semimartingale. Then the Stratonovich integral is well defined. It is related to the Itô integral by the formula:*

$$\int_0^t F(f_s, \circ ds) = \int_0^t F(f_s, ds) + \frac{1}{2}\sum_{j=1}^{d}\left\langle \int_0^t \frac{\partial F}{\partial x^j}(f_s, ds), f_t^j \right\rangle. \quad (21)$$

Proof Let $F(x, t) = M(x, t) + B(x, t)$ be the decomposition such that M is a localmartingale and B is a process of bounded variation. Since the local characteristic of M belongs to the class $B^{2,\delta}$, it is a C^2-localmartingale. Further B is a C^1-process of bounded variation. Now we may assume $t = t_n$ for some $t_n \in \Delta$. Then $\mathring{F}_t^{\Delta}(f)$ is written as

$$\mathring{F}_t^{\Delta}(f) = \sum_{k=0}^{n-1} \{F(f_{t_k}, t_{k+1}) - F(f_{t_k}, t_k)\}$$

$$+ \frac{1}{2}\sum_{k=0}^{n-1} \{F(f_{t_{k+1}}, t_{k+1}) - F(f_{t_k}, t_{k+1}) - (F(f_{t_{k+1}}, t_k) - F(f_{t_k}, t_k))\}.$$

Denote the first and the second terms of the right hand side by $F_t^{\Delta}(f)$ and $L_t^{\Delta}(f)$, respectively. We have

$$F_t^{\Delta}(f) \xrightarrow[|\Delta| \to 0]{} \int_0^t F(f_s, ds) \qquad \text{in probability.}$$

On the other hand, by the mean value theorem,

$$L_t^{\Delta}(f) = \frac{1}{2} \sum_{j=1}^{d} \sum_{k=0}^{n-1} \left\{ \frac{\partial F}{\partial x^j}(f_{t_k}, t_{k+1}) - \frac{\partial F}{\partial x^j}(f_{t_k}, t_k) \right\} (f_{t_{k+1}}^j - f_{t_k}^j)$$

$$+ \frac{1}{4} \sum_{i,j=1}^{d} \sum_{k=0}^{n-1} \left\{ \frac{\partial^2 M}{\partial x^i \partial x^j}(\xi_k, t_{k+1}) - \frac{\partial^2 M}{\partial x^i \partial x^j}(\eta_k, t_k) \right\} (f_{t_{k+1}}^i - f_{t_k}^i)(f_{t_{k+1}}^j - f_{t_k}^j)$$

$$+ \frac{1}{2} \sum_{j=1}^{d} \sum_{k=0}^{n-1} \left\{ \frac{\partial B}{\partial x^j}(\xi'_k, t_{k+1}) - \frac{\partial B}{\partial x^j}(\eta'_k, t_k) \right.$$

$$\left. - \left(\frac{\partial B}{\partial x^j}(f_{t_k}, t_{k+1}) - \frac{\partial B}{\partial x^j}(f_{t_k}, t_k) \right) \right\} (f_{t_{k+1}}^j - f_{t_k}^j),$$

where ξ_k, ξ'_k, $\eta_k \eta'_k$ are random variables such that $|\xi_k - f_{t_k}|$, $|\xi'_k - f_{t_k}|$, $|\eta_k - f_{t_k}|$, $|\eta'_k - f_{t_k}|$ are bounded by $|f_{t_{k+1}} - f_{t_k}|$. Denote by M_t^{Δ}, N_t^{Δ} and O_t^{Δ} the first, second and third terms of $L_t^{\Delta}(f)$, respectively. We wish to prove the following convergences in probability:

$$\lim_{|\Delta| \to 0} M_t^{\Delta} = \frac{1}{2} \sum_{j=1}^{d} \left\langle \int_0^t \frac{\partial F}{\partial x^j}(f_s, ds), f_t^j \right\rangle, \tag{22}$$

$$\lim_{|\Delta| \to 0} |N_t^{\Delta}| = 0, \tag{23}$$

$$\lim_{|\Delta| \to 0} |O_t^{\Delta}| = 0. \tag{24}$$

Let $(\partial F/\partial x^j)_t(f^{\Delta})$ and $(\partial F/\partial x^j)_t(f)$ be Itô's stochastic integrals of f_t^{Δ} and f_t based on the kernel $(\partial F/\partial x^j)(x, dt)$, respectively. Then $(\partial F/\partial x^j)_t(f^{\Delta})$ converges to $(\partial F/\partial x^j)_t(f)$ in probability as $|\Delta| \to 0$. Therefore we have by Corollary 2.2.19,

$$\sum_k \left(\frac{\partial F}{\partial x^j}(f_{t_k}, t_{k+1}) - \frac{\partial F}{\partial x^j}(f_{t_k}, t_k) \right) (f_{t_{k+1}}^j - f_{t_k}^j) \xrightarrow[|\Delta| \to 0]{} \left\langle \left(\frac{\partial F}{\partial x^j} \right)_t (f), f_t^j \right\rangle$$

in probability. Hence (22) is proved. We can prove (23) similarly as in the proof of Itô's formula (Theorem 2.3.11). The convergence (24) is obvious since $(\partial B/\partial x^j)(x, t)$ is a process of bounded variation. The proof is complete. □

Corollary 3.2.6 *The Stratonovich integral and the Itô integral are equal if the joint quadratic variation of the integrand f_t and the kernel $F(x, t)$ is 0 for* *every x. In particular if the integrand f_t is a process of bounded variation, both integrals are equal for any semimartingale F.*

Proof If $\langle F(x, t), f_t^j \rangle = 0$ is satisfied for any x, then $\langle (\partial F/\partial x^j)(x, t), f_t^j \rangle = 0$ is satisfied by Theorem 3.1.3. Therefore the second term of the right hand side of (21) is 0, proving that the integrals are equal. □

Remark The Stratonovich integral is well defined under a weaker condition for $F(x, t)$. Indeed, suppose the localmartingale part M is a continuous C^2-process with local characteristic belonging to the class $B^{1,0}$ and the bounded variation part B is a continuous C^0-process with local characteristic belonging to the class $B^{0,0}$. Then the Stratonovich integral is well defined. We have the relation

$$\int_0^t F(f_s, \circ ds) = \int_0^t F(f_s, ds) + \frac{1}{2}\sum_j \left\langle \int_0^t \frac{\partial M}{\partial x^j}(f_s, ds), f_t^j \right\rangle. \quad (25)$$

For the proof, show the convergences of $\{\mathring{M}_t^{\Delta_m}(f)\}$ and $\{\mathring{B}_t^{\Delta_m}(f)\}$ separately. Note that the last joint quadratic variation of (25) is equal to that of (21) since it depends only on the martingale part of the semimartingale (see Theorem 2.2.14).

Further if M is a C^1-localmartingale with local characteristic belonging to the class $B^{1,0}$, then the right hand side of (25) is well defined. Conventionally we could define the Stratonovich integral by the right hand side of (25) though the approximation such as in Theorem 3.2.5 might not be valid. Many of the results in this book could be extended to this kind of Stratonovich integral. However, in this book we will restrict our attention to semimartingales satisfying the conditions of Theorem 3.2.5.

Time change of stochastic integrals

Let B_t, $t \in [0, T]$ be a continuous strictly increasing process with $B_0 = 0$ adapted to $(\mathscr{F}_t)$. We define the *inverse function* τ_u, $0 \le u < \infty$ of B_t by

$$\tau_u = \sup\{s \in [0, T] : B_s \le u\}. \quad (26)$$

It is a continuous function of u, strictly increasing in the domain $[0, B_T)$ and is a constant on $[B_T, \infty)$. Further it is a stopping time for each u. We have the relation $B_{\tau_u} = u$ if $u < B_T$, $B_{\tau_u} = B_T$ if $u \ge B_T$ and $\tau_{B_u} = u$ if $u \le T$. Set

$$\tilde{\mathscr{F}}_u = \mathscr{F}_{\tau_u}, \qquad u \in [0, \infty). \quad (27)$$

It is a filtration. It is easily verified that for every t, B_t is an $(\tilde{\mathscr{F}}_u)$-stopping time.

We shall be concerned with the change of time of $C^{m,\delta}$-semimartingales by the random function τ_u. We prove a lemma.

Lemma 3.2.7 *Let M_t, $t \in [0, T]$ be a continuous localmartingale adapted to $(\mathscr{F}_t)$. Then $\tilde{M}_u = M_{\tau_u}$, $u \in [0, \infty)$ is a continuous localmartingale adapted to $(\tilde{\mathscr{F}}_u)$.*

Proof Suppose first that M_t is a martingale adapted to $(\mathscr{F}_t)$. Since $\{\tau_u\}$ is a family of stopping times adapted to $(\mathscr{F}_t)$ and $\tau_u \le \tau_v$ holds if $u < v$, we have $E[M_{\tau_v}|\mathscr{F}_{\tau_u}] = M_{\tau_u}$ by the optional stopping time theorem. Therefore $\tilde{M}_u$ is a martingale adapted to $(\tilde{\mathscr{F}}_u)$.

Suppose next that M_t is an arbitrary localmartingale. Let $\{\tau_n\}$ be an increasing sequence of stopping times such that $P(\tau_n < T) \to 0$ and each stopped process $M_t^{\tau_n}$ is a martingale. Define random times $\hat{t}_n$ by $\hat{t}_n = B_{\tau_n}$ if $\tau_n < T$, $\hat{t}_n = \infty$ if $\tau_n = T$. Then $\{\hat{t}_n\}$ is an increasing sequence of stopping times adapted to $(\tilde{\mathscr{F}}_u)$ such that $\hat{t}_n \uparrow \infty$ a.s. We have the equality $\tilde{M}_u^{\hat{t}_n} = M_{\tau_{(u \wedge \hat{t}_n)}} = M_{\tau_u \wedge \tau_n} = M_{\tau_u}^{\tau_n}$, so that $\tilde{M}_u^{\hat{t}_n}$ is a martingale adapted to $(\tilde{\mathscr{F}}_u)$ for every n. Therefore $\tilde{M}_u$ is a continuous localmartingale. The proof is complete. □

Theorem 3.2.8 *Let $F(x, t)$ be a continuous $C^{m,\delta}$-semimartingale adapted to $(\mathscr{F}_t)$. Set $\tilde{F}(x, u) = F(x, \tau_u)$, $u \in [0, \infty)$, where τ_u is the inverse function of a strictly increasing process B_t. Then it is a continuous $C^{m,\delta}$-semimartingale adapted to $(\tilde{\mathscr{F}}_u)$ with local characteristic $(\tilde{a}, \tilde{b}, \tilde{A})$ where*

$$\tilde{a}(x, y, u) = a(x, y, \tau_u), \qquad \tilde{b}(x, u) = b(x, \tau_u), \qquad \tilde{A}_u = A_{\tau_u}. \quad (28)$$

Proof For each x, $\tilde{F}(x, u)$ is decomposed as $\tilde{F}(x, u) = \tilde{M}(x, u) + \tilde{B}(x, u)$, where $\tilde{M}(x, u) = M(x, \tau_u)$ and $\tilde{B}(x, u) = B(x, \tau_u)$. The process $\tilde{M}(x, u)$ is a continuous $C^{m,\delta}$-localmartingale adapted to $(\tilde{\mathscr{F}}_u)$. The process $\tilde{B}(x, u)$ is a continuous $C^{m,\delta}$-process of bounded variation. Hence $\tilde{F}$ is a continuous $C^{m,\delta}$-semimartingale adapted to $(\tilde{\mathscr{F}}_u)$. Furthermore

$$(\tilde{M}(x, u) - \tilde{M}(x, 0))(\tilde{M}(y, u) - \tilde{M}(y, 0)) - A(x, y, \tau_u)$$

is a continuous localmartingale adapted to $(\tilde{\mathscr{F}}_u)$ by the previous lemma. Then the joint quadratic variation of $\tilde{M}(x, t)$ and $\tilde{M}(y, t)$ is given by $A(x, y, \tau_u) = \int_0^u \tilde{a}(x, y, r)\, d\tilde{A}_r$. We have further $B(x, \tau_u) = \int_0^u \tilde{b}(x, r)\, d\tilde{A}_r$. Therefore the local characteristic of $\tilde{F}$ is given by (28). □

Theorem 3.2.9

(i) *Let $F(x, t)$, $x \in \mathbb{D}$ be a continuous C-semimartingale with local characteristic belonging to the class $B^{0,0}$ and let f_t be a continuous process* *with values in $\mathbb{D}$ adapted to $(\mathscr{F}_t)$. Set $\tilde{f}_u = f_{\tau_u}$. Then the Itô integral $\int_0^u \tilde{F}(\tilde{f}_v, \mathrm{d}v)$ is well defined as a continuous $(\tilde{\mathscr{F}}_u)$-semimartingale. It satisfies*

$$\int_0^u \tilde{F}(\tilde{f}_r, \mathrm{d}r) = \int_0^{\tau_u} F(f_r, \mathrm{d}r). \tag{29}$$

(ii) *Assume next that $F(x, t)$ is a continuous C^1-semimartingale with local characteristic belonging to the class $(B^{2,\delta}, B^{1,0})$ for some $\delta > 0$ and f_t is a continuous semimartingale adapted to $(\mathscr{F}_t)$. Then $\tilde{f}_u$ is a continuous semimartingale adapted to $(\tilde{\mathscr{F}}_u)$ and the Stratonovich integral $\int_0^u \tilde{F}(\tilde{f}_r, \circ\mathrm{d}r)$ is well defined. It satisfies*

$$\int_0^u \tilde{F}(\tilde{f}_r, \circ\mathrm{d}r) = \int_0^{\tau_u} F(f_r, \circ\mathrm{d}r). \tag{30}$$

Proof If F is a continuous process of bounded variation, the above equalities are obvious. We shall prove them for a continuous localmartingale F. Denote the right hand side and the left hand side of (29) by M_t and N_t, respectively. Both are continuous localmartingales adapted to $(\tilde{\mathscr{F}}_u)$. We have

$$\langle M \rangle_u = \int_0^{\tau_u} a(f_r, f_r, r)\, \mathrm{d}A_r, \tag{31}$$

$$\langle N \rangle_u = \int_0^u \tilde{a}(\tilde{f}_r, \tilde{f}_r, r)\, \mathrm{d}\tilde{A}_r = \int_0^{\tau_u} a(f_r, f_r, r)\, \mathrm{d}A_r. \tag{32}$$

Further, since $\langle N_u, \tilde{F}(y, u)\rangle = \int_0^u \tilde{a}(\tilde{f}_r, y, r)\, \mathrm{d}\tilde{A}_r = \int_0^{\tau_u} a(f_r, y, r)\, \mathrm{d}A_r$ we have $\langle N_{B_t}, F(y, t)\rangle = \int_0^t a(f_r, y, r)\, \mathrm{d}A_r$. Then we obtain

$$\left\langle N_{B_t}, \int_0^t F(f_r, \mathrm{d}r) \right\rangle = \int_0^t a(f_r, f_r, r)\, \mathrm{d}A_r,$$

or equivalently

$$\langle N_u, M_u \rangle = \int_0^{\tau_u} a(f_r, f_r, r)\, \mathrm{d}A_r. \tag{33}$$

These three relations (31), (32) and (33) imply $\langle M - N \rangle_u = 0$, proving $M_u = N_u$. Equality (30) can be reduced to (29) by rewriting both sides of (30) using the Itô integrals. The proof is complete. □

Case of local semimartingales

A family of real random variables $F(x, t), x \in \mathbb{D}, t \in [0, \tau(x))$ is called a *local random field* if $\tau(x) = \tau(x, \omega)$ is an *accessible lower semicontinuous stopping time* in the following sense:

(a) for each x, $\tau(x, \omega)$ is a stopping time,
(b) for almost all ω, $\tau(x, \omega)$ is a positive lower semicontinuous function of x,
(c) there exists an increasing sequence of random fields $\tau_n(x, \omega)$ with properties (a) and (b) above such that $\tau_n(x) < \tau(x)$ and $\tau_n(x) \uparrow \tau(x)$ hold for any x a.s.

The random field $\tau(x)$ is often called the *terminal time* of $F(x, t)$. When the terminal time is a positive constant T for all x a.s., $F(x, t)$ is called a *global random field* or simply a *random field*.

In the following, the local random field $F(x, t)$ is assumed to be continuous in (x, t) and $(\mathscr{F}_t)$-adapted.

Let $F(x, t)$, $t \in [0, \tau(x))$ be a local random field. Set $\mathbb{D}_t = \mathbb{D}_t(\omega) = \{x | \tau(x) > t\}$. It is an open subset of $\mathbb{D}$ a.s. since $\tau(x)$ is lower semicontinuous a.s. Then $F(x, t)$ defines a mapping from $\mathbb{D}_t(\omega)$ into $\mathbb{R}$ for each ω. It is called a *local process with values in* $C^{m,\delta}$ or simply a *local* $C^{m,\delta}$*-process* if for almost all ω, the map $F(\cdot, t, \omega) : \mathbb{D}_t(\omega) \to \mathbb{R}$ is a $C^{m,\delta}$-function for any t. A *continuous local* $C^{m,\delta}$*-process* is defined similarly.

Now let $F(x, t)$, $x \in \mathbb{D}$, $t \in [0, \tau(x))$ be a local random field and let $\{\tau_n(x)\}$ be the associated sequence of stopping times increasing to $\tau(x)$. Then the stopped process $F^{\tau_n(x)}(x, t) = F(x, t \wedge \tau_n(x))$, $x \in \mathbb{D}$, $t \in [0, T]$ may be considered as a global random field. A continuous local $C^{m,\delta}$-process is called a *continuous local* $C^{m,\delta}$*-martingale* or *local* $C^{m,\delta}$*-semimartingale* if stopped processes $(D^\alpha F)^{\tau_n(x)}(x, t)$, $x \in \mathbb{D}$, $|\alpha| \le m$, $n = 1, 2, \dots$ are all martingales or semimartingales respectively. The local characteristic of a local $C^{m,\delta}$-martingale or a local $C^{m,\delta}$-semimartingale is defined in the obvious way.

Now let $F(x, t)$, $t \in [0, \tau(x))$ be a continuous local $C^{m,\delta}$-semimartingale with local characteristic $(a(x, y, t), b(x, t), A_t)$ belonging to the class $B^{m,\delta}$. Let f_t, $t \in [0, \sigma)$ be a continuous local process with values in $\mathbb{D}$ such that $f_t \in \{x | \tau(x) > t\}$ if $t < \sigma$. Then the Itô integral $\int_0^{t \wedge \sigma_n} F(f_s, ds)$ is well defined for any t, where $\{\sigma_n\}$ is a sequence of stopping times such that $\sigma_n < \sigma$ and $\sigma_n \uparrow \sigma$. Then there exists a continuous local semimartingale L_t, $t \in [0, \sigma)$ such that $L_t^{\sigma_n} = \int_0^{t \wedge \sigma_n} F(f_s, ds)$. We denote it as $\int_0^t F(f_s, ds)$, $t \in [0, \sigma)$.

Exercise 3.2.10 Let $\{M^n\}$ be an orthogonal basis of continuous martingales.

(i) Show that any continuous C-localmartingale $M(x, t)$ is represented as $M(x, t) = \sum_n \int_0^t f_n(x, s)\, dM_s^n$, where $f_n(x, s, \omega)$ are measurable random fields, predictable in s for each x.
(ii) Deduce that there exists a continuous increasing process A_t such that the joint quadratic variation $A(x, y, t)$ of $M(x, t)$ and $M(y, t)$ is absolutely continuous with respect to A_t for any x and y.

(iii) Suppose that the local characteristic (a, A_t) of $M(x, t)$ belongs to the class $B^{k,\delta}$. Show that we can choose the above $f_n(x, s)$ as predictable processes with values in $C^{k,\varepsilon}$ where $\varepsilon < \delta$.

(*Hint*: for the proof of (iii) see Lemma 3.2.3 and its proof.)

Exercise 3.2.11 Under the same condition as Exercise 3.2.10 show that

$$\int_0^t M(\varphi_s, ds) = \sum_n \int_0^t f_n(\varphi_s, s)\, dM_s^n$$

holds for any continuous predictable process φ_t. Here the right hand side is Itô's stochastic integral defined in Chapter 2.

3.3 Some formulas for stochastic integrals

Generalized Itô's formula

Let $F(x, t)$, $x \in \mathbb{D}$ be a continuous C^2-process and a continuous C^1-semimartingale. Let g_t be a continuous predictable process with values in $\mathbb{D}$. Then the composite process $F(g_t, t)$ is a continuous predictable process. We shall study the differential rule of $F(g_t, t)$ with respect to time t. If $F(x, t)$ is a deterministic function twice continuously differentiable in x and continuously differentiable in t, the differential rule of $F(g_t, t)$ is the Itô's formula introduced in the previous chapter. However if $F(x, t)$ is a continuous C^1-semimartingale, the differential formula is not the same as Itô's formula. It requires an additional term or a correction term as is stated below.

Theorem 3.3.1 (*Generalized Itô's formula* I). *Let $F(x, t)$, $x \in \mathbb{D}$ be a continuous C^2-process and a continuous C^1-semimartingale with local characteristic belonging to the class $B^{1,0}$ and let g_t be a continuous semimartingale with values in $\mathbb{D}$. Then $F(g_t, t)$ is a continuous semimartingale and satisfies*

$$\begin{aligned} F(g_t, t) - F(g_0, 0) = {} & \int_0^t F(g_s, ds) + \sum_{i=1}^d \int_0^t \frac{\partial F}{\partial x^i}(g_s, s)\, dg_s^i \\ & + \frac{1}{2} \sum_{i,j=1}^d \int_0^t \frac{\partial^2 F}{\partial x^i \partial x^j}(g_s, s)\, d\langle g_s^i, g_s^j \rangle \\ & + \sum_{i=1}^d \left\langle \int_0^t \frac{\partial F}{\partial x^i}(g_s, ds), g_t^i \right\rangle. \end{aligned} \tag{1}$$

Proof For a partition $\Delta = \{0 = t_0 < t_1 < \cdots < t_l = t\}$, we have

$$F(g_t, t) - F(g_0, 0) = \sum_{k=0}^{l-1} \{F(g_{t_k}, t_{k+1}) - F(g_{t_k}, t_k)\}$$

$$+ \sum_{k=0}^{l-1} \{F(g_{t_{k+1}}, t_{k+1}) - F(g_{t_k}, t_{k+1})\}. \tag{2}$$

The first term of the right hand side converges to $\int_0^t F(g_s, ds)$ by (19) of Section 3.2. The second term is written as the sum of the following three terms:

$$I^\Delta = \sum_{k=0}^{l-1} \sum_{i=1}^{d} \left\{\frac{\partial F}{\partial x^i}(g_{t_k}, t_{k+1}) - \frac{\partial F}{\partial x^i}(g_{t_k}, t_k)\right\}(g^i_{t_{k+1}} - g^i_{t_k}), \tag{3}$$

$$J^\Delta = \sum_{k=0}^{l-1} \sum_{i=1}^{d} \frac{\partial F}{\partial x^i}(g_{t_k}, t_k)(g^i_{t_{k+1}} - g^i_{t_k}), \tag{4}$$

$$K^\Delta = \frac{1}{2} \sum_{i,j=1}^{d} \sum_{k=0}^{l-1} \frac{\partial^2 F}{\partial x^i \partial x^j}(\xi_k, t_{k+1})(g^i_{t_{k+1}} - g^i_{t_k})(g^j_{t_{k+1}} - g^j_{t_k}), \tag{5}$$

where ξ_k are random variables such that $|\xi_k - g_{t_k}| \le |g_{t_{k+1}} - g_{t_k}|$. The following convergences in probability hold:

$$\lim_{|\Delta| \to 0} I^\Delta = \sum_{i=1}^{d} \left\langle \int_0^t \frac{\partial F}{\partial x^i}(g_s, ds), g^i_t \right\rangle, \tag{6}$$

$$\lim_{|\Delta| \to 0} J^\Delta = \sum_{i=1}^{d} \int_0^t \frac{\partial F}{\partial x^i}(g_s, s)\, dg^i_s, \tag{7}$$

$$\lim_{|\Delta| \to 0} K^\Delta = \frac{1}{2} \sum_{i,j=1}^{d} \int_0^t \frac{\partial^2 F}{\partial x^i \partial x^j}(g_s, s)\, d\langle g^i_s, g^j_s \rangle. \tag{8}$$

Indeed the convergence (6) can be shown similarly to the convergence (22) in the previous section. The convergence (7) is obvious from the definition of the stochastic integral. The convergence (8) can be shown similarly to the proof of Itô's formula (Theorem 2.3.11). The proof is complete. □

With an additional condition on $F(x, t)$, we can rewrite the generalized Itô's formula using the Stratonovich integrals. The new formula looks like the classical differential formula for composite functions.

Theorem 3.3.2 (*Generalized Itô's formula* II). *Let $F(x, t)$, $x \in \mathbb{D}$ be a continuous C^3-process and a continuous C^2-semimartingale with local characteristic belonging to the class $(B^{2,\delta}, B^{1,0})$ for some $\delta > 0$. Let g_t be a continuous* *semimartingale. Then the formula*

$$F(g_t, t) - F(g_0, 0) = \int_0^t F(g_s, \circ ds) + \sum_{i=1}^d \int_0^t \frac{\partial F}{\partial x^i}(g_s, s) \circ dg_s^i \tag{9}$$

is satisfied.

Proof Rewrite the right hand side using Itô integrals. By the generalized Itô's formula I, we have

$$\begin{aligned}&\frac{\partial F}{\partial x^i}(g_t, t) - \frac{\partial F}{\partial x^i}(g_0, 0)\\ &\quad = \int_0^t \frac{\partial F}{\partial x^i}(g_s, ds) + \sum_j \int_0^t \frac{\partial^2 F}{\partial x^i \partial x^j}(g_s, s)\, dg_s^j\\ &\qquad + \sum_j \left\langle \int_0^t \frac{\partial^2 F}{\partial x^i \partial x^j}(g_s, ds), g_t^j \right\rangle + \frac{1}{2}\sum_{j,k} \int_0^t \frac{\partial^3 F}{\partial x^i \partial x^j \partial x^k}(g_s, s)\, d\langle g_s^j, g_s^k \rangle.\end{aligned} \tag{10}$$

Therefore we have by Theorem 3.2.5

$$\begin{aligned}\int_0^t \frac{\partial F}{\partial x^i}(g_s, s) \circ dg_s^i &= \int_0^t \frac{\partial F}{\partial x^i}(g_s, s)\, dg_s^i + \frac{1}{2}\left\langle \frac{\partial F}{\partial x^i}(g_t, t), g_t^i \right\rangle\\ &= \int_0^t \frac{\partial F}{\partial x^i}(g_s, s)\, dg_s^i + \frac{1}{2}\left\langle \int_0^t \frac{\partial F}{\partial x^i}(g_s, ds), g_t^i \right\rangle\\ &\quad + \frac{1}{2}\sum_j \int_0^t \frac{\partial^2 F}{\partial x^i \partial x^j}(g_s, s)\, d\langle g_s^j, g_s^i \rangle.\end{aligned} \tag{11}$$

On the other hand by the same theorem,

$$\int_0^t F(g_s, \circ ds) = \int_0^t F(g_s, ds) + \frac{1}{2}\sum_i \left\langle \int_0^t \frac{\partial F}{\partial x^i}(g_s, ds), g_t^i \right\rangle.$$

The sum of the above two coincides with $F(g_t, t) - F(g_0, 0)$ by the previous theorem. Hence formula (9) is established. □

Integral and differential rules for stochastic integrals

We shall establish some integral formulas and differential formulas concerning the stochastic integrals. We first discuss the case of the Itô integral.

Theorem 3.3.3

(i) *Let $F(x, t)$, $x \in \mathbb{D}$, be a continuous $C^{m,\delta}$-semimartingale with local characteristic belonging to the class $B^{m,\delta}$ where $\delta > 0$. Let $f(\lambda, t)$, $\lambda \in \Lambda$,*

$t \in [0, T]$ be a continuous predictable process with values in $C^{k,\gamma} = C^{k,\gamma}(\Lambda, \mathbb{D})$ where $\gamma > 0$ and Λ is a domain in $\mathbb{R}^e$. Set

$$L(\lambda, t) = \int_0^t F(f(\lambda, s), ds). \tag{12}$$

Then $L(\lambda, t)$ has a modification of a continuous semimartingale with values in $C^{m\wedge k,\varepsilon} = C^{m\wedge k,\varepsilon}(\Lambda : \mathbb{R}^1)$ with local characteristic belonging to the class $B^{m\wedge k,\gamma\delta}$ where $\varepsilon < \gamma\delta$. Further if g_t is a continuous predictable process with values in Λ, then we have the equality:

$$\int_0^t L(g_s, ds) = \int_0^t F(f(g_s, s), ds). \tag{13}$$

(ii) *Let $F(x, \lambda, t)$, $x \in \mathbb{D}$, $\lambda \in \Lambda$ be a continuous $C^{m,\delta}(\mathbb{D} \times \Lambda : \mathbb{R})$-semimartingale with local characteristic belonging to the class $B^{m,\delta}$. Let f_t be a continuous predictable process with values in $\mathbb{D}$. Set*

$$K(\lambda, t) = \int_0^t F(f_s, \lambda, ds). \tag{14}$$

Then $K(\lambda, t)$ has a modification of a continuous $C^{m,\varepsilon}(\Lambda : \mathbb{R})$-valued semimartingale for $\varepsilon < \delta$ with local characteristic belonging to the class $B^{m,\delta}$. Further if g_t is a continuous predictable process with values in Λ, we have the equality:

$$\int_0^t K(g_s, ds) = \int_0^t F(f_s, g_s, ds). \tag{15}$$

(iii) *If $m \geq 1$ and $k \geq 1$, then we have the equality:*

$$\frac{\partial L}{\partial \lambda^i}(\lambda, t) = \sum_{l=1}^d \int_0^t \frac{\partial f^l}{\partial \lambda^i}(\lambda, s) \frac{\partial F}{\partial x^l}(f(\lambda, s), ds) \tag{16}$$

where $f(\lambda, s) = (f^1(\lambda, s), \ldots, f^d(\lambda, s))$. □

Remark The stochastic integral of the right hand side of (16) is defined precisely by $\int_0^t G(f(\lambda, s), (\partial f^l/\partial \lambda^i)(\lambda, s), ds)$, where $G(x, y, t) = y(\partial F/\partial x^l)(x, t)$. By (15), it is equal to $\int_0^t K((\partial f^l/\partial \lambda^i)(\lambda, s), ds)$, where $K(y, t) = y\int_0^t (\partial F/\partial x^l)(f(\lambda, s), ds)$.

For the application of formula (16), it is convenient to rewrite the formula compatible with a standard expression of the differential rule of a composite function. We will often write (16) as

$$\frac{\partial}{\partial \lambda^i}\left(\int_0^t F(f(\lambda, s), ds)\right) = \sum_{l=1}^d \int_0^t \frac{\partial F}{\partial x^l}(f(\lambda, s), ds) \frac{\partial f^l}{\partial \lambda^i}(\lambda, s).$$

Proof It is sufficient to prove the theorem in the case where F is a localmartingale, since the assertion is obvious for any F of bounded variation. We first consider (i). The local characteristic of $L(\lambda, t)$ is $(a(f(\lambda, t), f(\mu, t), t), A_t)$. A direct computation yields that it belongs to the class $B^{m \wedge k, \gamma\delta}$. Therefore $L(\lambda, t)$ has a modification of a continuous $C^{m \wedge k, \varepsilon}$-localmartingale by Theorem 3.1.2. We shall prove formula (13). Denote the left hand side and the right hand side of (13) by M_t and N_t, respectively. These are one dimensional localmartingales. Since two equalities

$$\langle L(\lambda, t), F(x, t)\rangle = \int_0^t a(f(\lambda, s), x, s)\, dA_s,$$

$$\langle L(\lambda, t), L(\lambda', t)\rangle = \int_0^t a(f(\lambda, s), f(\lambda', s), s)\, dA_s$$

hold, we have

$$\langle M\rangle_t = \langle N\rangle_t = \langle M, N\rangle_t = \int_0^t a(f(g(s), s), f(g(s), s), s)\, dA_s.$$

This implies $\langle M - N\rangle_t = 0$, proving $M = N$.

The second assertion (ii) is immediate from (i) replacing $F(x, t)$ and $f(\lambda, t)$ by $F(x, \lambda, t)$ and (f_t, λ), respectively.

We shall next prove (16). Denote the left hand side and the right hand side of (16) by $Y(\lambda, t)$ and $Z(\lambda, t)$, respectively. Then the local characteristic of $Y(\lambda, t)$ is (a^Y, A_t) where

$$a^Y(\lambda, \eta, t) = \frac{\partial^2}{\partial\lambda^i \partial\eta^i}\{a(f(\lambda, t), f(\eta, t), t)\}$$

by Theorem 3.1.3. It is equal to

$$\sum_{k,l} \frac{\partial^2 a}{\partial x^k \partial y^l}(f(\lambda, t), f(\eta, t), t)\frac{\partial f^k}{\partial\lambda^i}(\lambda, t)\frac{\partial f^l}{\partial\eta^i}(\eta, t).$$

Obviously the above with A_t is the local characteristic of $Z(\lambda, t)$. Furthermore the joint local characteristic of $Y(\lambda, t)$ and $Z(\lambda, t)$ denoted by (a^{YZ}, A_t) is given by the same quantity. Therefore we have $a^Y - 2a^{YZ} + a^Z = 0$, which yields $\langle Y - Z\rangle = 0$. Thus equality (16) holds. □

Theorem 3.3.4

(i) *Let $F(x, t)$, $x \in \mathbb{D}$ be a continuous $C^{m,\delta}$-semimartingale with local characteristic belonging to the class $(B^{m+1,\delta}, B^{m,\delta})$ where $m \geq 1$ and $\delta > 0$. Let $f(\lambda, t)$, $\lambda \in \Lambda$, $t \in [0, T]$ be a continuous $C^{k,\gamma}$-semimartingale with* *values in $\mathbb{D}$ with local characteristic belonging to the class $B^{k,\gamma}$ where $k \geq 2$ and $\gamma > 0$ and Λ is a domain in $\mathbb{R}^e$. Then the Stratonovich integral*

$$\mathring{L}(\lambda, t) = \int_0^t F(f(\lambda, s), \circ ds) \tag{17}$$

has a modification of a continuous semimartingale with values in $C^{m \wedge k, \varepsilon}$, $\varepsilon < \gamma\delta$ with local characteristic belonging to the class $(B^{(m+1)\wedge k, \varepsilon}, B^{m \wedge k, \varepsilon})$. Further if g_t is a continuous semimartingale with values in Λ, then $f(g_t, t)$ is a continuous semimartingale and satisfies

$$\int_0^t \mathring{L}(g_s, \circ ds) = \int_0^t F(f(g_s, s), \circ ds). \tag{18}$$

(ii) *Let $F(x, \lambda, t)$, $x \in \mathbb{D}$, $\lambda \in \Lambda$ be a continuous $C^{m,\delta}(\mathbb{D} \times \Lambda : \mathbb{R})$-semimartingale with local characteristic belonging to the class $(B^{m+1,\delta}, B^{m,\delta})$ where $m \geq 1$ and $\delta > 0$. Let f_t be a continuous semimartingale with values in $\mathbb{D}$. Then*

$$\mathring{K}(\lambda, t) = \int_0^t F(f_s, \lambda, \circ ds) \tag{19}$$

has a modification of a continuous $C^{m,\varepsilon}(\Lambda : \mathbb{R})$-semimartingale for any $\varepsilon < \delta$ with local characteristic belonging to the class $(B^{m+1,\delta}, B^{m,\delta})$. Further if g_t is a continuous semimartingale with values in Λ, the equality

$$\int_0^t \mathring{K}(g_s, \circ ds) = \int_0^t F(f_s, g_s, \circ ds) \tag{20}$$

holds.

(iii) *We have*

$$\frac{\partial \mathring{L}}{\partial \lambda^i}(\lambda, t) = \sum_{l=1}^d \int_0^t \frac{\partial f^l}{\partial \lambda^i}(\lambda, s) \frac{\partial F}{\partial x^l}(f(\lambda, s), \circ ds). \quad \square \tag{21}$$

Remark Formula (21) is often written as

$$\frac{\partial}{\partial \lambda^i}\left(\int_0^t F(f(\lambda, s), \circ ds)\right) = \sum_{l=1}^d \int_0^t \frac{\partial F}{\partial x^l}(f(\lambda, s), \circ ds) \frac{\partial f^l}{\partial \lambda^i}(\lambda, s).$$

Proof of Theorem 3.3.4 We first prove assertion (i). By Theorem 3.2.5 the Stratonovich integral is well defined and is written as

$$\mathring{L}(\lambda, t) = \int_0^t F(f(\lambda, s), ds) + \frac{1}{2}\sum_j \left\langle \int_0^t \frac{\partial M}{\partial x^j}(f(\lambda, s), ds), f^j(\lambda, t) \right\rangle, \tag{22}$$

where $M(x, t)$ is the localmartingale part of $F(x, t)$. The first Itô integral is a $C^{m\wedge k,\varepsilon}$-semimartingale by Theorem 3.3.3. The second joint quadratic variation is a $C^{m\wedge k,\varepsilon}$-process by Theorem 3.1.3. Therefore $\mathring{L}(\lambda, t)$ is a $C^{m\wedge k,\varepsilon}$-semimartingale with local characteristic belonging to the class $(B^{(m+1)\wedge k,\varepsilon}, B^{m\wedge k,\varepsilon})$. We shall prove formula (18). We rewrite the left hand side using the Itô integral. Then we have

$$\int_0^t \mathring{L}(g_s, \circ ds) = \int_0^t \mathring{L}(g_s, ds) + \frac{1}{2}\sum_i \left\langle \int_0^t \frac{\partial L}{\partial \lambda^i}(g_s, ds), g_t^i \right\rangle. \quad (23)$$

The first term of the right hand side is written as

$$\int_0^t F(f(g_s, s), ds) + \frac{1}{2}\sum_j \left\langle \int_0^t \frac{\partial F}{\partial x^j}(f(g_s, s), ds), \int_0^t f^j(g_s, ds) \right\rangle$$

by formula (22) and Theorem 3.2.4. The second term of the right hand side of (23) is written as

$$\frac{1}{2}\sum_{i,j} \left\langle \int_0^t \frac{\partial F}{\partial x^j}(f(g_s, s), ds)\frac{\partial f^j}{\partial \lambda^i}(g_s, s), g_t^i \right\rangle$$
$$= \frac{1}{2}\sum_{i,j} \left\langle \int_0^t \frac{\partial F}{\partial x^j}(f(g_s, s), ds), \int_0^t \frac{\partial f^j}{\partial \lambda^i}(g_s, s)\, dg_s^i \right\rangle$$

by Theorem 3.3.3 (iii). We next rewrite the right hand side of (18) using the Itô integral. Since $f^j(g_t, t)$ is a continuous semimartingale and is written as

$$f^j(g_t, t) = f^j(g_0, 0) + \int_0^t f^j(g_s, ds) + \sum_i \int_0^t \frac{\partial f^j}{\partial \lambda^i}(g_s, s)\, dg_s^i$$
$$+ \text{ a process of bounded variation}$$

using the generalized Itô's formula (Theorem 3.3.1), we have

$$\int_0^t F(f(g_s, s), \circ ds)$$
$$= \int_0^t F(f(g_s, s), ds) + \frac{1}{2}\sum_j \left\langle \int_0^t \frac{\partial F}{\partial x^j}(f(g_s, s), ds), f^j(g_t, t) \right\rangle$$
$$= \int_0^t F(f(g_s, s), ds) + \frac{1}{2}\sum_j \left\langle \int_0^t \frac{\partial F}{\partial x^j}(f(g_s, s), ds), \int_0^t f^j(g_s, ds) \right\rangle$$
$$+ \frac{1}{2}\sum_{i,j} \left\langle \int_0^t \frac{\partial F}{\partial x^j}(f(g_s, s), ds), \int_0^t \frac{\partial f^j}{\partial \lambda^i}(g_s, s)\, dg_s^i \right\rangle.$$

This is equal to $\int_0^t \mathring{L}(g_s, \circ ds)$ as we have shown above. Therefore formula (18) is established.

The second assertion (ii) is immediate from (i) replacing $F(x, t)$ and $f(\lambda, t)$ by $F(x, \lambda, t)$ and (f_t, λ). We shall prove (iii). Observe formula (22) and differentiate each term by λ_i. For the last term, we have

$$\frac{\partial}{\partial \lambda^i}\left(\left\langle \int_0^t \frac{\partial F}{\partial x^j}(f(\lambda, s), ds), f^j(\lambda, t)\right\rangle\right) = \left\langle \frac{\partial}{\partial \lambda^i}\int_0^t \frac{\partial F}{\partial x^j}(f(\lambda, s), ds), f^j(\lambda, t)\right\rangle + \left\langle \int_0^t \frac{\partial F}{\partial x^j}(f(\lambda, s), ds), \frac{\partial f^j}{\partial \lambda^i}(\lambda, t)\right\rangle$$

by Corollary 3.1.4. Therefore

$$\begin{aligned}\frac{\partial}{\partial \lambda^i}\int_0^t F(f(\lambda, s), \circ ds) &= \sum_l \int_0^t \frac{\partial f^l}{\partial \lambda^i}(\lambda, s)\frac{\partial F}{\partial x^l}(f(\lambda, s), ds)\\ &\quad + \frac{1}{2}\sum_{l,j}\left\langle \int_0^t \frac{\partial f^l}{\partial \lambda^i}(\lambda, s)\frac{\partial^2 F}{\partial x^l \partial x^j}(f(\lambda, s), ds), f^j(\lambda, t)\right\rangle\\ &\quad + \frac{1}{2}\sum_j \left\langle \int_0^t \frac{\partial F}{\partial x^j}(f(\lambda, s), ds), \frac{\partial f^j}{\partial \lambda^i}(\lambda, t)\right\rangle.\end{aligned}$$

The right hand side is equal to the Stratonovich integral:

$$\sum_l \int_0^t \frac{\partial f^l}{\partial \lambda^i}(\lambda, s)\frac{\partial F}{\partial x^l}(f(\lambda, s), \circ ds).$$

The proof is complete. □

Exercise 3.3.5 Let $F(x, t)$ be a continuous C^1-semimartingale with local characteristic belonging to the class $(B^{2,\delta}, B^{1,0})$. Suppose that it is represented by $\sum_n \int_0^t f_n(x, s) \circ dM_s^n$ where $\{f_n(x, t)\}$ are continuous C^2-processes and C^1-semimartingales with local characteristic belonging to the class $B^{1,0}$. Show that

$$\int_0^t F(\varphi_s, \circ ds) = \sum_n \int_0^t f_n(\varphi_s, s) \circ dM_s^n$$

holds for any continuous semimartingale φ_t.

Exercise 3.3.6. Let $F(x, t)$ be a continuous $C^{m,\delta}$-semimartingale with local characteristic belonging to $B^{m,\delta}$ and let $f(\lambda, t)$, $\lambda \in \Lambda \subset \mathbb{R}^e$ be a continuous $C^{m,\gamma}$-process adapted to $(\mathscr{F}_t)$ where $m \geq 1$, $\delta, \gamma > 0$. For a partition $\Delta = \{0 = t_0 < \cdots < t_l = t\}$, set

$$L^\Delta(\lambda, t) = \sum_{k=0}^{l-1}\{F(f(\lambda, t_k \wedge t), t_{k+1} \wedge t) - F(f(\lambda, t_k \wedge t), t_k \wedge t)\}.$$

Show that the $L^{\Delta}(\lambda, t)$ converge to $\int_0^t F(f(\lambda, s), ds)$ uniformly on compact sets in $\Lambda \times [0, T]$ in probability as $|\Delta| \to 0$. (*Hint*: consider first a special case where F is a martingale with local characteristic belonging to the class $B_{ub}^{m,\delta}$ where $A_t \equiv t$ and $f(\lambda, t)$ together with the derivatives are bounded by a constant. Show that for any $|\alpha| \le m$ $\sup_{\Delta,\lambda} E[\sup_t |D_\lambda^\alpha L^{\Delta}(\lambda, t)|^p] < \infty$ and $E[\sup_t |D_\lambda^\alpha L^{\Delta}(\lambda, t) - D_\lambda^\alpha \int_0^t F(f(\lambda, s), ds)|^p] \to 0$ as $|\Delta| \to 0$ for each λ. Then apply Sobolev's inequality.)

3.4 Stochastic differential equations

Itô's stochastic differential equation

Let $F(x, t) = (F^1(x, t), \ldots, F^d(x, t))$, $x \in \mathbb{R}^d$ be a continuous semimartingale with values in $C = C(\mathbb{R}^d : \mathbb{R}^d)$. The objective of this section is to discuss the solutions of the stochastic differential equation based on the semimartingale F. The stochastic differential equation is written as

$$d\varphi_t = F(\varphi_t, dt). \tag{1}$$

If $F(\cdot, t)$ has a density function $f(\cdot, t)$ with respect to dt, i.e. $f(\cdot, t) = dF(\cdot, t)/dt$, then the equation can be written as a stochastic ordinary differential equation

$$\frac{d\varphi_t}{dt} = f(\varphi_t, t). \tag{2}$$

However the density function $f(\cdot, t)$ does not exist except for a very special semimartingale. Instead we shall consider the right hand side of (1) as Itô's stochastic integral.

We shall first introduce assumptions for a continuous semimartingale F so that equation (1) is well defined. Let $F^i(x, t) = M^i(x, t) + B^i(x, t)$ be the decomposition such that $M^i(x, t)$ is a continuous localmartingale and $B^i(x, t)$ is a continuous process of bounded variation. Set $A^{ij}(x, y, t) = \langle M^i(x, t), M^j(y, t)\rangle$. As in Section 3.2 let A_t be a continuous strictly increasing process such that both $A^{ij}(x, y, t)$ and $B^i(x, t)$ are absolutely continuous with respect to A_t a.s. for any x, y of $\mathbb{R}^d$. Then there exist predictable processes $a^{ij}(x, y, t)$ and $b^i(x, t)$ with parameters x, y such that

$$A^{ij}(x, y, t) = \int_0^t a^{ij}(x, y, s)\, dA_s, \qquad B^i(x, t) = \int_0^t b^i(x, s)\, dA_s.$$

Set $a(x, y, t) = (a^{ij}(x, y, t))$, $i, j = 1, \ldots, d$ and $b(x, t) = (b^1(x, t), \ldots, b^d(x, t))$. Then $a(x, y, t)$ is a $d \times d$-matrix valued function with the following properties.

(a) **symmetric:** $a^{ij}(x, y, t) = a^{ji}(y, x, t)$ holds a.e. μ for any x, y and i, j.
(b) **non-negative definite:** $\sum_{i,j,p,q} a^{ij}(x_p, x_q, t)\xi_p^i \xi_q^j \geq 0$ holds a.e. μ for any x_p, $(\xi_p^1, \ldots, \xi_p^d)$, $p = 1, \ldots, n$.

The triple $(a(x, y, t), b(x, t), A_t)$ is called the *local characteristic* of $F(x, t)$.

The local characteristic $(a(x, y, t), b(x, t), A_t)$ is said to belong to the class $(B^{m,\delta}, B^{m',\delta'})$ if for every $i = 1, \ldots, d$ the local characteristic of $F^i(x, t)$ belongs to the class $(B^{m,\delta}, B^{m',\delta'})$. When $m = m'$ and $\delta = \delta'$ the triple is simply said to belong to the class $B^{m,\delta}$. The classes $B_b^{m,\delta}$ and $B_{ub}^{m,\delta}$ etc. are defined similarly to those in the previous section.

Now if F is a continuous C-semimartingale with the local characteristic belonging to the class $B^{0,\delta}$, then Itô's stochastic integral $\int_0^t F(\varphi_s, ds)$ is well defined for any continuous $\mathbb{R}^d$-valued predictable process φ_t. To be precise, we shall give the definition of the solution of the stochastic differential equation. Let $t_0 \in [0, T]$ and $x_0 \in \mathbb{R}^d$. A continuous $\mathbb{R}^d$-valued process φ_t, $t_0 \leq t \leq T$ adapted to $(\mathscr{F}_t)$ is called a *solution of Itô's stochastic differential equation based on* $F(x, t)$ *starting at* x_0 *at time* t_0 *if it satisfies*

$$\varphi_t = x_0 + \int_{t_0}^t F(\varphi_s, ds). \tag{1'}$$

Also φ_t is said to be *governed by Itô's stochastic differential equation based on* $F(x, t)$.

We shall study the existence and uniqueness of the solution of the above equation. In the case of an ordinary differential equation of the form (2), a well-known sufficient condition is a Lipschitz condition for the vector field $f(x, t)$ with respect to the spatial variable x. In our case of the stochastic differential equation a similar condition for the local characteristic will assert the existence and the uniqueness of the solution.

Theorem 3.4.1 *Let* $F(x, t)$ *be a continuous semimartingale with values in* $C(\mathbb{R}^d : \mathbb{R}^d)$ *with local characteristic belonging to the class* $B_b^{0,1}$. *Then for each* t_0 *and* x_0, *the equation* (1') *has a unique solution.* □

Existence and uniqueness of the solution

The proof of the theorem will be divided into two steps. In the first step, we discuss the case where the local characteristic (a, b, A_t) of F belongs to the class $B_{ub}^{0,1}$ and A_t is identical to t. Then the pair (a, b) is *uniformly Lipschitz continuous* and of *uniformly linear growth*, i.e. there exists a positive constant K such that

$$|b(x, t) - b(y, t)| \le K|x - y|,$$
$$\|a(x, x, t) - 2a(x, y, t) + a(y, y, t)\| \le K|x - y|^2,$$
$$|b(x, t)| \le K(1 + |x|),$$
$$\|a(x, y, t)\| \le K(1 + |x|)(1 + |y|),$$

hold for all $x, y \in \mathbb{R}^d$ a.s. Here $|\ |$ and $\|\ \|$ denote norms of vectors and matrices, respectively. The general case will be considered in the second step. We shall first construct a solution of equation (1′) by the method of the successive approximation.

Lemma 3.4.2 *Assume that the local characteristic (a, b, A_t) of F belongs to the class $B_{ub}^{0,1}$ and A_t is identical to t. Define a sequence of continuous $(\mathscr{F}_t)$-adapted processes $\{\varphi_t^n : n = 0, 1, 2, \ldots\}$ by*

$$\varphi_t^0 = x_0, \qquad \varphi_t^n = x_0 + \int_{t_0}^t F(\varphi_s^{n-1}, ds), \qquad n = 1, 2, \ldots \tag{3}$$

Then the sequence converges uniformly in t in L^p for any $p > 1$. The limit φ_t is a solution of equation (1′).

Proof Since

$$\varphi_t^{n+1} - \varphi_t^n = \left\{\int_{t_0}^t M(\varphi_s^n, ds) - \int_{t_0}^t M(\varphi_s^{n-1}, ds)\right\} + \int_{t_0}^t \{b(\varphi_s^n, s) - b(\varphi_s^{n-1}, s)\}\, ds,$$

we have for any $p > 1$

$$E\left[\sup_{t_0 \le u \le t} |\varphi_u^{n+1} - \varphi_u^n|^p\right] \le 2^p E\left[\sup_{t_0 \le u \le t}\left|\int_{t_0}^u M(\varphi_s^n, ds) - \int_{t_0}^u M(\varphi_s^{n-1}, ds)\right|^p\right] + 2^p E\left[\sup_{t_0 \le u \le t}\left|\int_{t_0}^u (b(\varphi_s^n, s) - b(\varphi_s^{n-1}, s))\, ds\right|^p\right]. \tag{4}$$

Using Doob's inequality and Burkholder's inequality, the first term of the right hand side is bounded by

$$CE\left[\left\{\sum_{i=1}^{d}\left\langle\int_{t_0}^{t} M^i(\varphi_s^n, ds) - \int_{t_0}^{t} M^i(\varphi_s^{n-1}, ds)\right\rangle\right\}^{p/2}\right]$$

with a positive constant C depending only on p. By Theorem 3.2.4 and the uniform Lipschitz condition for a, the above quadratic variation is estimated as

$$\sum_{i=1}^{d}\left\langle\int_{t_0}^{t} M^i(\varphi_s^n, ds) - \int_{t_0}^{t} M^i(\varphi_s^{n-1}, ds)\right\rangle$$
$$= \int_{t_0}^{t}\sum_{i=1}^{d}\{a^{ii}(\varphi_s^n, \varphi_s^n, s) - 2a^{ii}(\varphi_s^n, \varphi_s^{n-1}, s) + a^{ii}(\varphi_s^{n-1}, \varphi_s^{n-1}, s)\}\, ds$$
$$\leq K\int_{t_0}^{t}|\varphi_s^n - \varphi_s^{n-1}|^2\, ds$$
$$\leq K|t - t_0|^{1-(2/p)}\left(\int_{t_0}^{t}|\varphi_s^n - \varphi_s^{n-1}|^p\, ds\right)^{2/p},$$

where K is the Lipschitz constant for $a(x, y, t)$. Therefore the first term of the right hand side of (4) is bounded by

$$CK^{p/2}|t - t_0|^{(p/2)-1}E\left[\int_{t_0}^{t}|\varphi_s^n - \varphi_s^{n-1}|^p\, ds\right].$$

Furthermore the second term of (4) is bounded by

$$2^p|t - t_0|^{p-1}K^pE\left[\int_{t_0}^{t}|\varphi_s^n - \varphi_s^{n-1}|^p\, ds\right]$$

by the Lipschitz condition for $b(x, t)$ and Hölder's inequality. Consequently there exists a positive constant C_1 such that

$$E\left[\sup_{t_0\leq u\leq t}|\varphi_u^{n+1} - \varphi_u^n|^p\right] \leq C_1E\left[\int_{t_0}^{t}|\varphi_s^n - \varphi_s^{n-1}|^p\, ds\right] \tag{5}$$

holds for all n and $t \in [t_0, T]$. Denote the left hand side by $\rho_t^{(n)}$. The above inequality yields $\rho_t^{(n)} \leq C_1\int_{t_0}^{t}\rho_s^{(n-1)}\, ds$. By iteration we get

$$\rho_t^{(n)} \leq C_1^n\int_{t_0}^{t}\int_{t_0}^{t_1}\cdots\int_{t_0}^{t_{n-1}}\rho_{t_n}^{(0)}\, dt_n \ldots dt_1 \leq \frac{C_1^n}{n!}(t - t_0)^n\rho_t^{(0)},$$

since $\rho_t^{(0)}$ is an increasing function of t. Then we have

$$\sum_{n=0}^{\infty}E\left[\sup_{t_0\leq u\leq t}|\varphi_u^{n+1} - \varphi_u^n|^p\right]^{1/p} \leq \sum_{n=0}^{\infty}\left\{\frac{C_1^n}{n!}(t - t_0)^n\rho_t^{(0)}\right\}^{1/p} < \infty.$$

Therefore $E[(\sum_{n=0}^{\infty}\sup_{t_0\leq u\leq t}|\varphi_u^{n+1} - \varphi_u^n|)^p]^{1/p} < \infty$. This shows that $\{\varphi_t^n\}$

converges uniformly in $[t_0, T]$ a.s. and in L^p-norm. Denote the limit by φ_t. It is a continuous $(\mathscr{F}_t)$-adapted process.

We will show that this φ_t is a solution. Note that the integrals $\int_{t_0}^t F(\varphi_s^n, ds)$ converge to $\int_{t_0}^t F(\varphi_s, ds)$ in L^p. In fact

$$\sum_{i=1}^d \left\langle \int_{t_0}^t M^i(\varphi_s^n, ds) - \int_{t_0}^t M^i(\varphi_s, ds) \right\rangle \le K \int_{t_0}^t |\varphi_s^n - \varphi_s|^2 \, ds \xrightarrow[n\to\infty]{} 0$$

in $L^{p/2}$. Also we have

$$\left| \int_{t_0}^t b(\varphi_s^n, s) \, ds - \int_{t_0}^t b(\varphi_s, s) \, ds \right| \le K|t - t_0|^{1/2} \left(\int_{t_0}^t |\varphi_s^n - \varphi_s|^2 \, ds \right)^{1/2} \xrightarrow[n\to\infty]{} 0$$

in L^p. Then in view of the Burkholder inequality, $\int_{t_0}^t F(\varphi_s^n, ds)$ converges to $\int_{t_0}^t F(\varphi_s, ds)$ in L^p. Consequently, φ_t is a solution of equation (1′). We have thus proved Lemma 3.4.2. □

Before we proceed to the proof of the uniqueness of the solution, we need another lemma.

Lemma 3.4.3 *Let F and G be continuous semimartingales with values in $C(\mathbb{R}^d : \mathbb{R}^d)$ with local characteristics belonging to the class $B^{0,0}$. Assume that $F(x, \cdot) = G(x, \cdot)$ holds for all $x \in U$, where U is an open subset of $\mathbb{R}^d$. Let f_t and g_t be predictable processes with values in $\mathbb{R}^d$ such that for any $t < \sigma$ $f_t = g_t$ holds and belongs to U, where σ is a certain stopping time. Then we have*

$$\int_{t_0}^{t\wedge\sigma} F(f_s, ds) = \int_{t_0}^{t\wedge\sigma} G(g_s, ds) = \int_{t_0}^t F^\sigma(f_s, ds) = \int_{t_0}^t G^\sigma(g_s, ds), \quad (6)$$

where $F^\sigma(x, t) = F(x, t \wedge \sigma)$. □

The lemma is verified directly if f_t and g_t are simple processes. For general f_t and g_t, approximate them by sequences of simple processes. The details are left to the reader.

The uniqueness of the solution is established by the following lemma.

Lemma 3.4.4. *Let φ_t be the solution of equation (1′) constructed in Lemma 3.4.2. Let ψ_t be a solution of Itô's equation based on a continuous semimartingale G with values in $C(\mathbb{R}^d : \mathbb{R}^d)$ with local characteristic of the class $B^{0,1}$. Assume $G(x, \cdot) = F(x, \cdot)$ holds for all $x \in U$ where U is an open subset of $\mathbb{R}^d$. Then $\varphi_t = \psi_t$ holds for $t < \tau$, where $\tau = \inf\{t > t_0 : \psi_t \in \bar{U}^c\}$ ($= T$ if the set $\{\dots\}$ is empty).*

Proof We consider stopped processes $\varphi_t^\tau = \varphi_{t\wedge\tau}$ and $\psi_t^\tau = \psi_{t\wedge\tau}$. Since

$$\int_{t_0}^{t\wedge\tau} G(\psi_s^\tau, ds) = \int_{t_0}^{t\wedge\tau} F(\psi_s^\tau, ds)$$

holds by the previous lemma, we have

$$\varphi_t^\tau - \psi_t^\tau = \int_{t_0}^{t\wedge\tau} F(\varphi_s^\tau, ds) - \int_{t_0}^{t\wedge\tau} F(\psi_s^\tau, ds).$$

Then, by a similar computation as that used to obtain (5), we get

$$E\left[\sup_{t_0\le u\le t} |\varphi_u^\tau - \psi_u^\tau|^p\right] \le C_1 E\left[\int_{t_0}^{t\wedge\tau} |\varphi_s^\tau - \psi_s^\tau|^p \, ds\right], \tag{7}$$

where C_1 is a positive constant not depending on t. Denote the left hand side by ρ_t. Then inequality (7) shows $\rho_t \le C_1 \int_{t_0}^t \rho_s \, ds$. By Gronwall's inequality, we get $\rho_t = 0$. This proves $\varphi_t^\tau = \psi_t^\tau$, showing $\varphi_t = \psi_t$ for $t < \tau$.

For the second step, we shall consider the case where (a, b, A) belongs to $B_b^{0,1}$. We may assume that the local characteristic belongs to $B_{ub}^{0,1}$. We will reduce it to the previous case $A_t \equiv t$ by changing the scale of time. In the following we will show the existence and uniqueness of the solution starting at time 0 only. The existence and uniqueness of solutions starting at time t_0 can be shown similarly with obvious modifications.

Let τ_u be the inverse function of A_t and let $\tilde{F}(x, u) = F(x, \tau_u)$. It is a continuous $C^{0,\varepsilon}$-semimartingale adapted to $(\tilde{\mathscr{F}}_u) = (\mathscr{F}_{\tau_u})$ for any $\varepsilon < 1$. Since $A_{\tau_u} = u \wedge A_T$, the local characteristic of $\tilde{F}$ is $(a(x, y, \tau_u)\chi(u < A_T), b(x, \tau_u)\chi(u < A_T))$. It belongs to the class $B_{ub}^{0,1}$ with $A_u \equiv u$. Therefore Itô's stochastic differential equation governed by $\tilde{F}$ has a unique solution $\tilde{\varphi}_r$ starting from x_0 at time 0. Set $\varphi_t = \tilde{\varphi}_{A_t}$. Then by Theorem 3.2.9, it satisfies

$$\varphi_t = x_0 + \int_0^{A_t} \tilde{F}(\tilde{\varphi}_s, ds) = x_0 + \int_0^t F(\varphi_s, ds).$$

Therefore it is a solution of equation (1).

We will prove the uniqueness of the solution. Let φ_t' be an arbitrary solution. Set $\tilde{\varphi}_u' = \varphi_{\tau_u}'$. Then it satisfies

$$\tilde{\varphi}_u' = x_0 + \int_0^{\tau_u} F(\varphi_s', ds) = x_0 + \int_0^u \tilde{F}(\tilde{\varphi}_s', ds).$$

Then we have $\tilde{\varphi}_u' = \tilde{\varphi}_u$, proving $\varphi_u = \varphi_u'$. The proof of Theorem 3.4.1 is now complete. □

Example

As an example we shall consider an Itô's classical stochastic differential equation. Let $f_0(x, t), \ldots, f_m(x, t)$ be $\mathbb{R}^d$-valued functions continuous in (x, t) and Lipschitz continuous in x, i.e. there exists a positive constant K such that for any $l = 0, \ldots, m$

$$|f_l(x, t) - f_l(y, t)| \le K|x - y|, \qquad \text{for all } x, y \in \mathbb{R}^d, \quad t \in [0, T].$$

An Itô's classical stochastic differential equation is given by

$$\varphi_t = x_0 + \int_{t_0}^{t} f_0(\varphi_s, s)\, ds + \sum_{l=1}^{m} \int_{t_0}^{t} f_l(\varphi_s, s)\, dB_s^l, \tag{8}$$

where $B_t = (B_t^1, \ldots, B_t^m)$ is an m-dimensional standard Brownian motion. It is equivalent to stochastic differential equation (1) by setting

$$F(x, t) = \int_0^t f_0(x, s)\, ds + \sum_{l=1}^{m} \int_0^t f_l(x, s)\, dB_s^l, \tag{9}$$

(Exercise 3.2.11). It is a $C^{0,\gamma}$-valued Brownian motion for any $\gamma \in (0, 1)$. Its local characteristic is given by $(\sum_{l=1}^{m} f_l^i(x, t) f_l^j(y, t), f_0^i(x, t))$. It belongs to the class $B_{ub}^{0,1}$ since

$$\sum_i \{a^{ii}(x, x, t) - 2a^{ii}(x, y, t) + a^{ii}(y, y, t)\} = \sum_{l=1}^{m} |f_l(x, t) - f_l(y, t)|^2.$$

Therefore, for given initial data, the stochastic differential equation (8) has a unique solution.

Equation (8) can be extended naturally to that for an infinite number of Brownian motions. Let $B_t^l, l = 1, 2, \ldots$ be infinite independent copies of one dimensional standard Brownian motions and let $f_l(x, t), l = 0, 1, 2, \ldots$ be $\mathbb{R}^d$-valued continuous functions such that there exists a positive constant K satisfying

$$\sum_{l=0}^{\infty} |f_l(x, t)|^2 \le K(1 + |x|)^2,$$

$$\sum_{l=0}^{\infty} |f_l(x, t) - f_l(y, t)|^2 \le K|x - y|^2.$$

Consider the stochastic differential equation

$$\varphi_t = x_0 + \int_{t_0}^{t} f_0(\varphi_s, s)\, ds + \sum_{l=1}^{\infty} \int_{t_0}^{t} f_l(\varphi_s, s)\, dB_s^l. \tag{8'}$$

The corresponding $F(x, t)$ is given by

$$F(x, t) = \int_0^t f_0(x, s)\, ds + \sum_{l=1}^{\infty} \int_0^t f_l(x, s)\, dB_s^l. \tag{9'}$$

It has local characteristic $(\sum_{l=1}^{\infty} f_l^i(x,t) f_l^j(y,t), f_0^i(x,t))$ which belongs to the class $B_{ub}^{0,1}$. Therefore equation (8') also has a unique solution for any initial data.

Conversely let $F(x, t)$ be a C-valued Brownian motion with local characteristic (a, b) belonging to the class $B_{ub}^{0,1}$. Then there exist independent standard Brownian motions $B_t^1, B_t^2, \ldots$, and measurable functions $f_l(x, t)$, $l = 1, 2, \ldots$, uniformly Lipschitz continuous in x such that F is represented as (9') (see Exercise 3.2.10). Then the stochastic differential equation based on $F(x, t)$ is equation (8'). □

Local solution

In the sequel, we will not assume the uniform Lipschitz condition for the local characteristic. Then, as in the case of an ordinary differential equation, the equation may not have a global solution. The explosion may occur at a finite time. So we shall define a local solution of a stochastic differential equation.

Let φ_t, $t \in [t_0, \sigma_\infty)$ be a continuous local process with values in $\mathbb{R}^d$ adapted to $(\mathscr{F}_t)$. It is called a *local solution* of equation (1') if

$$\varphi_{t \wedge \sigma_N} = x_0 + \int_{t_0}^{t \wedge \sigma_N} F(\varphi_{s \wedge \sigma_N}, ds) \tag{10}$$

is satisfied for any N where $\{\sigma_N\}$ is a sequence of stopping times such that $\sigma_N < \sigma_\infty$ and $\sigma_N \uparrow \sigma_\infty$. Furthermore if $\lim_{t \uparrow \sigma_\infty} \varphi_t = \infty$ is satisfied when $\sigma_\infty < T$, it is called a *maximal solution* and σ_∞ is called the *explosion time.* If the explosion time is equal to T a.s., the solution φ_t, $t \in [t_0, T)$ is called a *global solution.* Further if equation (1') has a global solution for any initial condition, equation (1') or the corresponding C-semimartingale F is called *complete* (*to the forward*).

Theorem 3.4.5 *Let $F(x, t)$ be a continuous semimartingale with values in $C(\mathbb{R}^d : \mathbb{R}^d)$ with local characteristic belonging to the class $B^{0,1}$. Then for each t_0 and x_0 the stochastic differential equation* (1') *has a unique maximal solution.*

Proof We shall apply the method of truncation. For each positive integer N, take a C^∞-function $\psi_N(x)$, $x \in \mathbb{R}^d$ such that $\psi_N(x) = 1$ if $|x| \le N$, $0 \le \psi_N(x) \le 1$ if $N \le |x| \le N + 1$ and $\psi_N(x) = 0$ if $|x| \ge N + 1$. Define $F^N(x, t) \equiv F(x, t)\psi_N(x)$. Then its local characteristic is $(a(x, y, t)\psi_N(x)\psi_N(y), b(x, t)\psi_N(x), A_t)$, which belongs to $B_b^{0,1}$. Therefore Itô's stochastic differential equation based on F^N has a unique solution starting at x_0 at time t_0. We denote it by φ_t^N. Set $\sigma_N = \inf\{t \in [t_0, T] : |\varphi_t^N| > N\}$ $(= T$ if $\{\ldots\} = \emptyset)$. Now let φ_t^M be the solution of Itô's stochastic differential equation based on F^M starting at x_0 at time t_0. Then if $M < N$, $\varphi_t^M = \varphi_t^N$ holds for $t < \sigma_M$ by Lemma 3.4.4, which implies that $\{\sigma_N\}$ increases with N. Set $\sigma_\infty = \lim \sigma_N$ and define $\varphi_t, t < \sigma_\infty$ by $\varphi_t = \varphi_t^N$ if $t < \sigma_N$. Then $\lim_{t\to\sigma_\infty} \varphi_t = \infty$ holds if $\sigma_\infty < T$. Hence it is a maximal solution of equation (1′).

We shall prove the uniqueness of the solution. Let $\psi_t, t \in [s, \tau_\infty)$ be any maximal solution of equation (1′). Lemma 3.4.4 is valid for a local process ψ_t, obviously. Therefore we have $\varphi_t^N = \psi_t$ for $t < \sigma_N$. This proves $\varphi_t = \psi_t$ for $t < \sigma_\infty$ and hence $\sigma_\infty = \tau_\infty$. The proof is complete. □

A simple sufficient condition that equation (1′) has a global solution for any initial condition is that the local characteristic (a, b, A_t) is of $B^{0,1}$ and is of linear growth, i.e. there exists a positive predictable process K_t with $\int_0^T K_t\, dA_t < \infty$ such that

$$\|a(x, x, t)\| \le K_t(1 + |x|)^2, \tag{11}$$

$$|b(x, t)| \le K_t(1 + |x|). \tag{12}$$

Theorem 3.4.6 *Assume that the local characteristic (a, b, A_t) of a continuous C-semimartingale F belongs to the class $B^{0,1}$ and satisfies (11) and (12). Then for each t_0 and x_0 equation (1′) has a unique global solution. Furthermore, if the process K_t satisfies*

$$E\left[\exp\left\{\lambda \int_0^T K_u\, dA_u\right\}\right] < \infty, \qquad \textit{for all } \lambda > 0 \tag{13}$$

the global solution has finite moments of any order.

Proof We first assume that condition (13) is satisfied. Let $\varphi_t, t \in [t_0, \sigma_\infty)$ be the maximal solution of equation (1′). Set $\sigma_N = \inf\{t \in [t_0, T] : |\varphi_t| > N\}$ ($= T$ if $\{\ldots\} = \emptyset$) as before. Consider the stopped process $\varphi_t^N = \varphi_{t\wedge\sigma_N}$. We shall apply Itô's formula (Theorem 2.3.11) to the function $f(x) = |x|^{2p}$ and continuous semimartingale $X_t = \varphi_t^N$, where $p > 1$. Then we have

$$\begin{aligned}|\varphi_t^N|^{2p} - |x_0|^{2p} &= \sum_i \int_{t_0}^{t\wedge\sigma_N} \frac{\partial f}{\partial x^i}(\varphi_s^N) M^i(\varphi_s^N, ds)\\ &\quad + \sum_i \int_{t_0}^{t\wedge\sigma_N} \frac{\partial f}{\partial x^i}(\varphi_s^N) b^i(\varphi_s^N, s)\, dA_s\\ &\quad + \frac{1}{2}\sum_{i,j} \int_{t_0}^{t\wedge\sigma_N} \frac{\partial^2 f}{\partial x^i \partial x^j}(\varphi_s^N) a^{ij}(\varphi_s^N, \varphi_s^N, s)\, dA_s,\end{aligned} \tag{14}$$

where $M^i(x, t)$ is the martingale part of $F^i(x, t)$. The sum of the second and third terms of the right hand side is written as $\int_{t_0}^{t\wedge\sigma_N} g(\varphi_s^N, s)\,dA_s$, where

$$g(x,t) = 2p|x|^{2p-2}\left\{\sum_i b^i(x,t)x^i\right\}$$
$$+ p|x|^{2p-4}\left\{\sum_{i,j}(|x|^2\delta_{ij} + 2(p-1)x^ix^j)a^{ij}(x,x,t)\right\}.$$

From (11) and (12) there exists a positive constant C such that $|g(x,t)| \le CK_t(1+|x|^{2p})$. Therefore from (14) we have the equality

$$1 + |\varphi_t^N|^{2p} = 1 + |x_0|^{2p} + M_t^N - B_t^N + C\int_{t_0}^t K_u^N(1 + |\varphi_u^N|^{2p})\,dA_u, \quad (15)$$

where M_t^N is a continuous martingale representing the first term of the right hand side of (14), B_t^N is a continuous increasing process with $B_{t_0}^N = 0$ and $K_u^N = K_u\chi(\sigma_N > u)$. Then $1 + |\varphi_t^N|^{2p}$ can be regarded as a solution of the linear stochastic differential equation

$$\Phi_t = 1 + |x_0|^{2p} + M_t^N - B_t^N + C\int_{t_0}^t K_u^N\Phi_u\,dA_u. \quad (16)$$

It has a unique global solution by Theorem 3.4.1. The solution is written as

$$\Phi_t = \exp\left\{C\int_{t_0}^t K_u^N\,dA_u\right\}\left(1 + |x_0|^{2p} + \int_{t_0}^t \exp\left\{-C\int_{t_0}^u K_s^N\,dA_s\right\}d(M_u^N - B_u^N)\right). \quad (17)$$

Indeed, a direct computation yields that the above Φ_t satisfies equation (16). Consequently, we have

$$\exp\left\{-C\int_{t_0}^t K_u^N\,dA_u\right\}(1 + |\varphi_t^N|^{2p})$$
$$= 1 + |x_0|^{2p} + \int_{t_0}^t \exp\left\{-C\int_{t_0}^u K_s^N\,dA_s\right\}d(M_u^N - B_u^N). \quad (18)$$

The expectation of the last term is non-positive. Therefore we have

$$E\left[\exp\left\{-C\int_{t_0}^t K_u^N\,dA_u\right\}(1 + |\varphi_t^N|^{2p})\right] \le 1 + |x_0|^{2p}.$$

Then by Schwarz's inequality

$$E[(1 + |\varphi_t^N|^{2p})^{1/2}] \le (1 + |x_0|^{2p})^{1/2} E\left[\exp\left\{C\int_{t_0}^t K_u^N\,dA_u\right\}\right]^{1/2}. \quad (19)$$

Let N tend to infinity. Then we find that $E[|\varphi_{t \wedge \sigma_\infty}|^p]$ is finite by condition (13). This proves that $\sigma_\infty > t$ a.s. and φ_t has moments of any order.

Now suppose that condition (13) is not satisfied. Let τ_u be the inverse function of $\tilde{A}_t = \int_0^t K_u \, dA_u$ and consider the process $\tilde{F}(x, u) = F(x, \tau_u)$. Its local characteristic satisfies (11) and (12) with $K_t = 1$ and $A_t = t$. Therefore an equation governed by $\tilde{F}$ has a global solution $\tilde{\varphi}_u$ by the above argument. Then $\varphi_t \equiv \tilde{\varphi}_{\tilde{A}_t}$ is a global solution governed by $F(x, t)$. The proof is complete. □

Stratonovich's stochastic differential equation

Next we shall consider stochastic differential equations described in terms of Stratonovich integrals. As we will soon see, Stratonovich's stochastic differential equation can be rewritten as an Itô's equation. Hence most problems involving a Stratonovich's stochastic differential equation can be reduced to a problem involving an Itô's equation.

Let $\mathring{F}(x, t)$ be a continuous C-semimartingale with local characteristic belonging to the class $(B^{2,\delta}, B^{1,0})$ for some $\delta > 0$. A continuous $\mathbb{R}^d$-valued local semimartingale φ_t, $t \in [t_0, \sigma_\infty)$ is called a *local solution of Stratonovich's stochastic differential equation* based on $\mathring{F}(x, t)$ *starting at* x_0 *at time* t_0, if it satisfies

$$\varphi_{t \wedge \sigma_N} = x_0 + \int_{t_0}^{t \wedge \sigma_N} \mathring{F}(\varphi_{s \wedge \sigma_N}, \circ ds) \tag{20}$$

for any N where $\{\sigma_N\}$ is a sequence of stopping times such that $\sigma_N < \sigma_\infty$ and $\sigma_N \uparrow \sigma_\infty$. Also φ_t is said to be *governed by the Stratonovich's stochastic differential equation based on* $\mathring{F}(x, t)$. A maximal solution is defined similarly to the case of Itô's stochastic differential equation.

Theorem 3.4.7 *Let $\mathring{F}(x, t)$, $x \in \mathbb{R}^d$ be a continuous C^1-semimartingale with local characteristic (a, b, A_t) belonging to the class $(B^{2,\delta}, B^{1,0})$ for some $\delta > 0$. Then for each t_0 and x_0, the Stratonovich's equation* (20) *has a unique maximal solution. Further the solution satisfies Itô's equation based on $\mathring{F}(x, t) + C(x, t)$ where*

$$C(x, t) = \frac{1}{2} \int_0^t \left\{ \sum_{j=1}^d \frac{\partial a^{\cdot j}}{\partial x^j}(x, y, s) \bigg|_{y=x} \right\} dA_s. \tag{21}$$

Conversely let $F(x, t)$, $x \in \mathbb{R}^d$ be a continuous C^1-semimartingale with local characteristic (a, b, A_t) belonging to the class $(B^{2,\delta}, B^{1,0})$ for some $\delta > 0$. Then the solution of the Itô's equation based on F satisfies the Stratonovich's equation based on $F(x, t) - C(x, t)$.

Proof Consider Itô's stochastic differential equation based on the above $F \equiv \mathring{F} + C$. It has a unique maximal solution φ_t, $t \in [t_0, \sigma_\infty)$, as we have shown in Theorem 3.4.5. Since φ_t, $t \in [t_0, \sigma_\infty)$ is a continuous local semimartingale, the Stratonovich integral based on $\mathring{F}(x, \circ dt)$ is well defined. Using Theorem 3.2.5, it is computed as

$$\begin{aligned}\int_{t_0}^{t} \mathring{F}(\varphi_s, \circ ds) &= \int_{t_0}^{t} \mathring{F}(\varphi_s, ds) + \frac{1}{2}\sum_j \left\langle \int_{t_0}^{t} \frac{\partial \mathring{F}}{\partial x^j}(\varphi_s, ds), \varphi_t^j \right\rangle \\ &= \int_{t_0}^{t} \mathring{F}(\varphi_s, ds) + \frac{1}{2}\sum_j \left\langle \int_{t_0}^{t} \frac{\partial \mathring{F}}{\partial x^j}(\varphi_s, ds), \int_{t_0}^{t} \mathring{F}^j(\varphi_s, ds) \right\rangle \\ &= \int_{t_0}^{t} \mathring{F}(\varphi_s, ds) + \frac{1}{2}\int_{t_0}^{t} \left\{ \sum_{j=1}^{d} \frac{\partial a^{\cdot j}}{\partial x^j}(x, y, s)\bigg|_{x=y=\varphi_s} \right\} dA_s \\ &= \int_{t_0}^{t} F(\varphi_s, ds) = \varphi_t - x_0.\end{aligned}$$

Therefore the Stratonovich's stochastic differential equation has a maximal solution. The uniqueness of the solution is also reduced to the uniqueness of the corresponding Itô's equation. The latter assertion of the theorem can be shown similarly. □

The term $C(x, t)$ defined by (21) is often called the *correction term of the semimartingale* F or $\mathring{F}$.

Backward integrals and backward equations

Finally we shall introduce backward stochastic integrals and backward stochastic differential equations. The arguments are completely parallel to those of (forward) stochastic integral and stochastic differential equations. The only difference is that these are defined to the backward direction.

Let $\{\mathscr{F}_{s,t} : 0 \le s \le t \le T\}$ be a family of sub σ-fields of $\mathscr{F}$ which contain all null sets and satisfy $\mathscr{F}_{s,t} \subset \mathscr{F}_{s',t'}$ if $s' \le s \le t \le t'$, $\bigcap_{\varepsilon>0} \mathscr{F}_{s,t+\varepsilon} = \mathscr{F}_{s,t}$ and $\bigcap_{\varepsilon>0} \mathscr{F}_{s-\varepsilon,t} = \mathscr{F}_{s,t}$. It is called a *filtration* (*with two parameters*) *of sub σ-fields of* $\mathscr{F}$. A continuous process $\hat{M}_t$ is called a *backward martingale* adapted to $(\mathscr{F}_{s,t})$ if it is integrable, $\hat{M}_t - \hat{M}_s$ is $\mathscr{F}_{s,t}$-measurable and satisfies $E[\hat{M}_r - \hat{M}_t | \mathscr{F}_{s,t}] = \hat{M}_s - \hat{M}_t$ for any $r \le s \le t$. A *backward localmartingale* is defined similarly to the (forward) localmartingale. Let $\hat{X}_t$ be a continuous process such that $\hat{X}_t - \hat{X}_s$ is $(\mathscr{F}_{s,t})$-adapted. It is called a *backward semimartingale* if it can be written as the sum of a continuous backward localmartingale and a process of bounded variation.

Now let $F(\cdot, t)$ be a continuous backward semimartingale with values in C with local characteristic belonging to the class $B^{0,0}$. We will fix the time t for a moment. Let f_s, $s \in [0, t]$ be a continuous $(\mathscr{F}_{s,t})$-adapted process. Then the *backward Itô integral* of f_s based on $F(\cdot, s)$ is defined by

$$\int_s^t F(f_r, \hat{d}r) = \lim_{|\Delta|\to 0} \sum_{k=0}^{n-1} \{F(f_{t_{k+1}\vee s}, t_{k+1} \vee s) - F(f_{t_{k+1}\vee s}, t_k \vee s)\}, \quad (22)$$

where $\Delta = \{0 = t_0 < t_1 < \cdots < t_n = t\}$, $t \vee s = \max\{t, s\}$ if the right hand side converges in probability. It is a continuous backward semimartingale with respect to s. Suppose that $F(\cdot, t)$ is a continuous backward C-semimartingale with local characteristic belonging to the class $(B^{2,\delta}, B^{1,0})$ for some $\delta > 0$, and f_s is a continuous backward semimartingale. The *backward Stratonovich integral* is well defined:

$$\int_s^t F(f_r, \circ \hat{d}r) = \lim_{|\Delta|\to 0} \sum_{k=0}^{n-1} \frac{1}{2}\{F(f_{t_{k+1}\vee s}, t_{k+1} \vee s) + F(f_{t_k \vee s}, t_{k+1} \vee s)$$

$$- F(f_{t_{k+1}\vee s}, t_k \vee s) - F(f_{t_k\vee s}, t_k \vee s)\}, \quad (23)$$

since the right hand side converges in probability. These two integrals are related by

$$\int_s^t F(f_r, \circ \hat{d}r) = \int_s^t F(f_r, \hat{d}r) - \frac{1}{2}\sum_j \left\langle \int_s^t \frac{\partial F}{\partial x^j}(f_r, \hat{d}r), f_t^j \right\rangle, \quad (24)$$

where $\langle\ ,\ \rangle$ denotes the joint quadratic variation.

Now a continuous $(\mathscr{F}_{s,t_0})$-adapted process φ_s, $s \in [0, t_0]$ with values in $\mathbb{R}^d$ is called the solution of the *backward Itô's stochastic differential equation based on $F(x, t)$ starting at x_0 at time t_0* if it satisfies

$$\varphi_s = x_0 - \int_s^{t_0} F(\varphi_r, \hat{d}r). \quad (25)$$

The solution of the *backward Stratonovich's stochastic differential equation based on F starting at x_0 at time t_0* is defined similarly. The relation between the above two equations is parallel to the relation between the two corresponding forward equations stated in Theorem 3.4.7.

Exercise 3.4.8 Let $F(s)$ and $G(s)$ be $d \times d$ and $d \times m$ matrix-valued continuous functions, respectively, and B_s be a standard m-dimensional Brownian motion.

(i) Show that the solution of the equation

$$\varphi_t = x + \int_0^t F(s)\varphi_s\, ds + \int_0^t G(s)\, dB_s$$

exists uniquely and is written as

$$\varphi_t = \Phi(t, 0)x + \int_0^t \Phi(t, s)G(s)\, dB_s,$$

where $\Phi(t, s)$ is the fundamental solution of the linear deterministic differential equation $\mathrm{d}x/\mathrm{d}t = F(t)x$.

(ii) Show that its mean vector $m(t, x) = E[\varphi_t(x)]$ is equal to $\Phi(t, 0)x$ and that the covariance matrix $V(t, x, y) = E[(\varphi_t(x) - m(t, x))(\varphi_t(y) - m(t, y))^t]$ is independent of x, y and is given by

$$V(t, x, y) = \int_0^t \Phi(t, r)G(r)G(r)^t\Phi(t, r)^t \, \mathrm{d}r.$$

(iii) Suppose that the matrices F and G do not depend on t. Show that the matrix $V(t, x, y)$ is nonsingular for any $t > 0$ if and only if the rank of the $d \times md$ matrix $(G, FG, F^2G, \ldots, F^{d-1}G)$ is d.

Exercise 3.4.9 Suppose that φ_t is a time homogeneous continuous Gaussian Markov process such that the covariance of φ_t is nonsingular for any t.

(i) Show that there exists a continuous nonsingular matrix valued function $\Phi(t)$ with the semigroup property $\Phi(t + s) = \Phi(t)\Phi(s)$ such that $E[\varphi_t|\mathscr{F}_s] = \Phi(t - s)\varphi_s$ holds a.s. for any $t > s$, where $\{\mathscr{F}_t\}$ is the filtration generated by φ_t.

(ii) Show that $Y_t \equiv \Phi(t)^{-1}\varphi_t$ is a Brownian motion.

(iii) Set $F = (\mathrm{d}\Phi_t/\mathrm{d}t)|_{t=0}$ and $W_t = \int_0^t \Phi(s)\,\mathrm{d}Y_s$. Show that φ_t satisfies the stochastic differential equation $\mathrm{d}\varphi_t = F\varphi_t + \mathrm{d}W_t$.

Exercise 3.4.10 Let φ_t^0, $t \in [t_0, T]$ be a continuous $\mathbb{R}^d$-valued process adapted to $(\mathscr{F}_t)$ and let $F(x, t)$ be a continuous C-semimartingale with local characteristic belonging to $B_b^{0,1}$. Show that the following integral equation has a unique global solution φ_t:

$$\varphi_t = \varphi_t^0 + \int_{t_0}^t F(\varphi_s, \mathrm{d}s).$$

Further show that if φ_t^0 has moments of any order and condition (13) is satisfied, the solution has moments of any order. (*Hint*: apply the method of successive approximation.)

4

Stochastic flows

4.1 Preliminaries

This chapter is devoted to the study of stochastic flows, dividing it into nine sections. Since we discuss various problems, we shall first give the outline of contents of this chapter.

To begin with, we shall give the definition of the stochastic flow of homeomorphisms and of diffeomorphisms on the Euclidean space. Let $\varphi_{s,t}(x, \omega)$, $s, t \in [0, T]$, $x \in \mathbb{R}^d$ be a continuous $\mathbb{R}^d$-valued random field defined on the probability space $(\Omega, \mathscr{F}, P)$. Then for almost all ω, $\varphi_{s,t}(\omega) \equiv \varphi_{s,t}(\cdot, \omega)$ defines a continuous map from $\mathbb{R}^d$ into itself for any s, t. It is called a *stochastic flow of homeomorphisms* if there exists a null set N of Ω such that for any $\omega \in N^c$, the family of continuous maps $\{\varphi_{s,t}(\omega): s, t \in [0, T]\}$ defines a *flow of homeomorphisms* i.e. it satisfies the following properties:

(i) $\varphi_{s,u}(\omega) = \varphi_{t,u}(\omega) \circ \varphi_{s,t}(\omega)$ holds for all s, t, u, where $\circ$ denotes the composition of maps.
(ii) $\varphi_{s,s}(\omega) =$ identity map for all s.
(iii) the map $\varphi_{s,t}(\omega): \mathbb{R}^d \to \mathbb{R}^d$ is an onto homeomorphism for all s, t.

Further if $\varphi_{s,t}(\omega)$ satisfies (iv), it is called a *stochastic flow of C^k-diffeomorphisms.*

(iv) $\varphi_{s,t}(x, \omega)$ is k-times differentiable with respect to x for all s, t and the derivatives are continuous in (s, t, x).

Let $\varphi_{s,t}(\omega)^{-1}$ be the inverse map of $\varphi_{s,t}(\omega)$. Then (i) and (ii) imply $\varphi_{t,s}(\omega) = \varphi_{s,t}(\omega)^{-1}$. This fact and condition (iii) show that $\varphi_{s,t}(\omega)^{-1}(x)$ is also continuous in (s, t, x). Condition (iv) implies that $\varphi_{s,t}(\omega)^{-1}(x)$ is k-times continuously differentiable with respect to x. Hence $\varphi_{s,t}(\omega): \mathbb{R}^d \to \mathbb{R}^d$ is actually a C^k-diffeomorphism for all s, t if (iv) is satisfied. We can regard $\varphi_{s,t}(\omega)^{-1}(x)$ as a random field with parameter (s, t, x), we often denote it as $\varphi_{s,t}^{-1}(x, \omega)$. Therefore $\varphi_{s,t}^{-1}(x) = \varphi_{t,s}(x)$ holds for all s, t, x a.s.

Now let G be the set of all homeomorphisms of $\mathbb{R}^d$. Define the product of $\varphi, \psi \in G$ by the composite map $\varphi \circ \psi$. Then G is a group. We introduce a metric d on G by

$$d(\varphi, \psi) = \rho(\varphi, \psi) + \rho(\varphi^{-1}, \psi^{-1}), \tag{1}$$

where

$$\rho(\varphi, \psi) = \sum_{N=1}^{\infty} \frac{1}{2^N} \frac{\sup_{|x| \le N} |\varphi(x) - \psi(x)|}{1 + \sup_{|x| \le N} |\varphi(x) - \psi(x)|}. \tag{2}$$

Then G is a complete topological group. A stochastic flow of homeomorphisms can be considered as a continuous random field with values in G satisfying the flow properties (i) and (ii). We will call it a (continuous) *stochastic flow with values in G.*

Let G^k be the subset of G consisting of C^k-diffeomorphisms. It is a subgroup of G. Introducing the metric d_k:

$$d_k(\varphi, \psi) = \sum_{|\alpha| \le k} \rho(D^\alpha \varphi, D^\alpha \psi) + \sum_{|\alpha| \le k} \rho(D^\alpha \varphi^{-1}, D^\alpha \psi^{-1}), \tag{3}$$

G^k is a complete separable metric space. A stochastic flow of C^k-diffeomorphisms can be regarded as a continuous random field with values in G^k satisfying the flow property. It is called a (continuous) *stochastic flow with values in* G^k.

Let φ_t, $t \ge 0$ be a continuous process with values in G or G^k such that $\varphi_0 = e =$ identity a.s. Set $\varphi_{s,t} = \varphi_t \circ \varphi_s^{-1}$ where φ_s^{-1} is the inverse of φ_s. Then it is a stochastic flow with values in G or G^k. Conversely if $\varphi_{s,t}$ is a stochastic flow with values in G or G^k, then $\varphi_t \equiv \varphi_{0,t}$ is a continuous process with values in G or G^k such that $\varphi_0 = e$ a.s. It satisfies $\varphi_{s,t} = \varphi_t \circ \varphi_s^{-1}$ by the flow property (i). Therefore a continuous stochastic flow with values in G or G^k is equivalent to a continuous process φ_t with values in G or G^k such that $\varphi_0 = e$ a.e.

In the sequel we shall mainly consider a stochastic flow $\varphi_{s,t}$ with double parameters s and t. The first parameter s in $\varphi_{s,t}$ represents the initial time of the flow and $\varphi_{s,t}$ represents the state of the flow at time t.

For the analysis of a stochastic flow, it is convenient to divide the flow into the forward flow $\varphi_{s,t}$, $0 \le s \le t \le T$ and the backward flow $\varphi_{s,t}$, $0 \le t \le s \le T$ and discuss them separately. Here a continuous random field $\varphi_{s,t}$, $0 \le s \le t \le T$ with values in G or G^k satisfying (i) and (ii) is called a *forward stochastic flow. A backward stochastic flow* $\varphi_{s,t}$, $0 \le t \le s \le T$ is defined analogously. Given a forward stochastic flow $\varphi_{s,t}$, $0 \le s \le t \le T$ with values in G, there exists a unique flow $\tilde{\varphi}_{s,t}$, $s, t \in [0, T]$ such that its restriction to the forward time parameters $0 \le s \le t \le T$ coincides with the above $\varphi_{s,t}$. In fact its restriction to the backward time parameters $0 \le t \le s \le T$ is the inverse of $\varphi_{t,s}$, i.e. $\tilde{\varphi}_{s,t} = \varphi_{t,s}^{-1}$. Therefore when we consider the backward flow associated with the given forward flow $\varphi_{s,t}$, $0 \le s \le t \le T$, we denote it by $\varphi_{t,s}$, $0 \le s \le t \le T$ because of the property $\varphi_{t,s} = \varphi_{s,t}^{-1}$.

An important class of stochastic flows is that of Brownian flows. A continuous stochastic flow with values in G or G^k is called a *Brownian flow with values in* G or G^k if for any $0 \le t_0 < t_1 < \cdots < t_n \le T$, $\varphi_{t_i,t_{i+1}}$, $i = 0, \ldots, n-1$ are independent random variables. Hence a Brownian flow is a stochastic flow with independent increments with respect to the multiplicative operation in the group G or G^k. Further if the law of $\varphi_{s,t}$ and that of $\varphi_{s+h,t+h}$ coincide for any $h > 0$, it is called a *temporally homogeneous Brownian flow*. On the other hand, if the group G is replaced by a linear space $C = C(\mathbb{R}^d : \mathbb{R}^d)$ (or $C^k = C^k(\mathbb{R}^d : \mathbb{R}^d)$), a continuous process with values in C (or C^k) with independent increments with respect to the additivity of C (or C^k), is called a *Brownian motion with values in* C (or C^k). A Brownian motion with values in C (or C^k) has a Gaussian distribution, but a Brownian flow does not have a Gaussian distribution in general.

One of the main problems discussed in this chapter is the relation between stochastic flows and stochastic differential equations. For a given forward stochastic flow $\varphi_{s,t}$, $0 \le s \le t \le T$ with a suitable regularity condition, we can find a unique continuous C-valued semimartingale $F(x, t)$ with $F(x, 0) \equiv 0$ such that for every fixed s and x the process $\varphi_t \equiv \varphi_{s,t}(x)$ satisfies Itô's stochastic differential equation based on $F(x, t)$, i.e.

$$\varphi_t = x + \int_s^t F(\varphi_r, dr). \tag{4}$$

In particular if $\varphi_{s,t}$ is a Brownian flow, the associated $F(x, t)$ is a Brownian motion with values in C. The case of the Brownian flow will be discussed in Section 4.2 and the general case will be discussed in Section 4.4.

Conversely consider an Itô's stochastic differential equation based on a continuous C-valued semimartingale $F(x, t)$ with $F(x, 0) \equiv 0$ with local characteristic (a, b, A) satisfying the uniform Lipschitz condition (or belonging to the class $B_b^{0,1}$). We have seen in the previous chapter that equation (4) has a unique solution, which we denote by $\varphi_{s,t}(x)$, $s \le t \le T$. It will be shown in Section 4.5 that $\varphi_{s,t}(x)$ has a modification of a forward stochastic flow of homeomorphisms. Further, if the local characteristic of $F(x, t)$ belongs to the class $B_b^{k,\delta}$ for some $k \ge 1$ and $\delta > 0$, it will be shown in Section 4.6 that $\varphi_{s,t}(x)$ has a modification of a forward stochastic flow of C^k-diffeomorphisms.

Consequently there exists a one to one correspondence between forward stochastic flows and continuous semimartingales with values in C through Itô's stochastic differential equation (4). The latter will be called the *Itô's random infinitesimal generator* of the forward stochastic flow. We emphasize here that in this correspondence the classical Itô's stochastic differential equation stated in Section 3.4 is not sufficient: the correspondence requires stochastic differential equations based on infinitely many independent Brownian motions (or orthogonal martingales) or C-valued Brownian motions (or C-valued semimartingales) $F(x, t)$ mentioned above.

On the other hand, for the analysis of a backward stochastic flow associated with the forward flow $\varphi_{s,t}$, the backward stochastic integrals and the backward stochastic differential equations are needed. Indeed, there exists a one-to-one correspondence between the associated backward stochastic flow $\varphi_{t,s} \equiv \varphi_{s,t}^{-1}$, $0 \le s \le t \le T$ and a *backward Itô's stochastic differential equation based on* $\hat{F}(x, t)$: for every fixed t and x, $\varphi_{t,s}(x)$, $s \in [0, t]$ satisfies

$$\varphi_{t,s}(x) = x - \int_s^t \hat{F}(\varphi_{t,r}(x), \hat{\mathrm{d}}r), \tag{5}$$

where $\hat{F} = F - 2C$ and $C(x, t)$ is the *correction term* of $F(x, t)$ defined by

$$C(x, t) = \frac{1}{2}\int_0^t \left\{\sum_j \frac{\partial a^{\cdot j}}{\partial x^j}(x, y, r)|_{y=x}\right\} \mathrm{d}A_r. \tag{6}$$

The backward semimartingale $\hat{F}$ is called the *backward Itô's random infinitesimal generator*. If $\varphi_{s,t}$ is a deterministic flow differentiable with respect to s and t, the forward Itô's infinitesimal generator and the backward Itô's infinitesimal generator are identical. In order to obtain a similar result for a stochastic flow, we will have to consider the Stratonovich's random infinitesimal generators. Indeed, setting $\mathring{F} = F - C$, the forward flow is governed by a forward Stratonovich's equation based on $\mathring{F}$:

$$\varphi_{s,t}(x) = x + \int_s^t \mathring{F}(\varphi_{s,r}(x), \circ \mathrm{d}r), \tag{7}$$

and the backward flow is governed by a backward Stratonovich's equation based on $\mathring{F}$:

$$\varphi_{t,s}(x) = x - \int_s^t \mathring{F}(\varphi_{t,r}(x), \circ \hat{\mathrm{d}}r). \tag{8}$$

Hence $\mathring{F}$ is the *forward Stratonovich's random infinitesimal generator* and is the *backward Stratonovich's random infinitesimal generator* simultaneously (see Sections 4.2 and 4.4).

Section 4.3 is devoted to a certain asymptotic property of temporally homogeneous Brownian flow. Let $\varphi_t \equiv \varphi_{0,t}$, $t \in [0, \infty)$ be a temporally homogeneous forward Brownian flow. Then for each t and ω, the maps $\varphi_t(\omega)$ and $\varphi_t(\omega)^{-1} : \mathbb{R}^d \to \mathbb{R}^d$ transform a measure Π on $\mathbb{R}^d$ to the image measures $\varphi_t(\Pi)(\omega)$ and $\varphi_t^{-1}(\Pi)(\omega)$, respectively. The flow φ_t is called

Π-preserving if $\varphi_t(\Pi) = \Pi$ holds a.s. for any t. Assume that the measure Π is absolutely continuous with respect to the Lebesgue measure and its Radon–Nikodym derivative $\pi(x)$ is a positive smooth function. Then, similarly to the case of the flow determined by an ordinary differential equation, we show that the Brownian flow is Π-preserving if and only if $\text{div}(\pi\mathring{F})(x, t) \equiv 0$, where $\mathring{F}$ is the forward Stratonovich's infinitesimal generator of the Brownian flow. For non Π-preserving flow we discuss the asymptotic behavior of $\varphi_t(\Pi)$ and $\varphi_t^{-1}(\Pi)$ as $t \uparrow \infty$. It turns out that the asymptotic behavior is closely related to the ergodic property of the one point motion $\varphi_t(x)$ or $\varphi_t^{-1}(x)$.

If the local characteristic of a C-semimartingale $F(x, t)$ satisfies only a local Lipschitz condition (or belongs to the class $B^{0,1}$), equation (4) may not have a global solution, since the explosion can occur at finite time. In such a case the solution defines only a stochastic flow of *local homeomorphisms* or *local diffeomorphisms* (see Section 4.7). A criterion that it defines a stochastic flow of global homeomorphisms is that both the forward equation (4) and the backward equation (5) have global solutions for every initial (or terminal) condition a.s. It induces a clear criterion in the case of one dimensional stochastic flow.

Stochastic flows on manifolds are defined similarly to those on Euclidean space. Let M be a connected paracompact C^∞-manifold of dimension d. Let $\varphi_{s,t}(x)$, $s, t \in [0, T]$, $x \in M$ be a continuous random field with values in M. It is called a stochastic flow of homeomorphisms (or C^k-diffeomorphisms) if it satisfies conditions (i)–(iii) (or (i)–(iv)) stated at the beginning of this section, replacing $\mathbb{R}^d$ by M. Stochastic flows of local homeomorphisms or local diffeomorphisms are defined similarly to those in the case of Euclidian space.

In Section 4.8 we will consider a stochastic differential equation on a manifold. In order to define it invariantly under the change of local coordinates, it is more convenient to define it using the Stratonovich integral: let $F(t)$, $t \in [0, T]$ be a continuous semimartingale with values in the space of vector fields on M. The *Stratonovich's stochastic differential equation based on* $F(t)$ will be expressed as

$$f(\varphi_t) = f(x) + \int_s^t F(\circ\, dr) f(\varphi_r), \qquad \text{for all } f \in C^\infty(M). \tag{9}$$

With a local coordinate $(x^1, \ldots, x^d)$, $F(t)$ is given by

$$F(t)f(x) = \sum_i F^i(x, t) \frac{\partial f}{\partial x^i},$$

where $F = (F^1, \ldots, F^d)$ is an $\mathbb{R}^d$-valued continuous semimartingale. Then at each coordinate neighborhood, equation (9) is written as (7) using the above $F = (F^1, \ldots, F^d)$.

If the underlying manifold is compact, the solution always defines a stochastic flow of global diffeomorphisms. However if the manifold is non-compact, the solution only defines a stochastic flow of local diffeomorphisms. Although it is difficult to obtain a criterion that the solution defines a flow of global diffeomorphisms, we discuss the problem for Itô's classical stochastic differential equation.

A stochastic flow of (local) diffeomorphisms acts naturally on differential geometric objects such as vector fields, tensor fields etc. Let ψ be a diffeomorphism of M. For a vector field X, we define a new vector field $\psi_* X$ by $(\psi_* X)(f)(x) = X(f \circ \psi)(\psi^{-1}(x))$. Now let $\varphi_{s,t}$ be a stochastic flow. Then $(\varphi_{s,t})_* X$ is a continuous process with values in the space of vector fields. We study its differential rule with respect to time s and t in Section 4.9. The rule will be called Itô's formula for the flow $\varphi_{s,t}$ acting on the vector fields. A similar Itô's formula for $\varphi_{s,t}$ acting on tensor fields will also be discussed in the same section. Then several applications of these formulas will be discussed.

4.2 Brownian flows

n-point motions of Brownian flows

In this section we shall study the structure of Brownian flows. We shall first show the Markov property of a finite dimensional projection of a Brownian flow. Let $\varphi_{s,t}(x), 0 \le s \le t \le T$ be a forward Brownian flow of homeomorphisms. Let $\mathbf{x}^{(n)} = (x_1, \ldots, x_n)$ be n-point in $\mathbb{R}^d$ or an element of $\mathbb{R}^{nd}$. Set $\varphi_{s,t}(\mathbf{x}^{(n)}) = (\varphi_{s,t}(x_1), \ldots, \varphi_{s,t}(x_n))$. Then for each fixed s and $\mathbf{x}^{(n)}$, $\varphi_{s,t}(\mathbf{x}^{(n)})$, $t \in [s, T]$ is a stochastic process with values in $\mathbb{R}^{nd}$ starting at $\mathbf{x}^{(n)}$ at time s. It is called an *n-point motion* of the flow $\varphi_{s,t}$.

For $s < t$, let $\mathscr{F}_{s,t}$ be the least sub σ-field of $\mathscr{F}$ containing all null sets and $\bigcap_{\varepsilon > 0} \sigma(\varphi_{u,v} : s - \varepsilon \le u, v \le t + \varepsilon)$. Then $\{\mathscr{F}_{s,t} : 0 \le s \le t \le T\}$ is a filtration with two parameters. It is called the *filtration generated by the flow* $\varphi_{s,t}$.

Theorem 4.2.1 *The n-point motion has a Markov property with respect to* $\{\mathscr{F}_{s,t}\}$ *with transition probabilities*

$$P_{s,t}^{(n)}(\mathbf{x}^{(n)}, E) = P(\varphi_{s,t}(\mathbf{x}^{(n)}) \in E),$$

where the E's are Borel sets in $\mathbb{R}^{nd}$.

Proof Let $s < t < u$. Since $\varphi_{t,u}$ and $\mathscr{F}_{s,t}$ are independent and since $\varphi_{s,u} = \varphi_{t,u} \circ \varphi_{s,t}$, we have

$$E[f(\varphi_{s,u}(\mathbf{x}^{(n)}))|\mathscr{F}_{s,t}] = E[f(\varphi_{t,u}(\mathbf{y}^{(n)}))]_{\mathbf{y}^{(n)}=\varphi_{s,t}(\mathbf{x}^{(n)})}.$$

Therefore, the n-point motion is a Markov process with the transition probability $P_{s,t}^{(n)}(\mathbf{x}^{(n)}, \cdot)$. The proof is complete. □

Let $C_\infty^{(n)} = C_\infty(\mathbb{R}^{nd} : \mathbb{R})$ be the set of all bounded continous functions f on $\mathbb{R}^{nd}$ such that $\lim_{|x|\to\infty} f(x) = 0$. It is a Banach space with the supremum norm. For each $s \le t$ and $f \in C_\infty^{(n)}$ we define $T_{s,t}^{(n)}f$ by

$$T_{s,t}^{(n)}f(\mathbf{x}^{(n)}) = E[f(\varphi_{s,t}(\mathbf{x}^{(n)}))] = \int P_{s,t}^{(n)}(\mathbf{x}^{(n)}, d\mathbf{y}^{(n)})f(\mathbf{y}^{(n)}).$$

It is a bounded continuous function on $\mathbb{R}^{nd}$. Further since $|\varphi_{s,t}(\mathbf{x}^{(n)})| \to \infty$ holds a.s. as $|\mathbf{x}^{(n)}| \to \infty$, $E[f(\varphi_{s,t}(\mathbf{x}^{(n)}))]$ converges to 0 as $|\mathbf{x}^{(n)}| \to \infty$. Therefore $T_{s,t}^{(n)}$ defines a linear operator on $C_\infty^{(n)}$. The family of linear operators $\{T_{s,t}^{(n)} : 0 \le s < t \le T\}$ satisfies the semigroup property $T_{s,u}^{(n)}f = T_{t,u}^{(n)}T_{s,t}^{(n)}f$ for any $s \le t \le u$, because of the Markov property of the n-point motion.

The sequence of semigroups of linear operators $\{T_{s,t}^{(n)}\}$, $n = 1, 2, \ldots$ is consistent in the following sense. Let $n > m$ and $(x_{i_1}, \ldots, x_{i_m})$, $1 \le i_1 < i_2 < \cdots < i_m \le n$ be a subset of $(x_1, \ldots, x_n)$. Let f be an element of $C_\infty^{(n)}$ depending only on $(x_{i_1}, \ldots, x_{i_m})$. It may be regarded as a function on $\mathbb{R}^{md}$. Then $T_{s,t}^{(n)}f(\mathbf{x}^{(n)}) = T_{s,t}^{(m)}f(x_{i_1}, \ldots, x_{i_m})$ is satisfied. Now for each n, the law of the n-point motion is determined by the semigroup $\{T_{s,t}^{(n)}\}$. Then the sequence of semigroups $\{T_{s,t}^{(n)}\}$, $n = 1, 2, \ldots$ determines all finite dimensional distributions of the Brownian flow. Therefore $\{T_{s,t}^{(n)}\}$, $n = 1, 2, \ldots$ determines the law of the Brownian flow.

Infinitesimal means and covariances

In order to obtain the infinitesimal generators of the family of semigroups $\{T_{s,t}^{(n)}\}$, we shall introduce the infinitesimal mean and the infinitesimal covariance of the forward Brownian flow $\varphi_{s,t}$. We assume the following two conditions (A.1) and (A.2).

Condition (A.1) *The random variable $\varphi_{s,t}(x)$ is square integrable for each s, t, x. For any t, x, y the infinitesimal mean and the infinitesimal covariance exist and are given by:*

$$\lim_{h\to 0+} \frac{1}{h}\{E[\varphi_{t,t+h}(x)] - x\}, \tag{1}$$

the infinitesimal mean, and

$$\lim_{h\to 0+} \frac{1}{h} E[(\varphi_{t,t+h}(x) - x)(\varphi_{t,t+h}(y) - y)^t]. \tag{2}$$

the infinitesimal covariance.

Here $(\ldots)^t$ denotes the transpose of the vector. □

We denote by $b(x, t)$ the *infinitesimal mean* (1). It is a d-vector valued function $b(x, t) = (b^1(x, t), \ldots, b^d(x, t))$. The *infinitesimal covariance* (2) is denoted by $a(x, y, t)$. It is a $d \times d$-matrix valued function $(a^{ij}(x, y, t))$, symmetric and non-negative definite for each fixed t:

$$a^{ij}(x, y, t) = a^{ji}(y, x, t), \tag{3}$$

$$\sum_{p,q,i,j} a^{ij}(x_p, x_q, t)\xi_p^i \xi_q^j \geq 0 \tag{4}$$

holds for any x_p, $p = 1, \ldots, n$ of $\mathbb{R}^d$ and $\xi_p = (\xi_p^1, \ldots, \xi_p^d)$, $p = 1, \ldots, n$ of $\mathbb{R}^d$.

The second assumption, Condition (A.2), is rather technical. However it will be seen in Section 4.5 that most Brownian flows constructed by solving stochastic differential equations satisfy this condition.

Condition (A.2) *There exists a positive constant K such that*

$$|E[\varphi_{s,t}(x) - x]| \leq K(1 + |x|)|t - s|, \tag{5}$$

$$|E[(\varphi_{s,t}(x) - x)(\varphi_{s,t}(y) - y)^t]| \leq K(1 + |x|)(1 + |y|)|t - s| \tag{6}$$

holds for any s, t, x. y. □

By Condition (A.2), the infinitesimal mean and the infinitesimal covariance satisfy the linear growth property:

$$|b(x, t)| \leq K(1 + |x|), \tag{7}$$

$$\|a(x, y, t)\| \leq K(1 + |x|)(1 + |y|) \tag{8}$$

hold for any x, y, t.

We wish to show that the infinitesimal mean and the infinitesimal covariance determine the law of the Brownian flow. For this purpose we will show that the infinitesimal generator of the n-point diffusion process $(\varphi_{s,t}(x_1), \ldots, \varphi_{s,t}(x_n))$, $t \in [s, T]$ is completely characterized by the infinitesimal mean and the infinitesimal covariance. This will be done after proving two lemmas concerning the semimartingale property of the stochastic processes $\varphi_{s,t}(x)$, $t \in [s, T]$.

Lemma 4.2.2 *For each fixed s, x,*

$$M_{s,t}(x) \equiv \varphi_{s,t}(x) - x - \int_s^t b(\varphi_{s,r}(x), r)\, dr, \ t \geq s \tag{9}$$

is a square integrable martingale adapted to $(\mathscr{F}_{s,t})$.

Proof Set $m_{s,t}(x) = E[\varphi_{s,t}(x)]$. The flow property: $\varphi_{s,t+h} = \varphi_{t,t+h} \circ \varphi_{s,t}$ and the independence of $\varphi_{s,t}$ and $\varphi_{t,t+h}$ imply

$$m_{s,t+h}(x) = \int E[\varphi_{t,t+h}(y)]P(\varphi_{s,t}(x) \in \mathrm{d}y) = E[m_{t,t+h}(\varphi_{s,t}(x))].$$

Therefore we have

$$\frac{1}{h}\{m_{s,t+h}(x) - m_{s,t}(x)\} = E\left[\frac{1}{h}\{m_{t,t+h}(\varphi_{s,t}(x)) - \varphi_{s,t}(x)\}\right]. \tag{10}$$

From (A.2) we have the inequality

$$\left|\frac{1}{h}\{m_{t,t+h}(\varphi_{s,t}(x)) - \varphi_{s,t}(x)\}\right| \leq K(1 + |\varphi_{s,t}(x)|),$$

where the right hand side is integrable. Therefore, letting h tend to 0 at (10), we obtain

$$\frac{\partial m_{s,t}}{\partial t}(x) = E[b(\varphi_{s,t}(x), t)] \tag{11}$$

for any $t \geq s$. Integrating both sides with respect to t, we obtain

$$m_{s,t}(x) - x = \int_s^t E[b(\varphi_{s,r}(x), r)]\,\mathrm{d}r.$$

This proves $E[M_{s,t}(x)] = 0$ for any t, x.

Now $M_{s,t}(x)$ has the additive property:

$$M_{s,u}(x) = M_{s,t}(x) + M_{t,u}(\varphi_{s,t}(x)) \tag{12}$$

if $s < t < u$. Since $\mathscr{F}_{s,t}$ and $M_{t,u}$ are independent, we have

$$E[M_{s,u}(x)|\mathscr{F}_{s,t}] = M_{s,t}(x) + E[M_{t,u}(y)]_{y=\varphi_{s,t}(x)} = M_{s,t}(x),$$

proving that $M_{s,t}(x)$ is a martingale for each s, x. □

Lemma 4.2.3 *The joint quadratic variation of $M_{s,t}(x)$ is:*

$$\langle M^i_{s,t}(x), M^j_{s,t}(y)\rangle = \int_s^t a^{ij}(\varphi_{s,r}(x), \varphi_{s,r}(y), r)\,\mathrm{d}r. \tag{13}$$

Proof Define a matrix-valued function

$$V_{s,t}(x, y) \equiv E[M_{s,t}(x)M_{s,t}(y)']. \tag{14}$$

Since $M_{s,t}(x)$, $t \geq s$ is an L^2-martingale with the additive property (12), we have

$$
\begin{aligned}
V_{s,t+h}(x, y) &- V_{s,t}(x, y) \\
&= E[(M_{s,t+h}(x) - M_{s,t}(x))(M_{s,t+h}(y) - M_{s,t}(y))^t] \\
&= E[M_{t,t+h}(\varphi_{s,t}(x))M_{t,t+h}(\varphi_{s,t}(y))^t] \\
&= \int E[M_{t,t+h}(x')M_{t,t+h}(y')^t]P(\varphi_{s,t}(x) \in \mathrm{d}x', \varphi_{s,t}(y) \in \mathrm{d}y') \\
&= E[V_{t,t+h}(\varphi_{s,t}(x), \varphi_{s,t}(y))]. \qquad (15)
\end{aligned}
$$

On the other hand, a direct computation yields

$$
\begin{aligned}
V_{t,t+h}(x, y) = {} & E[(\varphi_{t,t+h}(x) - x)(\varphi_{t,t+h}(y) - y)^t] \\
& + E\left[(\varphi_{t,t+h}(x) - x)\left(\int_t^{t+h} b(\varphi_{t,r}(y), r)\,\mathrm{d}r\right)^t\right] \\
& + E\left[\left(\int_t^{t+h} b(\varphi_{t,r}(x), r)\,\mathrm{d}r\right)(\varphi_{t,t+h}(y) - y)^t\right] \\
& + E\left[\left(\int_t^{t+h} b(\varphi_{t,r}(x), r)\,\mathrm{d}r\right)\left(\int_t^{t+h} b(\varphi_{t,r}(y), r)\,\mathrm{d}r\right)^t\right]. \qquad (16)
\end{aligned}
$$

By Condition (A.2), the second, third and fourth terms on the right hand side of (16) are all bounded by $h^{3/2}K'(1 + |x|)(1 + |y|)$, where K' is a positive constant not depending on x, y, t, h. Therefore we have from (16) and Condition (A.1)

$$\lim_{h \to 0} \frac{1}{h} V_{t,t+h}(x, y) = a(x, y, t). \qquad (17)$$

Furthermore if $0 < h \leq 1$, we have

$$\left|\frac{1}{h} V_{t,t+h}(\varphi_{s,t}(x), \varphi_{s,t}(y))\right| \leq (K + 3K')(1 + |\varphi_{s,t}(x)|)(1 + |\varphi_{s,t}(y)|),$$

the right hand side being integrable. Divide both sides of (15) by h and let h tend to 0. Then we obtain

$$\frac{\partial V_{s,t}}{\partial t}(x, y) = E[a(\varphi_{s,t}(x), \varphi_{s,t}(y), t)]$$

for any t, x, y. Integrating the above with respect to t, we arrive at

$$V_{s,t}(x, y) = \int_s^t E[a(\varphi_{s,r}(x), \varphi_{s,r}(y), r)]\,\mathrm{d}r. \qquad (18)$$

We have thus shown that

$$N_{s,t}(x, y) \equiv M_{s,t}(x)M_{s,t}(y)^t - \int_s^t a(\varphi_{s,r}(x), \varphi_{s,r}(y), r)\,dr$$

is of mean 0 for any s, t, x, y.

Now the additivity of $M_{s,t}(x)$ yields for $s < t < u$

$$N_{s,u}(x, y) - N_{s,t}(x, y) = N_{t,u}(\varphi_{s,t}(x), \varphi_{s,t}(y)) + M_{s,t}(x)M_{t,u}(\varphi_{s,t}(y))^t + M_{t,u}(\varphi_{s,t}(x))M_{s,t}(y)^t.$$

We have

$$E[N_{t,u}(\varphi_{s,t}(x), \varphi_{s,t}(y))|\mathscr{F}_{s,t}] = E[N_{t,u}(x', y')]_{\substack{x'=\varphi_{s,t}(x)\\ y'=\varphi_{s,t}(y)}} = 0.$$

Conditional expectations of the second and the third terms with respect to $\mathscr{F}_{s,t}$ are also 0. Therefore $N_{s,t}(x, y)$ is a martingale for any s, x, y. Then $\int_s^t a(\varphi_{s,r}(x), \varphi_{s,r}(y), r)\,dr$ is the joint quadratic variation of $M_{s,t}(x)$ and $M_{s,t}(y)$ by Theorem 2.2.8. The proof is complete. □

Now we shall introduce an assumption to the infinitesimal mean and covariance. Let k be a non-negative integer and δ be a positive number less than or equal to 1.

Condition (A.3)$_{k,\delta}$ *The infinitesimal mean $b(x, t)$ belongs to the class $C_{ub}^{k,\delta}$ and the infinitesimal covariance $a(x, y, t)$ belongs to the class $\tilde{C}_{ub}^{k,\delta}$.* □

We shall define, for every positive integer n, second order linear differential operators over the space $\mathbb{R}^{nd}$ with parameter $t \in [0, T]$ by

$$L_t^{(n)}f(x_1, \ldots, x_n) \equiv \frac{1}{2}\sum_{p,q}\sum_{i,j} a^{ij}(x_p, x_q, t)\frac{\partial^2 f}{\partial x_p^i \partial x_q^j}(x_1, \ldots, x_n) + \sum_p \sum_i b^i(x_p, t)\frac{\partial f}{\partial x_p^i}(x_1, \ldots, x_n), \tag{19}$$

where $x_p = (x_p^1, \ldots, x_p^d)$, $p = 1, \ldots, n$. Then $L_t^{(n)} : t \in [0, T]$ is the infinitesimal generator of the n-point motion $(\varphi_{s,t}(x_1), \ldots, \varphi_{s,t}(x_n))$. Indeed, we have the following.

Theorem 4.2.4 *Let $\varphi_{s,t}$ be a forward Brownian flow satisfying Conditions* (A.1) *and* (A.2). *Let $T_{s,t}^{(n)}$ be the semigroup of linear operators of the n-point motion. Let f be a C^2-function on $\mathbb{R}^{nd}$ such that f and its derivatives are bounded. Then*

$$T_{s,t}^{(n)}f(x_1,\dots,x_n) - f(x_1,\dots,x_n) = \int_s^t T_{s,r}^{(n)}L_r^{(n)}f(x_1,\dots,x_n)\,dr \tag{20}$$

holds for any s, t and $x_1,\dots,x_n$. Furthermore, if Condition $(A.3)_{k,\delta}$ is satisfied for $k \ge 3$, $\delta > 0$ and f is a C^3-function, $T_{s,t}^{(n)}f$ is a C^2-function for each s, t and satisfies

$$T_{s,t}^{(n)}f(x_1,\dots,x_n) - f(x_1,\dots,x_n) = \int_s^t L_r^{(n)}T_{r,t}^{(n)}f(x_1,\dots,x_n)\,dr. \tag{21}$$

Proof Here we shall only prove (20). The proof of (21) will be given in Section 4.8 (see Theorem 4.8.11). We shall apply Itô's formula. Note that for each s, x, $\varphi_{s,t}(x)$, $t \in [s, T]$ is a continuous semimartingale with the martingale part $M(x, t) \equiv M_{s,t}(x)$ defined in Lemma 4.2.2. Then Theorem 2.3.11 tells us

$$\begin{aligned} &f(\varphi_{s,t}(x_1),\dots,\varphi_{s,t}(x_n)) - f(x_1,\dots,x_n) \\ &\quad = \sum_{i,p}\int_s^t \frac{\partial f}{\partial x_p^i} M^i(x_p, dr) + \sum_{i,p}\int_s^t \frac{\partial f}{\partial x_p^i} b^i(\varphi_{s,r}(x_p), r)\,dr \\ &\qquad + \frac{1}{2}\sum_{i,j,p,q}\int_s^t \frac{\partial^2 f}{\partial x_p^i \partial x_q^j} d\langle M^i(x_p, r), M^j(x_q, r)\rangle. \end{aligned} \tag{22}$$

Apply Lemma 4.2.3. Then the above is written as

$$\sum_{i,p}\int_s^t \frac{\partial f}{\partial x_p^i} M^i(x_p, dr) + \int_s^t L_r^{(n)}f(\varphi_{s,r}(x_1),\dots,\varphi_{s,r}(x_n))\,dr.$$

The first term is a localmartingale with the quadratic variation

$$\sum_{i,j,p,q}\int_s^t \frac{\partial f}{\partial x_p^i}\frac{\partial f}{\partial x_q^j} a^{ij}(\varphi_{s,r}(x_p), \varphi_{s,r}(x_q), r)\,dr.$$

Its expectation is finite. Therefore the localmartingale is actually a martingale with mean 0 by Corollary 2.2.7. Take the expectation of each term of (22). Then we obtain (20). □

The law of the forward Brownian flow is determined by the infinitesimal mean and covariance if these are bounded continuous functions having two bounded continuous spatial derivatives. Indeed, by Oleinik's theorem (see Stroock–Varadhan [118]), the semigroup $\{T_{s,t}^{(n)}\}$ is uniquely determined by the operators $\{L_t^{(n)}\}$. Therefore the law of the n-point motion is unique for each n. This proves the uniqueness of the law of the Brownian flow. However, as we will see later, for the uniqueness of the semigroup it is sufficient that the infinitesimal mean and covariance are continuous and uniformly Lipschitz continuous with respect to the spatial variables (see Theorem 4.2.5).

Conversely suppose that we are given a pair $(a(x, y, t), b(x, t))$ consisting of a symmetric non-negative definite matrix valued function $a(x, y, t)$ and a vector valued function $b(x, t)$ with parameter t. Define the operators $L_t^{(n)}$ by (19). If a and b are bounded continuous functions having two bounded continuous spatial derivatives, then there exists a unique Markovian semigroup $\{T_{s,t}^{(n)}\}$ satisfying (20) by Oleinik's theorem. The family of semigroups $\{\{T_{s,t}^{(n)}\} : n = 1, 2, \dots\}$ is consistent. Indeed, let $n > m$ and $(x_{i_1}, \dots, x_{i_m})$, $1 \le i_1 < \cdots < i_m \le n$ be a subset of $(x_1, \dots, x_n)$. Let f be a C^2-function on $\mathbb{R}^{nd}$ depending on $(x_{i_1}, \dots, x_{i_m})$ only. Then $T_{s,t}^{(n)}f = T_{s,t}^{(m)}f$ is satisfied since $L_t^{(n)}f = L_t^{(m)}f$ is satisfied.

A further problem is whether there exists a forward Brownian flow whose n-point motion has the above semigroup $\{T_{s,t}^{(n)}\}$. Since the infinitesimal generator $\{L_t^{(n)}\}$ is defined by (19), it is equivalent to the problem of whether there exists a Brownian flow whose infinitesimal covariance and mean coincide with the given pair (a, b). The answer is yes. Indeed, we have the assertion under a weaker assumption on the pair (a, b).

Theorem 4.2.5 *Let $a(x, y, t)$ be a continuous $d \times d$ matrix valued function satisfying* (3) *and* (4), *and let $b(x, t)$ be a continuous d-vector valued function. Suppose that $a(x, y, t)$ belongs to the class $\tilde{C}_{ub}^{0,1}$ (uniformly Lipschitz continuous) and $b(x, t)$, to the class $C_{ub}^{0,1}$ (uniformly Lipschitz continuous). Then there exists a forward Brownian flow of homeomorphisms such that its infinitesimal covariance and infinitesimal mean coincide with the pair $(a(x, y, t), b(x, t))$. The law of such a Brownian flow is unique and it satisfies Condition* (A.2). □

The theorem will be proved by solving a stochastic differential equation based on a C-valued Brownian motion $F(x, t)$ with mean vector $\int b(x, r)\, dr$ and the covariance matrix $\int a(x, y, r)\, dr$. Since our proof of the existence of the associated flow is very long, it will be postponed until Section 4.5. The uniqueness of the flow will be shown in Theorem 4.2.9 in a more refined form. Further, in Section 4.6 we show that if the pair (a, b) satisfies Condition $(A.3)_{k,\delta}$, the corresponding Brownian flow takes values in G^k (see Theorem 4.6.5).

Now suppose that we are given second order linear elliptic differential operators $\{L_t\}$ over $\mathbb{R}^d$ with parameter $t \in [0, T]$:

$$L_t f = \frac{1}{2} \sum_{i,j} a^{ij}(x, t) \frac{\partial^2 f}{\partial x^i \partial x^j} + \sum_i b^i(x, t) \frac{\partial f}{\partial x^i}. \tag{23}$$

An interesting problem is whether there exists a Brownian flow whose one point motion has the operator (23) as its infinitesimal generator. The problem is clearly reduced to that of finding a matrix valued function $a(x, y, t)$ which is symmetric, non-negative definite and Lipschitz continuous, such that its diagonal value $a(x, x, t)$ coincides with the coefficients $(a^{ij}(x, t))$ of the operator (23). We can find such an $a(x, y, t)$ if $a(x, t)$ has the Lipschitz continuous square root. Indeed, let $\sigma(x, t)$ be a $d \times r$-matrix valued function, uniformly Lipschitz continuous, satisfying $\sigma(x, t)\sigma(x, t)^t = a(x, t)$ where σ^t denotes the transpose of the matrix σ. Set

$$a(x, y, t) = \sigma(x, t)\sigma(y, t)^t. \tag{24}$$

Then it is symmetric and non-negative definite and belongs to the class $\tilde{C}^{0,1}$.

The existence of the symmetric square root of $a(x, t) = (a^{ij}(x, t))$ is known. We quote it from Stroock–Varadhan [118] without giving the proof. See also Ikeda–Watanabe [49].

Theorem 4.2.6 *Let $a = (a^{ij}(x, t))$ be a symmetric, non-negative definite matrix valued function and let $a^{1/2}(x, t)$ be the symmetric square root.*

(i) *Suppose there exists $c > 0$ such that*

$$\sum_{i,j} a^{ij}(x, t)\xi^i\xi^j \geq c|\xi|^2 \tag{25}$$

holds for any $\xi = (\xi^1, \ldots, \xi^d) \in \mathbf{R}^d$. Then if a is uniformly Lipschitz continuous, $a^{1/2}(x, t)$ is also uniformly Lipschitz continuous.

(ii) *Suppose $a(x, t)$ is twice continuously differentiable with respect to x and the derivatives $\partial^2 a/\partial x_i^2$, $i = 1, \ldots, d$ are bounded. Then $a^{1/2}(x, t)$ is again uniformly Lipschitz continuous.* □

The following corollary is immediate from Theorems 4.2.5 and 4.2.6.

Corollary 4.2.7 *Suppose that the coefficients of operator* (23) *are uniformly Lipschitz continuous and satisfy either property* (i) *or* (ii) *of Theorem* 4.2.6. *Then there exists a forward Brownian flow of homeomorphisms such that the infinitesimal generator of its one point motion coincides with the given operator* (23). □

Remark The law of a Brownian flow is determined by 2-point motion since its infinitesimal generator $L_t^{(2)}$ induces the infinitesimal mean and the infinitesimal covariance of the Brownian flow. However, there are many different Brownian flows having the same one point motion. In fact there are many square roots of the matrix $a(x, t) \equiv a(x, x, t)$. Each different square root $\sigma(x, t)$ defines a different matrix $a(x, y, t) = \sigma(x, t)\sigma(y, t)^t$. Then the forward Brownian flows with the infinitesimal covariances $a(x, y, t)$ are different from each other.

There exist still other types of Brownian flows which can not be obtained by the above method. As an example we shall consider Brownian flows on $\mathbb{R}$ following Harris [44]. Let $c(x)$, $x \in \mathbb{R}$ be any real non-negative definite function, twice continuously differentiable, such that $c(0) = 1$ and $c(x) = c(-x)$, for all x. Set $a(x, y) = c(x - y)$. Then it is symmetric and non-negative definite and satisfies the Lipschitz condition. Then by Theorem 4.2.5 there is a unique Brownian flow of homeomorphisms such that its infinitesimal mean is 0 and infinitesimal covariance is a. The infinitesimal generator of the one point motion is $L^1 = 2^{-1}\, \mathrm{d}^2/\mathrm{d}x^2$ since $a(x, x) = 1$. It is a standard Brownian motion. However, the laws of a Brownian flow corresponding to the different functions c are different from each other.

Forward random infinitesimal generators

We shall next consider another infinitesimal characteristic of a forward Brownian flow called the random infinitesimal generator. In order to get an idea of this, we shall first consider a special case: a Brownian flow whose infinitesimal covariance is 0. Then $M_{s,t}(x)$ of Lemma 4.2.2 is identically 0. Therefore $\varphi_{s,t}(x)$ is a deterministic flow and satisfies

$$\varphi_{s,t}(x) - x = \int_s^t b(\varphi_{s,r}(x), r)\, \mathrm{d}r.$$

Hence, $\varphi_{s,t}(x)$ is the solution of the ordinary differential equation $\mathrm{d}x/\mathrm{d}t = b(x, t)$ starting at x at time s. The vector function $b(x, t)$ is called the *infinitesimal generator* of the flow $\varphi_{s,t}$.

When we consider a Brownian flow whose infinitesimal covariance is not zero, the infinitesimal generator is not a deterministic function, but is a Brownian motion with values in $C = C(\mathbb{R}^d : \mathbb{R}^d)$. An ordinary differential equation is then replaced by a stochastic differential equation.

Theorem 4.2.8 *Let $\varphi_{s,t}$, $0 \le s \le t \le T$ be a forward Brownian flow with values in G^k satisfying Conditions* (A.1), (A.2) *and* $(\mathrm{A.3})_{k,\delta}$ *for some $k \ge 0$ and $\delta > 0$. Then there exists a unique $C^{k,\varepsilon}$-valued Brownian motion $F(x, t)$ with $F(x, 0) \equiv 0$ where $\varepsilon < \delta$ such that the flow is governed by Itô's stochastic differential equation based on $F(x, t)$:*

$$\varphi_{s,t}(x) = x + \int_s^t F(\varphi_{s,r}, \mathrm{d}r). \tag{26}$$

Further, the mean and the covariance of $F(x, t)$ coincide with $\int_0^t b(x, r)\,dr$ and $\int_0^t a(x, y, r)\,dr$, respectively, where b and a are the infinitesimal mean and covariance of the flow, respectively.

Proof Set

$$X_s(x, t) \equiv \varphi_{s,t}(x) - x. \tag{27}$$

For each s, it is a continuous C-semimartingale with local characteristic $(a(\varphi_{s,t}(x), \varphi_{s,t}(y), t), b(\varphi_{s,t}(x), t))$ by Lemmas 4.2.2 and 4.2.3. Then the stochastic integral

$$F_s(x, t) \equiv \int_s^t X_s(\varphi_{s,r}^{-1}(x), dr) \tag{28}$$

is well defined for each s and x. For each s, it is a family of continuous semimartingales with local characteristic $(a(x, y, t), b(x, t))$. Since the latter belongs to the class $B^{k,\delta}$, $F_s(x, t)$ has a modification of a continuous $C^{k,\varepsilon}$-semimartingale for each s by Theorem 3.1.2. We will show the relation:

$$F_s(x, u) = F_s(x, t) + F_t(x, u), \qquad s < t < u. \tag{29}$$

Let $\Delta = \{s = t_0 < \cdots < t_n = u\}$ be partitions of $[s, u]$ such that $t = t_m \in \Delta$. Then by formula (9) in Section 3.2,

$$\sum_{h=0}^{n-1} \{X_s(\varphi_{s,t_h}^{-1}(x), t_{h+1}) - X_s(\varphi_{s,t_h}^{-1}(x), t_h)\}$$

converges to $F_s(x, u)$ in probability as $|\Delta| \to 0$ for each x. Since $X_s(x, t)$ has the additive property $X_s(x, u) = X_s(x, t) + X_t(\varphi_{s,t}(x), u)$, the above is written as $\sum_{h=0}^{n-1} X_{t_h}(x, t_{h+1})$. It is divided into two:

$$\sum_{h=0}^{m-1} X_{t_h}(x, t_{h+1}) + \sum_{h=m}^{n-1} X_{t_h}(x, t_{h+1}).$$

Let $|\Delta| \to 0$. The first and second terms converge in probability to $F_s(x, t)$ and $F_t(x, u)$ respectively.

Now let us denote $F_0(x, t)$ by $F(x, t)$. Then $F_s(x, t) = F(x, t) - F(x, s)$ is satisfied. Hence the stochastic integral based on F_s coincides with that based on F. Theorem 3.3.3 (i) then gives us the equality:

$$X_s(x, t) = \int_s^t F(\varphi_{s,r}(x), dr).$$

Therefore $\varphi_{s,t}$ is governed by Itô's stochastic differential equation based on $F(x, t)$. Note further that $X_{t_h}(x, t_{h+1})$, $h = 0, \ldots, n - 1$ are independent. Then $F(x, t)$ has independent increments because of property (29). Therefore F is a $C^{k,\varepsilon}$-valued Brownian motion. Its mean and covariance are computed from the local characteristic $(a(x, y, t), b(x, t))$ immediately.

It remains to prove the uniqueness. Let $\tilde{F}(x, t)$ be another $C^{k,\varepsilon}$-valued Brownian motion such that the flow is governed by Itô's equation based on $\tilde{F}(x, t)$. Then, since $X_s(x, t) = \int_s^t \tilde{F}(\varphi_{s,r}(x), dr)$ holds, we have by Theorem 3.3.3 (i)

$$\tilde{F}(t) - \tilde{F}(s) = \int_s^t X_s(\varphi_{s,r}^{-1}(x), dr) = F(t) - F(s).$$

The proof is complete. □

The $C^{k,\varepsilon}$-valued Brownian motion $F(x, t)$ of the above theorem is called the *forward Itô's random infinitesimal generator* (or *forward random infinitesimal generator in the sense of the Itô integral*) of the Brownian flow $\varphi_{s,t}$.

As an application of the above theorem, we show that the law of the Brownian flow is determined by its infinitesimal mean $b(x, t)$ and the infinitesimal covariance $a(x, y, t)$ in the case where the pair (a, b) is uniformly Lipschitz continuous.

Theorem 4.2.9 *Let $\varphi_{s,t}$, $0 \le s \le t \le T$ be a forward Brownian flow satisfying Conditions* (A.1), (A.2) *and* $(A.3)_{0,1}$. *Then the law of the forward Brownian flow is uniquely determined by the pair (a, b). Further the joint law of the random infinitesimal generator F and the Brownian flow is also uniquely determined by the pair (a, b).*

Proof Consider Itô's stochastic differential equation based on the Brownian motion $F(x, t)$. The local characteristic of F belongs to the class $B_{ub}^{0,1}$. Then by Lemma 3.4.2, for each s and x, $\varphi_{s,t}(x)$ is approximated in $L^p(\Omega, \mathscr{F}, P)$ (for all $p > 1$) by a sequence of processes $\{\varphi_t^n\}$ defined successively by

$$\varphi_t^n = x + \int_0^t F(\varphi_r^{n-1}, dr), \qquad n = 1, 2, \ldots$$

where $\varphi_t^0 \equiv x$. Since the law of the Brownian motion $F(x, t)$ is determined by the pair (a, b), the joint law of the m_0-point motion $(F(y_1, t), \ldots, F(y_{m_0}, t))$ and n_0-point motion $(\varphi_{s,t}^n(x_1), \ldots, \varphi_{s,t}^n(x_{n_0}))$ is determined by the pair (a, b) for every n. Consequently the joint law of $(F(y_1, t), \ldots, F(y_{m_0}, t))$ and $(\varphi_{s,t}(x_1), \ldots, \varphi_{s,t}(x_{n_0}))$ is uniquely determined by the pair (a, b). The proof is complete. □

Backward random infinitesimal generators

Now let us consider a backward Brownian flow $\varphi_{t,s}$, $0 \le s \le t \le T$ associated with the given forward Brownian flow $\varphi_{s,t}$, $0 \le s \le t \le T$. We wish to obtain the random infinitesimal generator of the backward flow and find the relation between the forward and the backward random infinitesimal generators.

Theorem 4.2.10 *Let* $\varphi_{s,t}$, $0 \le s \le t \in T$ *be a forward Brownian flow with values in* G^k *satisfying* (A.1), (A.2) *and* $(A.3)_{k,\delta}$ *where* $k \ge 2$ *and* $\delta > 0$. *Let* $F(x, t)$ *be its forward Itô's random infinitesimal generator. Define the* C^{k-1}*-valued Brownian motion* $\hat{F}(x, t)$ *by* $F(x, t) - 2C(x, t)$ *where* C *is the correction term of* F *defined by* $C(x, t) = \int_0^t c(x, r)\, dr$ *where*

$$c(x, r) = \frac{1}{2}\left\{\sum_j \frac{\partial a^{\cdot j}}{\partial x^j}(x, y, r)|_{y=x}\right\}. \tag{30}$$

Then the associated backward flow $\varphi_{t,s} \equiv \varphi_{s,t}^{-1}$, $0 \le s \le t \le T$ *is governed by a backward Itô's stochastic differential equation based on* $\hat{F}$:

$$\varphi_{t,s}(y) = y - \int_s^t \hat{F}(\varphi_{t,r}(y), \hat{d}r). \tag{31}$$

Proof It is sufficient to prove the equality:

$$\int_s^t F(\varphi_{s,r}(x), dr)|_{x=\varphi_{t,s}(y)} = \int_s^t \hat{F}(\varphi_{t,r}(y), \hat{d}r). \tag{32}$$

In fact, substitute $x = \varphi_{t,s}(y)$ in equality (26) and apply the above. Then we obtain (31) immediately. Let $\Delta = \{s = t_0 < t_1 < \cdots < t_n = t\}$ be partitions of $[s, t]$. Then we have

$$\int_s^t F(\varphi_{s,r}(x), dr)|_{x=\varphi_{t,s}(y)}$$

$$= \lim_{|\Delta|\to 0} \sum_{h=0}^{n-1} \{F(\varphi_{s,t_h}(x), t_{h+1}) - F(\varphi_{s,t_h}(x), t_h)\}|_{x=\varphi_{t,s}(y)}$$

(convergence in probability; see Exercise 3.3.6). The latter is equal to

$$\lim_{|\Delta|\to 0} \sum_{h=0}^{n-1} \{F(\varphi_{t,t_h}(y), t_{h+1}) - F(\varphi_{t,t_h}(y), t_h)\}$$

by the flow property $\varphi_{s,t_h}(\varphi_{t,s}(y)) = \varphi_{t,t_h}(y)$. The above finite sum is written as

$$\sum_h \{F(\varphi_{t,t_{h+1}}(y), t_{h+1}) - F(\varphi_{t,t_{h+1}}(y), t_h)\} - \sum_h (\{F(\varphi_{t,t_{h+1}}(y), t_{h+1})$$

$$- F(\varphi_{t,t_h}(y), t_{h+1})\} - \{F(\varphi_{t,t_{h+1}}(y), t_h) - F(\varphi_{t,t_h}(y), t_h)\}).$$

The first term of the above converges to $\int_s^t F(\varphi_{t,r}(y), \hat{d}r)$ in probability as $|\Delta| \to 0$. The second term is written by the mean value theorem as

$$-\sum_j \sum_h \left(\frac{\partial F}{\partial x^j}(\varphi_{t,t_h}(y), t_{h+1}) - \frac{\partial F}{\partial x^j}(\varphi_{t,t_h}(y), t_h) \right) (\varphi^j_{t,t_{h+1}}(y) - \varphi^j_{t,t_h}(y))$$

$$-\frac{1}{2} \sum_{i,j,h} \xi^\Delta_{ijh} (\varphi^i_{t,t_{h+1}}(y) - \varphi^i_{t,t_h}(y))(\varphi^j_{t,t_{h+1}}(y) - \varphi^j_{t,t_h}(y)), \tag{33}$$

where ξ^Δ_{ijh} are random variables such that $\sup_h |\xi^\Delta_{ijh}| \to 0$ in probability as $|\Delta| \to 0$. Denote the first term of (33) by I^Δ. Then a simple computation yields

$$I^\Delta|_{y=\varphi_{s,t}(x)} = -\sum_j \sum_h \left(\frac{\partial F}{\partial x^j}(\varphi_{s,t_h}(x), t_{h+1}) - \frac{\partial F}{\partial x^j}(\varphi_{s,t_h}(x), t_h) \right) \times (\varphi^j_{s,t_{h+1}}(x) - \varphi^j_{s,t_h}(x)).$$

Then similarly to the proof of Theorem 3.2.5 it converges in probability to

$$-\sum_j \left\langle \int_s^t \frac{\partial F}{\partial x^j}(\varphi_{s,r}(x), dr), \int_s^t F^j(\varphi_{s,r}(x), dr) \right\rangle = -2 \int_s^t c(\varphi_{s,r}(x), r)\, dr.$$

Therefore we have

$$\lim_{|\Delta| \to 0} I^\Delta = -2 \int_s^t c(\varphi_{s,r}(x), r)\, dr|_{x=\varphi_{t,s}(y)} = -2 \int_s^t c(\varphi_{t,r}(y), r)\, dr.$$

The second term of (33) converges to 0, obviously. We have thus proved (32). The proof is complete. □

The function $c(x, r)$ of (30) is called the *correction term based on* $a = (a^{ij}(x, y, r))$.

The C^{k-1}-valued Brownian motion $\hat{F}$ of the above theorem is called the *backward Itô's random infinitesimal generator* (or *backward random infinitesimal generator in the sense of the Itô integral*) of the Brownian flow.

The above theorem indicates that the backward random infinitesimal generator requires the correction term $2C$. However if we rewrite both stochastic differential equations making use of Stratonovich's integrals, then the correction term disappears. Indeed, assume that the forward part of the Brownian flow $\varphi_{s,t}$ satisfies Conditions (A.1), (A.2) and $(A.3)_{k,\delta}$ with $k \geq 2$ and $\delta > 0$. Set $\mathring{F} = F - C$. Then it is the *forward Stratonovich's random infinitesimal generator* (or *forward random infinitesimal generator in the sense of the Stratonovich integral*), i.e. $\varphi_{s,t}$, $0 \leq s \leq t \leq T$ is governed by the forward Stratonovich's stochastic differential equation based on $\mathring{F}$. Further $\mathring{F}$ is the *backward Stratonovich's random infinitesimal generator* (or *backward random infinitesimal generator in the sense of the Stratonovich integral*), i.e. $\varphi_{s,t}$, $0 \leq t \leq s \leq T$ is governed by the backward Stratonovich's stochastic differential equaton based on $\mathring{F}$. We have thus shown the following corollary.

Corollary 4.2.11 *Let $\varphi_{s,t}$, $s, t \in [0, T]$ be a Brownian flow such that its forward part satisfies Conditions* (A.1), (A.2) *and* $(A.3)_{k,\delta}$ *for some* $k \geq 2$ *and* $\delta > 0$. *Then there exists a unique C^{k-1}-valued Brownian motion $\mathring{F}(x, t)$ with $F(x, 0) \equiv 0$ such that it is a forward and backward Stratonovich's random infinitesimal generator of the Brownian flow $\varphi_{s,t}$.* □

Remark

(i) In Theorem 4.2.10 and Corollary 4.2.11, Condition $(A.3)_{k,\delta}$ can be relaxed. It is sufficient to assume $(A.3)_{k,\delta}$, $k \geq 2, \delta > 0$ for $a(x, y, t)$ and $(A.3)_{k,\delta}$, $k \geq 1, \delta > 0$ for $b(x, t)$. Indeed, in the proof of Theorem 4.2.10, we have only to prove equality (32) for $N \equiv F - \int b(\cdot, r)\, dr$ instead of F, since the equality is clear for $\int b(\cdot, r)\, dr$. Since $N(x, t)$ is a $C^{2,\varepsilon}$-localmartingale for any $\varepsilon < \delta$ by Theorem 3.1.2, equality (32) can be shown for N in exactly the same way.

(ii) It is worth mentioning that the infinitesimal means of the forward part and the backward part are different from each other in general. The forward infinitesimal mean is $b(x, t)$, while the backward infinitesimal mean is $b(x, t) - 2c(x, t)$. However the forward and the backward infinitesimal covariances are identical.

Exercise 4.2.12 Let $a(x, y, t) = (a^{ij}(x, y, t))$, $x, y \in \mathbb{R}^d$, $t \in [0, T]$ be a continuous $d \times d$-matrix valued function, non-negative definite and symmetric. Let $b(x, t)$ be a continuous d-vector function. Suppose that a belongs to the class $\tilde{C}^{0,\delta}$ and b belongs to the class $C^{0,\delta}$ for some $\delta > 0$. Show that there exists a C-valued Brownian motion $F(x, t)$ such that

$$E[F(x, t)] = \int_0^t b(x, r)\, dr, \quad \operatorname{Cov}(F(x, t), F(y, s)) = \int_0^{t \wedge s} a(x, y, r)\, dr.$$

(*Hint*: show first the existence of a Gaussian random field with the above mean and covariance. Then apply Kolmogorov's theorem.)

Exercise 4.2.13 Let $f(x)$ be a Lipschitz continuous function and let $\xi(x, t)$ be the flow generated by $dx/dt = f(x)$. Let B_t be a one dimensional Brownian motion. Show that $\varphi_{s,t}(x) \equiv \xi(x, B_t - B_s)$ is a Brownian flow whose infinitesimal covariance and mean are $a(x, y) = f(x)f(y)$ and $b(x) = \frac{1}{2}f(x)f'(x)$ respectively. Show that the Stratonovich's random infinitesimal generator of φ_t is $f(x)B_t$. (*Hint*: apply Itô's formula to $\xi(x, B_t - B_s)$.)

Exercise 4.2.14 (*cf.* Exercise 3.4.9) Let $\varphi_{s,t}(x)$ be a temporally homogeneous Brownian flow and a Gaussian random field on $\mathbb{R}^d$.

(i) Show that its infinitesimal covariance and mean are given by $a(x, y) = V$ and $b(x) = Fx$ respectively, where V is a non-negative definite symmetric $d \times d$-matrix and F is a $d \times d$-matrix.
(ii) Show that the random infinitesimal generator is of the form $(Fx)t + W_t$, where W_t is a Brownian motion with mean 0 and covariance tV.

Exercise 4.2.15 A temporally homogeneous Brownian flow $\varphi_{s,t}$ is called *(spatially) homogeneous* if its infinitesimal mean is 0 and its infinitesimal covariance satisfies $a(x, y) = c(x - y)$, where $c(x)$ is a matrix valued function. For a homogeneous flow $\varphi_t = \varphi_{0,t}$ show the following.

(i) $\varphi_t(x)$ is a d-dimensional Brownian motion. Compute its covariance.
(ii) $\varphi_t(x) - \varphi_t(y)$ is a diffusion Markov process for every x, y. Determine its generator.

Exercise 4.2.16 A homogeneous flow $\varphi_{s,t}$ is called *isotropic* if $c(x) = G^*c(Gx)G$ holds for any orthogonal matrix G where $a(x, y) = c(x - y)$ is the infinitesimal covariance. For a homogeneous flow $\varphi_{s,t}$ show that the following three statements are equivalent.

(a) $\varphi_{s,t}$ is isotropic,
(b) $GF(x, t)$ and $F(G(x), t)$ have the same law for any orthogonal matrix G where F is the infinitesimal generator of $\varphi_{s,t}$,
(c) $\varphi_{s,t}(Gx)$ and $G\varphi_{s,t}(x)$ have the same law.

Exercise 4.2.17 (*Baxendale–Harris* [7]) Suppose that $\varphi_t \equiv \varphi_{0,t}$ is an isotropic Brownian flow. Show that $\rho_t \equiv |\varphi_t(x) - \varphi_t(y)|$ is a diffusion process on $\mathbb{R}$ with the generator

$$Lg(\rho) = (1 - B_L(\rho))g''(\rho) + (d - 1)\left(\frac{1 - B_N(\rho)}{\rho}\right)g'(\rho)$$

where $B_L(\rho) = a^{ii}(\rho e_i)$, $B_N(\rho) = a^{ii}(\rho e_j)$, $i \neq j$ (independent of i, j) and $e_i = (0, \ldots, 0, 1, 0, \ldots, 0)$ (1 is at the i-th component).

4.3 Asymptotic properties of Brownian flows

Measure preserving Brownian flow
Let $\varphi_{s,t}$, $0 \le s \le t < \infty$ be a forward Brownian flow. In this section we assume that it is temporally homogeneous. Then the pair of its infinitesimal covariance and infinitesimal mean (a, b) does not depend on t and the semigroup of the n-point motion satisfies $T_{s,t}^{(n)} = T_{0,t-s}^{(n)}$. We denote it simply by $T_{t-s}^{(n)}$. Then $\{T_t^{(n)}\}$ is a Feller semigroup. Its infinitesimal generator $L^{(n)}$

is represented by (19) of Section 4.2, where the coefficients a^{ij}, b^i do not depend on t. The flow $\varphi_{0,t}$ is written as φ_t as before.

We have seen in the previous section that for every t the inverse $\varphi_{s,t}^{-1}$, $s \in [0, t]$ is a backward Brownian flow with the infinitesimal covariance and the infinitesimal mean $(a^{ij}(x, y), -b^i(x) + 2c^i(x))$ for each t. Then the n-point motion of the inverse flow is a temporally homogeneous backward Markov process with the Feller semigroup $\{\hat{T}_t^{(n)}\}$ and infinitesimal generator $\hat{L}^{(n)}$, which are given by

$$\hat{T}_t^{(n)}f(\mathbf{x}^{(n)}) = E[f(\varphi_t^{-1}(\mathbf{x}^{(n)}))], \tag{1}$$

$$\hat{L}^{(n)}g = \frac{1}{2}\sum_{i,j,p,q} a^{ij}(x_p, x_q)\frac{\partial^2 g}{\partial x_p^i \partial x_q^j} - \sum_{i,p}(b^i(x_p) - 2c^i(x_p))\frac{\partial g}{\partial x_p^i}. \tag{2}$$

Let Π be a Borel measure on $\mathbb{R}^d$. We set $\Pi(f) = \int f(x)\Pi(dx)$ if the integral is well defined. Let $\mathscr{M}^1(\mathbb{R}^d)$ be the set of all probability measures on $\mathbb{R}^d$. It is a complete metric space by the weak topology. Next let $\mathscr{M}(\mathbb{R}^d)$ be the set of all Borel measures on $\mathbb{R}^d$. It is a complete metric space by the vague topology: here a sequence $\{\Pi_n\}$ of $\mathscr{M}(\mathbb{R}^d)$ is said to converge to Π *vaguely* if the sequence $\{\Pi_n(f)\}$ converges to $\Pi(f)$ for any f of $C_0(\mathbb{R}^d)$, where $C_0(\mathbb{R}^d)$ is the space of all continuous functions over $\mathbb{R}^d$ with compact supports equipped with the uniform topology.

Now let Π be an element of $\mathscr{M}(\mathbb{R}^d)$. For each t and ω we define the image measures of Π by the maps $\varphi_t(\cdot)(\omega)$ and $\varphi_t^{-1}(\cdot)(\omega)$ by

$$\varphi_t^{-1}(\Pi)(A) \equiv \Pi(\varphi_t(A)), \quad \varphi_t(\Pi)(A) \equiv \Pi(\varphi_t^{-1}(A)), \quad A \in \mathscr{B}(\mathbb{R}^d) \tag{3}$$

respectively (ω is suppressed). These can be regarded as stochastic processes with values in $\mathscr{M}(\mathbb{R}^d)$. The objective of this section is to study the asymptotic properties of the processes $\varphi_t^{-1}(\Pi)$ and $\varphi_t(\Pi)$ as $t \to \infty$.

A simple case is where $\varphi_t^{-1}(\Pi) = \Pi$ holds for every t or $\varphi_t^{-1}(\Pi)(A)$ increases (or decreases) with t a.s. for any A of $\mathscr{B}(\mathbb{R}^d)$. The flow is called Π-*expanding* if $\varphi_t^{-1}(\Pi)(A)$ increases with t a.s. for any A of $\mathscr{B}(\mathbb{R}^d)$. The flow is called Π-*shrinking* if $\varphi_t^{-1}(\Pi)(A)$ decreases with t a.s. for any A of $\mathscr{B}(\mathbb{R}^d)$. If it is Π-expanding and Π-shrinking, it is called Π-*preserving*. In particular it is called *incompressible* if it preserves the Lebesgue measure.

In the first part of this section we characterize the Π-preserving flow by means of the pair (a, b) or its random infinitesimal generator. In the second and the third parts, we discuss the asymptotic behavior of $\varphi_t^{-1}(\Pi)$ and $\varphi_t(\Pi)$ respectively when φ_t is not Π-preserving. It will be seen that the asymptotic behavior of these two is completely different.

In the following we will assume that the Brownian flow φ_t satisfies Conditions (A.1), (A.2) and $(A.3)_{k,\delta}$ for $k \geq 3$ and $\delta > 0$. For a moment we fix any sample ω for which $\{\varphi_t(\omega) : t \geq 0\}$ defines a flow of diffeomorphisms of $\mathbb{R}^d$ but ω is suppressed. Let $\partial\varphi_t(x)$ be the Jacobian matrix of $\varphi_t(x)$: $\partial\varphi_t(x) = \partial(\varphi_t^1, \ldots, \varphi_t^d)/\partial(x^1, \ldots, x^d)$. By the formula for the change of variables we have

$$\int_{\varphi_t(A)} f(y)\,\mathrm{d}y = \int_A f(\varphi_t(x))\,|\det \partial\varphi_t(x)|\,\mathrm{d}x \quad \text{a.s.} \tag{4}$$

for any continuous function f. Now let Π be a Borel measure on $\mathbb{R}^d$ having a strictly positive density function $\pi(x)$ with respect to the Lebesgue measure μ. Then the image measure $\varphi_t^{-1}(\Pi)$ satisfies

$$\varphi_t^{-1}(\Pi)(A) = \int_A \pi(\varphi_t(x))\,|\det \partial\varphi_t(x)|\,\mathrm{d}x \qquad \text{for all } A \in \mathscr{B}(\mathbb{R}^d). \tag{5}$$

Therefore

$$\alpha(x, t) = \pi(x)^{-1}\pi(\varphi_t(x))\,|\det \partial\varphi_t(x)| \tag{6}$$

is a Radon–Nikodym density of the measure $\varphi_t^{-1}(\Pi)$ with respect to Π. We shall compute it assuming that π is a C^3-function.

Observe that there exists a C^{k-1}-valued Brownian motion $\mathring{F}(x, t)$ and the flow is represented as the solution of Stratonovich's stochastic differential equation

$$\varphi_t(x) = x + \int_0^t \mathring{F}(\varphi_s(x), \circ\,\mathrm{d}s). \tag{7}$$

Denote by $\operatorname{div}(\pi\mathring{F})(x, t)$ the Brownian motion such that

$$\operatorname{div}(\pi\mathring{F})(x, t) = \sum_{i=1}^d \frac{\partial(\pi\mathring{F}^i)}{\partial x^i}(x, t). \tag{8}$$

The following is an analogue of the classical Liouville's theorem.

Lemma 4.3.1 *The density* $\alpha(x, t)$ *defined by* (6) *is represented by*

$$\alpha(x, t) = \exp\left\{\int_0^t \pi(\varphi_s(x))^{-1} \operatorname{div}(\pi\mathring{F})(\varphi_s(x), \circ\,\mathrm{d}s)\right\} \quad \text{a.s.} \tag{9}$$

Proof Apply Itô's formula (Theorem 2.3.11) to $\det \partial\varphi_t(x)$, where x is fixed. Then we have

$$\det \partial\varphi_t - 1 = \sum_\sigma \varepsilon(\sigma)\left(\sum_{k=1}^d \int_0^t \frac{\partial\varphi_s^1}{\partial x^{i_1}} \cdots \circ d\left(\frac{\partial\varphi_s^k}{\partial x^{i_k}}\right) \cdots \frac{\partial\varphi_s^d}{\partial x^{i_d}}\right) \tag{10}$$

where the first sum is taken for all permutations $\sigma = \begin{pmatrix} 1 \cdots d \\ i_1 \cdots i_d \end{pmatrix}$ and $\varepsilon(\sigma)$ is the sign of the permutation σ. Differentiate each term of (7) with respect to x and apply Theorem 3.3.4. Then,

$$\frac{\partial \varphi_t^k}{\partial x^l}(x) = \delta_{kl} + \sum_m \int_0^t \frac{\partial \mathring{F}^k}{\partial x^m}(\varphi_s(x), \circ ds) \frac{\partial \varphi_s^m}{\partial x^l}(x).$$

Substitute the above in (10). Then we obtain

$$\det \partial\varphi_t - 1 = \sum_k \sum_m \int_0^t \frac{\partial \mathring{F}^k}{\partial x^m}(\varphi_s, \circ ds) \det \begin{pmatrix} \frac{\partial \varphi_s^1}{\partial x^1} \cdots \frac{\partial \varphi_s^1}{\partial x^d} \\ \cdots\cdots\cdots\cdots \\ \frac{\partial \varphi_s^m}{\partial x^1} \cdots \frac{\partial \varphi_s^m}{\partial x^d} \\ \cdots\cdots\cdots\cdots \end{pmatrix}.$$

In the last determinant, $\left(\frac{\partial \varphi_s^m}{\partial x^1}, \ldots, \frac{\partial \varphi_s^m}{\partial x^d}\right)$ appears at the k-th row. The determinant equals $\det \partial\varphi_t(x)$ if $m = k$ and equals 0 if $m \neq k$. We have thus proved the equality

$$\det \partial\varphi_t - 1 = \int_0^t \operatorname{div} \mathring{F}(\varphi_s, \circ ds) \det \partial\varphi_s.$$

Finally by Itô's formula

$$\pi(\varphi_t) \det \partial\varphi_t - \pi(x)$$

$$= \sum_i \int_0^t \frac{\partial \pi}{\partial x^i}(\varphi_s)\mathring{F}^i(\varphi_s, \circ ds) \det \partial\varphi_s + \int_0^t \pi(\varphi_s) \operatorname{div} \mathring{F}(\varphi_s, \circ ds) \det \partial\varphi_s$$

$$= \int_0^t \{\pi(\varphi_s)^{-1} \operatorname{div}(\pi\mathring{F})(\varphi_s, \circ ds)\} \pi(\varphi_s) \det \partial\varphi_s.$$

It can be regarded as a linear integral equation for $\pi(\varphi_t) \det \partial\varphi_t$. The solution is represented by

$$\pi(\varphi_t) \det \partial\varphi_t = \pi(x) \exp\left\{\int_0^t \pi(\varphi_s(x))^{-1} \operatorname{div}(\pi\mathring{F})(\varphi_s(x), \circ ds)\right\}. \quad \square$$

We will give several equivalent characterizations of Π-preserving flow.

Theorem 4.3.2 (*c.f.* *Harris* [44]) *Let φ_t be a Brownian flow satisfying Conditions* (A.1), (A.2), (A.3)$_{k,\delta}$ *for some $k \geq 3$ and $\delta > 0$ and let $\mathring{F}(x, t)$ be its Stratonovich random infinitesimal generator. Let $\Pi(x)$ be a Borel measure on $\mathbb{R}^d$ having a strictly positive density function $\pi(x)$ of C^3-class with respect to the Lebesgue measure μ. The following statements are equivalent.*

(i) *φ_t is Π-preserving.*
(ii) *$\operatorname{div}(\pi \mathring{F})(x, t)$ is identically 0 a.s.*
(iii) *For any x and y*

$$\sum_i \frac{\partial}{\partial x^i}\{\pi(x)a^{ij}(x, y)\} = 0, \qquad j = 1, \ldots, d \tag{11}$$

and

$$\operatorname{div}\{\pi(b - c)\}(x) = 0. \tag{12}$$

(iv) *Let $\Pi^{(n)}$ be the n-product of the measures Π. Then for every n $\{\hat{T}_t^{(n)}\}$ is an adjoint semigroup of $\{T_t^{(n)}\}$ with respect to the measure $\Pi^{(n)}$, i.e.*

$$\int (T_t^{(n)}f)g \, d\Pi^{(n)} = \int (\hat{T}_t^{(n)}g)f \, d\Pi^{(n)} \tag{13}$$

holds for any bounded measurable functions f, g on $\mathbb{R}^{nd}$ with compact supports.
(v) *Π is an invariant measure of the one point motion and $\Pi \otimes \Pi$ is an invariant measure of the two point motion.*

Proof Note that φ_t is Π-preserving if and only if $\alpha(x, t) \equiv 1$ holds for any t a.s. for every x. Therefore (i) and (ii) are equivalent by Lemma 4.3.1. Since the infinitesimal covariance of $\operatorname{div}(\pi\mathring{F})$ and $\mathring{F}$ and the infinitesimal mean of $\operatorname{div}(\pi\mathring{F})$ are given by

$$\left(\sum_i \frac{\partial}{\partial x^i}\{\pi(x)a^{ij}(x, y)\}, \sum_i \frac{\partial}{\partial x^i}\{\pi(x)(b^i(x) - c^i(x))\}\right)$$

$\operatorname{div}(\pi\mathring{F}) = 0$ implies (11) and (12).

We next prove that (iii) implies (iv). Let $L^{(n)*}$ be the formal adjoint of $L^{(n)}$ with respect to the Lebesgue measure. Set $\pi^{(n)}(x_1, \ldots, x_n) = \pi(x_1)\ldots\pi(x_n)$. Then it is the density of the measure $\Pi^{(n)}$ with respect to the Lebesgue measure. A direct computation yields

$$\frac{1}{\pi^{(n)}} L^{(n)*}(\pi^{(n)}g)$$

$$= \frac{1}{2}\sum_{i,j,p,q} a^{ij}(x_p, x_q)\frac{\partial^2 g}{\partial x_p^i \partial x_q^j} - \sum_{i,p}(b^i(x_p) - 2c^i(x_p))\frac{\partial g}{\partial x_p^i}$$
$$+ \frac{1}{2}\sum_{j,q}\left(\sum_{i,p}\pi(x_p)^{-1}\frac{\partial}{\partial x_p^i}\{\pi(x_p)a^{ij}(x_p, y)\}|_{y=x_q}\right)\frac{\partial g}{\partial x_q^j}$$
$$-\left(\sum_p \pi(x_p)^{-1}\operatorname{div}\pi(b - c)(x_p)\right)g$$
$$+ \frac{1}{2}\left(\sum_{p\neq q}\{\pi(x_p)\pi(x_q)\}^{-1}\sum_{i,j}\frac{\partial^2}{\partial x_p^i \partial x_q^j}(\pi(x_p)\pi(x_q)a^{ij}(x_p, x_q))\right)g$$

$$+\frac{1}{2}\Bigg(\sum_p \pi(x_p)^{-1}\sum_{i,j}\frac{\partial^2}{\partial x_p^i \partial x_p^j}\{\pi(x_p)a^{ij}(x_p, y)\}|_{y=x_p}$$

$$+\sum_p \pi(x_p)^{-1}\sum_{i,j}\frac{\partial^2}{\partial x_p^i \partial y_p^j}\{\pi(x_p)a^{ij}(x_p, y_p)\}|_{y_p=x_p}\Bigg)g. \tag{14}$$

Consequently $(1/\pi^{(n)})L^{(n)*}(\pi^{(n)}g) = \hat{L}^{(n)}g$ holds if (iii) is satisfied. Now the above equality implies

$$\int (L^{(n)}f)g \, d\Pi^{(n)} = \int f\hat{L}^{(n)}g \, d\Pi^{(n)} \tag{15}$$

if f, g are smooth and $f, g, L^{(n)}f, L^{(n)}g$ are in $L^2(\Pi^{(n)})$. Set

$$U_\alpha^{(n)}f = \int_0^\infty e^{-\alpha t}T_t^{(n)}f \, dt, \qquad \hat{U}_\alpha^{(n)}f = \int_0^\infty e^{-\alpha t}\hat{T}_t^{(n)}f \, dt.$$

Since

$$T_t^{(n)}f = f + \int_0^t L^{(n)}T_r^{(n)}f \, dr, \qquad \hat{T}_t^{(n)}f = f + \int_0^t \hat{L}^{(n)}\hat{T}_r^{(n)}f \, dr$$

holds by Theorem 4.2.4, we have $(\alpha - L^{(n)})U_\alpha^{(n)}f = f$ and $(\alpha - \hat{L}^{(n)})\hat{U}_\alpha^{(n)}f = f$. Therefore, (15) yields

$$\int f\hat{U}_\alpha^{(n)}g \, d\Pi^{(n)} = \int ((\alpha - L^{(n)})U_\alpha^{(n)}f)(\hat{U}_\alpha^{(n)}g) \, d\Pi^{(n)}$$

$$= \int (U_\alpha^{(n)}f)(\alpha - \hat{L}^{(n)})\hat{U}_\alpha^{(n)}g \, d\Pi^{(n)} = \int (U_\alpha^{(n)}f)g \, d\Pi^{(n)}.$$

This implies (13) by taking the inverse Laplace transform.

Now set $g = 1$ in equality (13). Since $\hat{T}_t^{(n)}1 = 1$ holds, $\Pi^{(n)}$ is an invariant measure of the semigroup $\{T_t^{(n)}\}$ for every n. Therefore (iv) implies (v). Finally suppose that Π and $\Pi \otimes \Pi$ are invariant measures of $\{T_t^{(1)}\}$ and $\{T_t^{(2)}\}$, respectively. Then for every continuous function g on $\mathbb{R}^d$ with compact support, we have

$$E\left[\left(\int_{\mathbb{R}^d}\{g(\varphi_t(x)) - g(x)\}\Pi(dx)\right)^2\right]$$

$$= \iint_{\mathbb{R}^d\times\mathbb{R}^d}(T_t^{(2)}(g\otimes g)(x, y) - 2g(x)T_tg(y) + g(x)g(y))\Pi(dx)\Pi(dy)$$

$$= \iint_{\mathbb{R}^d\times\mathbb{R}^d}(T_t^{(2)}(g\otimes g)(x, y) - g\otimes g(x, y))\Pi(dx)\Pi(dy)$$

$$= 0$$

for every $t > 0$, where $g \otimes g$ is a function on $\mathbb{R}^d \times \mathbb{R}^d$ defined by

$g \otimes g(x, y) = g(x)g(y)$. This proves $\int g(\varphi_t(x))\Pi(dx) = \int g(x)\Pi(dx)$ a.s. Therefore φ_t is Π-preserving. □

Asymptotic behavior of $\varphi_t^{-1}(\Pi)$

We shall consider the asymptotic behavior of $\varphi_t^{-1}(\Pi)$ as t tends to infinity in the case where φ_t is not Π-preserving. The asymptotic property depends crucially on the ergodic property of the one point motion. In order to study the problem we introduce another assumption for the one point motion.

Condition (A.4) *The matrix $(a^{ij}(x, x))$ is positive definite, i.e. $\sum a^{ij}(x, x)\xi^i\xi^j > 0$ holds for any x if $(\xi^1, \ldots, \xi^d)$ is not zero.* □

Under the above assumption the transition probability $P_t(x, \cdot)$ has a strictly positive continuous density function $p_t(x, y)$ with respect to the Lebesgue measure (see Friedman [33]). Therefore, the recurrence–transience dichotomy discussed in Section 1.3 is valid.

We first consider the case where φ_t is Π-expanding or Π-shrinking. The recurrent and transient cases will be discussed separately.

Theorem 4.3.3 *Assume the same conditions as in Theorem 4.3.2. Then the flow φ_t is Π-expanding (or π-shrinking) if and only if (11) holds and the following (16) is non-negative (or non-positive).*

$$l(x) = \operatorname{div}\{\pi(b - c)\}(x)/\pi(x). \tag{16}$$

(i) *Assume further that Condition (A.4) is satisfied and that the one point motion is recurrent in the sense of Harris.*
  (a) *If φ_t is Π-expanding but is not Π-preserving, then $\varphi_t^{-1}(\Pi)(A)$ increases to infinity a.s. as t tends to infinity whenever $\Pi(A) > 0$.*
  (b) *If φ_t is Π-shrinking but is not Π-preserving, then $\varphi_t^{-1}(\Pi)(A)$ decreases to 0 a.s. as t tends to infinity whenever $\Pi(A) < \infty$.*

(ii) *Assume further that Condition (A.4) is satisfied and the one point motion is transient.*
  (c) *If φ_t is Π-expanding and $\underline{\lim}_{x\to\infty} l(x) > 0$ holds, then $\varphi_t^{-1}(\Pi)(A)$ increases to infinity a.s. as t tends to infinity whenever $\Pi(A) > 0$. If φ_t is Π-expanding and $l(x)$ is of compact support, $\varphi_t^{-1}(\Pi)(A)$ increases to a finite valued random variable whenever $\Pi(A) < \infty$.*
  (d) *If φ_t is Π-shrinking and $\overline{\lim}_{x\to\infty} l(x) < 0$ holds, then $\varphi_t^{-1}(\Pi)(A)$ decreases to 0 a.s. whenever $\Pi(A) < \infty$. If φ_t is Π-shrinking and $l(x)$ is of compact support, it decreases to a strictly positive random variable a.s. whenever $\Pi(A) > 0$.*

Proof Note that φ_t is Π-expanding if and only if $\operatorname{div}(\pi \mathring{F})$ is an increasing process by Lemma 4.3.1. The latter property is valid if and only if (11) holds and $l(x)$ of (16) is non-negative. A similar fact is valid for Π-shrinking flow. Therefore the first assertion is verified. If φ_t is Π-expanding but is not Π-preserving, then the function l of (16) is non-negative but is not identically 0. Since $\varphi_t(x)$ is recurrent in the sense of Harris, we have $\int_0^\infty l(\varphi_s(x))\,ds = \infty$ a.s. for any x. Since their density $\alpha(x, t)$ is written as

$$\alpha(x, t) = \exp\left\{\int_0^t l(\varphi_s(x))\,ds\right\}, \tag{17}$$

it increases to infinity for any x as t tends to infinity. Therefore we have proved (a). Assertion (b) can be proved similarly.

We next consider the case where one point motion is transient. We have $\int_0^\infty l(\varphi_s(x))\,ds = \infty$ a.s. for any x if $\underline{\lim}_{x\to\infty} l(x) > 0$, and $\int_0^\infty l(\varphi_s(x))\,ds < \infty$ a.s. for any x if $l(x)$ is of compact support. Then assertions (c) and (d) can be shown easily. □

We will next consider the case where φ_t is neither Π-expanding nor Π-shrinking. We begin by rewriting the density function $\alpha(x, t)$ of Lemma 4.3.1 using the Itô integral.

Lemma 4.3.4 *Set $M(x, t) = F(x, t) - tb(x)$ and define*

$$G(x, t) = \frac{1}{\pi(x)}\operatorname{div}(\pi M)(x, t). \tag{18}$$

Then the density $\alpha(x, t)$ given by (6) can be written as

$$\alpha(x, t) = \exp\left\{\int_0^t G(\varphi_s(x), ds) - \frac{1}{2}\int_0^t a^G(\varphi_s(x), \varphi_s(x))\,ds\right\}$$
$$\times \exp\left\{\int_0^t \pi(\varphi_s(x))^{-1}\hat{L}^{(1)*}\pi(\varphi_s(x))\,ds\right\}, \tag{19}$$

where $a^G(x, y)$ is the local characteristic of G and $\hat{L}^{(1)*}$ is the adjoint operator of $\hat{L}^{(1)}$.

Proof We shall rewrite the stochastic integral of (9) using the Itô integral. Then

$$\int_0^t \frac{1}{\pi(\varphi_s)}\operatorname{div}(\pi \mathring{F})(\varphi_s, \circ ds)$$
$$= \int_0^t G(\varphi_s, ds) + \int_0^t \frac{1}{\pi(\varphi_s)}\operatorname{div}\pi(b - c)(\varphi_s)\,ds$$
$$+ \frac{1}{2}\int_0^t \sum_j \frac{\partial}{\partial x^j}\left\{\frac{1}{\pi(x)}\sum_i \frac{\partial}{\partial x^i}(\pi(x)a^{ij}(x, y))\right\}\bigg|_{x=y=\varphi_s} ds \tag{20}$$

The first integral is a localmartingale. The last term is written as

$$-\frac{1}{2}\int_0^t \sum_{i,j}\left(\frac{1}{\pi^2}\frac{\partial \pi}{\partial x^j}\right)(\varphi_s)\frac{\partial}{\partial x^i}\{\pi(x)a^{ij}(x,y)\}|_{x=y=\varphi_s}\,ds$$
$$+\frac{1}{2}\int_0^t \frac{1}{\pi(\varphi_s)}\sum_{i,j}\frac{\partial^2}{\partial x^i \partial x^j}\{\pi(x)a^{ij}(x,y)\}|_{x=y=\varphi_s}\,ds.$$

On the other hand we have similarly to (14),

$$\frac{1}{\pi}\hat{L}^{(1)*}\pi = \frac{1}{\pi}\operatorname{div}\pi(b-c) + \frac{1}{2\pi}\sum_{i,j}\frac{\partial^2}{\partial x^i \partial x^j}\{\pi(x)a^{ij}(x,y)\}|_{y=x}$$
$$+\frac{1}{2\pi}\sum_{i,j}\frac{\partial^2}{\partial x^i \partial y^j}\{\pi(x)a^{ij}(x,y)\}|_{y=x}.$$

The last term can be written as

$$\frac{1}{2\pi^2}\left(\sum_{i,j}\frac{\partial^2}{\partial x^i \partial y^j}\{\pi(x)\pi(y)a^{ij}(x,y)\}|_{y=x} - \sum_j \frac{\partial \pi}{\partial y^j}(y)\sum_i \frac{\partial}{\partial x^i}\{\pi(x)a^{ij}(x,y)\}|_{y=x}\right).$$

Consequently the sum of the second and third terms of the right hand side of (20) is equal to

$$-\frac{1}{2}\int_0^t a^G(\varphi_s(x),\varphi_s(x))\,ds + \int_0^t \pi(\varphi_s(x))^{-1}\hat{L}^{(1)*}\pi(\varphi_s(x))\,ds.$$

This proves equality (19). □

Corollary 4.3.5

(i) *If $\hat{L}^{(1)*}\pi \le 0$, then $\alpha(x,t)$ is a continuous supermartingale for any x.*

(ii) *If $\hat{L}^{(1)*}\pi = 0$, it is a continuous localmartingale.*

(iii) *If $\hat{L}^{(1)*}\pi \ge 0$, it is a continuous local-submartingale.*

Proof Apply Itô's formula to (19). Then

$$\alpha(\cdot,t) = 1 + \int_0^t \alpha(\cdot,s)G(\varphi_s,ds) + \int_0^t \alpha(\cdot,s)\pi(\varphi_s)^{-1}\hat{L}^{(1)*}\pi(\varphi_s)\,ds.$$

Then the assertion of the corollary is immediate. □

We shall further consider the case where the density function $\pi(x)$ satisfies $\hat{L}^{(1)*}\pi \le 0$.

Theorem 4.3.6 *Assume the same conditions as in Theorem 4.3.2. Assume further $\hat{L}^{(1)*}\pi \le 0$. Then an $\mathcal{M}(\mathbf{R}^d)$-valued stochastic process $\varphi_t^{-1}(\Pi)$ con-*

verges vaguely to an $\mathcal{M}(\mathbb{R}^d)$-valued random variable Π_∞ a.s. as $t \to \infty$. Further if (11) *is not satisfied, for almost all ω the measure $\Pi_\infty(\omega)$ is singular to the Borel measure Π provided that $\Pi_\infty(\omega)$ is not a zero measure.*

Proof Let f be a non-negative function of $C_0(\mathbb{R}^d)$. Then $\varphi_t^{-1}(\Pi)(f)$, $t \geq 0$ is a positive supermartingale by Corollary 4.3.5. Therefore $\lim_{t\to\infty} \varphi_t^{-1}(\Pi)(f)$ exists a.s. by Theorem 1.2.3. Now let $D_0(\mathbb{R}^d)$ be a countable dense subset of $C_0(\mathbb{R}^d)$ such that $c_1 f_1 + c_2 f_2 \in D_0(\mathbb{R}^d)$ if $f_1, f_2 \in D_0(\mathbb{R}^d)$ and c_1, c_2 are rationals, and $|f| \in D_0(\mathbb{R}^d)$ if $f \in D_0(\mathbb{R}^d)$. Then $\Pi_\infty(f) \equiv \lim_{t\to\infty} \varphi_t^{-1}(\Pi)(f)$ exists a.s. for all f of $D_0(\mathbb{R}^d)$. We can choose a family of modifications $\{\Pi_\infty(f)(\omega) : f \in D_0(\mathbb{R}^d)\}$ (denoted by the same notation) such that it is a positive linear functional on $D_0(\mathbb{R}^d)$ for almost all ω. Then for such ω, $\Pi_\infty(f)(\omega)$ can be extended to a positive linear functional on $C_0(\mathbb{R}^d)$. Therefore there exists a Borel measure $\Pi(\omega)$ on $\mathbb{R}^d$ such that $\Pi_\infty(f)(\omega) = \int f(x)\Pi_\infty(\omega)(dx)$ holds for any f. Consequently $\varphi_t^{-1}(\Pi)(\omega)$ converges to $\Pi_\infty(\omega)$ vaguely for almost all ω.

Let $\Pi_\infty^{ac}(\omega)$ be the absolutely continuous part of $\Pi_\infty(\omega)$ with respect to the Borel measure Π and let $\alpha_\infty(x, \omega)$ be the Radon–Nikodym density. We can choose it so that it is $\mathcal{B}(\mathbb{R}^d) \otimes \mathcal{F}_{0,\infty}$-measurable. Since $E[\Pi_\infty^{ac}(A)|\mathcal{F}_{0,t}] \leq \varphi_t^{-1}(\Pi)(A)$ holds for any Borel set A, we have $E[\alpha_\infty(x)|\mathcal{F}_{0,t}] \leq \alpha(x, t)$ a.e. (x, ω) with respect to $\Pi \times P$. Let t tend to infinity. Then we obtain $\alpha_\infty(x) \leq \alpha(x, \infty)$ a.e. with respect to $\Pi \times P$. Now assume that φ_t does not satisfy (11). Then the function a^G in Corollary 4.3.5 is not identical to 0. Since $\varphi_t(x)$ is recurrent in the sense of Harris, we have

$$\int_0^\infty a^G(\varphi_r(x), \varphi_r(x))\, dr = \infty \quad \text{a.s.} \quad \text{for all } x.$$

Therefore $\alpha(x, \infty) = 0$ holds a.s. for any x (see Exercise 2.3.20). Then $\alpha_\infty(x) = 0$, a.e. (x, ω) with respect to $\Pi \times P$. Therefore for almost all ω $\Pi_\infty(\omega)$ is singular to the Borel measure Π if $\Pi_\infty(\omega)$ is not a zero measure. □

If the semigroup $\{\hat{T}_t^{(1)}\}$ has an invariant probability, Theorem 4.3.6 can be modified to the following form.

Theorem 4.3.7 (*c.f.* Le Jan [88, 89]). *Let φ_t be a Brownian flow satisfying Conditions* (A.1), (A.2), (A.3)$_{k,\delta}$ *for some $k \geq 1$, $\delta > 0$ and Condition* (A.4). *Assume that the semigroup $\{\hat{T}_t^{(1)}\}$ has an invariant probability Λ. Then the $\mathcal{M}^1(\mathbb{R}^d)$-valued stochastic process $\varphi_t^{-1}(\Lambda)$ converges weakly to an $\mathcal{M}^1(\mathbb{R}^d)$-valued random variable Λ_∞ a.s. as $t \to \infty$. It satisfies $E[\Lambda_\infty(A)] = \Lambda(A)$ for any Borel set A. Further if* (11) *is not satisfied, the measures Λ_∞ are singular to Λ a.s.*

Proof We first show that Λ has a strictly positive density function $\lambda(x)$ of C^3-class with respect to the Lebesgue measure and it satisfies $\hat{L}^{(1)*}\lambda = 0$. Note that the semigroup $\{\hat{T}_t^{(1)}\}$ satisfies the equation $(d/dt)\hat{T}_t^{(1)}f = \hat{T}_t^{(1)}\hat{L}^{(1)}f$. The coefficients of the operator $\hat{L}^{(1)}$ are non-degenerate and Hölder continuous. Then the transition probability $\hat{P}_t^{(1)}(x, dy)$ has a density function $\hat{p}_t^{(1)}(x, y)$ which is a strictly positive continuous function and is three times continuously differentiable in y (Friedmann [33]). Define the function $\lambda(y)$ by $\int \Lambda(dx)\hat{p}_t^{(1)}(x, y)$. It does not depend on t and $\Lambda(dy) = \lambda(y)\,dy$ is satisfied. Hence λ is the desired density function. Now since $\Lambda(\hat{T}_t^{(1)}f) = \Lambda(f)$ is satisfied, we have $\Lambda(\hat{L}^{(1)}f) = 0$ if f belongs to the domain of the generator $\hat{L}^{(1)*}$. This implies $\hat{L}^{(1)*}\lambda = 0$.

Now by Theorem 4.3.6 there exists an $\mathscr{M}(\mathbb{R}^d)$-valued random variable Λ_∞ such that $\Lambda_\infty(f) \equiv \lim_{t\to\infty} \varphi_t^{-1}(\Lambda)(f)$ holds for any $f \in C_0(\mathbb{R}^d)$. We show that Λ_∞ is an $\mathscr{M}^1(\mathbb{R}^d)$-valued random variable. Since Λ is a probability, we have $0 \le \Lambda_\infty(f) \le 1$ if $0 \le f \le 1$. Further since Λ is $\{\hat{T}_t^{(1)}\}$-invariant, equality $E[\varphi_t^{-1}(\Lambda)(f)] = \Lambda(f)$ holds. Letting t tend to infinity, we obtain $E[\Lambda_\infty(f)] = \Lambda(f)$. This implies $E[\Lambda_\infty(\mathbb{R}^d)] = 1$. Hence Λ_∞ is an $\mathscr{M}^1(\mathbb{R}^d)$-valued random variable. Therefore the convergence takes place with respect to the weak topology. The last assertion of the theorem is obvious from the previous theorem. □

Finally we consider the case where the semigroup $\{\hat{T}_t^{(1)}\}$ does not have an invariant probability. In the following theorem, we do not have to assume Conditions (A.1)–(A.4).

Theorem 4.3.8 *Assume that the semigroup $\{\hat{T}_t^{(1)}\}$ does not have an invariant probability. Then for any probability measure Π on $\mathbb{R}^d$,*

$$\lim_{t\to\infty} \varphi_t^{-1}(\Pi)(A) = 0 \qquad \text{in } L^1 \tag{21}$$

holds for any bounded Borel set A.

Proof Note the relation

$$\varphi_t^{-1}(\Pi)(A) = \int \chi(A)(\varphi_t^{-1}(x))\, d\Pi(x). \tag{22}$$

Since $\chi(A)(\varphi_t^{-1}(x))$ converges to 0 in L^1 for any x by Theorem 1.3.12, the right hand side of (22) converges to 0 in L^1. □

Asymptotic behavior of $\varphi_t(\Pi)$

In this subsection, we shall consider the asymptotic behavior of $\varphi_t(\Pi)$ as $t \to \infty$ in the case where φ_t^{-1} is not Π-preserving. The discussion will be divided into two cases. The first is the case where the one point motion has an invariant probability.

Theorem 4.3.9 *Let φ_t be a Brownian flow satisfying Conditions* (A.1), (A.2), (A.3)$_{k,\delta}$ *for some* $k \geq 0$, $\delta > 0$ *and* (A.4). *If its one point motion has an invariant probability* Λ, *the following assertions hold.*

(i) *For any probability* Π,

$$\lim_{T\to\infty} \frac{1}{T}\int_0^T \varphi_s(\Pi)(A)\,\mathrm{d}s = \Lambda(A) \quad a.s. \tag{23}$$

holds for any Borel set A.

(ii) *For any infinite Borel measure* Π,

$$\lim_{T\to\infty} \frac{1}{T}\int_0^T \varphi_s(\Pi)(A)\,\mathrm{d}s = \infty \quad a.s. \tag{24}$$

holds for any Borel set A *such that* $\Lambda(A) > 0$.

Proof Note the relation

$$\frac{1}{T}\int_0^T \varphi_s(\Pi)(A)\,\mathrm{d}s = \int_{\mathbb{R}^d}\left\{\frac{1}{T}\int_0^T \chi(A)(\varphi_s(y))\,\mathrm{d}s\right\}\Pi(\mathrm{d}y). \tag{25}$$

The integrand $\{\dots\}$ on the right hand side coverges to $\Lambda(A)$ a.s. for any y by an ergodic theorem (see Theorem 1.3.12). Then if Π is a probability equation (25) converges to $\Lambda(A)$ by Fubini's theorem. Suppose next $\Pi(\mathbb{R}^d) = \infty$. Let K be any compact subset of $\mathbb{R}^d$. Then we have

$$\frac{1}{T}\int_0^T \Pi(\varphi_s^{-1}(A)\cap K)\,\mathrm{d}s = \int_K\left\{\frac{1}{T}\int_0^T \chi(A)(\varphi_s(y))\,\mathrm{d}s\right\}\Pi(\mathrm{d}y).$$

It converges to $\Lambda(A)\Pi(K)$ a.s. Since $\Pi(K)\uparrow\infty$ as $K\uparrow\mathbb{R}^d$, we obtain the second assertion. The proof is complete. □

We next consider the case where the one point motion does not have an invariant probability. In the next theorem, Conditions (A.1)–(A.4) are not assumed.

Theorem 4.3.10 *Assume that the one point motion does not have an invariant probability. Then for any probability* Π,

$$\lim_{t\to\infty} \varphi_t(\Pi)(A) = 0 \qquad in\ L^1(P) \tag{26}$$

holds for any bounded Borel set A.

Proof Note the relation $\varphi_t(\Pi)(A) = \int \chi(A)(\varphi_t(x))\, d\Pi(x)$. Since $\chi(A)(\varphi_t(x))$ converges to 0 in L^1 for any x, we have (26). □

We emphasize here that the asymptotic behavior of $\varphi_t(\Pi)$ is strikingly different from that of $\varphi_t^{-1}(\Lambda)$. In Theorem 4.3.9 we showed that if $\Pi \in \mathcal{M}^1(\mathbb{R}^d)$ for almost all ω the time average of $\varphi_t(\Pi)(\omega)$ converges to a single element Λ of $\mathcal{M}^1(\mathbb{R}^d)$, which is absolutely continuous with respect to the Lebesgue measure, no matter whether the measure Π is absolutely continuous or not. On the other hand, Lemma 4.3.4 indicates that $\varphi_t^{-1}(\Lambda)(\omega)$ converges to an $\mathcal{M}^1(\mathbb{R}^d)$-valued random variable $\Lambda_\infty(\omega)$ such that for almost all ω, $\Lambda_\infty(\omega)$ is singular to the Lebesgue measure. However its expectation coincides with Λ which is absolutely continuous with respect to the Lebesgue measure.

The rate of the convergence or divergence of $\varphi_t^{-1}(\Pi)(A)$ (or $\varphi_t(\Pi)(A)$) is defined by

$$\overline{\lim_{t\to\infty}} \frac{1}{t} \log \varphi_t^{-1}(\Pi)(A), \quad \left(\text{or } \overline{\lim_{t\to\infty}} \frac{1}{t} \log \varphi_t(\Pi)(A)\right). \tag{27}$$

It is called the *Lyapunov exponent* of the masses $\{\varphi_t^{-1}(\Pi)(A)\}$ (or $\{\varphi_t(\Pi)(A)\}$). It is 0 if $\varphi_t^{-1}(\Pi)(A)$ converges to a positive value. However if it converges to 0 or diverges to infinity, the Lyapunov exponent may take a non-zero value. See Exercise 4.3.12 for such an example.

Similarly we can define the Lyapunov exponent of the density function $\alpha(x, t)$ of $\varphi_t^{-1}(\Pi)$ with respect to Π by the quantity $\overline{\lim}_{t\to\infty} t^{-1} \log \alpha(x, t)$. Note the expression of Corollary 4.3.5. If $a^G(x, x)$ is a bounded function, the growth order of $\int_0^t G(\varphi_s(x), ds)$ as $t \to \infty$ is $o(t^{1/2+\varepsilon})$ for any $\varepsilon > 0$ (see Exercise 2.3.19). Therefore the Lyapunov exponent of $\alpha(x, t)$ is computed as

$$\overline{\lim_{t\to\infty}} \frac{1}{t} \log \alpha(x, t) = \lim_{t\to\infty} \frac{1}{t} \int_0^t \left(-2^{-1}a^G + \frac{1}{\pi} \hat{L}^{(1)*}\pi\right)(\varphi_s(x))\, ds \quad \text{a.s.}$$

It is equal to $\displaystyle\int \left(-2^{-1}a^G + \frac{1}{\pi} \hat{L}^{(1)*}\pi\right) \Lambda(dx)$ if the one point motion has an invariant probability Λ. It is equal to $\displaystyle\lim_{|x|\to\infty} \left\{-2^{-1}a^G(x, x) + \frac{1}{\pi(x)} \hat{L}^{(1)*}\pi(x)\right\}$ if it exists and the one point motion does not have an invariant probability (see Exercise 1.3.18).

There are extensive studies of Lyapunov exponents of the type

$$\overline{\lim_{t\to\infty}} \frac{1}{t} \log \|\partial\varphi_t \xi\|,$$

where ξ are vectors of the tangent space. Some of the works are connected with Oseledec's multiplicative ergodic theorem. See Arnold–Wihstutz [3], Carverhill [17], Le Jan [88].

Exercise 4.3.11 (*c.f.* Harris [44]) Let φ_t be a homogeneous Brownian flow. Show that it is incompressible if and only if for each n $\{T_t^{(n)}\}$ is symmetric with respect to the product measure $dx_1 \ldots dx_n$, i.e. $\int g T_t^{(n)} f \, dx_1 \ldots dx_n = \int f T_t^{(n)} g \, dx_1 \ldots dx_n$ holds for any functions f, g with compact supports.

Exercise 4.3.12 Let $\varphi_{s,t}(x)$ be a stochastic flow satisfying the bilinear Itô's stochastic differential equation

$$\varphi_{s,t}(x) = x + \int_s^t F\varphi_{s,u}(x)\, du + \sum_{k=1}^m \int_s^t G_k \varphi_{s,u}(x)\, dB_u^k$$

where $F, G_1, \ldots, G_m$ are constant matrices and $(B_t^1, \ldots, B_t^m)$ is a standard Brownian motion.

(i) Show that $\varphi_{s,t}$ is incompressible if and only if $\mathrm{tr} G_k = 0$, $k = 1, \ldots, m$ and $\mathrm{tr} F = \frac{1}{2}\sum_k \mathrm{tr}(G_k^2)$.
(ii) Show that the Lyapunov exponent of $\{\varphi_t^{-1}(\mu)(A)\}$ (μ is the Lebesgue measure) is given by $\lambda = \mathrm{tr} F - \frac{1}{2}\sum_k \mathrm{tr}(G_k^2)$ if $0 < \mu(A) < \infty$.

(*Hint*: show that $\mathrm{div}\, \mathring{F}(x, t) = \sum_k (\mathrm{tr} G_k) B_t^k + \mathrm{tr}(F - \frac{1}{2}\sum_k G_k^2)t$, where $\mathring{F}$ is the Stratonovich's infinitesimal generator of $\varphi_{s,t}$.)

4.4 Semimartingale flows

Random infinitesimal generators

In Section 4.2 we studied the structure of Brownian flows. In particular we derived the forward and backward infinitesimal generators of the Brownian flow. In the course of the discussion, an important role was played by the semimartingale property of the process $\varphi_{s,t}(x)$, $t \in [s, T]$ for each fixed s. In this section we shall study non-Brownian stochastic flows having the semimartingale property. It will be shown that such stochastic flows have many properties similar to those of Brownian flows.

Let $\varphi_{s,t}$, $s, t \in [0, T]$ be a stochastic flow with values in G^k, where k is a non-negative integer. Let $\{\mathscr{F}_{s,t} : 0 \le s \le t \le T\}$ be the filtration generated by the flow $\varphi_{s,t}$. The forward part $\varphi_{s,t}$, $0 \le s \le t \le T$ is called a *forward $C^{k,\delta}$-semimartingale flow* if for every s, $\varphi_{s,t}$, $t \in [s, T]$ is a continuous $C^{k,\delta}$-semimartingale adapted to $(\mathscr{F}_{s,t} : t \in [s, T])$. The backward part $\varphi_{s,t}$, $0 \le t \le s \le T$ is called a *backward $C^{k,\delta}$-semimartingale flow* if for every s, $\varphi_{s,t}$, $t \in [0, s]$ is a continuous backward $C^{k,\delta}$-semimartingale adapted to $(\mathscr{F}_{s,t})$.

The stochastic flow $\varphi_{s,t}$ is called a *forward–backward $C^{k,\delta}$-semimartingale flow* if its forward part is a forward $C^{k,\delta}$-semimartingale flow and its backward part is a backward $C^{k,\delta}$-semimartingale flow, simultaneously.

The *forward Itô's and Stratonovich's random infinitesimal generators* (or *forward random infinitesimal generators in the sense of the Itô and the Stratonovich integrals*, respectively) of the forward semimartingale flows are defined similarly to those of forward Brownian flows. The *backward random infinitesimal generators* of the backward semimartingale flows are also defined in the same way.

We shall first restrict our attention to a forward $C^{k,\delta}$-semimartingale flow $\varphi_{s,t}$, $0 \le s \le t \le T$. The proof of the following theorem is similar to the proof of Theorem 4.2.8.

Theorem 4.4.1 *Let $\varphi_{s,t}$, $0 \le s \le t \le T$ be a forward $C^{k,\delta}$-semimartingale flow such that for every s the local characteristic belongs to the class $B^{k,\delta}$ where $k \ge 0$ and $\delta > 0$. Then there exists a unique continuous $C^{k,\varepsilon}$-semimartingale $F(x, t)$ with $F(x, 0) \equiv 0$ (for all $\varepsilon < \delta$) with local characteristic belonging to the class $B^{k,\delta}$ such that it is the forward Itô's random infinitesimal generator of $\varphi_{s,t}$. Further if $k \ge 2$, $\mathring{F} \equiv F - C$ is the unique forward Stratonovich's random infinitesimal generator, where C is the correction term of F.* □

In the case of Brownian flows, both the forward and backward infinitesimal generators are Brownian motions, so that those are forward–backward semimartingales adapted to the filtration generated by the flow. However, in the present case the above property is not immediately obvious. In the sequel we will prove this fact by showing that $\mathring{F}$ becomes Stratonovich's random infinitesimal generator of the backward flow. For this purpose, the forward semimartingale property of the inverse flow will be needed.

Theorem 4.4.2 *Let $\varphi_{s,t}$, $0 \le s \le t \le T$ be a forward $C^{k,\delta}$-semimartingale flow such that for every s, its local characteristic belongs to the class $B^{k,\delta}$ where $k \ge 3$ and $\delta > 0$. Then the inverse flow $\varphi_{t,s}(x) \equiv \varphi_{s,t}^{-1}$ is a forward semimartingale with respect to t for any s and x. Furthermore it satisfies the Stratonovich's stochastic differential equation:*

$$\varphi_{t,s}(x) = x - \int_s^t \partial\varphi_{s,r}(\varphi_{r,s}(x))^{-1}\mathring{F}(x, \circ dr), \tag{1}$$

where $\partial\varphi_{s,r}(x)^{-1}$ is the inverse matrix of the Jacobian matrix $\partial\varphi_{s,r}(x)$ of the map $\varphi_{s,r}(x)$. □

Before we proceed to the proof, we have to show that the Jacobian matrix $\partial\varphi_{s,t}(x)$ is invertible for any s, t, x.

Lemma 4.4.3 *For any s, t, x, the Jacobian matrix $\partial\varphi_{s,t}(x)$ has the inverse $\partial\varphi_{s,t}(x)^{-1}$. It satisfies*

$$\partial\varphi_{s,t}(x)^{-1} = I - \int_s^t \partial\varphi_{s,r}(x)^{-1}\partial\mathring{F}(\varphi_{s,r}(x), \circ\, dr), \tag{2}$$

where I is the identity matrix and $\partial\mathring{F}$ is the matrix $(\partial\mathring{F}^i/\partial x^j)$. Further for each s $\partial\varphi_{s,t}(x)^{-1}$ is a continuous forward $C^{k-1,\varepsilon}$-semimartingale for some $\varepsilon > 0$.

Proof Since the flow $\varphi_{s,t}$ satisfies the Stratonovich's equation based on $\mathring{F}(x, t)$, the Jacobian matrix satisfies

$$\partial\varphi_{s,t}(x) = I + \int_s^t \partial\mathring{F}(\varphi_{s,r}(x), \circ\, dr)\partial\varphi_{s,r}(x)$$

by Theorem 3.3.4. Consider now a Stratonovich's equation for a matrix-valued semimartingale J_t, $t \geq s$:

$$J_t = I - \int_s^t J_r\partial\mathring{F}(\varphi_{s,r}(x), \circ\, dr),$$

where s and x are fixed. It has a unique solution. We will show $J_t\partial\varphi_{s,t}(x) = I$, which will ensure the existence of the inverse map $\partial\varphi_{s,t}(x)^{-1}$. Apply Itô's formula. Then

$$\begin{aligned} J_t\partial\varphi_{s,t} - I &= \int_s^t J_r \circ d\partial\varphi_{s,r} + \int_s^t \circ\, dJ_r\partial\varphi_{s,r} \\ &= \int_s^t J_r\partial\mathring{F}(\varphi_{s,r}(x), \circ\, dr)\partial\varphi_{s,r}(x) - \int_s^t J_r\partial\mathring{F}(\varphi_{s,r}(x), \circ\, dr)\partial\varphi_{s,r}(x) \\ &= 0. \end{aligned}$$

The last assertion is obvious if we write the inverse matrix explicitly. □

Proof of Theorem 4.4.2 Consider a Stratonovich's stochastic differential equation based on

$$G(y, t) = -\int_s^t \partial\varphi_{s,r}(y)^{-1}\mathring{F}(\varphi_{s,r}(y), \circ\, dr).$$

Note that $\partial\varphi_{s,r}(y)^{-1}$ is a $C^{k-1,\varepsilon}$-semimartingale for some $\varepsilon > 0$, $\varphi_{s,r}(y)$ is a

$C^{k,\varepsilon'}$-semimartingale and $\mathring{F}(x, t)$ is a $C^{k-1,\varepsilon'}$-semimartingale for any $\varepsilon' < \delta$. Then $G(y, t)$ is a $C^{k-2,\varepsilon''}$-semimartingale for some $\varepsilon'' > 0$ by Theorem 3.3.4. Its local characteristic belongs to the class $(B^{k-1,\varepsilon''}, B^{k-2,\varepsilon''})$. Then for any initial condition the Stratonovich's equation based on $G(y, t)$ has a unique maximal solution by Theorem 3.4.7 since we assume $k \geq 3$. We denote the solution starting at x at time s by $\eta_{s,t}(x)$, $t \in [s, \sigma_\infty(x))$.

We claim $\eta_{s,t}(x) = \varphi_{t,s}(x)$ for $t \in [s, \sigma_\infty(x))$. We can apply Theorem 3.3.2 (generalized Itô's formula) for $\varphi_{s,t}(\eta_{s,t}(x))$, since $\varphi_{s,t}$ is a forward $C^{k,\delta}$-semimartingale with a local characteristic belonging to the class $B^{k,\delta}$, $k \geq 3$ by the assumption. Set $\varphi_{s,t}(x) = \varphi(x, t)$. Then we have

$$\begin{aligned}\varphi_{s,t}(\eta_{s,t}(x)) &= x + \int_s^t \varphi(\eta_{s,r}(x), \circ dr) + \sum_j \int_s^t \frac{\partial \varphi}{\partial x^j}(\eta_{s,r}(x), r) \circ d\eta_{s,r}^j(x) \\ &= x + \int_s^t \mathring{F}(\varphi_{s,r}(\eta_{s,r}(x)), \circ dr) \\ &\quad - \int_s^t \partial\varphi_{s,r}(\eta_{s,r}(x))\partial\varphi_{s,r}(\eta_{s,r}(x))^{-1}\mathring{F}(\varphi_{s,r}(\eta_{s,r}(x)), \circ dr) \\ &= x.\end{aligned}$$

Therefore we have $\eta_{s,t}(x) = \varphi_{s,t}^{-1}(x) = \varphi_{t,s}(x)$ if $t < \sigma_\infty(x)$. The relation also implies $\sigma_\infty(x) = T$ a.s. for any x. Hence equation (1) is verified. □

Theorem 4.4.4 *Let $\varphi_{s,t}$, $s, t \in [0, T]$ be a forward–backward $C^{k,\delta}$-semimartingale flow where $k \geq 3$ and $\delta > 0$. Suppose that for every s both of the forward and backward local characteristics belong to the class $B^{k,\delta}$. Let $\mathring{F}(x, t)$ be the forward Stratonovich's random infinitesimal generator. Then it is a backward $C^{k-1,\varepsilon}$-semimartingale for some $\varepsilon > 0$ and is the backward Stratonovich's random infinitesimal generator.*

Proof Set $X_{s,t}(x) = \varphi_{s,t}(x) - x$ and $X_t(x, s) = X_{s,t}(x)$. For each t, it is a backward C-semimartingale. Define

$$\hat{\mathring{F}}(x, s) = -\int_s^T X_T(\varphi_{T,r}^{-1}(x), \hat{d}r).$$

Then the backward flow $\varphi_{t,s}$ is governed by the backward Itô's equation based on $\hat{\mathring{F}}$. Let $\Delta = \{s = t_0 < \cdots < t_n = t\}$ be partitions. Then,

$$\sum_{h=0}^{n-1} (X_T(\varphi_{T,t_{h+1}}^{-1}(x), t_{h+1}) - X_T(\varphi_{T,t_{h+1}}^{-1}(x), t_h)) = -\sum_{h=0}^{n-1} X_{t_{h+1}}(x, t_h),$$

and it converges to $\hat{\mathring{F}}(x, t) - \hat{\mathring{F}}(x, s)$ in probability as $|\Delta| \to 0$ simlarly to the forward case. We have by Theorem 4.4.2

$$
\begin{aligned}
-\sum_h X_{t_{h+1}}(x, t_h) &= \sum_h \int_{t_h}^{t_{h+1}} \partial\varphi_{t_h,r}(\varphi_{r,t_h}(x))^{-1}\mathring{F}(x, \circ \mathrm{d}r) \\
&= \sum_h \int_{t_h}^{t_{h+1}} \partial\varphi_{t_h,r}(\varphi_{r,t_h}(x))^{-1}\mathring{F}(x, \mathrm{d}r) \\
&\quad + \frac{1}{2}\sum_h \langle \partial\varphi_{t_h,t_{h+1}}(\varphi_{t_{h+1},t_h}(x))^{-1}, \mathring{F}(x, t_{h+1}) - \mathring{F}(x, t_h)\rangle \\
&= I_1^\Delta + I_2^\Delta, \qquad \text{say.}
\end{aligned}
$$

Since $\partial\varphi_{t_h,r}(\varphi_{r,t_h}(x))^{-1}$ converges to the identity matrix a.s. as $r \to t_h$, we have $\lim_{|\Delta|\to 0} I_1^\Delta = \mathring{F}(x, t) - \mathring{F}(x, s)$ in probability.

In order to compute I_2^Δ, we shall apply the generalized Itô's formula. Note the relation (2). Then we have

$$
\begin{aligned}
\partial\varphi_{t_h,t_{h+1}}(\varphi_{t_{h+1},t_h}(x))^{-1} &= I - \int_{t_h}^{t_{h+1}} \partial\varphi_{t_h,r}(\varphi_{r,t_h}(x))^{-1}\partial\mathring{F}(x, \circ \mathrm{d}r) \\
&\quad - \sum_i \int_{t_h}^{t_{h+1}} \left.\frac{\partial(\partial\varphi_{t_h,r}(y)^{-1})}{\partial y^i}\right|_{y=\varphi_{r,t_h}(x)} \mathrm{d}\varphi_{r,t_h}^i(x).
\end{aligned}
$$

Therefore,

$$
\begin{aligned}
I_2^\Delta &= -\frac{1}{2}\sum_h \int_{t_h}^{t_{h+1}} (\partial\varphi_{t_h,r})(\varphi_{r,t_h}(x))^{-1} d\langle \partial\mathring{F}(x, r), \mathring{F}(x, r)\rangle \\
&\quad -\frac{1}{2}\sum_h\sum_i \left\langle \int_{t_h}^{t_{h+1}} \left.\frac{\partial(\partial\varphi_{t_h,r}(y)^{-1})}{\partial y^i}\right|_{y=\varphi_{r,t_h}} \mathrm{d}\varphi_{r,t_h}^i(x), \mathring{F}(x, t_{h+1}) - \mathring{F}(x, t_h)\right\rangle.
\end{aligned}
$$

Note that

$$
\sum_j \left\langle \frac{\partial\mathring{F}}{\partial x^j}(x, r), \mathring{F}^j(x, r)\right\rangle = 2C(x, r).
$$

Then the first term of I_2^Δ converges to $-(C(x, t) - C(x, s))$ in probability as $|\Delta| \to 0$. The second term converges to 0 since $\partial(\partial\varphi_{t_h,r})^{-1}/\partial y^i$ converges to 0 uniformly on compact sets with respect to y a.s. We have thus proved

$$
\hat{F}(x, t) - \hat{F}(x, s) = \mathring{F}(x, t) - \mathring{F}(x, s) - (C(x, t) - C(x, s)).
$$

Then the backward Stratonovich's random infinitesimal generator is equal to $\mathring{F}$. The proof is complete. □

Itô's formulas for stochastic flows

Let $\varphi_{s,t}$, $s, t \in [0, T]$ be a stochastic flow such that its forward part is a forward semimartingale flow satisfying the condition of Theorem 4.4.1. Let $F(x, t)$ be its Itô's forward random infinitesimal generator with local characteristic $(a^{ij}(x, y, t), b^i(x, t), A_t)$. Then if f is a C^2-function on $\mathbb{R}^d$ the forward part $\varphi_{s,t}$, $0 \le s \le t \le T$ satisfies

$$f(\varphi_{s,t}(x)) = f(x) + \sum_i \int_s^t \frac{\partial f}{\partial x^i}(\varphi_{s,r}(x))F^i(\varphi_{s,r}(x), \mathrm{d}r)$$

$$+ \frac{1}{2}\sum_{i,j}\int_s^t \frac{\partial^2 f}{\partial x^i \partial x^j}(\varphi_{s,r}(x))a^{ij}(\varphi_{s,r}(x), \varphi_{s,r}(x), r)\,\mathrm{d}A_r, \quad (3)$$

by Itô's formula. Further suppose that it is a forward–backward semimartingale flow satisfying the condition of Theorem 4.4.4. Then $\hat{F}(x, t) \equiv F(x, t) - 2C(x, t)$ is the backward Itô's random infinitesimal generator. Therefore, applying Itô's formula (to the backward variable), the backward part satisfies

$$f(\varphi_{t,s}(x)) = f(x) - \sum_i \int_s^t \frac{\partial f}{\partial x^i}(\varphi_{t,r}(x))\hat{F}^i(\varphi_{t,r}(x), \hat{\mathrm{d}}r)$$

$$+ \frac{1}{2}\sum_{i,j}\int_s^t \frac{\partial^2 f}{\partial x^i \partial x^j}(\varphi_{t,r}(x))a^{ij}(\varphi_{t,r}(x), \varphi_{t,r}(x), r)\,\mathrm{d}A_r. \quad (4)$$

We will call the above *Itô's second formula* for the stochastic flow, since it gives the differential rule of the flow for the second variable.

We shall obtain a differential rule of the flow for the first variable. We denote by $\mathring{F}(x, t)$ the Stratonovich random infinitesimal generator of the flow $\varphi_{s,t}(x)$.

Theorem 4.4.5 *Let $\varphi_{s,t}$, $s, t \in [0, T]$ be a stochastic flow satisfying the condition of Theorem 4.4.4. Then the forward flow $\varphi_{s,t}$, $0 \le s \le t \le T$ satisfies*

$$f(\varphi_{s,t}(x)) = f(x) + \sum_i \int_s^t \frac{\partial(f\circ\varphi_{r,t})}{\partial x^i}(x)\mathring{F}^i(x, \circ\hat{\mathrm{d}}r) \quad (5)$$

$$= f(x) + \sum_i \int_s^t \frac{\partial(f\circ\varphi_{r,t})}{\partial x^i}(x)F^i(x, \hat{\mathrm{d}}r)$$

$$+ \frac{1}{2}\sum_{i,j}\int_s^t \frac{\partial^2(f\circ\varphi_{r,t})}{\partial x^i \partial x^j}(x)a^{ij}(x, x, r)\mathrm{d}A_r, \quad (6)$$

and the backward flow $\varphi_{t,s}$, $0 \le s \le t \le T$ satisfies

$$f(\varphi_{t,s}(x)) = f(x) - \sum_i \int_s^t \frac{\partial(f\circ\varphi_{r,s})}{\partial x^i}(x)\mathring{F}^i(x, \circ\mathrm{d}r) \quad (7)$$

$$= f(x) - \sum_i \int_s^t \frac{\partial(f\circ\varphi_{r,s})}{\partial x^i}(x)F^i(x, \mathrm{d}r)$$

$$+ \frac{1}{2}\sum_{i,j}\int_s^t \frac{\partial^2(f\circ\varphi_{r,s})}{\partial x^i \partial x^j}(x)a^{ij}(x, x, r)\,\mathrm{d}A_r. \quad (8)$$

Proof We first show (7) by applying Theorem 4.4.2. Since $\varphi_{s,t}\circ\varphi_{t,s}(x)=x$, Jacobian matrices $\partial\varphi_{s,t}$ and $\partial\varphi_{t,s}$ are related by $\partial\varphi_{s,t}(\varphi_{t,s}(x))\partial\varphi_{t,s}(x)=$ identity. This means $\partial\varphi_{s,t}(\varphi_{t,s}(x))^{-1}=\partial\varphi_{t,s}(x)$. Then equation (1) is written as

$$\varphi_{t,s}(x)=x-\int_s^t \partial\varphi_{r,s}(x)\mathring{F}(x,\circ dr). \tag{9}$$

Then Itô's formula implies

$$\begin{aligned} f(\varphi_{t,s}(x))-f(x) &= -\sum_{i,j}\int_s^t \frac{\partial f}{\partial x^i}(\varphi_{r,s}(x))\frac{\partial\varphi_{r,s}^i}{\partial x^j}(x)\mathring{F}^j(x,\circ dr)\\ &= -\sum_j\int_s^t \frac{\partial(f\circ\varphi_{r,s})}{\partial x^j}(x)\mathring{F}^j(x,\circ dr). \end{aligned} \tag{10}$$

This shows formula (7). We will rewrite the last term using the Itô integral. Note the relation

$$\begin{aligned} &\sum_j\int_s^t \frac{\partial(f\circ\varphi_{r,s})}{\partial x^j}(x)\mathring{F}^j(x,\circ dr)\\ &\quad=\sum_j\int_s^t \frac{\partial(f\circ\varphi_{r,s})}{\partial x^j}(x)\mathring{F}^j(x,dr)+\frac{1}{2}\sum_j\left\langle\frac{\partial(f\circ\varphi_{t,s})}{\partial x^j}(x),\mathring{F}^j(x,t)\right\rangle. \end{aligned}$$

From (10) we have

$$\begin{aligned} &\frac{\partial(f\circ\varphi_{t,s})}{\partial x^j}(x)-\frac{\partial f}{\partial x^j}(x)\\ &\quad=-\sum_i\int_s^t \frac{\partial^2(f\circ\varphi_{r,s})}{\partial x^j\partial x^i}(x)\mathring{F}^i(x,\circ dr)-\sum_i\int_s^t\frac{\partial(f\circ\varphi_{r,s})}{\partial x^i}(x)\frac{\partial\mathring{F}^i}{\partial x^j}(x,\circ dr). \end{aligned}$$

Therefore

$$\begin{aligned} \sum_j\left\langle\frac{\partial(f\circ\varphi_{t,s})}{\partial x^j}(x),\mathring{F}^j(x,t)\right\rangle &= -\sum_{i,j}\int_s^t\frac{\partial^2(f\circ\varphi_{r,s})}{\partial x^j\partial x^i}(x)a^{ij}(x,x,r)\,dA_r\\ &\quad-2\sum_i\int_s^t\frac{\partial(f\circ\varphi_{r,s})}{\partial x^i}(x)c^i(x,r)\,dA_r, \end{aligned}$$

where $c^i(x,t)$ is the correction term based on $\mathring{F}(x,t)$. Consequently we obtain

$$\begin{aligned} &\sum_j\int_s^t\frac{\partial(f\circ\varphi_{r,s})}{\partial x^j}(x)\mathring{F}^j(x,\circ dr)\\ &\quad=\sum_j\int_s^t\frac{\partial(f\circ\varphi_{r,s})}{\partial x^j}(x)\hat{F}^j(x,dr)-\frac{1}{2}\sum_{i,j}\int_s^t\frac{\partial^2(f\circ\varphi_{r,s})}{\partial x^i\partial x^j}(x)a^{ij}(x,x,r)\,dA_r. \end{aligned}$$

Therefore (8) is proved.

For the proof of (5), we apply Theorem 4.4.2 to the backward flow $\varphi_{t,s}$. The theorem states that $\varphi_{t,s} \equiv \varphi_{s,t}^{-1}$ is a backward semimartingale with respect to s for any t and x. Equation (9) induces the following:

$$\varphi_{s,t}(x) = x + \int_s^t \partial\varphi_{r,t}(x)\mathring{F}(x, \circ \hat{d}r).$$

Then Itô's formula proves formula (5) similarly to the proof of formula (7). Formula (6) is derived from (5) similarly. □

The formulas of the above theorem are called *Itô's first formulas* of the stochastic flow.

Finally set $u(x, t) = f(\varphi_{t,0}(x))$. Formula (7) tells us that it satisfies the first order stochastic partial differential equation:

$$u(x, t) = f(x) - \sum_i \int_0^t \mathring{F}^i(x, \circ dr)\frac{\partial u}{\partial x^i}(x, r). \tag{11}$$

In other words, for any initial function f, the first order equation has at least one solution which is represented by $f(\varphi_{t,0}(x))$. Thus $\varphi_{t,0}$ can be regarded as a characteristic curve associated with equation (11). We will discuss a similar problem more systematically in Chapter 6.

4.5 Homeomorphic property of solutions of SDE

Construction of stochastic flows

In the previous section we have shown that a continuous semi-martingale flow with some regularity property can be represented as a solution of a stochastic differential equation. In this section we shall discuss the converse problem: we wish to show that a system of solutions of a stochastic differential equation defines a stochastic flow of homeomorphisms provided that the local characteristic of $F(x, t)$ governing the stochastic differential equation belongs to $B_b^{0,1}$.

Let $F(x, t) = (F^1(x, t), \ldots, F^d(x, t))$, $x \in \mathbf{R}^d$ be a continuous $C(\mathbf{R}^d : \mathbf{R}^d)$-valued semimartingale with local characteristic belonging to $B_b^{0,1}$. Consider an Itô's stochastic differential equation

$$\varphi_t = x + \int_s^t F(\varphi_r, dr). \tag{1}$$

We have seen in Theorem 3.4.1 that equation (1) has a unique global solution for any s, x. We denote its solution by $\varphi_{s,t}(x)$, $t \geq s$. One of the main objectives of this section is to prove the following.

Theorem 4.5.1

(i) *Assume that the local characteristic of $F(x, t)$ belongs to the class $B_b^{0,1}$. Then there exists a modification of the system of solutions denoted by $\varphi_{s,t}(x), 0 \le s \le t \le T$ such that it is a forward stochastic flow of homeomorphisms. Further for every s, $\varphi_{s,t}$, $t \in [s, T]$ is a $C^{0,\gamma}$-semimartingale flow for any $\gamma < 1$.*

(ii) *Assume that $F(x, t)$ is a Brownian motion with values in $C^{0,\gamma}$ with mean vector $\int_0^t b(x, r)\,dr$ and covariance $\int_0^t a(x, y, r)\,dr$ where a belongs to the class $\tilde{C}_{ub}^{0,1}$ and b belongs to the class $C_{ub}^{0,1}$. Then the associated flow is a Brownian flow with infinitesimal mean b and infinitesimal covariance a.* □

For the proof of the theorem, we need a number of L^p-estimates for the random field $\varphi_{s,t}(x)$. Throughout Lemmas 4.5.2–4.5.8, we assume that the local characteristic $(a^{ij}(x, y, t), b^i(x, t), A_t)$ of F belongs to the class $B_{ub}^{0,1}$ and the increasing process A_t is identical to t. Hence the pair (a, b) is uniformly Lipschitz continuous and of uniformly linear growth.

To begin with, we shall prove a lemma concerning a martingale property for a function of $\varphi_{s,t}(x)$, which is a direct consequence of Itô's formula. Let n be any positive integer. Associated with the local characteristic $(a^{ij}(x, y, t), b^i(x, t), t)$ of the C-semimartingale $F(x, t)$, we define a second order linear partial differential operator of random coefficients over the space $\mathbb{R}^{nd}$:

$$L_t^{(n)} f(x_1, \ldots, x_n, \omega) = \frac{1}{2} \sum_{i,j,p,q} a^{ij}(x_p, x_q, t, \omega) \frac{\partial^2 f}{\partial x_p^i \partial x_q^j}(x_1, \ldots, x_n)$$
$$+ \sum_{i,p} b^i(x_p, t, \omega) \frac{\partial f}{\partial x_p^i}(x_1, \ldots, x_n), \tag{2}$$

where $x_p = (x_p^1, \ldots, x_p^d)$, $p = 1, \ldots, n$ are elements of $\mathbb{R}^d$ and $f(x_1, \ldots, x_n)$ is a C^2-function on the space $\mathbb{R}^{nd}$.

Lemma 4.5.2 *Let f be a C^2-function over $\mathbb{R}^{nd}$ such that f and its derivatives are of polynomial growth. Then for each fixed $x_1, \ldots, x_n, s_1, \ldots, s_n$ and $s \ge \max_{1 \le i \le n}\{s_i\}$, the stochastic process*

$$f(\varphi_{s_1,t}(x_1), \ldots, \varphi_{s_n,t}(x_n)) - f(\varphi_{s_1,s}(x_1), \ldots, \varphi_{s_n,s}(x_n))$$
$$- \int_s^t L_r^{(n)} f(\varphi_{s_1,r}(x_1), \ldots, \varphi_{s_n,r}(x_n))\,dr \tag{3}$$

with time parameter $t \in [s, T]$ is a martingale with mean 0.

Proof Our arguments are close to those of Theorem 4.2.4. Let $N^i(x, t)$ be the martingale part of $F^i(x, t)$. Then by Itô's formula the process (3) equals

$$\sum_{i,p} \int_s^t \frac{\partial f}{\partial x_p^i} N^i(\varphi_{s_p,r}(x_p), dr).$$

It is a localmartingale with the quadratic variation

$$\sum_{i,j,p,q} \int_s^t \frac{\partial f}{\partial x_p^i} \frac{\partial f}{\partial x_q^j} a^{ij}(\varphi_{s_p,r}(x_p), \varphi_{s_q,r}(x_q), r)\, dr.$$

Since $\varphi_{s_p,r}(x_p)$ belongs to L^p for any $p > 1$ (Lemma 3.4.2) and $a(x, y, r)$, $\partial f/\partial x_p^i$ are functions of polynomial growth, the above is integrable. Therefore the localmartingale is actually a martingale with mean 0 (*cf.* Corollary 2.2.7). The lemma is thus proved. □

L^p-estimates of solutions

We now proceed to L^p-estimates of $\varphi_{s,t}(x)$.

Lemma 4.5.3 *For each real p, there exists a positive constant $C = C(p)$ such that*

$$E[(1 + |\varphi_{s,t}(x)|^2)^p] \le C(1 + |x|^2)^p \tag{4}$$

holds for any x, y, s, t.

Proof Let $L_t^{(1)}$ be the second order operator of (2) when $n = 1$. Set $g(x) = 1 + |x|^2$ and $f(x) = g(x)^p$. Then we have

$$L_t^{(1)} f(x) = 2pg(x)^{p-1} \sum_i b^i(x, t)x^i$$

$$+ pg(x)^{p-2} \sum_{i,j} \{g(x)\delta_{ij} + 2(p-1)x^i x^j\} a^{ij}(x, x, t).$$

Since $a(x, y, t)$ and $b(x, t)$ are of linear growth, there exists $C_i > 0$, $i = 1, 2$ such that

$$\left| \sum_i b^i(x, t)x^i \right| \le C_1(1 + |x|^2),$$

$$\left| \sum_{i,j} \{g(x)\delta_{ij} + 2(p-1)x^i x^j\} a^{ij}(x, x, t) \le C_2(1 + |x|^2)^2. \right.$$

Therefore there exists $C_3 > 0$ such that $|L_t^{(1)} f| \le C_3 f$. Now apply Lemma 4.5.2. Since $\varphi_{s,s}(x) = x$, we obtain

$$E[f(\varphi_{s,t}(x))] \le f(x) + C_3 \int_s^t E[f(\varphi_{s,r}(x))]\, dr.$$

Gronwall's inequality yields

$$E[f(\varphi_{s,t}(x))] \le e^{C_3(t-s)} f(x).$$

This proves inequality (4) by setting $C = e^{C_3 T}$. The proof is complete. □

Lemma 4.5.4 *For each $p > 1$ there exists a positive constant $C = C(p)$ such that*

$$E[|\varphi_{s,t}(x) - \varphi_{s,t'}(x)|^{2p}] \le C(1 + |x|^2)^p |t - t'|^p \tag{5}$$

holds for any x, s, t, t'.

Proof Assume $t' < t$. Then

$$\varphi_{s,t}(x) - \varphi_{s,t'}(x) = \int_{t'}^{t} N(\varphi_{s,r}(x), dr) + \int_{t'}^{t} b(\varphi_{s,r}(x), r)\, dr.$$

By Burkholder's inequality, there exists $C_1 > 0$ such that

$$E\left[\left|\int_{t'}^{t} N(\varphi_{s,r}(x), dr)\right|^{2p}\right] \le C_1 E\left[\left|\int_{t'}^{t} tr(a)(\varphi_{s,r}(x), r)\, dr\right|^{p}\right].$$

Since $|tr(a)(x, t)| \le C_2(1 + |x|^2)$ holds, the right hand side is bounded by

$$C_1 C_2^p E\left[\left|\int_{t'}^{t} (1 + |\varphi_{s,r}(x)|^2)\, dr\right|^{p}\right]$$
$$\le C_1 C_2^p |t - t'|^{p-1} E\left[\int_{t'}^{t} (1 + |\varphi_{s,r}(x)|^2)^p\, dr\right].$$

The above is bounded by $C_3 |t - t'|^p (1 + |x|^2)^p$ by Lemma 4.5.3. Similarly we have the estimate

$$E\left[\left|\int_{t'}^{t} b(\varphi_{s,r}(x), r)\, dr\right|^{2p}\right] \le C_4 |t - t'|^{2p} (1 + |x|^2)^p.$$

These two estimates imply (5). The proof is complete. □

Lemma 4.5.5 *For each real p these exists a positive constant $C = C(p)$ such that*

$$E[(\varepsilon + |\varphi_{s,t}(x) - \varphi_{s',t}(x')|^2)^p] \le CE[(\varepsilon + |\varphi_{s,s'}(x) - x'|^2)^p] \tag{6}$$

holds for any s, s', t, x, x', and $\varepsilon > 0$. We have in particular

$$E[(\varepsilon + |\varphi_{s,t}(x) - \varphi_{s,t}(x')|^2)^p] \le C(\varepsilon + |x - x'|^2)^p. \tag{7}$$

Proof Let $L_t^{(2)}$ be the second order partial differential operator of (2) when $n = 2$. Set $g(x, y) = \varepsilon + |x - y|^2$ and $f(x, y) = g(x, y)^p$. Then we have

$$L_t^{(2)}f(x, y) = 2pg(x, y)^{p-1} \sum_i (b^i(x, t) - b^i(y, t))(x^i - y^i)$$
$$+ pg(x, y)^{p-2} \sum_{i,j} \{g(x, y)\delta_{ij} + 2(p-1)(x^i - y^i)(x^j - y^j)\}$$
$$\times \{a^{ij}(x, x, t) - 2a^{ij}(x, y, t) + a^{ij}(y, y, t)\}.$$

Since $a(x, y, t)$ and $b(x, t)$ are uniformly Lipschitz continuous, there exist $C_i > 0$, $i = 1, 2$ such that

$$\left|\sum_i (b^i(x, t) - b^i(y, t))(x^i - y^i)\right|$$
$$\leq C_1 g(x, y) \left|\sum_{i,j} \{g(x, y)\delta_{ij} + 2(p-1)(x^i - y^i)(x^j - y^j)\}\right.$$
$$\left. \times \{a^{ij}(x, x, t) - 2a^{ij}(x, y, t) + a^{ij}(y, y, t)\}\right| \leq C_2 g(x, y)^2.$$

Therefore there exists $C_3 > 0$ such that $|L_t^{(2)}f(x, y)| \leq C_3 f(x, y)$. Now apply Lemma 4.5.2. Then we obtain

$$E[f(\varphi_{s,t}(x), \varphi_{s',t}(x'))]$$
$$= E[f(\varphi_{s,s'}(x), x')] + E\left[\int_{s'}^t L_r^{(2)} f(\varphi_{s,r}(x), \varphi_{s',r}(x'))\, dr\right]$$
$$\leq E[f(\varphi_{s,s'}(x), x')] + C_3 \int_{s'}^t E[f(\varphi_{s,r}(x), \varphi_{s',r}(x'))]\, dr.$$

Gronwall's inequality yields

$$E[f(\varphi_{s,t}(x), \varphi_{s',t}(x'))] \leq e^{C_3(t-s')} E[f(\varphi_{s,s'}(x), x')].$$

This proves (6). In particular if $s = s'$, the right hand side is $e^{C_3(t-s)} \times (\varepsilon + |x - x'|^2)^p$. The proof is complete. □

Lemma 4.5.6 *For each $p > 1$ there exists a positive constant $C = C(p)$ such that*

$$E[|\varphi_{s,t}(x) - \varphi_{s',t'}(x')|^{2p}]$$
$$\leq C\{|x - x'|^{2p} + (1 + |x|^2 + |x'|^2)^p(|t - t'|^p + |s - s'|^p)\} \quad (8)$$

holds for any s, s', t, t' and x, x'.

Proof Let ε tend to 0 in (6). Then we have

$$E[|\varphi_{s,t}(x) - \varphi_{s',t}(x')|^{2p}] \leq 2^{2p} C\{E[|\varphi_{s,s'}(x) - x|^{2p}] + |x - x'|^{2p}\}.$$

Apply Lemma 4.5.4 to $E[|\varphi_{s,s'}(x) - x|^{2p}]$. Then we obtain

$$E[|\varphi_{s,t}(x) - \varphi_{s',t}(x')|^{2p}] \leq C_2\{(1 + |x|^2)^p |s - s'|^p + |x - x'|^{2p}\}. \quad (9)$$

We have, again from Lemma 4.5.4

$$E[|\varphi_{s',t}(x') - \varphi_{s',t'}(x')|^{2p}] \le C_3(1 + |x'|^2)^p|t - t'|^p. \tag{10}$$

These two inequalities (9) and (10) establish the lemma. □

Homeomorphic property

Now Lemma 4.5.6 tells us that the solution $\varphi_{s,t}(x)$ has a modification which is continuous in three variables (s, t, x) a.s. by Kolmogorov's theorem (Theorem 1.4.1). Then for almost all ω, $\varphi_{s,t}(\omega) \equiv \varphi_{s,t}(\cdot, \omega)$ defines a continuous map $\mathbb{R}^d \to \mathbb{R}^d$. We will prove that the map is actually a homeomorphism of $\mathbb{R}^d$ onto itself a.s.

We will first consider the 'one to one' property of the map $\varphi_{s,t}(\omega)$. Lemma 4.5.5 implies the inequality

$$E[|\varphi_{s,t}(x) - \varphi_{s,t}(y)|^{2p}] \le C'|x - y|^{2p} \tag{11}$$

for negative p. This shows that if $x \ne y$, then $\varphi_{s,t}(x) \ne \varphi_{s,t}(y)$ a.s. for any $s < t$. But this does not imply immediately that the map $\varphi_{s,t}(\omega)$ is one to one a.s. To prove the latter assertion, we require a lemma.

Lemma 4.5.7 *Set*

$$\eta_{s,t}(x, y) = \frac{1}{|\varphi_{s,t}(x) - \varphi_{s,t}(y)|}. \tag{12}$$

Then for each $p > 1$, there exists a positive constant $C = C(p)$ such that for every $\delta > 0$

$$\begin{aligned} &E[|\eta_{s,t}(x, y) - \eta_{s',t'}(x', y')|^{2p}] \\ &\quad \le C\delta^{-4p}\{|x - x'|^{2p} + |y - y'|^{2p} \\ &\qquad + (1 + |x| + |x'| + |y| + |y'|)^{2p}(|t - t'|^p + |s - s'|^p)\} \end{aligned} \tag{13}$$

holds for any $s < t$ and *x, y, x', y' such that $|x - y| \ge \delta$ and $|x' - y'| \ge \delta$.*

Proof A simple computation yields

$$\begin{aligned} &|\eta_{s,t}(x, y) - \eta_{s',t'}(x', y')|^{2p} \\ &\quad \le 2^p \eta_{s,t}(x, y)^{2p} \eta_{s',t'}(x', y')^{2p} \\ &\qquad \times \{|\varphi_{s,t}(x) - \varphi_{s',t'}(x')|^{2p} + |\varphi_{s,t}(y) - \varphi_{s',t'}(y')|^{2p}\}. \end{aligned}$$

Take expectations for both sides and use Hölder's inequality. Then,

$$\begin{aligned} &E[|\eta_{s,t}(x, y) - \eta_{s',t'}(x', y')|^{2p}] \\ &\quad \le 2^p E[|\eta_{s,t}(x, y)|^{8p}]^{1/4} E[|\eta_{s',t'}(x', y')|^{8p}]^{1/4} \\ &\qquad \times \{E[|\varphi_{s,t}(x) - \varphi_{s',t'}(x')|^{4p}]^{1/2} + E[|\varphi_{s,t}(y) - \varphi_{s',t'}(y')|^{4p}]^{1/2}\}. \end{aligned}$$

From (11) we have

$$E[|\eta_{s,t}(x, y)|^{8p}]^{1/4} \leq C'|x - y|^{-2p} \leq C'\delta^{-2p},$$

where $|x - y| \geq \delta$. Then we get inequality (13) by applying Lemma 4.5.6. □

We can prove the 'one to one' property of the continuous map $\varphi_{s,t}(\omega)$ for almost all ω. Take p as large as $p > 2(d + 1)$ in Lemma 4.5.7. Kolmogorov's theorem states that $\eta_{s,t}(x, y)$ has a modification which is continuous in (s, t, x, y) in domain $\{(s, t, x, y) : s < t, |x - y| \geq \delta\}$. Since δ is arbitrary, it is also continuous in the domain $\{(s, t, x, y) : s < t, x \neq y\}$. This proves that for almost all ω, the map $\varphi_{s,t}(\omega) : \mathbb{R}^d \to \mathbb{R}^d$ is one to one for all $0 < s < t < T$.

We will next consider the onto property of the map $\varphi_{s,t}$. We need a lemma.

Lemma 4.5.8 *Set $\hat{x} = |x|^{-2}x$ if $x \neq 0$ and define*

$$\eta_{s,t}(\hat{x}) = \begin{cases} \dfrac{1}{1 + |\varphi_{s,t}(x)|} & \text{if } \hat{x} \neq 0, \\ 0 & \text{if } \hat{x} = 0. \end{cases} \tag{14}$$

Then for each $p > 1$, there exists a positive constant $C = C(p)$ such that

$$E[|\eta_{s,t}(\hat{x}) - \eta_{s',t'}(\hat{x}')|^{2p}] \leq C\{|\hat{x} - \hat{x}'|^{2p} + |t - t'|^p + |s - s'|^p\} \tag{15}$$

holds for any s, s', t, t' and $\hat{x}, \hat{x}'$.

Proof Suppose $\hat{x}$ and $\hat{x}'$ are not 0. Since

$$|\eta_{s,t}(\hat{x}) - \eta_{s',t'}(\hat{x}')|^{2p} \leq \eta_{s,t}(\hat{x})^{2p}\eta_{s',t'}(\hat{x}')^{2p}|\varphi_{s,t}(x) - \varphi_{s',t'}(x')|^{2p},$$

we have by Hölder's inequality

$$\begin{aligned} E[|\eta_{s,t}(\hat{x}) - \eta_{s',t'}(\hat{x}')|^{2p}] &\leq E[|\eta_{s,t}(\hat{x})|^{8p}]^{1/4}E[|\eta_{s',t'}(\hat{x}')|^{8p}]^{1/4} \\ &\quad \times E[|\varphi_{s,t}(x) - \varphi_{s',t'}(x')|^{4p}]^{1/2}. \end{aligned}$$

Apply Lemmas 4.5.3 and 4.5.6. Then the right hand side is dominated by

$$\begin{aligned} &C(1+|x|)^{-2p}(1+|x'|)^{-2p}\{|x-x'|^{2p}+(1+|x|+|x'|)^{2p}(|t-t'|^p+|s-s'|^p)\} \\ &\leq C\{|\hat{x} - \hat{x}'|^{2p} + |t - t'|^p + |s - s'|^p\}, \end{aligned}$$

if $\hat{x}$ and $\hat{x}'$ are not 0. Here we have used the inequality $(1 + |x|)^{-1} \times (1 + |x'|)^{-1}|x - x'| \leq |\hat{x} - \hat{x}'|$. In the case where $\hat{x} = 0$, we have by Lemma 4.5.3,

$$E[|\eta_{s',t'}(\hat{x}')|^{2p}] \le C'(1+|x'|)^{-2p} \le C'|\hat{x}'|^{2p}.$$

Therefore the inequality of the lemma follows. □

The 'onto' property of the map $\varphi_{s,t}(\omega)$ follows from Lemma 4.5.8. Take p greater than $d+3$. Then by Kolmogorov's theorem, for almost all ω $\eta_{s,t}(\hat{x}, \omega)$ is continuous at $\hat{x}=0$ for all $s<t$. We will fix such ω for a moment. Then $\varphi_{s,t}(\omega)$ can be extended to a continuous map from $\hat{\mathbb{R}}^d$ into itself for all $s<t$, where $\hat{\mathbb{R}}^d \equiv \mathbb{R}^d \cup \{\infty\}$ is the one point compactification of $\mathbb{R}^d$. Further, the extension $\tilde{\varphi}_{s,t}(x, \omega)$ is continuous in (s, t, x). The map $\hat{\phi}_{s,t}(\omega): \hat{\mathbb{R}}^d \to \hat{\mathbb{R}}^d$ is then homotopic to the identity map $\hat{\phi}_{s,s}(\omega)$, so that it is an onto map by a well known theorem in homotopic theory. The restriction of $\varphi_{s,t}(\omega)$ to $\mathbb{R}^d$ is again an 'onto' map since $\hat{\phi}_{s,t}(\infty, \omega)=\infty$.

The map $\varphi_{s,t}(\omega): \mathbb{R}^d \to \mathbb{R}^d$ is therefore one-to-one and onto for all $s<t$. Hence the inverse map $\varphi_{s,t}(\omega)^{-1}: \mathbb{R}^d \to \mathbb{R}^d$ is also one-to-one and onto for all $s<t$. We claim that $\varphi_{s,t}(\omega)^{-1}(x)$ is continuous in (s, t, x). Let $\{(s_n, t_n, x_n)\}$ be a sequence converging to a point (s, t, x). Set $y_n = \varphi_{s_n,t_n}^{-1}(x_n)$. Since $\hat{\mathbb{R}}^d$ is compact, a subsequence $\{y_n\}$ converges to y in $\hat{\mathbb{R}}^d$. Then we have $x=\hat{\phi}_{s,t}(y, \omega)$. Hence y belongs to $\mathbb{R}^d$ and it does not depend on the choice of the subsequence $\{y_{n_k}\}$. This means that $y_n = \varphi_{s_n,t_n}(\omega)^{-1}(x_n)$ converges to $y=\varphi_{s,t}^{-1}(x)$ proving that $\varphi_{s,t}(\omega)^{-1}(x)$ is continuous in (s, t, x).

We can now complete the proof of our main theorem in this section.

Proof of Theorem 4.5.1 Assume first that the local characteristic of F belongs to the class $B_{ub}^{0,1}$ and $A_t \equiv t$. It remains to prove that the continuous modification $\varphi_{s,t}(x)$ of the solution has the property $\varphi_{s,s}=$ identity and the flow property $\varphi_{s,u}=\varphi_{t,u}\circ\varphi_{s,t}$ for all $s<t<u$. The first property is obvious since $\varphi_{s,s}(x)=x$ holds a.s. for any s. Now $\varphi_{t,u}(y)$ satisfies

$$\varphi_{t,u}(y) = y + \int_t^u F(\varphi_{t,r}(y), dr). \tag{16}$$

The last integral has a modification continuous in (t, u, y). Further we have

$$\int_t^u F(\varphi_{t,r}(y), dr)|_{y=\varphi_{s,t}(x)} = \int_t^u F(\varphi_{t,r}(\varphi_{s,t}(x)), dr)$$

by Theorem 3.3.3. Therefore setting $y=\varphi_{s,t}(x)$ in (16), we obtain

$$\varphi_{t,u}(\varphi_{s,t}(x)) = \varphi_{s,t}(x) + \int_t^u F(\varphi_{t,r}(\varphi_{s,t}(x)), dr).$$

Define $\varphi'_{s,r}(x)$ by $\varphi'_{s,r}(x)=\varphi_{s,r}(x)$ if $r<t$ and $\varphi'_{s,r}(x)=\varphi_{t,r}(\varphi_{s,t}(x))$ if $r>t$.

Then the above equality proves that $\varphi'_{s,r} = \varphi_{s,r}$ a.s. for any $r > s$. Therefore the flow property is satisfied.

We have thus proved that the solution of equation (1) has a modification of a forward stochastic flow of homeomorphisms. For each s, the flow $\varphi_{s,t}$, $t \in [s, T]$ can be regarded as a continuous $C^{0,\gamma}$-semimartingale.

We shall next prove the theorem for the general case by changing the scale of the time and reducing it to the case $A_t \equiv t$. Let τ_t be the inverse function of A_t. Set $\tilde{\mathscr{F}}_t = \mathscr{F}_{\tau_t}$ and $\tilde{F}(x, t) = F(x, \tau_t)$. Then $\tilde{F}(x, t)$ is a $C^{0,\gamma}$-semimartingale for any $\gamma < 1$ and its local characteristic belongs to $B_{ub}^{0,1}$. Then the solution of the stochastic differential equation based on $\tilde{F}(x, t)$ has a modification of a stochastic flow of homeomorphisms generated by $\tilde{F}(x, t)$. Define $\varphi_{s,t} = \tilde{\varphi}_{A_s, A_t}(x)$. Then it is a forward stochastic flow of homeomorphisms generated by $F(x, t)$. Obviously it is a continuous $C^{0,\gamma}$-semimartingale flow.

Next, assume that $F(x, t)$ is a Brownian motion. Since the associated flow $\varphi_{s,t}$ is measurable with respect to $\sigma(F(\cdot, v) - F(\cdot, u) : s \le u < v \le t)$, the flow has independent increments. Therefore it is a Brownian flow. It satisfies

$$E[\varphi_{s,t}(x)] - x = \int_s^t E[b(\varphi_{s,r}(x), r)]\, dr. \tag{17}$$

Here $b(\varphi_{s,r}(x), r)$ converges to $b(x, s)$ a.s. and in L^1 as $r \downarrow s$. Therefore, the infinitesimal mean of the flow is $b(x, s)$. We have further

$$\begin{aligned}
&\lim_{h \to 0+} \frac{1}{h} E[(\varphi_{s,s+h}(x) - x)(\varphi_{s,s+h}(y) - y)^t] \\
&= \lim_{h \to 0+} \frac{1}{h} E\left[\left\langle \int_s^{s+h} F(\varphi_{s,r}(x), dr), \int_s^{s+h} F(\varphi_{s,r}(y), dr) \right\rangle\right] \\
&= \lim_{h \to 0+} \frac{1}{h} \int_s^{s+h} E[a(\varphi_{s,r}(x), \varphi_{s,r}(y), r)]\, dr \\
&= a(x, y, s).
\end{aligned}$$

Therefore a is the infinitesimal covariance. The proof is complete. □

We can now give the proof of a theorem stated in Section 4.2.

Proof of Theorem 4.2.5. On a suitable probability space we can define a C-valued Brownian motion $F(x, t)$ with mean vector $\int b(x, r)\, dr$ and covariance matrix $\int a(x, y, r)\, dr$ (Exercise 4.2.12). Let $\varphi_{s,t}$ be the Brownian flow generated by $F(x, t)$. It has infinitesimal mean b and infinitesimal covariance a. We show that it satisfies Condition (A.2). Since the function b is of linear growth inequality (5) of Section 4.2 follows from (17) and Lemma 4.5.3. Inequality (6) of Section 4.2 follows from Lemma 4.5.4, using Schwarz's inequality. The proof is complete. □

Exercise 4.5.9 Assume the same condition as in Lemmas 4.5.3–4.5.8. Let $0 < \varepsilon < 1$ and $\hat{x} = |x|^{-(1+\varepsilon)}x$ if $x \neq 0$. Define

$$\eta_{s,t}(\hat{x}) = \begin{cases} \dfrac{|\varphi_{s,t}(x)|}{1 + |x|^{1+\varepsilon}} & \text{if } \hat{x} \neq 0, \\ 0 & \text{if } \hat{x} = 0. \end{cases}$$

(i) Show that

$$E[|\eta_{s,t}(\hat{x}) - \eta_{s',t'}(\hat{x}')|^{2p}] \leq C\{|\hat{x} - \hat{x}'|^{2p} + |t - t'|^p + |s - s'|^p\}$$

holds for all t, t', s, s' and $\hat{x}$, $\hat{x}'$.

(ii) Deduce from this

$$\lim_{|x|\to\infty} \frac{|\varphi_{s,t}(x)|}{(1 + |x|)^{1+\varepsilon}} = 0 \qquad \text{uniformly in } s, t.$$

(iii) Show that $\sup_{x,t,s} |\varphi_{s,t}(x)|(1 + |x|)^{-(1+\varepsilon)}$ has moments of any order.

(*Hint*: if $|x'| \leq |x|$ then $|x - x'| \geq \varepsilon |x - x'|(|x||x'|^{\varepsilon})^{-1}$ and

$$|\eta_{s,t}(\hat{x}) - \eta_{s',t'}(\hat{x}')| \leq \frac{|\varphi_{s,t}(x) - \varphi_{s',t'}(x')|}{(1 + |x|)^{1+\varepsilon}} + \frac{(1 + \varepsilon)|x - x'|}{(1 + |x|)(1 + |x'|)^{1+\varepsilon}} |\varphi_{s',t'}(x')|.$$

Then apply Lemmas 4.5.3 and 4.5.6.)

Exercise 4.5.10 (*cf.* Lemma 4.5.7) Assume the same condition as in Exercise 4.5.9. Let $0 < \varepsilon < 1$ and $\hat{x} = |x|^{-(2-\varepsilon)}x$ if $x \neq 0$. Define

$$\eta_{s,t}(\hat{x}) = \begin{cases} = \dfrac{(1 + |x|)^{\varepsilon}}{(1 + |\varphi_{s,t}(x)|)} & \text{if } \hat{x} \neq 0, \\ = 0 & \text{if } \hat{x} = 0. \end{cases}$$

(i) Show that

$$E[|\eta_{s,t}(\hat{x}) - \eta_{s',t'}(\hat{x}')|^{2p}] \leq C[|\hat{x} - \hat{x}'|^{2p} + |t - t'|^p + |s - s'|^p\}$$

holds for any t, t', s, s' and $\hat{x}$, $\hat{x}'$.

(ii) Deduce from this that

$$\lim_{|x|\to\infty} \frac{(1 + |x|)^{\varepsilon}}{1 + |\varphi_{s,t}(x)|} = 0 \qquad \text{uniformly in } s, t.$$

(iii) Show that $\sup_{x,t,s}(1+|x|)^{\varepsilon}(1+|\varphi_{s,t}(x)|)^{-1}$ has moments of any order.

(*Hint*: if $|x'| \le |x|$ then $|\hat{x}-\hat{x}'| \ge (1-\varepsilon)|x-x'|(|x||x'|^{1-\varepsilon})^{-1}$ and

$$|\eta_{s,t}(\hat{x})-\eta_{s',t'}(\hat{x}')| \le \frac{(1+|x'|)^{\varepsilon}|\varphi_{s,t}(x)-\varphi_{s',t'}(x')|}{(1+|\varphi_{s,t}(x)|)(1+|\varphi_{s',t'}(x')|)} + \frac{\varepsilon|x-x'|}{(1+|x'|)^{1-\varepsilon}}\frac{1}{(1+|\varphi_{s,t}(x)|)}.$$

Then apply Lemmas 4.5.3 and 4.5.6.)

4.6 Diffeomorphic property of solutions of SDE

Differentiability of solutions

We shall next discuss the smoothness of the solution $\varphi_{s,t}(x)$ with respect to the spatial variable x. It will be shown that the problem is reduced to the differentiability of the solution with respect to a suitable parameter. For this purpose we introduce a continuous semimartingale with a parameter and consider the stochastic differential equation based on it.

Let $G(\lambda,\tau,t)$, $(\lambda,\tau)\in\mathbb{R}^e\times[0,T]$ be a family of continuous $\mathbb{R}^d$-semimartingales with parameter (λ,τ) with local characteristic $(a(\lambda,\tau,\lambda',\tau',t), b(\lambda,\tau,t), t)$. Let $0<\delta\le 1$, $0<\gamma\le\frac{1}{2}$ and $p\ge 1$. We assume that both a and b are continuous random fields and continuously differentiable with respect to λ and λ'. Let $a' = D_\lambda^\alpha D_{\lambda'}^\alpha a$, $b' = D_\lambda^\alpha b$ for $|\alpha|\le 1$. Set

$$L_1^{\alpha,\delta,\gamma}(\lambda,\tau,\lambda',\tau',t) = \frac{\|a'(\lambda,\tau,\lambda,\tau,t)-2a'(\lambda,\tau,\lambda',\tau',t)+a'(\lambda',\tau',\lambda',\tau',t)\|}{|\lambda-\lambda'|^{2\delta}+|\tau-\tau'|^{2\gamma}},$$

$$L_2^{\alpha,\delta,\gamma}(\lambda,\tau,\lambda',\tau',t) = \frac{|b'(\lambda,\tau,t)-b'(\lambda',\tau',t)|}{|\lambda-\lambda'|^{\delta}+|\tau-\tau'|^{\gamma}}.$$

These are called *L^p-bounded* if $E[|L_i^{\alpha,\delta,\gamma}(\lambda,\tau,\lambda',\tau',t)|^p]$ are bounded with respect to $(\lambda,\tau,\lambda',\tau',t)$. Now the local characteristic (a,b) is said to belong to the class $B_p^{1,\delta,\gamma}$ if a, b, $L_i^{0,1,\gamma}$ and $L_i^{\alpha,\delta,\gamma}$ for $|\alpha|=1$ are all L^p-bounded.

Now let $G_1(\lambda,\tau,t)$, $(\lambda,\tau)\in\mathbb{D}\times[0,T]$ be a family of continuous $\mathbb{R}^d$-valued semimartingales with parameter (λ,τ) and let $G_2(\lambda,\tau,t)$ be a family of continuous $\mathbb{R}^d\otimes\mathbb{R}^d$-valued semimartingales with parameter (λ,τ). Let $G_3(y,t)$ be a continuous $C(\mathbb{R}^d:\mathbb{R}^d)$-valued semimartingale. For these three semimartingales, we assume the following.

Condition (A.5)$_{\delta,\gamma}$

(1) *The local characteristic (a_1,b_1,t) of G_1 belongs to the class $B_p^{1,\delta,\gamma}$ for any $p\ge 1$.*

(2) *The local characteristic (a_2, b_2, t) of G_2 belongs to the class $B_p^{1,\delta,\gamma}$ for any $p \geq 1$. Further, a_2, b_2 are uniformly bounded.*
(3) *The local characteristic (a_3, b_3, t) of G_3 belongs to the class $B_{ub}^{1,\delta}$.*

Define

$$G(y, \lambda, \tau, t) = G_1(\lambda, \tau, t) + G_2(\lambda, \tau, t)y + G_3(y, t). \tag{1}$$

It is a family of continuous $C(\mathbb{R}^d : \mathbb{R}^d)$-valued semimartingales with parameter (λ, τ). Consider Itô's stochastic differential equation with parameter (λ, τ):

$$\eta_t = y + \int_s^t G(\eta_u, \lambda, \tau, du). \tag{2}$$

For each y, λ, τ and s it has a unique solution. We denote it by $\eta_{s,t}(y, \lambda, \tau)$. For a given C^∞-function $q(\lambda)$ with values in $\mathbb{R}^d$, we set $\eta_{s,t}(\lambda) \equiv \eta_{s,t}(q(\lambda), \lambda, s)$. We will study its continuity with respect to (s, t, λ) and its differentiability with respect to λ. For this we prove Lemmas 4.6.1–4.6.3. In these lemmas Condition $(A.5)_{\delta,\gamma}$ for some $\delta > 0$, $0 < \gamma \leq \frac{1}{2}$ is always assumed.

Lemma 4.6.1 *$\eta_{s,t}(\lambda)$ has a modification which is continuous in (s, t, λ) a.s. Further assume that $q(\lambda)$ and its first derivatives are bounded. Then for every $p > 1$ there exists a positive constant $c > 0$ such that*

$$E[|\eta_{s,t}(\lambda) - \eta_{s',t'}(\lambda')|^{2p}] \leq c\{|s - s'|^{2p\gamma} + |t - t'|^p + |\lambda - \lambda'|^{2p}\} \tag{3}$$

holds for any s, s', t, t' and λ, λ'. Furthermore $\eta_{s,t}(\lambda)$ is L^{2p}-bounded.

Proof The existence of a continuous modification of $\eta_{s,t}(\lambda)$ follows immediately from inequality (3) by Kolmogorov's theorem (Theorem 1.4.1). Our proof of inequality (3) is similar to those of Lemmas 4.5.3–4.5.6. We first show that $\eta_{s,t}(\lambda)$ is L^{2p}-bounded. Note that the generator of the one point motion of $\eta_{s,t}(y, \lambda, \tau)$ where (λ, τ) is fixed, is given by

$$L_t^{(1)}f = \frac{1}{2}\sum_{i,j} a^{ij}(y, \lambda, \tau, y, \lambda, \tau, t)\frac{\partial^2 f}{\partial y^i \partial y^j} + \sum_i b^i(y, \lambda, \tau, t)\frac{\partial f}{\partial y^i}. \tag{4}$$

From Condition $(A.5)_{\delta,\gamma}$, we have the estimate

$$|a^{ij}(y, \lambda, \tau, y, \lambda, \tau, t)| \leq c_1(1 + |y|)^2 + K_1,$$
$$|b^i(y, \lambda, \tau, t)| \leq c_1(1 + |y|) + K_1.$$

Here and in the following c_i, $i = 1, 2, \ldots$ are suitable positive constants and $K_i(\lambda, \tau, t)$ or $K_i(\lambda, \tau, \lambda', \tau', t)$, $i = 1, 2, \ldots$ are random fields which are $L^{p'}$-bounded for any $p' > 1$. For a given $p > 1$ set $f(y) = |y|^{2p}$. Then, similarly to the proof of Lemma 4.5.3, we have the estimate

$$|L_t^{(1)}f| \le c_2 f + K_2(\lambda, \tau, t). \tag{5}$$

Then Itô's formula implies

$$E[|\eta_{s,t}(\lambda)|^{2p}] \le |q(\lambda)|^{2p} + E\left[\int_s^t L_u f(\eta_{s,u}(\lambda))\,du\right]$$
$$\le |q(\lambda)|^{2p} + c_3 + c_2 \int_s^t E[|\eta_{s,u}(\lambda)|^{2p}]\,du. \tag{6}$$

Gronwall's inequality proves that $\eta_{s,t}(\lambda)$ is L^{2p}-bounded if q is bounded.

Now the generator of the n-point motion $(\eta_{s,t}(y_1, \lambda_1, \tau_1), \ldots, \eta_{s,t}(y_n, \lambda_n, \tau_n))$ is given by

$$L_t^{(n)}f = \frac{1}{2}\sum_{p,q}\sum_{i,j} a^{ij}(y_p, \lambda_p, \tau_p, y_q, \lambda_q, \tau_q, t)\frac{\partial^2 f}{\partial y_p^i \partial y_q^j} + \sum_{p,i} b^i(y_p, \lambda_p, \tau_p, t)\frac{\partial f}{\partial y_p^i}. \tag{7}$$

Consider the case $n = 2$. Set $f(y, y') = |y - y'|^{2p}$. Then we have

$$|a^{ij}(y, \lambda, s, y, \lambda, s, t) - 2a^{ij}(y, \lambda, s, y', \lambda', s', t) + a^{ij}(y', \lambda', s', y', \lambda', s', t)|$$
$$\le c_4|y - y'|^2 + K_4(\lambda, \lambda', s, s', t)(1 + |y| + |y'|)^2(|\lambda - \lambda'|^2 + |s - s'|^{2\gamma}) \tag{8}$$

and

$$|b^i(y, \lambda, s, t) - b^i(y', \lambda', s', t)|$$
$$\le c_4|y - y'| + K_4(\lambda, \lambda', s, s', t)(1 + |y| + |y'|)(|\lambda - \lambda'| + |s - s'|^{\gamma}), \tag{9}$$

by Condition $(A.5)_{\delta,\gamma}$. Therefore we have

$$|L_t^{(2)}f| \le c_5 f + K_5(\lambda, \lambda', s, s', t)(1 + |y| + |y'|)^{2p}(|\lambda - \lambda'|^{2p} + |s - s'|^{2p\gamma}), \tag{10}$$

similarly to Lemma 4.5.5.

Now if $s < s' < t = t'$ we have by Lemma 4.5.2,

$$E[|\eta_{s,t}(\lambda) - \eta_{s',t}(\lambda')|^{2p}]$$
$$= E[|\eta_{s,s'}(\lambda) - q(\lambda')|^{2p}] + \int_{s'}^t E[L_u^{(2)}f]\,du$$
$$\le 2^{2p}|q(\lambda) - q(\lambda')|^{2p} + 2^{2p}E\left[\left|\int_s^{s'} G(\eta_{s,r}(\lambda), \lambda, s, dr)\right|^{2p}\right]$$
$$+ \int_{s'}^t E[L_u^{(2)}f]\,du. \tag{11}$$

The second term of the right hand side is bounded by

$$c_6\left\{E\left[\left(\int_s^{s'} \operatorname{tr}(a(\eta_{s,r}(\lambda), \lambda, s, r))\,dr\right)^p\right]+E\left[\left(\int_s^{s'} |b(\eta_{s,r}(\lambda), \lambda, s, r)|\,dr\right)^{2p}\right]\right\}$$
$$\le c_6\left\{E\left[\left(\int_s^{s'} K_1 + c_1(1+|\eta_{s,r}(\lambda)|)^2\,dr\right)^p\right]\right.$$
$$\left.+E\left[\left(\int_s^{s'} K_1 + c_1(1+|\eta_{s,r}(\lambda)|)\,dr\right)^{2p}\right]\right\}.$$

Note that $K_1, |\eta_{s,r}(\lambda)|$ are L^p-bounded for any p. Then the above is bounded by $c_7|s'-s|^p$. The last term of the right hand side of (11) is bounded by

$$c_8\left\{|\lambda-\lambda'|^{2p}+|s-s'|^{2p\gamma}+\int_{s'}^{t} E[|\eta_{s,u}(\lambda)-\eta_{s',u}(\lambda')|^{2p}]\,du\right\}$$

by (10). Therefore (11) implies

$$E[|\eta_{s,t}(\lambda)-\eta_{s',t}(\lambda')|^{2p}]$$
$$\le c_9\{|\lambda-\lambda'|^{2p}+|s'-s|^{2p\gamma}\}+c_8\int_{s'}^{t} E[|\eta_{s,u}(\lambda)-\eta_{s',u}(\lambda')|^{2p}]\,du.$$

Gronwall's inequality implies (3) in the case where $t=t'$.

Finally if $t<t'$, we have

$$E[|\eta_{s',t}(\lambda')-\eta_{s',t'}(\lambda')|^{2p}] = E\left[\left|\int_t^{t'} G(\eta_{s',r}(\lambda'), \lambda', s', dr)\right|^{2p}\right]$$
$$\le c_{10}|t'-t|^p.$$

Combining this with the case $s\neq s'<t=t'$, we obtain (3). The proof is complete. □

We next discuss the differentiability of $\eta_{s,t}(\lambda)$. We need a numerical lemma.

Lemma 4.6.2 *Let $L_t^{(4)}$ be the linear operator defined in equation* (7) *with $n=4$ and $\tau_1=\tau_2, \tau_3=\tau_4$. For a given $p>1$ set $f=g^p$, where*

$$g(y_1, y_2, y_3, y_4) = \left|\frac{1}{h}(y_1-y_2)-\frac{1}{h'}(y_3-y_4)\right|^2. \tag{12}$$

Then there exists a positive constant c and a positive random field $K=K(\lambda_1,\ldots,\lambda_4,\tau_1,\tau_3,t)$ which is $L^{p'}$-bounded for any $p'>1$ such that

$$|L_t^{(4)}f| \le cf+\tilde f K\left(1+\sum|y_i|\right)^{2p}$$
$$+K\left(1+\sum|y_i|\right)^{2p}\left(\left|\frac{1}{h'}(y_3-y_4)\right|+\left|\frac{1}{h'}(\lambda_3-\lambda_4)\right|\right)^{2p}$$
$$\times\{|y_1-y_3|^{2\delta}+|y_2-y_4|^{2\delta}+|\lambda_1-\lambda_3|^{2\delta}$$
$$+|\lambda_2-\lambda_4|^{2\delta}+|\tau_1-\tau_3|^{2\gamma}\}^p \tag{13}$$

holds for any $y_1,\ldots,y_4,\lambda_1,\ldots,\lambda_4$ and τ_1,τ_3, where $\tilde f=f(\lambda_1,\ldots,\lambda_4)$.

Further if $F_2 \equiv 0$ *the terms* $(1 + \sum |y_i|)^{2p}$ *on the right hand side of* (13) *can be replaced by* 1.

Proof Set $(x^1, \ldots, x^{d+e}) = (y^1, \ldots, y^d, \lambda^1, \ldots, \lambda^e)$ and

$$u^i = \frac{1}{h}(x_1^i - x_2^i), \quad v^i = \frac{1}{h'}(x_3^i - x_4^i), \quad w^i = u^i - v^i.$$

A direct computation yields $L_t^{(4)} f = I_1 + I_2 + I_3$, say, where

$$I_1 = pg^{p-1} \sum_{i=1}^{d} \left\{ (b^i(x_1) - b^i(x_2))\frac{1}{h} - (b^i(x_3) - b^i(x_4))\frac{1}{h'} \right\} w^i$$

$$\begin{aligned} I_2 = pg^{p-1} \Bigg(& \sum_{i=1}^{d} \{a^{ii}(x_1, x_1) - 2a^{ii}(x_1, x_2) + a^{ii}(x_2, x_2)\} \frac{1}{h^2} \\ & - 2 \sum_{i=1}^{d} \{a^{ii}(x_1, x_3) - a^{ii}(x_1, x_4) - a^{ii}(x_2, x_3) + a^{ii}(x_2, x_4)\} \frac{1}{hh'} \\ & + \sum_{i=1}^{d} \{a^{ii}(x_3, x_3) - 2a^{ii}(x_3, x_4) + a^{ii}(x_4, x_4)\} \frac{1}{h'^2} \Bigg) \end{aligned}$$

$$\begin{aligned} I_3 = 2p(p-1)g^{p-2} \sum_{i \neq j} \Bigg(& \{a^{ij}(x_1, x_1) - 2a^{ij}(x_1, x_2) + a^{ij}(x_2, x_2)\} \frac{1}{h^2} \\ & - 2\{a^{ij}(x_1, x_3) - a^{ij}(x_1, x_4) - a^{ij}(x_2, x_3) + a^{ij}(x_2, x_4)\} \frac{1}{hh'} \\ & + \{a^{ij}(x_3, x_3) - 2a^{ij}(x_3, x_4) + a^{ij}(x_4, x_4)\} \frac{1}{h'^2} \Bigg) w^i w^j. \end{aligned}$$

The first term is written as

$$\begin{aligned} I_1 = {} & pg^{p-1} \sum_{i=1}^{d} \sum_{j=1}^{d+e} \left\{ \int_0^1 \frac{\partial b^i}{\partial x^j}(x_1 + \theta(x_2 - x_1), \tau_1)\, d\theta \right\} w^i w^j \\ & + pg^{p-1} \sum_{i=1}^{d} \sum_{j=1}^{d+e} \left\{ \int_0^1 \left(\frac{\partial b^i}{\partial x^j}(x_1 + \theta(x_2 - x_1), \tau_1) \right.\right. \\ & \left.\left. - \frac{\partial b^i}{\partial x^j}(x_3 + \theta(x_4 - x_3), \tau_3) \right) d\theta \right\} w^i v^j. \end{aligned}$$

Note the inequalities

$$\left| \frac{\partial b^i}{\partial x^j}(y, \lambda, \tau, t) \right| \le \begin{cases} c_1 & \text{if } j = 1, \ldots, d, \\ K_1(\lambda, \tau, t)(1 + |y|) & \text{if } j = d+1, \ldots, d+e, \end{cases}$$

$$\left| \frac{\partial b^i}{\partial x^j}(y, \lambda, \tau, t) - \frac{\partial b^i}{\partial x^j}(y', \lambda', \tau', t) \right|$$

$$\le K_2(\lambda, \tau, \lambda', \tau', t)(1 + |y| + |y'|)(|y - y'|^\delta + |\lambda - \lambda'|^\delta + |\tau - \tau'|^\gamma)$$

if $j = 1, \ldots, d+e$. Here and in the sequel, c_i, $i = 1, 2, \ldots$ are positive constants and $K_i(\lambda, \ldots)$, $i = 1, 2, \ldots$ are positive random fields, which are $L^{p'}$-bounded for any $p' > 1$. Then the first term of I_1 is bounded by

$$c_3 f + g^{p-(1/2)} \tilde{g}^{1/2} K_3 (1 + \sum |y_i|)$$

where $\tilde{g} = g(\lambda_1, \ldots, \lambda_4)$. The second term of I_1 is bounded by

$$\begin{aligned} &g^{p-(1/2)} K_4 (1 + \sum |y_i|)(|v| + |\tilde{v}|) \\ &\qquad \times (|y_1 - y_3|^\delta + |y_2 - y_4|^\delta + |\lambda_1 - \lambda_3|^\delta + |\lambda_2 - \lambda_4|^\delta + |\tau_1 - \tau_3|^\gamma) \end{aligned}$$

where $\tilde{v} = (\lambda_3 - \lambda_4)/h'$. Summing these two bounds and using the inequality $ab \le a^{\tilde{p}} \tilde{p}^{-1} + b^{\tilde{q}} \tilde{q}^{-1}$ $(a, b \ge 0, \tilde{p}, \tilde{q} > 1, \tilde{p}^{-1} + \tilde{q}^{-1} = 1)$, we obtain the estimate

$$\begin{aligned} |I_1| \le c_5 f + \tilde{f} K_5 (1 + \sum |y_i|)^{2p} &+ \{K_5 (1 + \sum |y_i|)(|v| + |\tilde{v}|) \\ &\times (|y_1 - y_3|^\delta + |y_2 - y_4|^\delta + |\lambda_1 - \lambda_3|^\delta \\ &+ |\lambda_2 - \lambda_4|^\delta + |\tau_1 - \tau_3|^\gamma)\}^{2p}. \end{aligned}$$

We shall next estimate I_2. Set $a^{ii} = a$ for simplicity. By the mean value theorem, we have

$$\begin{aligned} &a(x_k, x_m) - a(x_k, x_n) - a(x_l, x_m) + a(x_l, x_n) \\ &\quad = \sum_{i,j=1}^{d+e} \left(\int_0^1 \int_0^1 \xi_{ln}^{ij}(\theta, \tau)\, d\theta\, d\tau \right) (x_k^i - x_l^i)(x_m^j - x_n^j), \end{aligned}$$

where

$$\xi_{ln}^{ij}(\theta, \tau) = \frac{\partial^2 a}{\partial x^i \partial x^j} (x_l + \theta(x_k - x_l), x_n + \sigma(x_m - x_n)).$$

Then I_2 is written as the sum of the following:

$$\begin{aligned} &g^{p-1} \int_0^1 \int_0^1 \sum_{i,j=1}^{d+e} (\xi_{22}^{ij} u^i u^j - 2 \xi_{24}^{ij} u^i v^j + \xi_{44}^{ij} v^i v^j)\, d\theta\, d\sigma \\ &\quad = g^{p-1} \int_0^1 \int_0^1 \sum_{i,j=1}^{d+e} \xi_{22}^{ij} w^i w^j\, d\theta\, d\sigma \\ &\qquad + g^{p-1} \int_0^1 \int_0^1 \sum_{i,j=1}^{d+e} (\xi_{22}^{ij} - \xi_{24}^{ij} - \xi_{42}^{ij} + \xi_{44}^{ij}) v^i v^j\, d\theta\, d\sigma \\ &\qquad + g^{p-1} \int_0^1 \int_0^1 \sum_{i,j=1}^{d+e} (2\xi_{22}^{ij} - \xi_{24}^{ij} - \xi_{42}^{ij}) w^i v^j\, d\theta\, d\sigma \\ &\quad = J_1 + J_2 + J_3, \qquad \text{say}. \end{aligned}$$

Here the relation

$$\int_0^1 \int_0^1 \xi_{24}^{ij}\, d\theta\, d\sigma = \int_0^1 \int_0^1 \xi_{42}^{ij}\, d\theta\, d\sigma$$

is used. Note the inequality

$$\begin{aligned} |\xi_{22}^{ij}| &\le c_6 \text{ if } 0 \le i, j \le d, \\ &\le K_6(\lambda_1, \dots, \lambda_4, \tau_1, \tau_3, t)(1 + |y_1| + |y_2|)^2 \\ &\qquad \text{if } d+1 \le i \le d+e, \quad d+1 \le j \le d+e \\ &\le K_6(\lambda_1, \dots, \lambda_4, \tau_1, \tau_3, t)(1 + |y_1| + |y_2|) \\ &\qquad \text{if } 1 \le i \le d, \quad d+1 \le j \le d+e. \end{aligned}$$

Then $|J_1|$ is bounded by

$$\begin{aligned} &c_7 g^p + g^{p-1}\tilde{g}K_7(1 + \sum |y_i|)^2 + g^{p-(1/2)}\tilde{g}^{1/2}K_7(1 + \sum |y_i|) \\ &\le c_8 g^p + \tilde{g}^p K_8(1 + \sum |y_i|)^{2p}. \end{aligned}$$

For the estimate of $|J_2|$, note that

$$\begin{aligned} |\xi_{22}^{ij} - \xi_{24}^{ij} - \xi_{42}^{ij} + \xi_{44}^{ij}| \le K_9(1 + \sum |y_i|)^2(&|y_1 - y_3|^{2\delta} + |y_2 - y_4|^{2\delta} \\ &+ |\lambda_1 - \lambda_3|^{2\delta} + |\lambda_2 - \lambda_4|^{2\delta} + |\tau_1 - \tau_3|^{2\gamma}). \end{aligned}$$

Then $|J_2|$ is bounded by the above multiplied by $g^{p-1}(|v| + |\tilde{v}|)^2$. It is dominated by

$$\begin{aligned} &c_{10}g^p + K_{10}(1 + \sum |y_i|)^{2p}(|v| + |\tilde{v}|)^{2p}(|y_1 - y_3|^{2\delta} + |y_2 - y_4|^{2\delta} \\ &\quad + |\lambda_1 - \lambda_3|^{2\delta} + |\lambda_2 - \lambda_4|^{2\delta} + |\tau_1 - \tau_3|^{2\gamma})^p. \end{aligned}$$

We can show similarly that $|J_3|$ is also bounded by the above. Consequently $|I_2|$ is bounded by a similar quantity to $|I_1|$ and a similar estimation is valid for $|I_3|$. Summing up all these estimations for I_1, I_2, I_3, we get inequality (13).

The last assertion will be obvious from the above computations. The proof is complete. □

We now proceed to the proof of the differentiability of $\eta_{s,t}(\lambda)$.

Lemma 4.6.3 *Assume that the first derivatives of $q(\lambda)$ are bounded and uniformly δ-Hölder continuous. Set for $h \neq 0$:*

$$\zeta_{s,t}(\lambda, h) = \frac{1}{h}\{\eta_{s,t}(\lambda + he) - \eta_{s,t}(\lambda)\}, \tag{14}$$

where e is a unit vector. Then it is L^p-bounded for every $p \ge 1$. Further for *every $p > 1$ there exists a positive constant c such that*

$$E[|\zeta_{s,t}(\lambda, h) - \zeta_{s',t'}(\lambda', h')|^{2p}]$$
$$\le c\{|\lambda - \lambda'|^{2p\delta} + |h - h'|^{2p\delta} + |s - s'|^{2p\gamma} + |t - t'|^{p}\} \tag{15}$$

holds for any $s, t, s', t', \lambda, \lambda'$ and $h, h' \neq 0$.

Proof The L^p-boundedness of $\zeta_{s,t}(\lambda, h)$ is immediate from inequality (3). For the proof of (15), we apply Lemma 4.6.2. Let f be the function of Lemma 4.6.2. Suppose $s < s' < t$. Then we have by Lemma 4.5.2,

$$E[|\zeta_{s,t}(\lambda, h) - \zeta_{s',t}(\lambda', h')|^{2p}]$$
$$= E[f(\eta_{s,s'}(\lambda), \eta_{s,s'}(\lambda + he), q(\lambda'), q(\lambda' + h'e))] + \int_{s'}^{t} E[L_u^{(4)} f]\, du. \tag{16}$$

The first term of the right hand side is bounded by the sum of two quantities. The first is

$$2^{2p}\left|\frac{1}{h}\{q(\lambda) - q(\lambda + he)\} - \frac{1}{h'}\{q(\lambda') - q(\lambda' + h'e)\}\right|^{2p} \tag{17}$$

which is bounded by $c_1(|\lambda - \lambda'|^{2p\delta} + |h - h'|^{2p\delta})$. The second is

$$2^{2p}E\left[\left|\frac{1}{h}\left\{\int_s^{s'} G(\eta_{s,u}(\lambda), \lambda, s, du) - \int_s^{s'} G(\eta_{s,u}(\lambda + he), \lambda + he, s, du)\right\}\right|^{2p}\right]. \tag{18}$$

Now, using Burkholder's inequality, (18) divided by 2^{2p} is bounded by

$$\frac{1}{h^{2p}}\left\{E\left[\left|\int_s^{s'} \operatorname{tr}\{a(\eta_{s,u}(\lambda), \lambda, s, \eta_{s,u}(\lambda), \lambda, s, u)\right.\right.\right.$$
$$- 2a(\eta_{s,u}(\lambda), \lambda, s, \eta_{s,u}(\lambda + he), \lambda + he, s, u)$$
$$\left.\left. + a(\eta_{s,u}(\lambda + he), \lambda + he, s, \eta_{s,u}(\lambda + he), \lambda + he, s, u)\}\, du\right|^{p}\right]$$
$$\left. + E\left[\left|\int_s^{s'} \{b(\eta_{s,u}(\lambda), \lambda, s, u) - b(\eta_{s,u}(\lambda + he), \lambda + he, s, u)\}\, du\right|^{2p}\right]\right\}.$$

Apply inequalities (8) and (9). Then the above is bounded by

$$\frac{c_2}{h^{2p}} E\left[\left(\int_s^{s'} |\eta_{s,u}(\lambda) - \eta_{s,u}(\lambda + he)|^2\, du\right)^p\right]$$
$$+ c_2 E\left[\left(\int_s^{s'} K_1(1 + |\eta_{s,u}(\lambda)| + |\eta_{s,u}(\lambda + he)|)^2\, du\right)^p\right].$$

The first term is bounded by $c_3|s-s'|^p$ by Lemma 4.6.1. The second term is bounded by the same quantity since K_1, $\eta_{s,u}(\lambda)$, $\eta_{s,u}(\lambda+he)$ are L^{2p}-bounded. We next consider the last term of (16). By Lemma 4.6.2, it is bounded by

$$c_4\int_{s'}^{t} E[|\zeta_{s,u}(\lambda,h)-\zeta_{s',u}(\lambda',h')|^{2p}]\,du$$

$$+\int_{s'}^{t} E\Bigg[K_3\Big(1+\sum_i|\eta_{s_i,u}(\lambda_i)|\Big)^{2p}(|\zeta_{s',u}(\lambda',h')|+1)^{2p}$$

$$\times\{|\eta_{s,u}(\lambda)-\eta_{s',u}(\lambda')|^{2\delta}+|\eta_{s,u}(\lambda+he)-h_{s',u}(\lambda'+h'e)|^{2\delta}$$

$$+|\lambda-\lambda'|^{2\delta}+|h-h'|^{2\delta}+|s-s'|^{2\gamma}\}^p\Bigg]\,du,$$

where $\lambda_1=\lambda$, $\lambda_2=\lambda+he$, $\lambda_3=\lambda'$, $\lambda_4=\lambda'+h'e$, $s_1=s_2=s$, $s_3=s_4=s'$. Random fields K_3, $\eta_{s_i,u}(\lambda_i)$, $\zeta_{s',u}(\lambda',h')$ etc. are L^{2p}-bounded. $\zeta_{s',u}(\lambda',h)$ is also L^{2p}-bounded by (3) in Lemma 4.6.1. Apply Lemma 4.6.1. Then the second integral of the above is bounded by

$$c_5(|\lambda-\lambda'|^{2p\delta}+|h-h'|^{2p\delta}+|s-s'|^{2p\gamma}).$$

Consequently (16) implies the functional inequality

$$E[|\zeta_{s,t}(\lambda,h)-\zeta_{s',t}(\lambda',h')|^{2p}]\le c_5(|\lambda-\lambda'|^{2p\delta}+|h-h'|^{2p\delta}+|s-s'|^{2p\gamma})$$
$$+c_4\int_{s'}^{t} E[|\zeta_{s,u}(\lambda,h)-\zeta_{s',u}(\lambda',h')|^{2p}]\,du.$$

Then Gronwall's inequality establishes (15) in the case where $t=t'$.

Next we consider the case $t<t'$. Note that

$$E[|\zeta_{s',t}(\lambda',h')-\zeta_{s',t'}(\lambda',h')|^{2p}]$$

$$\le\frac{1}{h'^{2p}}E\Bigg[\Bigg|\int_t^{t'}G(\eta_{s',u}(\lambda'+h'e),\lambda'+h'e,s',du)$$

$$-\int_t^{t'}G(\eta_{s',u}(\lambda'),\lambda',s',du)\Bigg|^{2p}\Bigg].$$

It is bounded by $c_6|t-t'|^p$. See the estimate of (18). Then we obtain (15) in the case where $t<t'$. The case where $t'<t$ can be shown similarly. The proof is complete. □

Theorem 4.6.4 *Assume that $G(y,\lambda,\tau,t)$ of (1) satisfies Condition $(A.5)_{\delta,\gamma}$ for some $\delta,\gamma>0$. Let $\eta_{s,t}(y,\lambda,\tau)$ be the solution of equation (2). Set $\eta_{s,t}(\lambda)=\eta_{s,t}(q(\lambda),\lambda,s)$ where $q(\lambda)$ is a smooth function. Then $\eta_{s,t}(\lambda)$ has a modification* *which is continuous in (s, t, λ). Any continuous modification is differentiable with respect to λ for any s, t and the derivatives are continuous in (s, t, λ) a.s. Further if q and its first derivatives are bounded and the latter is uniformly δ-Hölder continuous, then for every $p > 1$ there exists a positive constant c such that the modification $\eta_{s,t}(\lambda)$ satisfies*

$$E\left[\left|\frac{\partial \eta_{s,t}}{\partial \lambda^i}(\lambda) - \frac{\partial \eta_{s',t'}}{\partial \lambda^i}(\lambda')\right|^{2p}\right] \le c\{|\lambda - \lambda'|^{2p\delta} + |s - s'|^{2p\gamma} + |t - t'|^{p}\}, \tag{19}$$

$$E\left[\left|\frac{\partial \eta_{s,t}}{\partial \lambda^i}(\lambda)\right|^{2p}\right] \le c \tag{20}$$

for any $s, t, s', t', \lambda, \lambda'$. Furthermore, for every s it is a continuous $C^{1,\varepsilon}$-semimartingale for any $\varepsilon < \delta$.

Proof It is sufficient to prove the theorem in the case where q and its first derivatives are bounded and the latter is uniformly δ-Hölder continuous. The existence of a continuous modification of $\eta_{s,t}(\lambda)$ is shown in Lemma 4.6.1. We denote any modification again by $\eta_{s,t}(\lambda)$. For the proof of the differentiability with respect to λ, we apply Lemma 4.6.3. Since $\zeta_{s,t}(\lambda, h)$ of (14) satisfies (15), $\zeta_{s,t}(\lambda, h, \omega)$ can be extended continuously at $h = 0$ for almost all ω by Kolmogorov's theorem (Theorem 1.4.1). For such ω, $\eta_{s,t}(\lambda, \omega)$ is differentiable with respect to λ for any s, t and the derivatives are continuous in (s, t, λ). Inequality (19) follows from (15) immediately. Therefore the derivative is a continuous $C^{0,\varepsilon}$-process for every $\varepsilon < \delta$ when s is fixed, by Kolmogorov's theorem. Then $\eta_{s,t}(\lambda)$, $t \in [s, T]$ is a continuous semimartingale with values in $C^{1,\varepsilon}$ for every $\varepsilon < \delta$ by Theorem 3.3.3. Inequality (20) follows from the L^{2p}-boundedness of $\zeta_{s,t}(\lambda, h)$. The proof is complete. □

Diffeomorphic property

We now establish the diffeomorphic property of the solution of Itô's stochastic differential equation based on $F(x, t)$.

Theorem 4.6.5 *Assume that the local characteristic of the continuous C-semimartingale $F(x, t)$ belongs to the class $B_b^{k,\delta}$ for some $k \ge 1$ and $\delta > 0$. Then the solution of Itô's stochastic differential equation based on F has a modification $\varphi_{s,t}$, $0 \le s \le t \le T$ such that it is a forward stochastic flow of C^k-diffeomorphisms. Further it is a $C^{k,\varepsilon}$-semimartingale for any $\varepsilon < \delta$.*

Proof Let (a, b, A_t) be the local characteristic of F. It is enough to prove the theorem in the case where the local characteristic belongs to the class

$B_{ub}^{k,\delta}$ and $A_t = t$, since the general case can be reduced to this by a change of time. Let $\varphi_{s,t}(x)$ be any continuous modification of the solution. We apply Theorem 4.6.4 by setting $G_1 \equiv G_2 \equiv 0$, $G_3 \equiv F$ and $q(\lambda) \equiv \lambda \equiv x$. Then $\varphi_{s,t}(x)$ is continuously differentiable in x for any s, t and the derivatives are continuous in (s, t, x) a.s. The derivative satisfies

$$\frac{\partial \varphi_{s,t}^j}{\partial x^i}(x) = \delta_{ij} + \sum_l \int_s^t \frac{\partial F^j}{\partial x^l}(\varphi_{s,u}(x), du)\frac{\partial \varphi_{s,u}^l}{\partial x^i}(x) \tag{21}$$

by Theorem 3.3.3. It is a continuous $C^{0,\gamma}$-semimartingale.

Now suppose next $k \geq 2$ and we shall show that $\partial \varphi_{s,t}^j/\partial x^i$ is also differentiable. Define a continuous $\mathbb{R}^d \otimes \mathbb{R}^d$-valued semimartingale with parameter (λ, τ) by

$$G_2(\lambda, \tau, t) = \left(\int_0^t \frac{\partial F^j}{\partial x^l}(\varphi_{\tau,\tau\vee u}(\lambda), du)\right). \tag{22}$$

Its local characteristic is given by

$$\left(\frac{\partial^2 a^{ij}}{\partial x^l \partial y^m}(\varphi_{\tau,\tau\vee t}(\lambda), \varphi_{\tau,\tau\vee t}(\lambda'), t), \frac{\partial b^j}{\partial x^l}(\varphi_{\tau,\tau\vee t}(\lambda), t)\right), \tag{23}$$

where (a, b) is the local characteristic of F which belongs to the class $B_{ub}^{k,\delta}$. It is continuously differentiable with respect to λ. Further since

$$\frac{|\varphi_{s,t}(\lambda) - \varphi_{s',t}(\lambda')|}{|\lambda - \lambda'| + |s - s'|^{1/2}}, \qquad \frac{|\eta_{s,t}(\lambda) - \eta_{s',t}(\lambda')|}{|\lambda - \lambda'|^\delta + |s - s'|^{1/2}} \qquad \left(\eta_{s,t} = \left(\frac{\partial \varphi^j}{\partial x^i}\right)\right)$$

are L^p-bounded for any $p \geq 1$ by Lemma 4.5.6 and Theorem 4.6.4, respectively, the local characteristic (23) belongs to the class $B_p^{1,\delta,1/2}$ for any $p > 1$. Therefore G_2 satisfies Condition $(A.5)_{\delta,\gamma}(2)$. ($G_1 \equiv G_3 \equiv 0$ in this case.) Now equality (21) shows that $\eta_{s,t} = (\partial \varphi^j/\partial x^i)$ satisfies

$$\eta_{s,t}(x) = I + \int_s^t G_2(x, s, du)\eta_{s,u}(x). \tag{24}$$

Consequently $\eta_{s,t}(x)$ is again continuously differentiable with respect to x by Theorem 4.6.4. The derivative satisfies

$$\frac{\partial^2 \varphi_{s,t}^j}{\partial x^l \partial x^i}(x) = \sum_{m,m'} \int_s^t \frac{\partial^2 F^j}{\partial x^m \partial x^{m'}}(\varphi_{s,u}(x), du)\frac{\partial \varphi_{s,u}^{m'}}{\partial x^l}(x)\frac{\partial \varphi_{s,u}^m}{\partial x^i}(x)$$
$$+ \sum_m \int_s^t \frac{\partial F^j}{\partial x^m}(\varphi_{s,u}(x), du)\frac{\partial^2 \varphi_{s,u}^m}{\partial x^l \partial x^i}(x).$$

Hence $\varphi_{s,t}$ is a continuous $C^{2,\varepsilon}$-semimartingale.

Suppose next $k \geq 3$. Set

$$G^{jli}(y, \lambda, \tau, t) = \sum_{m,m'} \int_0^t \frac{\partial^2 F^j}{\partial x^m \partial x^{m'}}(\varphi_{\tau,\tau\vee u}(\lambda), du) \frac{\partial \varphi^{m'}_{\tau,\tau\vee u}}{\partial x^l}(\lambda) \frac{\partial \varphi^m_{\tau,\tau\vee u}}{\partial x^i}(\lambda)$$

$$+ \sum_m \int_0^t \frac{\partial F^j}{\partial x^m}(\varphi_{\tau,\tau\vee u}(\lambda), du) y^{mli},$$

where $y = (y^{mli})$. We can show similarly to the case of (22) that $G = (G^{mli})$ satisfies Condition $(A.5)_{\delta,\gamma}$ (1), (2). ($G_3 \equiv 0$ in this case.) Since $\eta_{s,t} = (\partial^2 \varphi^j_{s,t}/\partial x^l \partial x^i)$ satisfies

$$\eta_{s,t}(x) = \int_s^t G(\eta_{s,u}(x), x, s, du),$$

it is continuously differentiable by the same theorem and $\varphi_{s,t}$ is in fact a continuous $C^{3,\varepsilon}$-semimartingale if $\varepsilon < \delta$.

Repeating the above arguments inductively, we can prove that $\varphi_{s,t}(x)$ is a continuous $C^{k,\varepsilon}$-semimartigale if $\varepsilon < \delta$.

Now as in Lemma 4.4.3 we can prove that the Jacobian matrix $\partial \varphi_{s,t}(x)$ is nonsingular for any (s, t, x) a.s. (see Exercise 4.6.8). Then the implicit function theorem shows that $\varphi^{-1}_{s,t}(x)$ is also k-times differentiable in x and the derivatives are continuous in (s, t, x) a.s. The proof is complete.

□

Corollary 4.6.6 *The forward flow $\varphi_{s,t}: 0 \leq s \leq t \leq T$ of Theorem 4.6.5 is a $C^{k,\varepsilon}$-semimartingale flow for any $\varepsilon < \delta$. Suppose further that $k \geq 2$, $\delta > 0$ and that F is a backward C-semimartingale with local characteristic belonging to the class $B_b^{k,\delta}$. Then the backward flow $\varphi_{t,s} \equiv \varphi^{-1}_{s,t}$; $0 \leq s \leq t \leq T$ is a backward $C^{k-1,\varepsilon}$-semimartingale flow for any $\varepsilon < \delta$. Its backward Itô's random infinitesimal generator is given by $\hat{F} \equiv F - 2C$.*

Proof The first assertion is obvious from Theorem 4.6.5. Assuming $k \geq 2$, $\delta > 0$ and the backward semimartingale property to $F(\cdot, t)$, we can show equality (31) of Section 4.2 in exactly the same way as in the case of Brownian flow (see the proof of Theorem 4.2.10). As a consequence, we see that $\varphi_{s,t}$ is a backward $C^{k-1,\varepsilon}$-semimartingale flow. The proof is complete.

□

Corollary 4.6.7 *Assume that the local characteristic (a, b, A_t) of $F(x, t)$ belongs to the class $B_{ub}^{k,\delta}$ and $A_t = t$. Then for each α with $1 \leq |\alpha| \leq k$ $D^\alpha \varphi_{s,t}(x)$ is L^p-bounded for any $p \geq 1$. Further*

$$E\left[\sup_{|x|\le N} |D^\alpha \varphi_{s,t}(x)|^{2p}\right] < \infty, \qquad \textit{for all } p > 1 \tag{25}$$

holds for any $N > 0$.

Proof Apply Theorem 4.6.4 repeatedly. Then for any p there exists $c > 0$ such that

$$E[|D^\alpha \varphi_{s,t}(x)|^{2p}] \le c, \qquad \text{for all } x,$$
$$E[|D^\alpha \varphi_{s,t}(x) - D^\alpha \varphi_{s,t}(x')|^{2p}] \le c|x - y|^{2p\delta'}, \qquad \text{for all } x, x',$$

where $\delta' = 1$ if $|\alpha| < k$ and $\delta' = \delta$ if $|\alpha| = k$. Then Theorem 1.4.1 implies the assertion. □

Exercise 4.6.8 Under the condition of Theorem 4.6.5, show that the Jacobian matrix $\partial\varphi_{s,t}(x)$ is invertible and the inverse $J_t = (\partial\varphi_{s,t})^{-1}(x)$ satifies

$$J_t = I - \int_s^t J_r \partial F(\varphi_{s,r}(x), dr) - \int_s^t \tilde{a}(\varphi_{s,r}(x), \varphi_{s,r}(x), r)\, dr$$

where $\partial F(x, t) = (\partial F^i(x, t)/\partial x^j)$ and $\tilde{a}^{ij}(x, y, t) = \sum_l (\partial^2 a^{il}/\partial x^j \partial y^l)(x, y, t)$.

Exercise 4.6.9 Under the condition of Theorem 4.6.5, show that for any $0 < \varepsilon < 1$

$$\lim_{|x|\to\infty} \frac{|D^\alpha \varphi_{s,t}(x)|}{(1 + |x|)^\varepsilon} = 0 \quad \text{uniformly in } s, t \text{ a.s.}$$

if $1 \le |\alpha| \le k$. (*Hint*: *c.f.* Exercise 4.5.9.)

4.7 Stochastic flows of local diffeomorphisms

Construction of stochastic flows of local diffeomorphisms

In Sections 4.5 and 4.6 we have shown that solutions of a stochastic differential equation based on a suitable continuous C-semimartingale $F(x, t)$ define a stochastic flow of homeomorphisms or diffeomorphisms. A basic assumption for the semimartingale $F(x, t)$ was that its local characteristic belonged to the class $B_b^{0,1}$ or $B_b^{k,\delta}$, $k \ge 1$, $\delta > 0$. In this section we shall discuss the case where the local characteristic belongs to the class $B^{0,1}$ or $B^{k,\delta}$, $k \ge 1$, $\delta > 0$. Then the equation may not have a global solution since the explosion can occur at a finite time, so that the solution can not define a stochastic flow of homeomorphisms or diffeomorphisms. The first objective of this section is to show that the solution defined up to the explosion time (the maximal solution) has a modification of a stochastic flow of *local homeomorphisms* or *local diffeomorphisms*.

Let us begin with the definition of a forward stochastic flow of local homeomorphisms. Let $\sigma(s, x)$, $0 \le s \le T$, $x \in \mathbb{R}^d$ be a random field with values in $(s, T]$ such that $\sigma(s, x)$ is lower semicontinuous with respect to s and x. Let $\varphi_{s,t}(x)$, $x \in \mathbb{R}^d$, $0 \le s \le t < \sigma(s, x)$ be a continuous random field with values in $\mathbb{R}^d$ with random domain of parameter (s, t, x). For each $s < t$, we set

$$\mathbb{D}_{s,t}(\omega) = \{x : \sigma(s, x, \omega) > t\}.$$

For almost all ω, it is an open subset of $\mathbb{R}^d$ and $\varphi_{s,t}(\omega) = \varphi_{s,t}(\cdot, \omega)$ can be regarded as a continuous map from $\mathbb{D}_{s,t}(\omega)$ into $\mathbb{R}^d$. Denote the range of the map $\varphi_{s,t}(\omega)$ by $\mathbb{R}_{s,t}(\omega)$. We will assume there exists a null set N of Ω such that for any ω of N^c, the family of maps $\{\varphi_{s,t}(\omega) : s < t\}$ defines a *forward flow of local diffeomorphisms*, i.e. it satisfies the following three properties.

(i) $\varphi_{s,u}(\omega) = \varphi_{t,u}(\omega) \circ \varphi_{s,t}(\omega)$ holds on $\mathbb{D}_{s,u}(\omega)$ for all $s < t < u$.
(ii) $\varphi_{s,s}(\omega) =$ identity for all s.
(iii) $\varphi_{s,t}(\omega) : \mathbb{D}_{s,t}(\omega) \to \mathbb{R}_{s,t}(\omega)$ is a homeomorphism for all $s < t$ and the inverse $\varphi_{s,t}^{-1}(x, \omega) \equiv \varphi_{s,t}(\omega)^{-1}(x)$ is continuous in (s, t, x).

Then $\varphi_{s,t}$ is called a *forward stochastic flow of local homeomorphisms*. Further if $\varphi_{s,t}(x, \omega)$ and $\varphi_{s,t}^{-1}(x, \omega)$ are k-times continuously differentiable with respect to x, and their derivatives are continuous in (s, t, x), it is called a *forward stochastic flow of local C^k-diffeomorphisms*. Further if $\varphi_{s,t}(x)$, $t \in [s, \sigma(s, x))$ is a local $C^{k,\delta}$-semimartingale for every s, it is called a *forward local $C^{k,\delta}$-semimartingale flow*. We should note that the range $\mathbb{R}_{s,t}(\omega)$ is an open set a.s. since it is homeomorphic to the open set $\mathbb{D}_{s,t}(\omega)$ a.s.

We will show that the maximal solution of a stochastic differential equation defines a forward stochastic flow of local homeomorphisms.

Let $F(x, t) = (F^1(x, t), \ldots, F^d(x, t))$, $x \in \mathbb{R}^d$ be a continuous $C(\mathbb{R}^d : \mathbb{R}^d)$-semimartingale with local characteristic belonging to the class $B^{0,1}$. Consider Itô's stochastic differential equation based on $F(x, t)$: we have seen in Theorem 3.4.5 that it has a unique maximal solution.

Theorem 4.7.1 *Assume that the local characteristic of F belongs to the class $B^{0,1}$. There exists a unique forward stochastic flow of local homeomorphisms $\varphi_{s,t}(x)$, $x \in \mathbb{R}^d$, $0 \le s \le t < \sigma(s, x)$ such that for each s, x, it is the maximal solution of Itô's equation based on $F(x, t)$ starting at x at time s, where $\sigma(s, x)$ is the explosion time. Further it is a continuous local $C^{0,\gamma}$-semimartingale flow for any $\gamma < 1$.*

Proof Let us recall the argument in Theorem 3.4.5. We will use the same notation. Since the local characteristic of the truncated semimartingale F^N belongs to the class $B_b^{0,1}$, there exists a forward stochastic flow of homeomorphisms $\varphi_{s,t}^N(x)$, $0 \le s \le t \le T$ generated by F^N. Set for each s and x, $\sigma_N(s, x) = \inf\{t > s : |\varphi_{s,t}^N(x)| \ge N\}$ $(= T$ if $\{\dots\} = \emptyset)$. Then $\varphi_{s,t}^N(x) = \varphi_{s,t}^M(x)$ holds for $t < \sigma_N(s, x)$ if $N < M$ by the uniqueness of the solution. Set $\sigma(s, x) = \lim_{N\to\infty} \sigma_N(s, x)$. We can define $\varphi_{s,t}(x)$, $s \le t < \sigma(s, x)$ by $\varphi_{s,t}(x) = \varphi_{s,t}^N(x)$ if $t < \sigma_N(s, x)$. It is a maximal solution of Itô's equation based on $F(x, t)$ starting at x at time s.

We shall prove that the above $\varphi_{s,t}$ is a forward stochastic flow of local homeomorphisms. Take any sample ω such that $\{\varphi_{s,t}^N(\omega) : 0 \le s \le t \le T\}$ defines a forward flow of homeomorphisms for every N. Probability of the set of all such samples is 1. We first note that each stopping time $\sigma_N(s, x, \omega)$ is lower semicontinuous in (s, x) i.e. $\{(s, x) : \sigma_N(s, x, \omega) > c\}$ is open for any $c > 0$. In fact, if (s_0, x_0) belongs to this set, $|\varphi_{s_0,t}^N(x_0, \omega)| < N$ for all $t \le c$. Then the same inequality holds for any (s, x) belonging to a suitable neighborhood of (s_0, x_0). Now since $\sigma(s, x, \omega)$ is the upper limit of $\{\sigma_N(s, x, \omega)\}$, it is also lower semicontinuous. Set $\mathbb{D}_{s,t}^N(\omega) = \{x : \sigma_N(s, x, \omega) > t\}$. It is an open set and $\bigcup_N \mathbb{D}_{s,t}^N(\omega) = \mathbb{D}_{s,t}(\omega)$. Since $\varphi_{s,t}(\omega) = \varphi_{s,t}^N(\omega)$ holds on $\mathbb{D}_{s,t}^N(\omega)$, the map $\varphi_{s,t}(\omega) : \mathbb{D}_{s,t}^N(\omega) \to \mathbb{R}^d$ is an into homeomorphism and $\varphi_{s,t}^{-1}(x, \omega)$ is continuous in (s, t, x) by Theorem 4.5.1. Consequently the map is a homeomorphism from $\mathbb{D}_{s,t}(\omega)$ into $\mathbb{R}^d$. We have the relation $\varphi_{s,u}(\omega) = \varphi_{t,u}(\omega) \circ \varphi_{s,t}(\omega)$ on $\mathbb{D}_{s,u}^N(\omega)$ for $s < t < u$ a.s. since $\varphi_{s,u}^N(\omega)$ satisfies the same property on $\mathbb{D}_{s,u}^N(\omega)$. Therefore $\varphi_{s,t}(\omega)$ is a forward stochastic flow of local homeomorphisms. The uniqueness of the flow follows from the uniqueness of the maximal solution of the stochastic differential equation. The last assertion of the theorem will be obvious. □

If the local characteristic of $F(x, t)$ is smooth with respect to the spatial variable, then the corresponding local flow is also smooth. In fact, we have the following.

Theorem 4.7.2 *Assume that the local characteristic of $F(x, t)$ belongs to the class $B^{k,\delta}$ for some $k \ge 1$ and $\delta > 0$. Then $\varphi_{s,t}(x)$, $x \in \mathbb{R}^d$, $0 \le s \le t < \sigma(s, x)$ of Theorem 4.7.1 defines a forward stochastic flow of local C^k-diffeomorphisms. Further it is a forward local $C^{k,\varepsilon}$-semimartingale flow for any $\varepsilon < \delta$.* □

The proof can be reduced to Theorem 4.6.5 similarly to the proof of Theorem 4.7.1. Details are left to the reader.

We next consider the Stratonovich's stochastic differential equation based on $\mathring{F}(x, t)$, where $\mathring{F}(x, t)$ is a continuous C-semimartingale with local characteristic belonging to the class $(B^{k+1,\delta}, B^{k,\delta})$ for some $k \geq 1$ and $\delta > 0$. For each s and x, the equation has a unique maximal solution $\varphi_{s,t}(x)$, $s \leq t < \sigma(s, x)$. It satisfies Itô's stochastic differential equation based on $F = \mathring{F} + C$ where C is the correction term of $\mathring{F}$ (see Theorem 3.4.7). Then the above theorem implies the following immediately.

Theorem 4.7.3 *Assume that the local characteristic of* $\mathring{F}$ *belongs to the class* $(B^{k+1,\delta}, B^{k,\delta})$ *for some* $k \geq 1$ *and* $\delta > 0$. *Then the system of maximal solutions of Stratonovich's equation based on* $\mathring{F}$ *defines a forward stochastic flow of local* C^k*-diffeomorphisms. Further it is a continuous local* $C^{k,\varepsilon}$*-semimartingale flow for any* $\varepsilon < \delta$. □

We shall extend the forward flow of local homeomorphisms obtained above to the backward direction, making use of the inverse map. Take any sample ω such that $\{\varphi_{s,t}(\omega) : s \leq t\}$ is a forward flow of local diffeomorphisms. Let $\mathbb{D}_{s,t}(\omega)$ and $\mathbb{R}_{s,t}(\omega)$ be the domain and the range of the map $\varphi_{s,t}(\omega)$ as before. These are open subsets of $\mathbb{R}^d$. For each t, the family of open sets $\mathbb{R}_{s,t}(\omega)$, $s \in [0, t]$ is increasing, i.e. $\mathbb{R}_{s',t}(\omega) \subset \mathbb{R}_{s,t}(\omega)$ holds if $s' < s$ since the relation $\varphi_{s',t}(\omega) = \varphi_{s,t}(\omega) \circ \varphi_{s',s}(\omega)$ is satisfied. Now given t, x, we define

$$\tau(t, x, \omega) = \inf\{s \in [0, t] : x \in \mathbb{R}_{s,t}(\omega)\} \qquad (= t \text{ if } \{\ldots\} = \emptyset). \qquad (1)$$

Then we have the relation $\{x : \tau(t, x, \omega) < s\} = \mathbb{R}_{s,t}(\omega)$. Hence $\tau(t, x, \omega)$ is an upper semicontinuous function of (t, x) and is a backward stopping time for each t and x. The inverse $\varphi_{s,t}^{-1}(x, \omega) = \varphi_{s,t}(\omega)^{-1}(x)$ is then well defined if $s > \tau(t, x, \omega)$ or $x \in \mathbb{R}_{s,t}(\omega)$. We define the random field $\varphi_{t,s}(x, \omega)$, $x \in \mathbb{R}^d$, $\tau(t, x, \omega) < s \leq t \leq T$ by $\varphi_{t,s}(x, \omega) = \varphi_{s,t}^{-1}(x, \omega)$. Then it has the flow property similar to that of the forward flow. We will call it the *backward flow of local homeomorphisms associated with the forward flow* $\varphi_{s,t}$.

We shall obtain a backward stochastic differential equation governing the associated backward flow. The following can be shown similarly to Theorem 4.2.10. See also the remark after Corollary 4.2.11.

Theorem 4.7.4 *Assume that* $\mathring{F}(\cdot, t)$ *is a forward–backward* C^k*-semimartingale such that both the forward and the backward local characteristics belong to the class* $(B^{k+1,\delta}, B^{k,\delta})$ *for some* $k \geq 1$ *and* $\delta > 0$. *Let* $\varphi_{s,t}(x)$ *be the forward stochastic flow of Theorem 4.7.3. Then the associated backward flow* $\varphi_{t,s}(x)$ *satisfies the backward Stratonovich stochastic differential equation based on* $\mathring{F}$. □

Global stochastic flows

The second objective of this section is to find a necessary and sufficient condition that the system of maximal solutions $\{\varphi_{s,t}(x) : x \in \mathbb{R}^d, s \leq t < \sigma(s, x)\}$ of the equation based on a C-semimartingale $F(x, t)$ defines a forward stochastic flow of (global) homeomorphisms or diffeomorphisms. Consider first a time homogeneous ordinary differential equation $dx/dt = f(x)$. It is well known that the system of its solutions defines a flow of diffeomorphisms if and only if the equation has global solutions both to the forward and backward direction for any initial conditions. The associated vector field $f(x)$ (infinitesimal generator) is called *complete* in such case. In the following we shall obtain a similar criterion for the stochastic differential equation.

Let $F(x, t)$ be a C-semimartingale satisfying the condition of Theorem 4.7.1 and let $\varphi_{s,t}(x)$, $0 \leq s \leq t < \sigma(s, x)$ be the forward stochastic flow of local homeomorphisms stated in the same theorem. Obviously, it becomes a forward flow of global homeomorphisms if and only if both the domain and the range of the map $\varphi_{s,t}$ are the whole space $\mathbb{R}^d$, or equivalently, the explosion time $\sigma(s, x)$ is equal to T for all s, x a.s. and the time $\tau(s, x)$ defined by (1) is 0 for all s, x a.s.

In the first step we shall discuss the case where the explosion time $\sigma(s, x)$ is equal to T for all x a.s. The corresponding C-semimartingale F is called *strictly complete to the forward.* In this case, $\varphi_{s,t}$ is called a *stochastic flow of into homeomorphisms* and trajectory $\varphi_{s,t}(x)$, $t \in [s, T]$ is called *strictly conservative.* On the other hand if the explosion time $\sigma(s, x)$ is equal to T a.s. for any x, i.e. $P(\sigma(s, x) = T) = 1$ for any x, the C-semimartingale F is called *complete to the forward* and the trajectories $\varphi_{s,t}(x)$, $t \in [s, T)$ are called *conservative.* In this case $\mathbb{D}_{s,t}$ are open dense subsets of $\mathbb{R}^d$ a.s. Obviously if $\varphi_{s,t}$ is strictly conservative it is conservative. The converse is, however, not valid in general. A counter example was given by Elworthy [29]. Although the example is stated in terms of manifold, we will give it here as an illustration.

Example 4.7.5 Let $M = \mathbb{R}^d - \{0\}$ be the punctured space. Let B_t be a d-dimensional Brownian motion and let $\varphi_{s,t}(x) = B_t - B_s + x$. It is conservative for any s, x since it does not hit the origin with probability 1. However, it is not strictly conservative. In fact, for any ω, $\sigma(s, x, \omega) < T$ holds for some x, since for any $t \in [0, T]$, there is x such that $B_t(\omega) - B_s(\omega) + x = 0$.

On the other hand, in the case of one dimensional stochastic differential equations, conservativeness implies strict conservativeness as is shown below. Let $\varphi_{s,t}(x)$, $x \in \mathbb{R}$, $0 \le s \le t < \sigma(s, x)$ be a stochastic flow of local diffeomorphisms on $\mathbb{R}$ such that it is a maximal solution of a certain one dimensional equation. Then $\mathbb{D}_{s,t}(\omega)$ is an open interval for any $s < t$ a.s. In fact if $y_1, y_2 \in \mathbb{D}_{s,t}(\omega)$ and $y_1 < y_2$, $\varphi_{s,t}(y_1) < \varphi_{s,t}(y) < \varphi_{s,t}(y_2)$ is satisfied for any y of the interval (y_1, y_2) a.s., since $\varphi_{s,t} : \mathbb{D}_{s,t}(\omega) \to \mathbb{R}$ is a homeomorphism for any $s < t$ a.s: therefore the interval (y_1, y_2) is included in $\mathbb{D}_{s,t}(\omega)$ if $y_1, y_2 \in \mathbb{D}_{s,t}(\omega)$. Now suppose that the trajectory of the maximal solution is conservative for any (s, x). Then $\mathbb{D}_{s,t}(\omega) = \mathbb{R}$ holds a.s. for any $s < t$ since $\mathbb{D}_{s,t}(\omega)$ is an open dense interval. Noting the continuity of $\varphi_{s,t}(x)$ with respect to (s, t, x), we see that $\mathbb{D}_{s,t}(\omega) = \mathbb{R}$ holds for any $s < t$ a.s.

A necessary and sufficient condition for the conservativeness of a one dimensional diffusion process was given by W. Feller, Itô-McKean [61]. We will use it in order to get the strict completeness (to the forward) of $C(\mathbb{R} : \mathbb{R})$-valued Brownian motion.

Theorem 4.7.6 *Let $F(x, t)$ be a $C(\mathbb{R} : \mathbb{R})$-valued Brownian motion with local characteristic $(a(x, y), b(x))$ such that $a(x, x) \neq 0$ everywhere. Set $a(x) = a(x, x)$ and $c(x) = \exp\left\{2 \int_0^x \frac{b(y)}{a(y)} dy\right\}$ and define*

$$K(x) = \int_0^x \frac{2}{c(z)} \int_0^z \frac{c(y)}{a(y)} dy\, dz. \tag{2}$$

Then $F(x, t)$ is strictly complete to the forward if and only if $K(+\infty) = K(-\infty) = \infty$ is satisfied.

Proof We follow Ikeda–Watanabe [49], p. 365. Since the local characteristic (a, b) is temporally homogeneous, we may assume that the maximal solution $\varphi_{s,t}(x)$ of an equation based on F is defined for any t until the explosion happens. Then we have $\lim_{t \uparrow \sigma(s,x)} |\varphi_{s,t}(x)| = \infty$ whenever $\sigma(s, x) < \infty$. Now set

$$Lu = \frac{1}{2} a u'' + b u'.$$

Then there exists a solution of equation $(L - 1)u = 0$ satisfying $u(0) = 1$, $1 + K(x) \le u(x) \le \exp K(x)$. By Itô's formula,

$$e^{-(t-s)} u(\varphi_{s,t}(x)) = u(x) + \int_s^t e^{-(r-s)} u'(\varphi_{s,r}(x)) M(\varphi_{s,r}(x), dr),$$

where M is the martingale part of F. Hence $e^{-(t-s)} u(\varphi_{s,t}(x))$ is a positive martingale. Therefore $e^{-(\sigma(s,x)-s)} \lim_{t \uparrow \sigma(s,x)} u(\varphi_{s,t}(x))$ exists and is finite a.s. If

$K(+\infty) = K(-\infty) = \infty$, then $\lim_{t\uparrow\sigma(s,x)} u(\varphi_{s,t}(x)) = \infty$, so that we have have $e^{-(\sigma(s,x)-s)} = 0$ a.s. This proves $P(\sigma(s, x) = \infty) = 1$ for any s, x.

Next suppose $K(+\infty) < \infty$. Let $x > 0$ and $\tau = \tau(s, x)$ be the hitting time of $\varphi_{s,t}(x)$ to the interval $(-\infty, 0]$. Since $u(+\infty) \le \exp K(\infty) < \infty$, $e^{-(t\wedge\tau-s)}u(\varphi_{s,t\wedge\tau}(x))$ is a bounded martingale by Doob's optional stopping time theorem. Therefore

$$E\left[e^{-(\sigma(s,x)\wedge\tau-s)} \lim_{t\uparrow\sigma(x,s)} u(\varphi_{s,t\wedge\tau}(x))\right] = u(x) > 0.$$

If $P(\sigma(s, x) = \infty) = 1$ is satisfied, then we have $E[e^{-(\tau-s)}u(0)] = u(x)$ and $u(0) > u(x)$, which is a contradiction. We have thus proved $P(\sigma(s, x) < \infty) > 0$ if $K(+\infty) < +\infty$. The proof is complete. □

The one point motion of the associated flow of the above theorem is a one dimensional temporally homogeneous diffusion process. According to Feller the point $+\infty$ or $-\infty$ is called a *non-exit boundary point* if $K(+\infty) = \infty$ or $K(-\infty) = \infty$ is satisfied respectively (see Itô–Mckean [61]).

Now we shall obtain a criterion for which maximal solutions of a given stochastic differential equation define a stochastic flow of diffeomorphisms.

Theorem 4.7.7. *Assume that $\mathring{F}(x, t)$ is a continuous forward–backward C^k-semimartingale such that both the forward and backward local characteristics belong to the class $(B^{k+1,\delta}, B^{k,\delta})$ for some $k \ge 1$ and $\delta > 0$. Then the system of the maximal solutions of Stratonovich's equation based on $\mathring{F}$ defines a stochastic flow of C^k-diffeomorphisms if and only if $\mathring{F}$ is strictly complete to both the forward and the backward direction.*

Proof. If the system of the maximal solutions $\{\varphi_{s,t}\}$ defines a stochastic flow of onto C^k-diffeomorphisms, then the associated backward flow $\varphi_{t,s}$, $s \le t$ is strictly conservative to the backward direction. Therefore $\mathring{F}$ is strictly complete to the backward.

Suppose conversely that $\mathring{F}$ is strictly complete to the backward. Then the maximal solution of Stratonovich's backward equation based on $\mathring{F}$ defines a flow of into diffeomorphisms $\hat{\varphi}_{t,s}(x)$, $s < t$. Clearly it is an extension of $\varphi_{s,t}^{-1} = \varphi_{t,s}$ by Theorem 4.7.4 and the uniqueness of the solution of the equation. Now take any sample ω such that both $\{\varphi_{s,t}(\omega)\}$ and $\{\hat{\varphi}_{s,t}(\omega)\}$ define flows of into diffeomorphisms. In order to prove the onto property of the map $\varphi_{s,t}(\omega)$, it is sufficient to prove that $\mathbb{R}_{s,t}(\omega)$ is closed, since $\mathbb{R}_{s,t}(\omega)$ is a non-void open set. Let $\overline{\mathbb{R}_{s,t}(\omega)}$ be the closure of $\mathbb{R}_{s,t}(\omega)$ and $y \in \overline{\mathbb{R}_{s,t}(\omega)}$. Choose a sequence $\{y_n\}$ from $\mathbb{R}_{s,t}(\omega)$ converging to y. Then we have $y_n = \varphi_{s,t}(\omega) \circ \varphi_{s,t}^{-1}(\omega)(y_n) = \varphi_{s,t}(\omega) \circ \hat{\varphi}_{t,s}(\omega)(y_n)$ for all n. Making n

tend to infinity, we see $y = \varphi_{s,t}(\omega) \circ \hat{\varphi}_{t,s}(\omega)(y)$, since $\hat{\varphi}_{t,s}(\omega)\colon \mathbb{R}^d \to \mathbb{R}^d$ is a continuous map. Therefore $y \in \mathbb{R}_{s,t}(\omega)$. □

As an example we consider a one dimensional Stratonovich's stochastic differential equation based on a $C(\mathbb{R} : \mathbb{R})$-valued Brownian motion. Let $\mathring{F}(x, t)$ be a continuous C-Brownian motion with local characteristic $(a(x, y), b(x))$ belonging to the class $(B^{k+1,\delta}, B^{k,\delta})$ for some $k \geq 1$ and $\delta > 0$, where $a(x, x) = a(x)$ is a strictly positive function. Consider Stratonovich's equation based on $\mathring{F}(x, t)$. The trajectory of the maximal solution is strictly conservative or equivalently $\mathring{F}$ is strictly complete to the forward if $K(+\infty) = K(-\infty) = \infty$, where K is defined by (2). In the present case, we have

$$c(x) = \exp\left\{\int_0^x \frac{(2b(y) + a'(y))}{a(y)} \mathrm{d}y\right\} = \left(\frac{a(x)}{a(0)}\right)^{1/2} \exp\left\{\int_0^x \frac{2b(y)}{a(y)} \mathrm{d}y\right\}.$$

Next consider the backward Stratonovich's equation based on $\mathring{F}$. The trajectory of the maximal solution is strictly conservative if $\hat{K}(+\infty) = \hat{K}(-\infty) = \infty$, where $\hat{K}$ is defined by (2) using

$$\hat{c}(x) = \exp\left\{\int_0^x \frac{-2b(y) + a'(y)}{a(y)} \mathrm{d}y\right\} = \left(\frac{a(x)}{a(0)}\right)^{1/2} \exp\left\{-\int_0^x \frac{2b(y)}{a(y)} \mathrm{d}y\right\}$$

$$= \frac{a(x)}{a(0)} c(x)^{-1}.$$

Hence $\mathring{F}$ is strictly complete to the backward if and only if

$$\int_0^\infty \frac{c(z)}{a(z)} \int_0^z \frac{1}{c(y)} \mathrm{d}y\, \mathrm{d}z = \infty \quad \text{and} \quad \int_{-\infty}^0 \frac{c(z)}{a(z)} \int_0^z \frac{1}{c(y)} \mathrm{d}y\, \mathrm{d}z = \infty$$

hold. According to Feller the above conditions state that $+\infty$ and $-\infty$ are *non-entrance boundary points* of the maximal solution. Combining this with Theorem 4.7.6, we obtain the following.

Theorem 4.7.8 *The system of the maximal solutions of the one dimensional stochastic differential equation defines a Brownian flow of C^k-diffeomorphisms if and only if $+\infty$ and $-\infty$ are natural (non-exit and non-entrance) boundary points.* □

Remark Set

$$s(x) = \int_0^x \frac{\mathrm{d}y}{c(y)}, \qquad m(x) = \int_0^x \frac{2c(y)}{a(y)} \mathrm{d}y.$$

These are called the canonical scale and the speed measure of the one dimensional diffusion process, respectively: its generator is written as $L^{(1)} = (d/dm)(d/ds)$ (Itô–McKean [61]). Now let $\hat{s}$ and $\hat{m}$ be the canonical scale and the speed measure of the one point motion of the inverse flow $\varphi_t^{-1}(x)$, respectively. The above argument shows that $\hat{s}(x) = (a(0)/2)m(x)$ and $\hat{m}(x) = (2/a(0))s(x)$. Therefore the canonical scale and the speed measures are reciprocal (up to constants) between these two diffusions: the generator of the inverse flow is then written as $\hat{L}^{(1)} = (d/ds)(d/dm)$.

The above theorem gives us a complete set of criteria for solutions of a one dimensional stochastic differential equation to define a Brownian flow of homeomorphisms or diffeomorphisms. As to multi-dimensional stochastic differential equations, we do not yet have such criteria other than the uniform Lipschitz condition for the local characteristic.

Random infinitesimal generators

In the remainder of this section we will consider the converse problem, i.e. for a given stochastic flow of local diffeomorphisms, we will find a stochastic differential equation governing the stochastic flow, under the condition that the flow is a local semimartingale with some regularity property.

Let $\varphi_{s,t}(x)$, $0 \le s \le t < \sigma(x, x)$ be a forward stochastic flow of local homeomorphisms. Let $\{\mathscr{F}_{s,t}\}$ be the filtration generated by $\varphi_{s,t}$. The flow is called a *forward local semimartingale flow* if for each s, x, $\varphi_{s,t}(x)$ is a continuous local semimartingale adapted to $(\mathscr{F}_{s,t})$.

Theorem 4.7.9 Let $\varphi_{s,t}(x)$, $x \in \mathbb{R}^d$, $0 \le s \le t < \sigma(s, x)$ *be a forward local semimartingale flow of local C^k-diffeomorphisms. Assume that for every s the local characteristic belongs to the class $B^{k,\delta}$ for some $k \ge 0$ and $\delta > 0$. Then for any $\varepsilon < \delta$, there exists a unique continuous $C^{k,\varepsilon}$-semimartingale $F(x, t)$ with $F(x, 0) = 0$ adapted to $(\mathscr{F}_{0,t})$ such that for every s and x, $\varphi_{s,t}(x)$, $s \le t < \sigma(s, x)$ is a solution of Itô's stochastic differential equation based on $F(x, t)$. Further if $k \ge 2$ $\varphi_{s,t}(x)$ is represented as a solution of Stratonovich's stochastic differential equation based on $\mathring{F}(x, t) = F(x, t) - 2C(x, t)$.* □

The above F and $\mathring{F}$ are called *Itô's random infinitesimal generator* and *Stratonovich's random infinitesimal generator* of the stochastic flow $\varphi_{s,t}$. The theorem is obviously an analogue of Theorem 4.4.1.

Finally we will study the random infinitesimal generator of a backward local semimartingale flow, which is defined in the obvious way.

Theorem 4.7.10 *Let $\varphi_{s,t}(x)$, $0 \le s \le t < \sigma(s, x)$ be a forward $C^{k,\delta}$-semimartingale flow with local characteristic belonging to the class $B^{k,\delta}$ where*

$k \geq 3$ and $\delta > 0$. Assume that the associated backward flow is also a backward $C^{k,\delta}$-semimartingale flow with local characteristic belonging to the same class. Let $\mathring{F}(x, t)$ be the forward Stratonovich's random infinitesimal generator. Then it is a forward–backward $C^{k-1,\varepsilon}$-semimartingale for some $\varepsilon > 0$ and is the backward Stratonovich's random infinitesimal generator. □

The proof can be carried out similarly to that of Theorem 4.4.4. We omit the details.

4.8 Stochastic flows on manifolds

Preliminaries

In this section we shall study stochastic flows on manifolds. The main objective is to discuss the relationship between stochastic flows and stochastic differential equations (or random infinitesimal generators). This relationship is similar to that in the case of Euclidean space which we have discussed in detail in previous sections. In order to define stochastic differential equations without using local coordinates, we will introduce semimartingales with values in the space of vector fields and stochastic differential equations based on them. The necessity of introducing the vector field valued semimartingales will be obvious once we recall the relation of the deterministic flow and the associated vector field.

Let M be a connected paracompact C^∞-manifold of dimension d. Let k be a non-negative integer. A continuous random field $\varphi_{s,t}(x)$, $s, t \in [0, T]$, $x \in M$ with values in M is called a *stochastic flow of C^k-diffeomorphisms* if it satisfies the following three properties:

(i) $\varphi_{s,u} = \varphi_{t,u} \circ \varphi_{s,t}$ holds for any s, t, u a.s.
(ii) $\varphi_{s,s} =$ identity map for any s, a.s.
(iii) $\varphi_{s,t}(x)$ is k-times differentiable with respect to x and the derivatives are continuous in (s, t, x). Further, the map $\varphi_{s,t} : M \to M$ is an onto C^k-diffeomorphism for any s, t a.s.

The *forward* and the *backward flows* are defined similarly to the case of Euclidean space. A stochastic flow $\varphi_{s,t}$ on the manifold M is called a *Brownian flow* if for any $t_1 < t_2 < \cdots < t_n$, $\varphi_{t_i,t_{i+1}}$, $i = 1, \ldots, n-1$ are independent. A *stochastic flow of local C^k-diffeomorphisms* on the manifold M is defined similarly to that on a Euclidean space.

Now let $\mathfrak{X}^k(M)$ be the set of all C^k-vector fields (= first order linear partial differential operators with C^k-coefficients). It is a linear topological space. Let $F(t)$, $t \in [0, T]$ be a continuous stochastic process with values in $\mathfrak{X}^k(M)$. Then for each $f \in C^\infty(M)$, $F(t)f$ is a $C^k(M : \mathbb{R})$-valued process. The process

$F(t)$ is called a *continuous localmartingale with values in* $\mathscr{X}^k(M)$ if $F(t)f$ is a continuous localmartingale with values in $C^k(M:\mathbb{R})$ for any f. *A continuous semimartingale with values in* $\mathscr{X}^k(M)$ is defined similarly.

In the following we will only consider the continuous semimartingale $F(t)$ such that for any $f \in C^\infty$ $F(t)f$ has a local characteristic of the form $(a(f, g), b(f), A_t)$ belonging to the class $B^{k,\delta}$, where A_t is a continuous strictly increasing process defined commonly for every f. We wish to define the local characteristic of $F(t)$ itself. For this we shall express $F(t)$ using a local coordinate. Let (U, η) be a coordinate neighborhood of M, i.e. U is an open subset of M and η is a C^∞-diffeomorphism from U to an open subset E in $\mathbb{R}^d$. Let $\eta = (x^1, \dots, x^d)$ be the local coordinate. Set $F^i(x, t) = F(t)x^i, i = 1, \dots, d$. Then $F^i(x, t)$ are continuous semimartingales with values in $C^k(U:\mathbb{R})$, and $F(t)f$ is represented by

$$F(t)f(x) = \sum_i F^i(x, t)\frac{\partial f}{\partial x^i}(x), \qquad x \in U. \tag{1}$$

Let now $(a^{ij}(x, y, t), b^i(x, t), A_t)$ be the local characteristic of $(F^1, \dots, F^d)$. We shall see how the local characteristics are transformed by the change of the local coordinates. Let $(\bar{x}^1, \dots, \bar{x}^d)$ be another local coordinate. Then $F(t)f$ is represented by $F(t)f = \sum_i \bar{F}^i(x, t)\partial f/\partial \bar{x}^i$. Let $(\bar{a}^{ij}(x, y, t), \bar{b}^i(x, t), A_t)$ be the local characteristic of $(\bar{F}^1, \dots, \bar{F}^d)$. Since F^i and $\bar{F}^i$ are related by $F^i(x, t) = \sum_j \bar{F}^j(x, t)\partial x^i/\partial \bar{x}^j$, we have

$$a^{ij}(x, y, t) = \sum_{k,l} \bar{a}^{kl}(x, y, t)\frac{\partial x^i}{\partial \bar{x}^k}(x)\frac{\partial x^j}{\partial \bar{x}^l}(y), \tag{2}$$

$$b^i(x, t) = \sum_j \bar{b}^j(x, t)\frac{\partial x^i}{\partial \bar{x}^j}(x). \tag{3}$$

Therefore, at each coordinate neighborhood,

$$B_t f(x) = \sum_i b^i(x, t)\frac{\partial f}{\partial x^i} \tag{4}$$

is defined not depending on the choice of local coordinates. Thus B_t defines a stochastic process with values in $\mathscr{X}^k(M)$. Set

$$H(t)f \equiv F(t)f - \int_0^t B_r f \, dA_r. \tag{5}$$

It is a localmartingale with values in $\mathscr{X}^k(M)$. At each coordinate neighborhood, the local characteristic of $(H(t)f, H(t)g)$ is given by

$$\langle f, g\rangle_t(x, y) \equiv \sum_{i,j} a^{ij}(x, y, t)\frac{\partial f}{\partial x^i}\frac{\partial g}{\partial x^j}. \tag{6}$$

Obviously it does not depend on the choice of the local coordinate because of property (2).

The following is easily verified.

Lemma 4.8.1 *For any t and almost all ω, $\langle\ ,\ \rangle_t$ of* (6) *defines a bilinear form from $C^\infty(M \times M)$ into $C^k(M \times M)$ satisfying the following property*:

(i) $\langle f, g\rangle_t(x, y) = \langle g, f\rangle_t(y, x)$.

(ii) $\langle f_1 f_2, g\rangle_t(x, y) = f_1(x)\langle f_2, g\rangle_t(x, y) + f_2(x)\langle f_1, g\rangle_t(x, y)$ *holds for any* f_1, f_2, g *of* $C^\infty(M)$ *and* x, y.

(iii) $\sum_{i,j=1}^m \langle f_i, f_j\rangle(x^i, x^j) \geq 0$ *for any positive integer* m *and* $x^i \in M$, $f_i \in C^\infty(M)$, $i = 1, \ldots, m$. □

The triple $(\langle,\rangle_t, B_t, A_t)$ is called a *local characteristic* of the $\mathscr{X}^k(M)$*-valued semimartingale* $F(t)$. It is said to belong to the *class* $(B^{k,\delta}, B^{k',\delta'})$ if at every coordinate neighborhood, the local characteristic of $(F^1, \ldots, F^d)$ in equation (1) belongs to the class $(B^{k,\delta}, B^{k',\delta'})$.

Now let f_t be a continuous $(\mathscr{F}_t)$-adapted process with values in $C^2(M : \mathbb{R})$ and let φ_t be a continuous $(\mathscr{F}_t)$-adapted process with values in M. The Itô integral is well defined by

$$\int_s^t F(dr)f_r(\varphi_r) \equiv \lim_{|\Delta|\to 0} \sum_{k=0}^{n-1} (F(t_{k+1}) - F(t_k))f_{t_k}(\varphi_{t_k}) \tag{7}$$

where $\Delta = \{s = t_0 < t_1 < \cdots < t_n = t\}$. The backward Itô integral can be defined similarly. Assume that the local characteristic of F belongs to $(B^{2,\delta}, B^{1,\delta})$ for some $\delta > 0$ and $Xf_t(\varphi_t)$ is a continuous semimartingale for any C^∞-vector field X. Then the Stratonovich integral is well defined by

$$\int_s^t F(\circ dr)f_r(\varphi_r) \equiv \lim_{|\Delta|\to 0} \frac{1}{2} \sum_{k=0}^{n-1} (F(t_{k+1}) - F(t_k))(f_{t_{k+1}}(\varphi_{t_{k+1}}) + f_{t_k}(\varphi_{t_k})). \tag{8}$$

The backward Stratonovich integral can be defined similarly.

The above definition of stochastic integrals is extended to a continuous local process φ_t, $t \in [0, \sigma)$ with values in M adapted to $(\mathscr{F}_t)$. Let $\{\sigma_n\}$ be an increasing sequence of stopping times such that $\sigma_n < \sigma$ and $\sigma_n \uparrow \sigma$. For each t, $\varphi_t^n \equiv \varphi_{t\wedge\sigma_n}$ is a global process. Then the Itô integral $g_t^n = \int_0^t F(dr)f(\varphi_r^n)$ is well defined. It satisfies $g_{t\wedge\sigma_m}^n = g_t^m = g_{t\wedge\sigma_m}^m$ if $n > m$. Therefore there exists a continuous local semimartingale g_t, $t \in [0, \sigma)$ such that for any n $g_t = g_t^n$ holds for $t < \sigma_n$. We denote it by $\int_0^t F(dr)f_r(\varphi_r)$. The Stratonovich integral $\int_0^t F(\circ dr)f_r(\varphi_r)$ is defined similarly if $Xf_t(\varphi_t)$ is a continuous local semimartingale for any C^∞-vector field X.

Stochastic differential equations on manifolds

Suppose that we are given a continuous semimartingale $F(t)$ with values in $\mathscr{X}^k(M)$ with local characteristic $(\langle\ ,\ \rangle_t, B_t, A_t)$ belonging to the class $(B^{k+1,\delta}, B^{k,\delta})$ where $k \geq 1$ and $0 < \delta \leq 1$. We wish to show the existence of a stochastic flow of local C^k-diffeomorphisms $\varphi_{s,t}$ satisfying

$$f(\varphi_{s,t}(x)) - f(x) = \int_s^t F(\circ \mathrm{d}r) f(\varphi_{s,r}(x)) \tag{9}$$

for any $f \in C^\infty(M)$ by solving a stochastic differential equation of the type

$$\mathrm{d}\varphi_t = F(\varphi_t, \circ \mathrm{d}t). \tag{10}$$

We first give a precise definition of the maximal solution of the above equation. Given $s \in [0, T]$ let φ_t, $t \in [s, \sigma)$ be a continuous local process with values in M adapted to $(\mathscr{F}_t)$. It is called a *local solution* of the Stratonovich equation based on $F(t)$ starting at x at time s, if $f(\varphi_t)$ is a local semimartingale for any f of $C^\infty(M)$ and satisfies

$$f(\varphi_t) = f(x) + \int_s^t F(\circ \mathrm{d}r) f(\varphi_r) \tag{10$'$}$$

for any $t < \sigma$. Further it is called a *maximal solution* if $\lim_{t \to \sigma} \varphi_t = \infty$ holds if $\sigma < T$ a.s. in the case where the manifold M is non-compact, and if $\sigma = T$ holds a.s. in the case where M is compact. Here ∞ is the infinity of M adjoined as the one point compactification. The random time σ is called the explosion time.

We wish to show that equation (10) has a unique maximal solution starting at x at time s for any s and x. In the first step, we shall construct a solution of equation (10) in a coordinate neighborhood. Then we shall obtain a maximal solution by piecing together the system of solutions in coordinate neighborhoods.

Let $\eta = (x^1, \ldots, x^d)$ be a local coordinate in a coordinate neighborhood (U, η). Then $F(t)f(x)$ is represented as in (1). Let V be an open subset of M such that $\bar{V} \subset U$ and let ψ be a C^∞-function on M such that $\psi(x) = 1$ for $x \in V$ and its support is included in U. We define $\tilde{F}^i(\tilde{x}, t)$, $\tilde{x} \in \mathbb{R}^d$ by

$$\tilde{F}^i(\tilde{x}, t) \equiv \begin{cases} \psi(\eta^{-1}(\tilde{x}))F^i(\eta^{-1}(\tilde{x}), t), & \text{if } \tilde{x} \in E \\ 0, & \text{if } \tilde{x} \in E^c, \end{cases} \tag{11}$$

where E is an open subset of $\mathbb{R}^d$ such that $E = \eta(U)$. Then $\tilde{F}(\tilde{x}, t) = (\tilde{F}^1(\tilde{x}, t), \ldots, \tilde{F}^d(\tilde{x}, t))$ is a continuous semimartingale with values in $C^k(\mathbb{R}^d, \mathbb{R}^d)$ with local characteristic belonging to the class $(B_b^{k+1,\delta}, B_b^{k,\delta})$. Consider Stratonovich's stochastic differential equation based on $\tilde{F}$ on the Euclidean space $\mathbb{R}^d$. It generates a stochastic flow of C^k-diffeomorphisms

$\tilde{\varphi}_{s,t}(\tilde{x})$. Set $\tilde{\tau}_V(s, \tilde{x}) = \inf\{t > s : \tilde{\varphi}_{s,t}(\tilde{x}) \notin \overline{\eta(V)}\}$ $(= T$ if $\{\ldots\} = \emptyset)$. Then $\tilde{\varphi}_{s,t}(\tilde{x})$, $0 \le s \le t \le \tilde{\tau}_V(s, \tilde{x})$ does not depend on the choice of the function ψ by Lemma 3.4.4. Define $\tau_V(s, x) = \tilde{\tau}_V(s, \eta(x))$ and $\varphi_{s,t}(x) = \eta^{-1}(\tilde{\varphi}_{s,t}(\eta(x)))$ for $x \in V$ and $0 \le s \le t \le \tau_V(s, x)$. It satisfies

$$x^i(\varphi_{s,t}(x)) - x^i = \int_s^t F^i(\varphi_{s,r}(x), \circ dr), \tag{12}$$

for $t < \tau_V(s, x)$. Therefore by Itô's formula $\varphi_{s,t}(x), x \in V, 0 \le s \le t \le \tau_V(s, x)$ satisfies (9) for any $f \in C^\infty(M)$. This implies that $\varphi_{s,t}(x)$, $x \in V, 0 \le s \le t \le \tau_V(s, x)$ does not depend on the choice of the local coordinate and in fact it is a solution of equation (10) at the coordinate neighborhood (V, η).

Next we shall piece together the above solutions to get the maximal solution on the manifold M. Let $\{U_l\}$, $\{V_l\}$ and $\{W_l\}$ be countable families of coordinate neighborhoods of M satisfying the following properties. For each l, U_l, V_l and W_l are balls with the same center and radius $3r$, $2r$, r, respectively, where r is a sufficiently small positive number; $\{W_l\}$ is a covering of M, i.e. $\bigcup_l W_l = M$. Then for each l, there exists a unique solution $\varphi_{s,t}^{(l)}(x)$, $t \in [s, \tau_{U_l}(s, x))$ of equation (10) starting at $x \in U_l$ at time s such that it is continuous in (s, t, x), where $\tau_{U_l}(s, x)$ is the hitting time of $\varphi_{s,t}^{(l)}(x)$ to $\bar{U}_l^c$. If $U_l \cap U_m \neq \emptyset$, then $\varphi_{s,t}^{(l)}(x) = \varphi_{s,t}^{(m)}(x)$ holds for $x \in U_l \cap U_m$ and $t < \tau_{U_l}(s, x) \wedge \tau_{U_m}(s, x)$.

For every $s \in [0, T]$ and $x \in M$ we define $\varphi_{s,t}(x)$, $x \in M$, $s \le t \le \sigma_1(s, x)$ by

$$\sigma_1(s, x) = \tau_{V_m}(s, x) \qquad \text{if } x \in W_m - \bigcup_{l<m} W_l, \tag{13}$$

$$\varphi_{s,t}(x) = \varphi_{s,t}^{(m)}(x) \qquad \text{if } x \in W_m - \bigcup_{l<m} W_l \text{ and } t \le \sigma_1(s, x). \tag{14}$$

It satisfies (9) for any $f \in C^\infty(M)$.

We will extend the solution so as to get a maximal solution. Define a local random field $\varphi_{s,t}(x)$, $x \in M$, $0 \le s \le t \le \sigma_n(s, x)$ by induction as follows.

$$\begin{aligned}\varphi_{s,t}(x) &= \varphi_{\sigma_1(s,x),t} \circ \varphi_{s,\sigma_1(s,x)}(x) \qquad \text{if } \sigma_1(s, x) < t \le \sigma_2(s, x),\\ &\qquad \text{where } \sigma_2(s, x) = \sigma_1(\sigma_1(s, x), \varphi_{s,\sigma_1(s,x)}(x))\\ &= \varphi_{\sigma_{n-1}(s,x),t} \circ \varphi_{s,\sigma_{n-1}(s,x)}(x) \qquad \text{if } \sigma_{n-1}(s, x) < t \le \sigma_n(s, x),\\ &\qquad \text{where } \sigma_n(s, x) = \sigma_1(\sigma_{n-1}(s, x), \varphi_{s,\sigma_{n-1}(s,x)}(x)).\end{aligned} \tag{15}$$

Then it again satisfies (9) for any f of $C^\infty(M)$. Set $\sigma(s, x) = \lim_{n\to\infty} \sigma_n(s, x)$. Then equality (9) holds for any $t < \sigma(s, x)$. We claim that it is the unique maximal solution.

Lemma 4.8.2 *For every s and x, the local process $\varphi_{s,t}(x)$, $t \in [s, \sigma(s, x))$ defined by* (15) *is a maximal solution of equation* (10). *Further it is the unique maximal solution.*

Proof In the following argument we fix any s and x. We first consider the case where M is non-compact. Let $\varphi_{s,t}(x)$, $t \in [s, \sigma(s, x))$ be the local process defined by (15). Let f be a C^∞-function with compact support. Then equality (9) is valid for $t < \sigma(s, x)$. Let t tend to $\sigma(s, x)$ at (9). Then the right hand side of (9) has a limit a.s. by the martingale convergence theorem (Theorem 1.2.3). Therefore the limit of $f(\varphi_{s,t}(x))$ as $t \uparrow \sigma(s, x)$ exists a.s. Now take a countable family $\{f_n\}$ of such functions separating any two points of M. Then we see that for almost all ω one of the following alternatives holds.

(a) $\lim_{t\uparrow\sigma(s,x,\omega)} \varphi_{s,t}(x, \omega)$ exists and equals ∞,
(b) $\lim_{t\uparrow\sigma(s,x,\omega)} \varphi_{s,t}(x, \omega)$ exists and belongs to M.

But (b) implies $\sigma(s, x, \omega) = T$. Indeed, suppose on the contrary that $\sigma(s, x) < T$ and $\lim_{t\uparrow\sigma(s,x,\omega)} \varphi_{s,t}(x, \omega)$ exists in M. Then there are positive integers m and n such that $\varphi_{s,t}(x, \omega) \in W_m - \bigcup_{l<m} W_l$ for all $t \geq \sigma_n(s, x, \omega)$. This implies $\sigma_{n+1}(s, x, \omega) = T$ which is a contradiction. We have thus seen that only (a) can occur if $\sigma(s, x, \omega) < T$. This proves that the solution is maximal. Next if M is compact the above argument shows that $\sigma(s, x) = T$ holds a.s. Therefore $\varphi_{s,t}(x)$, $t \in [s, \sigma(s, x))$ is a maximal solution.

We will next show the uniqueness of the maximal solution of equation (10). Let $\psi_{s,t}(x)$, $t \in [s, \tau(s, x))$ be another maximal solution of equation (10) starting at x at time s. Let (U, η) be a coordinate neighborhood containing x. Then the equality $\varphi_{s,t}(x) = \psi_{s,t}(x)$ holds for $t \in [s, \tau_U(s, x))$, since both $\tilde{\varphi}_{s,t}(\tilde{x}) \equiv \eta \circ \varphi_{s,t} \circ \eta^{-1}(\tilde{x})$ and $\tilde{\psi}_{s,t}(\tilde{x}) \equiv \eta \circ \psi_{s,t} \circ \eta^{-1}(\tilde{x})$ are solutions of the same stochastic differential equation on Euclidean space with the same initial condition. Therefore we have $\varphi_{s,t}(x) = \psi_{s,t}(x)$ for $t \in [s, \sigma_1(s, x))$ where σ_1 is the stopping time defined by (13). We can easily show by induction that $\varphi_{s,t}(x) = \psi_{s,t}(x)$ holds for $t \in [s, \sigma_n(s, x))$, where σ_n is the stopping time defined by (15). Consequently the equality $\varphi_{s,t}(x) = \psi_{s,t}(x)$ holds for any $t \in [s, \sigma(s, x))$. The proof is complete. □

We will show that the system of maximal solutions $\varphi_{s,t}(x)$ defined by (15) is a forward stochastic flow of local C^k-diffeomorphisms.

Lemma 4.8.3 *There exists a null set N of Ω such that for any $\omega \in N^c$ $\varphi_{s,t}(x, \omega)$ defined by* (15) *satisfies the following properties.*

(i) *It has the flow property $\varphi_{s,u}(x, \omega) = \varphi_{t,u}(\varphi_{s,t}(x, \omega), \omega)$ if $s < t < u < \sigma(s, x, \omega)$.*

(ii) *Set* $\mathbb{D}_{s,t}(\omega) = \{x : \sigma(s, x, \omega) > t\}$. *Then* $\mathbb{D}_{s,t}(\omega)$ *is an open subset of* $\mathbb{R}^d$ *for every* $s < t$. *The map* $\varphi_{s,t}(\omega) : \mathbb{D}_{s,t}(\omega) \to \mathbb{R}^d$ *is an into* C^k-*diffeomorphism for every* $s < t$.

(iii) *The derivatives* $D_x^\alpha \varphi_{s,t}(x, \omega)$ *and* $D_x^\alpha \varphi_{s,t}^{-1}(x, \omega)$, $|\alpha| \le k$ *are continuous with respect to* (s, t, x).

Proof Take any ω *such that* $\varphi_{s,t}(x, \omega)$ defined by (15) is continuous in (s, t, x). Assertion (i) is obvious from the definition of $\varphi_{s,t}(x, \omega)$. For the proof of (ii), we will first show the local diffeomorphism of the map $\varphi_{s,t}(\omega) : \mathbb{D}_{s,t}(\omega) \to \mathbb{R}^d$. Take any x_0 from $\mathbb{D}_{s,t}(\omega)$. Consider a sample function $\{\varphi_{s,r}(x_0, \omega) : r \in [s, t]\}$. We may choose a chain of coordinate neighborhoods $V_0, \dots, V_n$ such that $x_0 \in V_0$ and for any $i = 1, \dots, n$,

$$\bigcup_{x \in V_0} \{\varphi_{s,r}(x, \omega) : r \in [t_{i-1}, t_i]\} \subset V_i,$$

where $t_i = (i/n)(t - s) + s$. By the flow property (i), we have

$$\varphi_{s,t}(\omega) = \varphi_{t_{n-1},t}(\omega) \circ \varphi_{t_{n-2},t_{n-1}}(\omega) \circ \cdots \circ \varphi_{s,t_1}(\omega).$$

Since $\varphi_{t_{i-1},t_i}(x, \omega)$ defines a local C^k-diffeomorphism in V_i, we see that $\varphi_{s,t}(x, \omega)$ is a local C^k-diffeomorphism at a neighborhood of x_0. This implies that $\mathbb{D}_{s,t}(\omega)$ is open.

The 'one to one' of the map $\varphi_{s,t}(\omega)$ is a rather local property as we will see below. Let $\{U_n : n = 1, 2, \dots\}$ be coordinate neighborhoods of M such that $\bigcup_n U_n = M$. Let $\{S_m : m = 1, 2, \dots\}$ be a set of open time intervals generating all open sets in (s, T). We denote by $N_{n,m}$ the set of all ω such that there exist x, y $(x \neq y)$ of M and $\sigma(\omega) \in S_m$ such that $\varphi_{s,t}(x, \omega) = \varphi_{s,t}(y, \omega)$ for $t \ge \sigma(\omega)$, $\varphi_{s,t}(x, \omega) \neq \varphi_{s,t}(y, \omega)$ for $t < \sigma(\omega)$ and $\varphi_{s,\sigma}(x, \omega) = \varphi_{s,\sigma}(y, \omega)$ is in U_n. In the coordinate neighborhood, we see by Theorem 4.7.1 that $N_{n,m}$ is a null set. Therefore, $\bigcup_{n,m} N_{n,m}$ is a null set. Note that if $\varphi_{s,t}(\omega)$ is not a one to one map for some t, then ω belongs to some $N_{n,m}$.

Now the derivatives $D_x^\alpha \varphi_{s,t}(x, \omega)$, $|\alpha| \le k$ are continuous with respect to (s, t, x) since each $D_x^\alpha \varphi_{t_{i-1},t_i}(x, \omega)$ is continuous with respect to (t_{i-1}, t_i, x) in the corresponding coordinate neighborhood. Further since the inverse $\varphi_{s,t}^{-1}(x)$ is given by $\varphi_{s,t_1}^{-1} \circ \cdots \circ \varphi_{t_{n-1},t_n}^{-1}(x)$, derivatives $D_x^\alpha \varphi_{s,t}^{-1}(x, \omega)$ are also continuous with respect to (s, t, x) by the same reasoning. The proof is complete. □

Summing up these two lemmas we obtain the following.

Theorem 4.8.4 *Let* $F(t)$ *be a continuous semimartingale with values in* $\mathfrak{X}^k(M)$ *such that its local characteristic belongs to the class* $(B^{k+1,\delta}, B^{k,\delta})$ *for some*

$k \geq 1$ and $\delta > 0$. Consider the Stratonovich's stochastic differential equation based on $F(t)$. Then the system of the maximal solutions $\varphi_{s,t}(x)$, $x \in M$, $0 \leq s \leq t < \sigma(s, x)$ has a modification of a stochastic flow of local C^k-diffeomorphisms of M. □

Diffeomorphic property of solutions

We shall study the case where the system of maximal solutions of equation (10) defines a stochastic flow of global diffeomorphisms. An $\mathscr{X}^k(M)$-valued semimartingale $F(t)$ is called *strictly complete to the forward* if the domain $\mathbb{D}_{s,t}(\omega)$ of the flow $\varphi_{s,t}(x, \omega)$ of Lemma 4.8.3 is the whole space M a.s. for any $s < t$. The following theorem corresponds to Theorem 4.7.7.

Theorem 4.8.5 *Assume that $F(t)$ is a continuous forward–backward semimartingale with values in $\mathscr{X}^k(M)$ such that both the forward and backward local characteristics belong to the class $(B^{k+1,\delta}, B^{k,\delta})$ for some $k \geq 1$ and $\delta > 0$. Then the system of maximal solutions $\varphi_{s,t}(x)$ of the previous theorem defines a stochastic flow of C^k-diffeomorphisms of M if and only if $F(t)$ is strictly complete both to the forward and the backward direction. Furthermore, if $\varphi_{s,t}$ is a diffeomorphism, the inverse $\varphi_{t,s}(x) = \varphi_{s,t}^{-1}(x)$ satisfies*

$$f(\varphi_{t,s}(x)) = f(x) + \int_s^t F(\circ \hat{d}r) f(\varphi_{t,r}(x)) \tag{16}$$

for any f of $C^\infty(M)$. □

The proof of the above theorem is similar to the proof of Theorem 4.7.7 and it is left to the reader.

Corollary 4.8.6 *Assume the same condition as in Theorem 4.8.5. If M is a compact manifold, the system of solutions defines a stochastic flow of C^k-diffeomorphisms of M a.s.*

Proof Let $\varphi_{s,t}(x)$, $t \in [s, \sigma(s, x))$ be the maximal solution defined by (15), starting at x at time s. We will prove that it is strictly conservative i.e. $\sigma(s, x) = T$ holds for all x a.s. for every s. For this we will define the solution in a different manner. We may assume that the number of coordinate neighborhoods $\{W_l\}$ which covers M is finite, say $\{W_l\}_{l=1}^m$. Set

$$\sigma_1(s) = \inf_{x \in \overline{W}_1} \tau_{V_1}(s, x) \wedge \inf_{x \in \overline{W_2 - W_1}} \tau_{V_2}(s, x) \wedge \cdots \wedge \inf_{x \in \overline{W_m - \bigcup_{l<m} W_l}} \tau_{V_m}(s, x),$$

$$\sigma_n(s) = \sigma_1(\sigma_{n-1}(s)).$$

Then $\{\sigma_n(s)\}$ is an increasing sequence of stopping times for each s. Define the limit by $\sigma_\infty(s)$.

We wish to prove $\sigma_\infty(s) = T$ a.s. P for any s. Observe that $\sigma_1(s) - s$ is strictly positive and lower semicontinuous in s a.s. Indeed, for each $\omega \in \Omega$

$$\{s : \sigma_1(s) > b\} = \bigcap_{l=1}^{m} \left\{ s : \left\{ \varphi_{s,t}(x) : s \le t \le b, x \in \overline{W_l - \bigcup_{i<l} W_i} \right\} \subset V_l \right\}.$$

Since the V_l are open sets, the set $\{s : \sigma_1(s) > b\}$ is open for any b. Therefore $\sigma_1(s)$ is lower semicontinuous. Consequently for any $\varepsilon > 0$ there exists s_0 in $[0, T-\varepsilon]$ such that $\inf_{s\in[0,T-\varepsilon]} \sigma_1(s) - s \ge \sigma_1(s_0) - s_0$. This implies $\sigma_{n+1}(s) - \sigma_n(s) = \sigma_1(\sigma_n(s)) - \sigma_n(s) \ge \sigma_1(s_0) - s_0 > 0$ if $\sigma_n(s) \le T - \varepsilon$. Therefore we have

$$\sigma_\infty(s) - s = \sum_{n=0}^{\infty} (\sigma_{n+1}(s) - \sigma_n(s)) \ge T - \varepsilon - s \quad \text{a.s.}$$

where $\sigma_0(s) = s$. Since $\varepsilon > 0$ is arbitrary we have $\sigma_\infty \ge T$ a.s.

We will now define $\tilde{\varphi}_{s,t}(x)$, $s \le t < T$ by

$$\tilde{\varphi}_{s,t}(x) = \begin{cases} \varphi_{s,t}(x) & \text{if } s < t \le \sigma_1(s), \\ \varphi_{\sigma_{n-1}(s),t} \circ \varphi_{s,\sigma_{n-1}(s)}(x) & \text{if } \sigma_{n-1}(s) < t \le \sigma_n(s). \end{cases}$$

Then $\tilde{\varphi}_{s,t}$ is the maximal solution. It is strictly conservative, i.e. $F(t)$ is strictly complete to the forward. By the same reasoning, $F(t)$ is strictly complete to the backward. Consequently the system of solutions defines a stochastic flow of C^k-diffeomorphisms. □

A vector field X is called *complete* if the maximal solution of the ordinary differential equation based on X defines a one parameter group of diffeomorphisms. In the present context, it is complete if and only if $F(t) \equiv tX$ is strictly complete both to the forward and backward directions. Hence if the manifold is compact any vector field on it is complete by the above corollary. Now suppose that $F(t)$ is a continuous semimartingale taking values of complete vector fields on a noncompact manifold. The question naturally arises of whether or not this $F(t)$ is strictly complete to the forward and the backward. The answer is not always affirmative. A counter example will be found in Exercise 4.8.12. In the sequel we will give a sufficient condition that the answer is yes, in the case of the classical Itô's stochastic differential equation.

Let $X_0, \ldots, X_m$ be complete C^∞-vector fields and let $(B_t^1, \ldots, B_t^m)$ be a standard Brownian motion. Consider the Stratonovich stochastic differential equation based on $\mathscr{X}^\infty(M)$-valued Brownian motion

$$F(t) = tX_0 + \sum_{l=1}^{m} B_t^l X_l,$$

that is to say,

$$f(\varphi_t) = f(x) + \int_s^t X_0 f(\varphi_r)\,\mathrm{d}r + \sum_{l=1}^m \int_s^t X_l f(\varphi_r) \circ \mathrm{d}B_r^l. \tag{17}$$

There exists a stochastic flow of local C^∞-diffeomorphisms $\varphi_{s,t}(x)$, $0 \le s \le t < \sigma(s, x)$ such that for every s and x it is a maximal solution of the above equation.

Before we proceed further, we need some facts from differential geometry. For two vector fields X and Y on the manifold M, we define the *Lie bracket* $[X, Y]$ by $XY - YX$. It is again a vector field. Now let $X_0, \ldots, X_m$ be complete vector fields defining equation (17). The real vector space spanned by all vector fields consisting of Lie brackets $[\ldots[X_{i_1}, X_{i_2}], \ldots, X_{i_n}]$, $i_1, \ldots, i_n \in \{0, 1, \ldots, m\}$ is called the *Lie algebra generated by* $X_0, \ldots, X_m$ and is denoted by $\mathscr{L}(X_0, \ldots, X_m)$ or simply by $\mathscr{L}$.

Suppose that the Lie algebra $\mathscr{L}$ is a finite dimensional vector space. A remarkable fact is that every element of $\mathscr{L}$ is a complete vector field. Furthermore, there exists a Lie group G with properties (i)–(iii) below.

(i) G is a Lie transformation group of M, i.e. there exists a C^∞-map ψ from the product manifold $G \times M$ into M such that
   (a) for each g, $\psi(g, \cdot)$ is a diffeomorphism of M and
   (b) $\psi(e, \cdot) =$ identity and $\psi(gh, \cdot) = \psi(g, \psi(h, \cdot))$ for any g, h of G.

(ii) The map $g \to \psi(g, \cdot)$ is an isomorphism from G into the group of all diffeomorphisms of M.

(iii) Let $\mathscr{G}$ be the Lie algebra of G (=right invariant vector fields on G). For any X of $\mathscr{L}$ there exists $\hat{X}$ of $\mathscr{G}$ such that

$$\hat{X}(f \circ \psi_x)(g) = Xf(\psi(g, x)) \tag{18}$$

holds for any f of $C^\infty(M)$. Here $f \circ \psi_x$ is a C^∞-function on G such that $f \circ \psi_x(g) = f \circ \psi(g, x)$.

We will call G the *Lie (transformation) group associated with* $\mathscr{L}$. For details see Palais [107].

Theorem 4.8.7 *Assume that $X_0, \ldots, X_m$ defining equation* (17) *are complete C^∞-vector fields and that the Lie algebra $\mathscr{L}$ generated by $X_0, \ldots, X_m$ is of finite dimension. Then the system of maximal solutions defines a forward Brownian flow of diffeomorphisms. Furthermore, let G be the Lie group associated with the Lie algebra $\mathscr{L}$. Then the solution takes values in the set of diffeomorphisms $\{\phi(g, \cdot) : g \in G\}$.*

Proof Let $\hat{X}_0, \ldots, \hat{X}_m$ be right invariant vector fields on the Lie group G associated with the Lie algebra $\mathscr{L}$ such that these are related to $X_0, \ldots, X_m$ by formula (18), respectively. Consider a Stratonovich stochastic differential equation on G based on $\hat{F}(t) = t\hat{X}_0 + \sum_{l=1}^m B_t^l \hat{X}_l$. We may assume that $\hat{F}(t)$ is defined for $t \in [0, \infty)$. Let ϕ_t be the solution starting at e at time 0. We show that its explosion time is infinite a.s. For an open neighborhood of e, define a sequence of stopping times τ_n by induction:

$$\tau_1 = \inf\{t > 0 : \phi_t \notin \bar{U}\} \qquad (= \infty \text{ if } \{\dots\} = \emptyset),$$

$$\vdots$$

$$\tau_n = \inf\{t > \tau_{n-1} : \phi_t \phi_{\tau_{n-1}}^{-1} \notin \bar{U}\} \qquad (= \infty \text{ if } \{\dots\} = \emptyset),$$

where ϕ_t^{-1} is the inverse of ϕ_t. Then $\tau_1, \tau_2 - \tau_1, \dots, \tau_n - \tau_{n-1}$ are independent identically distributed random variables such that $E[\tau_1] > 0$. Therefore τ_n diverges to $+\infty$ as $n \to \infty$ by the law of large numbers. Hence the explosion time is infinite a.s.

Now set $\varphi_{s,t}(x) = \psi(\phi_t \phi_s^{-1}, x)$. It is a Brownian flow of diffeomorphisms. Indeed we have the flow property by

$$\varphi_{t,u} \circ \varphi_{s,t}(x) = \psi(\phi_u \phi_t^{-1}, \psi(\phi_t \phi_s^{-1}, x)) = \psi(\phi_u \phi_s^{-1}, x) = \varphi_{s,u}(x)$$

for any $s < t < u$. Further,

$$\begin{aligned} f(\varphi_{s,t}(x)) &= f \circ \psi(e, x) + \sum_{l=0}^m \int_s^t \hat{X}_l(f \circ \psi_x)(\phi_r \phi_s^{-1}) \circ dB_r^l \\ &= f(x) + \sum_{l=0}^m \int_s^t X_l f(\varphi_{s,r}(x)) \circ dB_r^l, \end{aligned} \tag{19}$$

where $B_t^0 \equiv t$. Therefore $\varphi_{s,t}(x)$ is the solution of the given equation (17). The proof is complete. □

Random infinitesimal generators

Suppose that we are given a stochastic flow of C^k-diffeomorphisms $\varphi_{s,t}(x)$ on the manifold M where $k \ge 2$ and let $0 \le \delta \le 1$. It is called a *forward $C^{k,\delta}$-semimartingale flow* if $f(\varphi_{s,t}(x))$ is a forward $C^{k,\delta}$-semimartingale for any $f \in C^\infty(M)$. The *backward semimartingale flow* is defined similarly. A continuous semimartingale $F(t)$ with values in $\mathscr{X}^k(M)$ such that $F(0) \equiv 0$ is called a *forward random C^k-infinitesimal generator of $\varphi_{s,t}$ (in the sense of the Stratonovich integral)* if (9) is satisfied for any $f \in C^\infty(M)$.

We shall rewrite the relation (9) using the Itô integral assuming the existence of a forward random infinitesimal generator $F(t)$. We first represent relation (9) using a local coordinate. Let (U, η) be a coordinate neighborhood of M where $\eta = (x^1, \dots, x^d)$ is a local coordinate. Set $\varphi_{s,t}^i(x) = x^i(\varphi_{s,t}(x))$. Then at the coordinate neighborhood U (9) is written as

$$\varphi_{s,t}^i(x) - x^i = \int_s^t F^i(\varphi_{s,r}(x), \circ dr), \qquad t \le \tau_U(s, x) \tag{20}$$

where $\tau_U(s, x) = \inf\{t > s : \varphi_{s,t}(x) \in \bar{U}^c\}$ ($= T$ if $\{\dots\} = \emptyset$). Then by Itô's formula, we have for $s \le t \le \tau_U(s, x)$

$$f(\varphi_{s,t}(x)) - f(x) = \int_s^t F(\mathrm{d}r)f(\varphi_{s,r}(x)) + \int_s^t L_r^0 f(\varphi_{s,r}(x))\,\mathrm{d}A_r, \tag{21}$$

where L_t^0 is a second order partial differential operator on U defined by

$$L_t^0 f(x) = \frac{1}{2}\sum_{i,j} a^{ij}(x, x, t)\frac{\partial^2 f}{\partial x^i \partial x^j}(x) + \sum_i c^i(x, t)\frac{\partial f}{\partial x^i}(x). \tag{22}$$

Here $(a^{ij}(x, y, t), b^i(x, t), A_t)$ is the local characteristic of $F(x, t)$ associated with the local coordinate $(x^1, \dots, x^d)$ and $c^i(x, t)$ is the correction term based on (a^{ij}). The operator L_t^0 does not depend on the choice of the local coordinates, since the first term of the right hand side of (21) does not depend on the choice of the local coordinate. (Note that both the second order term involving a^{ij} and the first order term involving c^i of the operator L_t^0 may depend on the choice of the local coordinate.) Consequently there exists a second order differential operator on M denoted by the same symbol L_t^0 such that it is represented as (22) at each coordinate neighborhood. Then, relation (21) is satisfied for all (s, t, x) such that $0 \le s \le t \le \sigma_1(s, x)$, where σ_1 is the stopping time defined by (13). Therefore by induction relation (21) is extended to all (s, t, x) such that $0 \le s \le t < \sigma(s, x)$, where σ is the explosion time of the flow $\varphi_{s,t}$.

Now equation (21) can be written as

$$f(\varphi_{s,t}(x)) - f(x) = \int_s^t H(\mathrm{d}r)f(\varphi_{s,r}(x)) + \int_s^t (L_r^0 + B_r)f(\varphi_{s,r}(x))\,\mathrm{d}A_r, \tag{23}$$

where $H(t)$ and $F(t)$ are defined by (4) and (5) respectively. The first term of the right hand side is a local martingale. Le Jan–Watanabe [90] calls the pair $\{L_t^0 + B_t, \langle\,,\,\rangle_t\}$ the *L.C.-system* of the stochastic flow.

Theorem 4.8.8 *Let $\varphi_{s,t}$ be a forward $C^{k,\delta}$-semimartingale flow such that for every s the local characteristic of $f(\varphi_{s,t}(x))$ belongs to the class $B^{k,\delta}$ for any $f \in C^\infty(M)$ where $k \ge 2$ and $\delta > 0$. Then there exists a unique forward random infinitesimal generator $F(t)$ of $\varphi_{s,t}$ (in the sense of the Stratonovich integral) with values in $\mathscr{X}^{k-1}(M)$ with local characteristic belonging to the class $(B^{k,\delta}, B^{k-1,\delta})$.*

Further assume that $\varphi_{s,t}$ is a forward–backward $C^{k,\delta}$-semimartingale flow such that both the forward and backward local characteristics of $f(\varphi_{s,t}(x))$ belong to the class $B^{k,\delta}$ for any s and $f \in C^\infty(M)$ where $k \ge 3$ and $0 < \delta \le 1$. Then the random infinitesimal generator is a forward–backward semimartingale with values in $\mathscr{X}^{k-1}(M)$ with local characteristic belonging to

$(B^{k,\delta}, B^{k-1,\delta})$ both to the forward and backward direction. The backward flow satisfies

$$f(\varphi_{t,s}) - f(x) = -\int_s^t F(\circ \hat{\mathrm{d}}r)f(\varphi_{t,r}(x)) \tag{24}$$

$$= -\int_s^t F(\hat{\mathrm{d}}r)f(\varphi_{t,r}(x)) + \int_s^t L_r^0 f(\varphi_{t,r}(x))\, \hat{\mathrm{d}}A_r. \tag{25}$$

Proof Set $X_s(t)f(x) = f(\varphi_{s,t}(x)) - f(x)$. For each s, it may be regarded as a continuous semimartingale with values in $\mathscr{L}$, where $\mathscr{L}$ is the space of all linear maps from $C^\infty(M)$ into $C^k(M)$. Define

$$G_s(t)f(x) = \int_s^t X_s(\mathrm{d}r)f(\varphi_{s,r}^{-1}(x)).$$

Then similarly to Theorem 4.2.8, we can show that it is a continuous $C^{k,\varepsilon}$-semimartingale for $\varepsilon < \delta$ for each s, f and satisfies $G_s(t)f = G_0(t)f - G_0(s)f$. Therefore, $G(t)f \equiv G_0(t)f$ is a continuous $C^{k,\varepsilon}$-semimartingale with local characteristic belonging to the class $B^{k,\varepsilon}$. Further it satisfies

$$f(\varphi_{s,t}(x)) - f(x) = \int_s^t G(\mathrm{d}r)f(\varphi_{s,r}(x)). \tag{26}$$

Now rewrite (26) at the coordinate neighborhood (U, η). Set $\varphi_{s,t}^i(x) = x^i(\varphi_{s,t}(x))$ and $G^i(x, t) = G(t)x^i(x)$. Let $(a_G^{ij}(x, y, t), b_G^i(x, t), A_t)$ be the local characteristic of $G = (G^1, \ldots, G^d)$. It belongs to the class $B^{k,\varepsilon}$ for some $\varepsilon < \delta$. Equation (26) implies

$$\varphi_{s,t}^i(x) - x^i = \int_s^t H_G^i(\varphi_{s,r}(x), \mathrm{d}r) + \int_s^t b_G^i(\varphi_{s,r}(x), r)\, \mathrm{d}A_r, \quad t < \tau_U(s, x)$$

where $H_G^i(x, t) = G^i(x, t) - \int_0^t b_G^i(x, r)\, \mathrm{d}A_r$. Then by Itô's formula we have

$$\begin{aligned} f(\varphi_{s,t}(x)) - f(x) &= \sum_i \int_s^t H_G^i(\varphi_{s,r}(x), \mathrm{d}r)\frac{\partial f}{\partial x^i}(\varphi_{s,r}(x)) \\ &+ \int_s^t \left\{\frac{1}{2}\sum_{i,j} a_G^{ij}(\varphi_{s,r}(x), \varphi_{s,r}(x), r)\frac{\partial^2 f}{\partial x^i \partial x^j}(\varphi_{s,r}(x)) \right. \\ &\left. + \sum_i b_G^i(\varphi_{s,r}(x), r)\frac{\partial f}{\partial x^i}(\varphi_{s,r}(x))\right\} \mathrm{d}A_r. \end{aligned} \tag{27}$$

The martingale part and the bounded variation part of (27) do not depend on the choice of the local coordinate. Therefore

$$H_G(t)f \equiv \sum_i H_G^i(\cdot, t)\frac{\partial f}{\partial x^i}$$

is a $\mathscr{X}^k(U)$-valued localmartingale. Then piecing together the system of these $\mathscr{X}(U)$-valued localmartingales, we get an $\mathscr{X}^k(M)$-valued localmartingale $H_G(t)$ such that at each coordinate neighborhood it is represented as the above using a local coordinate. Further, we can define a second order partial differential operator L_t on M such that

$$L_t^G f = \frac{1}{2}\sum_{i,j} a_G^{ij}(x, x, t)\frac{\partial^2 f}{\partial x^i \partial x^j} + \sum_i b_G^i(x, t)\frac{\partial f}{\partial x_i} \tag{28}$$

holds at each coordinate neighborhood, since the right hand side does not depend on the choice of the local coordinate. Then $G(t)f$ is decomposed as the sum of $H_G(t)f$ and $\int_0^t L_s^G f\, dA_s$. We can define a second order partial differential operator L_t^0 on M such that at each coordinate neighborhood it is given by (22) using $a^{ij}(x, y, t) \equiv a_G^{ij}(x, y, t)$. Set $F(t) \equiv G(t) - \int_0^t L_s^0\, dA_s$. It is a first order random differential operator and in fact an $\mathscr{X}^{k-1}(M)$-valued semimartingale with local characteristic belonging to $(B^{k,\delta}, B^{k-1,\delta})$. Further, equality (26) yields equality (21). Therefore $F(t)$ is the infinitesimal generator of $\varphi_{s,t}$. The second assertion of the theorem can be proved similarly. □

The forward (or backward) local semimartingale flow is defined with the obvious change. The following theorem corresponds to Theorem 4.7.9.

Theorem 4.8.9 *Let $\varphi_{s,t}(x)$, $x \in M$, $0 \le s \le t < \sigma(s, x)$ be a forward local $C^{k,\delta}$-semimartingale flow of local diffeomorphisms such that for every s the local characteristic of $f(\varphi_{s,t})$ belongs to the class $B^{k,\delta}$ for any $f \in C^\infty(M)$ where $k \ge 2$ and $\delta > 0$. Then there exists a unique continuous semimartingale $F(t)$ with values in $\mathscr{X}^{k-1}(M)$ with local characteristic belonging to $(B^{k,\delta}, B^{k-1,\delta})$ such that $\varphi_{s,t}$ is represented by (9).*

Further assume that $\varphi_{s,t}$ is a forward–backward local $C^{k,\delta}$-semimartingale flow such that both the forward and backward local characteristics of $f(\varphi_{s,t}(x))$ belong to the class $B^{k,\delta}$ for any s and $f \in C^\infty(M)$ where $k \ge 3$ and $0 < \delta \le 1$. Then the random infinitesimal generator F is a forward–backward semimartingale with values in $\mathscr{X}^{k-1}(M)$ with local characteristic belonging to $(B^{k,\delta}, B^{k-1,\delta})$ both to the forward and backward direction. The backward flow satisfies (24) and (25). □

Itô's formulas for stochastic flows on manifolds

In Section 4.4 we obtained Itô's first and second formulas for forward–backward semimartingale flow of diffeomorphisms on Euclidean space. We shall obtain similar formulas for stochastic flow on manifolds.

Theorem 4.8.10 *Let $F(t)$ be a forward–backward semimartingale with values in the space of vector fields on a manifold M such that both the forward and backward local characteristics belong to $(B^{k+1,\delta}, B^{k,\delta})$ for some $k \geq 3$ and $\delta > 0$. Let $\varphi_{s,t}(x)$, $x \in M$, $0 \leq s \leq t < \sigma(s, x)$ be the forward stochastic flow of local C^k-diffeomorphisms generated by $F(t)$ in the sense of the Stratonovich integral and $\varphi_{t,s}(x)$, $x \in M$, $\tau(t, x) < s \leq t \leq T$ be the associated backward flow of local C^k-diffeomorphisms. Then the forward flow $\varphi_{s,t}$, $s < t$ satisfies the following differential rule with respect to the variable s:*

$$f(\varphi_{s,t}(x)) - f(x) = \int_s^t F(\circ \hat{d}r)(f \circ \varphi_{r,t})(x) \tag{29}$$

$$= \int_s^t F(\hat{d}r)(f \circ \varphi_{r,t})(x) + \int_s^t L_r^0(f \circ \varphi_{r,t})(x)\, \hat{d}A_r, \tag{30}$$

where $f \in C^\infty(M)$, $x \in \mathbb{D}_{s,t} = \{x' : \sigma(s, x') > t\}$ and L_r^0 is a second order differential operator defined by (22).

Further the backward flow $\varphi_{t,s}$, $s < t$ satisfies the following differential rule with respect to the variable t:

$$f(\varphi_{t,s}(x)) - f(x) = -\int_s^t F(\circ dr)(f \circ \varphi_{r,s})(x) \tag{31}$$

$$= -\int_s^t F(dr)(f \circ \varphi_{r,s})(x) + \int_s^t L_r^0(f \circ \varphi_{r,s})(x)\, dA_r, \tag{32}$$

where $f \in C^\infty(M)$ and $x \in \mathbb{R}_{s,t}$.

Proof At each coordinate neighborhood, the forward flow $\varphi_{s,t}$ satisfies Itô's second formula

$$f(\varphi_{s,t}(x)) - f(x) = \sum_i \int_s^t \frac{\partial f}{\partial x^i}(\varphi_{s,r}(x))F^i(\varphi_{s,r}(x), \circ dr) \tag{33}$$

because of (9). Therefore, Itô's first formula (29) is valid by Theorem 4.4.5. Let $\{U_n\}$ be a countable set of coordinate neighborhoods covering M. Fix the time t and define a decreasing sequence of backward stopping times $\tau_n(t, x)$ as

$$\tau_1(t, x) = \sup\{r < t : \varphi_{r,t}(x) \notin \bar{U}_1\} \qquad (=0 \text{ if } \{\dots\} = \emptyset)$$

$$\tau_n(t, x) = \sup\{r < t : \varphi_{r,t}(x) \notin \bar{U}_1 \cup \dots \cup \bar{U}_n\} \qquad (=0 \text{ if } \{\dots\} = \emptyset)$$

and $\tau_\infty(t, x) = \lim_{t \to \infty} \tau_n(t, x)$. Formula (29) is valid for any s of $(\tau_1(t, x), t)$.

We can show by induction that (29) is valid for any s of $(\tau_n(t, x), t)$ and hence of $(\tau_\infty(t, x), t)$. Since $\{x : \tau_\infty(t, x) < s\} = \mathbb{D}_{s,t}$, formula (29) is valid if $x \in \mathbb{D}_{s,t}$.

Further for each coordinate neighborhood, the forward flow satisfies formula (5) of Section 4.4. Then the formula is written as (30) at the coordinate neighborhood. We can verify that formula (30) is valid for any $x \in \mathbb{R}_{s,t}$ by a similar argument to the above. Formulas (31) and (32) can be proved similarly. □

Itô's first formula can be applied to obtain Kolmogorov's backward equation. Let $\varphi_{s,t}(x)$ be a Brownian flow on $\mathbb{R}^d$ with infinitesimal mean $b(x, t)$ and covariance $a(x, y, t)$ satisfying Conditions (A.1) (A.2) and $(A.3)_{k,\delta}$ for $k \geq 3$ and $\delta > 0$. Let $\{T_{s,t}\}$ be the semigroup of the one point motion $\varphi_{s,t}(x)$. Set

$$L_t = \frac{1}{2}\sum_{i,j} a^{ij}(x, x, t)\frac{\partial^2}{\partial x^i \partial x^j} + \sum_i b^i(x, t)\frac{\partial}{\partial x^i}. \tag{34}$$

Theorem 4.8.11 *Let f be a C^3-function such that its derivatives are of polynomial growth. Then $T_{s,t}f(x)$ is differentiable with respect to s and twice differentiable with respect to x and satisfies*

$$\frac{\partial}{\partial s}T_{s,t}f(x) + L_s T_{s,t}f(x) = 0, \qquad \textit{for all } s < t. \tag{35}$$

Proof Let $F(x, t)$ be Itô's infinitesimal generator of the flow and let $M(x, t) = F(x, t) - \int_0^t b(x, r)\,dr$. It is a Brownian motion with mean 0. Itô's first formula yields

$$f(\varphi_{s,t}(x)) - f(x) = \sum_i \int_s^t \frac{\partial (f \circ \varphi_{r,t})}{\partial x^i}(x) M^i(x, \hat{d}r) + \int_s^t L_r(f \circ \varphi_{r,t})(x)\,dr.$$

Note that $\sup_{|x| \leq N} |D^\alpha \varphi_{s,t}(x)|$ has finite moments of any order for any $N > 0$ and $|\alpha| \leq 2$ (see Corollary 4.6.7). Take the expectation of each term of the above. The first term of the right hand side is a martingale with mean 0. Then we have

$$T_{s,t}f(x) - f(x) = \int_s^t E[L_r(f \circ \varphi_{r,t})(x)]\,dr.$$

We can interchange the order of the expectation and the differentiation. Therefore

$$T_{s,t}f(x) - f(x) = \int_s^t L_r T_{r,t} f(x)\, dr.$$

This yields the theorem. □

Exercise 4.8.12 (*cf.* Palais [107]) Let $X_1 = x^2(\partial/\partial x^1)$ and $X_2 = ((x^1)^2/2) \times (\partial/\partial x^2)$ be two vector fields on $\mathbb{R}^2$ and let (B_t^1, B_t^2) be a standard Brownian motion. Show the following.

(i) X_1 and X_2 are complete vector fields but the Lie bracket $[X_1, X_2]$ is not complete.
(ii) The Lie algebra generated by X_1 and X_2 is infinite dimensional.
(iii) $F(t) \equiv B_t^1 X_1 + B_t^2 X_2$ is not strictly complete to the forward.

(*Hint*: for the proof of (iii), note that the equation based on the above $F(t)$ is $d\varphi_t^1 = \varphi_t^2\, dB_t^1$ and $d\varphi_t^2 = ((\varphi_t^1)^2/2)\, dB_t^2$.)

Exercise 4.8.13 Let $\varphi_{s,t}$ be the stochastic flow governed by the classical Stratonovich's stochastic differential equation (17). Show that the operator L_t^0 of (22) is given by $\frac{1}{2}\sum_{i=1}^m X_i^2$.

4.9 Itô's formulas for stochastic flows and their applications

Itô's formula for $\varphi_{s,t}$ acting on vector fields

The stochastic flow of diffeomorphisms on a manifold acts naturally on tensor fields. We shall discuss the differential rule or Itô's formula governing the tensor field valued process. Although we can extend our discussion to stochastic flows of local diffeomorphisms with appropriate modification, we restrict our attention to the stochastic flow of diffeomorphisms for simplicity.

We shall first deal with vector fields: the case of general tensor fields will be discussed later.

We begin by introducing the differential of the map. Let ψ be a diffeomorphism of the manifold M. The *differential* $(\psi_*)_x$ of the map ψ is by definition the linear map from the tangent space $T_x(M)$ to the tangent space $T_{\psi(x)}(M)$ such that $(\psi_*)_x X_x f = X_x(f \circ \psi)$ for all $X_x \in T_x(M)$. Given a vector field X on M, we denote by X_x the restriction of X at the point x. We define a new vector field $\psi_* X$ by $(\psi_* X)_x = (\psi_*)_{\psi^{-1}(x)} X_{\psi^{-1}(x)}$. Then for any f of $C^\infty(M)$, $\psi_* X f(x) = X(f \circ \psi)(\psi^{-1}(x))$ for every $x \in M$ for any f of $C^\infty(M)$.

We shall express $\psi_* X$ using a local coordinate $(x^1, \ldots, x^d)$. Let $X = \sum X^i(x)(\partial/\partial x^i)$ be the coordinate expression. Then the vector field $\psi_* X$ is expressed as $\psi_* X = \sum (\psi_* X)^i(x)(\partial/\partial x^i)$, where

$$(\psi_* X)^i(x) = \sum_j X^j(\psi^{-1}(x)) \frac{\partial \psi^i}{\partial x^j}(\psi^{-1}(x))$$

and $\psi^i(x) = x^i(\psi(x))$. Thus denoting the d-vector $(X^1, \dots, X^d)$ by X and Jacobian matrix $(\partial \psi^i/\partial x^j)$ by $\partial \psi(x)$, the vector $\psi_* X(x) = ((\psi_* X)^1(x), \dots, (\psi_* X)^d(x))$ satisfies $\psi_* X(x) = \partial \psi(\psi^{-1}(x)) X(\psi^{-1}(x))$.

Let Y be a vector field and ψ_t the flow of Y, i.e. the local one parameter group of local transformations of M generated by Y:

$$\frac{d}{dt}(f \circ \psi_t)(x)|_{t=0} = Yf(x), \qquad \text{for all } f \in C^\infty(M).$$

The *Lie derivative* of the vector field X with respect to Y is the vector field $\mathscr{L}(Y)X$ defined by

$$\begin{aligned}(\mathscr{L}(Y)X)_x &= -\lim_{t\to 0} \frac{1}{t}\{(\psi_{t*}X)_x - X_x\} \\ &= -\lim_{t\to 0} \frac{1}{t}\{(\psi_{t*})_{\psi_t^{-1}(x)} X_{\psi_t^{-1}(x)} - X_x\}. \qquad (1)\end{aligned}$$

The relation $\mathscr{L}(Y)X = [Y, X]$ is well known.

Let $F(t)$ be a continuous forward–backward semimartingale with values in $\mathfrak{X}^k(M)$ such that both the forward and backward local characteristics belong to the class $(B^{k+1,\delta}, B^{k,\delta})$ where $k \geq 4$ and $\delta > 0$. In the remainder of this section we assume that it generates a *global* stochastic flow of C^k-diffeomorphisms $\varphi_{s,t}$, $s, t \in [0, T]$, though most results are extended easily to stochastic flow of local diffeomorphisms with the obvious changes.

The Lie derivative $\mathscr{L}(F(t))X$ of X with respect to $F(t)$ is a continuous semimartingale with values in $\mathfrak{X}^{k-1}(M)$. Let $X(t)$ be a continuous $(\mathscr{F}_t)$-adapted process with values in $\mathfrak{X}^k(M)$. The stochastic integrals based on $\mathscr{L}(F(t))$ such as $\int_s^t \mathscr{L}(F(\circ dr))X(r)$ etc. are defined similarly to $\int_s^t F(\circ dr)f(r)$ in the previous section.

Now the differential $(\varphi_{s,t})_*$ is well defined for any s, t a.s. The forward stochastic process $\int_s^t (\varphi_{s,r})_* \mathscr{L}(F(\circ dr))X(r)$ is defined by a continuous process $G(t)$ with values in $\mathfrak{X}^{k-1}(M)$ satisfying

$$G(t)f = \int_s^t \mathscr{L}(F(\circ dr))X(r)(f \circ \varphi_{s,r})(\varphi_{s,r}^{-1}) \qquad (2)$$

for any smooth function f on M. The backward integrals are defined similarly.

Theorem 4.9.1 *The differentials $(\varphi_{s,t})_* X$ and $(\varphi_{t,s})_* X$, $0 \le s \le t \le T$ satisfy the following version of Itô's second formula for any C^3-vector field X.*

$$(\varphi_{s,t})_* X - X = -\int_s^t \mathscr{L}(F(\circ \mathrm{d}r))((\varphi_{s,r})_* X), \tag{3}$$

$$(\varphi_{t,s})_* X - X = \int_s^t \mathscr{L}(F(\circ \hat{\mathrm{d}}r))((\varphi_{t,r})_* X). \tag{4}$$

Further they satisfy the following version of Itô's first formula.

$$(\varphi_{s,t})_* X - X = -\int_s^t (\varphi_{r,t})_*(\mathscr{L}(F(\circ \hat{\mathrm{d}}r))X), \tag{5}$$

$$(\varphi_{t,s})_* X - X = \int_s^t (\varphi_{r,s})_*(\mathscr{L}(F(\circ \mathrm{d}r))X). \tag{6}$$

Proof We prove the theorem in the case where M is a Euclidean space only, since the theorem can be reduced to this case by utilizing local coordinates. We shall prove (3). Let $f \in C^\infty(M)$. Then

$$(\varphi_{s,t})_* X f(x) = X(f \circ \varphi_{s,t})(\varphi_{t,s}(x)).$$

We fix s, x and consider $\Phi(y, t) = X(f \circ \varphi_{s,t})(y)$ and $g_t = (\varphi_{t,s}^1(x), \dots, \varphi_{t,s}^d(x))$ where $t \ge s$. Then $\Phi(g_t, t) = (\varphi_{s,t})_* X f(x)$ and $\Phi(g_s, s) = Xf(x)$ are satisfied. The generalized Itô's formula (Theorem 3.3.2) tells us

$$\Phi(g_t, t) - \Phi(x, s) = \int_s^t \Phi(g_r, \circ \mathrm{d}r) + \sum_{i=1}^d \int_s^t \frac{\partial \Phi}{\partial x^i}(g_r, r) \circ \mathrm{d}g_r^i,$$

where $g_r^i = \varphi_{r,s}^i(x)$. Since $\Phi(y, t) = Xf(y) + \int_s^t X\{F(\circ \mathrm{d}r) f(\varphi_{s,r})\}(y)$ holds, we have

$$\int_s^t \Phi(g_r, \circ \mathrm{d}r) = \int_s^t X\{F(\circ \mathrm{d}r) f(\varphi_{s,r})\}(\varphi_{r,s}(x)) = \int_s^t (\varphi_{s,r})_* X F(\circ \mathrm{d}r) f(x).$$

Further, by Itô's first formula (31) of the previous section, we have

$$g_t^i - x^i = -\int_s^t F(\circ \mathrm{d}r)(\varphi_{r,s}^i)(x) = -\int_s^t (\varphi_{r,s})_* F(\circ \mathrm{d}r)^i(\varphi_{r,s}(x)).$$

Therefore,

$$\begin{aligned}\sum_{i=1}^d \int_s^t \frac{\partial \Phi}{\partial x^i}(g_r, r) \circ \mathrm{d}g_r^i &= -\int_s^t (\varphi_{r,s})_* F(\circ \mathrm{d}r)\{X(f \circ \varphi_{s,r})\}(\varphi_{r,s}(x)) \\ &= -\int_s^t F(\circ \mathrm{d}r)(\varphi_{s,r})_* X f(x).\end{aligned}$$

The above two relations imply

$$(\varphi_{s,t})_* Xf - Xf = -\int_s^t [F(\circ dr), (\varphi_{s,r})_* X]f = -\int_s^t \mathscr{L}(F(\circ dr))(\varphi_{s,r})_* Xf.$$

This proves (3). Formula (4) can be proved similarly.

For the proof of (5), we will again apply Itô's first formula. Let t and x be fixed and set $\Psi(y, s) = X(f \circ \varphi_{s,t})(y)$ and $h_s = (\varphi_{t,s}^1(x), \ldots, \varphi_{t,s}^d(x))$ where $s < t$. Then $\Psi(h_s, s) = (\varphi_{s,t})_* Xf(x)$ and $\Phi(h_t, t) = Xf(x)$ are satisfied. By Theorem 4.8.10, we have $\Psi(y, s) = Xf(y) + \int_s^t X\{F(\circ \hat{d}r)(f \circ \varphi_{r,t})\}(y)$. Therefore,

$$\int_s^t \Psi(h_r, \circ \hat{d}r) = -\int_s^t X\{F(\circ \hat{d}r)(f \circ \varphi_{r,t})\}(\varphi_{t,r}(x)).$$

Further since h_s satisfies the backward Stratonovich's equation $h_s - x = -\int_s^t F^i(\varphi_{t,r}(x), \circ \hat{d}r)$, we have

$$\sum_i \int_s^t \frac{\partial \Psi}{\partial x^i}(h_r, r) \circ \hat{d}h_r = \int_s^t F(\circ \hat{d}r)\{X(f \circ \varphi_{r,t})\}(\varphi_{t,r}(x)).$$

Consequently applying the generalized Itô's formula to the backward direction, we obtain

$$\begin{aligned} Xf(x) - \Psi(h_s, s) &= \int_s^t (F(\circ \hat{d}r)X - XF(\circ \hat{d}r))\{(f \circ \varphi_{r,t})\}(\varphi_{t,r}(x)) \\ &= \int_s^t (\varphi_{r,t})_*[F(\circ \hat{d}r), X]f(x) \\ &= \int_s^t (\varphi_{r,t})_* \mathscr{L}(F(\circ \hat{d}r))Xf(x). \end{aligned}$$

This proves (5). Formula (6) can be proved similarly. □

Itô's formula for $\varphi_{s,t}$ acting on tensor fields

In the previous theorem, we obtained Itô's formula for $\varphi_{s,t}$ acting on vector fields. In the sequel, we will obtain a similar formula for $\varphi_{s,t}$ acting on general tensor fields.

We begin with the differential form. Let ψ be a diffeomorphism of M and let $(\psi_*)_x$ be the differential of ψ, which is a linear map from $T_x(M)$ into $T_{\psi(x)}(M)$. We denote the dual map by $(\psi^*)_x$: it is a linear map from $T_{\psi(x)}(M)^*$ (cotangent space) to $T_x(M)^*$ such that $\langle \theta_{\psi(x)}, (\psi_*)_x X_x \rangle = \langle (\psi^*)_x \theta_{\psi(x)}, X_x \rangle$ holds for any $\theta_{\psi(x)} \in T_{\psi(x)}(M)^*$ and $X_x \in T_x(M)$. Given a (differential) 1-form θ, we denote by $\psi^*\theta$ a 1-form such that $(\psi^*\theta)_x = (\psi^*)_x \theta_{\psi(x)}$. Then $\langle \psi^*\theta, X \rangle_x = \langle \theta, \psi_* X \rangle_{\psi(x)}$. Let X be a complete vector field and let ψ_t be the one parameter group of transformations generated by X. The *Lie derivative of* 1-*form* θ is defined by

$$\mathscr{L}(X)\theta = \lim_{t\to 0}\frac{1}{t}\{\psi_t{}^*\theta - \theta\}. \tag{7}$$

The following relation is well known,

$$\langle \mathscr{L}(X)\theta, Y\rangle + \langle \theta, \mathscr{L}(X)Y\rangle = X(\langle \theta, Y\rangle). \tag{8}$$

The definition of stochastic integrals based on $\mathscr{L}(F(t))$ is similar to the case of the vector field.

Theorem 4.9.2 *The differentials $(\varphi_{s,t})^*$ and $(\varphi_{t,s})^*$, $0 \le s \le t \le T$ satisfy the following version of the Itô's second formula for any C^3-1-form θ.*

$$(\varphi_{s,t})^*\theta - \theta = \int_s^t (\varphi_{s,r})^*(\mathscr{L}(F(\circ\, dr))\theta), \tag{9}$$

$$(\varphi_{t,s})^*\theta - \theta = -\int_s^t (\varphi_{t,r})^*(\mathscr{L}(F(\circ\, \hat{d}r))\theta). \tag{10}$$

Further they satisfy Itô's first formula.

$$(\varphi_{s,t})^*\theta - \theta = \int_s^t \mathscr{L}(F(\circ\, \hat{d}r))((\varphi_{r,t})^*\theta), \tag{11}$$

$$(\varphi_{t,s})^*\theta - \theta = -\int_s^t \mathscr{L}(F(\circ\, dr))((\varphi_{r,s})^*\theta). \tag{12}$$

Proof We shall prove (9). Other formulas will be proved by similar methods. From (3) we have

$$\langle \theta, (\varphi_{s,t})_* Y\rangle_y - \langle \theta, Y\rangle_y = -\int_s^t \langle \theta, \mathscr{L}(F(\circ\, dr))(\varphi_{s,r})_* Y\rangle_y.$$

Apply the generalized Itô's formula to $\langle \theta, (\varphi_{s,t})_* Y\rangle_y$ and $y = \varphi_{s,t}(x)$ where s and x are fixed. Then

$$\begin{aligned}\langle (\varphi_{s,t})^*\theta, Y\rangle_x - \langle \theta, Y\rangle_x &= -\int_s^t \langle \theta, \mathscr{L}(F(\circ\, dr))(\varphi_{s,r})_* Y\rangle_{\varphi_{s,r}(x)} \\ &\quad + \int_s^t F(\circ\, dr)\langle \theta, (\varphi_{s,r})_* Y\rangle_{\varphi_{s,r}(x)} \\ &= \int_s^t \langle (\varphi_{s,r})^*\mathscr{L}(F(\circ\, dr))\theta, Y\rangle_x.\end{aligned}$$

The last equality follows from (8) and the relation $\langle \psi^*\theta, X\rangle_x = \langle \theta, \psi_* X\rangle_{\psi(x)}$. □

A tensor field K of type (p, q) is, by definition, an assignment of a tensor K_x of $T_q^p(x) \equiv T_x(M) \times \cdots \times T_x(M) \times T_x(M)^* \times \cdots \times T_x(M)^*$ ($T_x(M)$ p times and $T_x(M)^*$ q times) to each point x of M. Hence for each x, K_x is a multilinear form on the product space $T_x(M)^* \times \cdots \times T_x(M)^* \times T_x(M) \times \cdots \times T_x(M)$. Thus, for given 1-forms $\theta^1, \ldots, \theta^p$ and vector fields $Y_1, \ldots, Y_q$,

$$K_x(\theta^1, \ldots, \theta^p, Y_1, \ldots, Y_q) \qquad (\equiv K_x(\theta_x^1, \ldots, \theta_x^p, Y_{1x}, \ldots, Y_{qx}))$$

is a scalar field. In the sequel, we assume that it is a C^3-function.

Let ψ be a diffeomorphism of M. Given a tensor field K of type (p, q), we define a tensor field ψ^*K by the relation

$$\begin{aligned}(\psi^*K)_x(\theta^1, \ldots, \theta^p, Y_1, \ldots, Y_q)& \\ = K_{\psi(x)}((\psi^*)^{-1}\theta^1, \ldots, (\psi^*)^{-1}\theta^p, \psi_* Y_1, \ldots, \psi_* Y_q).& \qquad (13)\end{aligned}$$

Then if K is a vector field, $\psi^*K = \psi_*^{-1}K$ and if K is a 1-form, it is equal to ψ^*K defined before.

Remark The definition of the above ψ^* is not the same as that of $\tilde{\psi}$ in Kobayashi–Nomizu [70], p. 28. The relation of these is $\tilde{\psi}^{-1} = \psi^*$ or $\tilde{\psi} = (\psi^{-1})^*$.

Let X be a complete vector field and ψ_t, $t \in (-\infty, \infty)$ be the one parameter group of transformations generated by X. The *Lie derivative of the tensor field K* with respect to X is defined by

$$\mathscr{L}(X)K = \lim_{t \downarrow 0} \frac{1}{t}\{\psi_t^*K - K\}. \qquad (14)$$

If K is a tensor field of type (p, q), then

$$\begin{aligned}(\mathscr{L}(X)K)_x&(\theta^1, \ldots, \theta^p, Y_1, \ldots, Y_q) \\ &= X(K_x(\theta^1, \ldots, \theta^p, Y_1, \ldots, Y_q)) \\ &\quad - \sum_{i=1}^{p} K_x(\theta^1, \ldots, \mathscr{L}(X)\theta^i, \ldots, \theta^p, Y_1, \ldots, Y_q) \\ &\quad - \sum_{j=1}^{q} K_x(\theta^1, \ldots, \theta^p, Y_1, \ldots, \mathscr{L}(X)Y_j, \ldots, Y_q). \qquad (15)\end{aligned}$$

We can now state Itô's formula for $\varphi_{s,t}^*$ acting on a tensor field.

Theorem 4.9.3 *The differentials* $(\varphi_{s,t})^*$ *and* $(\varphi_{t,s})^*$, $0 \le s \le t \le T$ *satisfy Itô's second formula for any* C^3*-tensor field* K *of type* (p, q).

$$(\varphi_{s,t})^* K - K = \int_s^t (\varphi_{s,r})^* (\mathscr{L}(F(\circ\, \mathrm{d}r))K), \tag{16}$$

$$(\varphi_{t,s})^* K - K = -\int_s^t (\varphi_{t,r})^* (\mathscr{L}(F(\circ\, \hat{\mathrm{d}}r))K. \tag{17}$$

Further they satisfy Itô's first formula.

$$(\varphi_{s,t})^* K - K = \int_s^t \mathscr{L}(F(\circ\, \hat{\mathrm{d}}r))((\varphi_{r,t})^* K), \tag{18}$$

$$(\varphi_{t,s})^* K - K = -\int_s^t \mathscr{L}(F(\circ\, \mathrm{d}r))((\varphi_{r,s})^* K). \tag{19}$$

Proof We shall prove (16) in the case where $s < t$. We will write $\varphi_{s,t}$ as φ_t for convenience. Apply the generalized Itô's formula to the multilinear form K. Then, using Theorems 4.9.1 and 4.9.2, we have

$$\begin{aligned}
&K_x((\varphi_t^*)^{-1}\theta^1, \dots, (\varphi_t^*)^{-1}\theta^p, \varphi_{t*} Y_1, \dots, \varphi_{t*} Y_q) - K_x(\theta^1, \dots, \theta^p, Y_1, \dots, Y_p) \\
&\quad = -\sum_{i=1}^p \int_s^t K_x((\varphi_r^*)^{-1}\theta^1, \dots, \mathscr{L}(F(\circ\, \mathrm{d}r))(\varphi_r^*)^{-1}\theta^i, \dots, \varphi_{r*} Y_1, \dots, \varphi_{r*} Y_q) \\
&\qquad - \sum_{j=1}^q \int_s^t K_x((\varphi_r^*)^{-1}\theta^1, \dots, \varphi_{r*} Y_1, \dots, \mathscr{L}(F(\circ\, \mathrm{d}r))\varphi_{r*} Y_j, \dots, \varphi_{r*} Y_q).
\end{aligned}$$

Set

$$\Phi(y, t) = K_y((\varphi_t^*)^{-1}\theta^1, \dots, (\varphi_t^*)^{-1}\theta^p, \varphi_{t*} Y_1, \dots, \varphi_{t*} Y_q)$$

and apply the generalized Itô formula to $\Phi(\varphi_t(x), t)$ again. Then

$$\begin{aligned}
&\Phi(\varphi_t(x), t) - \Phi(x, s) \\
&\quad = -\sum_{i=1}^p \int_s^t K_{\varphi_r(x)}((\varphi_r^*)^{-1}\theta^1, \dots, \mathscr{L}(F(\circ\, \mathrm{d}r))(\varphi_r^*)^{-1}\theta^i, \dots, \varphi_{r*} Y_1, \dots, \varphi_{r*} Y_q) \\
&\qquad - \sum_{j=1}^q \int_s^t K_{\varphi_r(x)}((\varphi_r^*)^{-1}\theta^1, \dots, \varphi_{r*} Y_1, \dots, \mathscr{L}(F(\circ\, \mathrm{d}r))\varphi_{r*} Y_j, \dots, \varphi_{r*} Y_q) \\
&\qquad + \int_s^t F(\circ\, \mathrm{d}r)\Phi(\varphi_r(x), r).
\end{aligned} \tag{20}$$

Noting relation (15), we see that the right hand side of (20) is

$$\int_s^t \varphi_r^* \mathscr{L}(F(\circ\, \mathrm{d}r))K_x(\theta^1, \ldots, \theta^p, Y_1, \ldots, Y_q).$$

We have thus proved (16). □

In the case where $\varphi_{s,t}$ is a stochastic flow generated by a classical stochastic differential equation, we can rewrite the above Itô's formula using the Itô integral explicitly.

Corollary 4.9.4 *Let $\varphi_{s,t}$, $0 \le s \le t \le T$ be a forward Brownian flow determined by the classical stochastic differential equation* (17) *of the previous section. Then we have the following versions of Itô's second and first formulas.*

$$(\varphi_{s,t})^* K - K = \sum_l \int_s^t (\varphi_{s,r})^*(\mathscr{L}(X_l)K)\, \mathrm{d}B_r^l + \int_s^t (\varphi_{s,r})^* \left[\left\{ \frac{1}{2} \sum_l \mathscr{L}(X_l)^2 + \mathscr{L}(X_0) \right\} K \right] \mathrm{d}r, \tag{21}$$

$$(\varphi_{s,t})^* K - K = \sum_l \int_s^t \mathscr{L}(X_l)((\varphi_{r,t})^* K)\, \hat{\mathrm{d}}B_r^l + \int_s^t \left\{ \frac{1}{2} \sum_l \mathscr{L}(X_l)^2 + \mathscr{L}(X_0) \right\} ((\varphi_{r,t})^* K)\, \mathrm{d}r. \quad \square \tag{22}$$

Composition and decomposition of the solution

Let $\varphi_{s,t}(x)$ and $\eta_{s,t}(x)$ be the forward stochastic flows of diffeomorphisms of the same manifold M defined on the same probability space. We assume that these have the forward Stratonovich infinitesimal generators $F(x, t)$ and $G(x, t)$, respectively, such that the local characteristic of $F(x, t)$ (of $G(x, t)$ respectively) belongs to the class $(B^{4,\delta}, B^{3,\delta})$ $((B^{2,\delta}, B^{1,\delta})$ respectively) for some $\delta > 0$. We shall obtain the stochastic differential equation governing the composite process $\zeta_{s,t}(x) = \varphi_{s,t} \circ \eta_{s,t}(x)$.

Theorem 4.9.5 *For each fixed s and x, the composite process $\zeta_{s,t}(x) = \varphi_{s,t} \circ \eta_{s,t}(x)$, $t \in [s, T]$ satisfies a Stratonovich's stochastic differential equation based on $F(t) + (\varphi_{s,t})_* G(t)$.*

Proof We prove the theorem in the case where M is a Euclidean space only. For a C^3-function f on M, we set $\Phi(y, t) = f \circ \varphi_{s,t}(y)$ and $g_t = (\eta_{s,t}^1(x), \ldots, \eta_{s,t}^d(x))$, where s and x are fixed. Since $\Phi(g_t, t) = f(\zeta_{s,t}(x))$ and $\Phi(g_s, s) = f(x)$, we have by the generalized Itô's formula

$$f(\zeta_{s,t}(x)) - f(x) = \sum_i \int_s^t F^i(\varphi_{s,r} \circ \eta_{s,r}(x), \circ \mathrm{d}r) \frac{\partial f}{\partial x^i}(\varphi_{s,r} \circ \eta_{s,r}(x))$$

$$+ \sum_i \int_s^t G^i(\eta_{s,r}(x), \circ \mathrm{d}r) \frac{\partial (f \circ \varphi_{s,r})}{\partial x^i}(\eta_{s,r}(x))$$

$$= \int_s^t F(\circ \mathrm{d}r) f(\zeta_{s,r}(x)) + \int_s^t (\varphi_{s,r})_* G(\circ \mathrm{d}r) f(\zeta_{s,r}(x)).$$

The proof is complete. □

Now, Theorem 4.9.1 tells us that $(\varphi_{s,t})_* G(t) = G(t)$ holds for $s < t$ if and only if $[F(r), G(t)] = 0$ holds for any $r \in [s, t]$. Therefore we have the following.

Corollary 4.9.6 *Suppose that $[F(s), G(t)] = 0$ holds for any $s \le t$. Then the composite process $\zeta_{s,t} = \varphi_{s,t} \circ \eta_{s,t}$ is governed by the Stratonovich stochastic differential equation based on $F + G$.* □

We shall next consider the problem of decomposing the flow $\varphi_{s,t}$.

Theorem 4.9.7 *Let $G(t)$ and $H(t)$ be continuous semimartingales with values in $\mathscr{X}^k(M)$ with local characteristics belonging to the class $(B^{k+1,\delta}, B^{k,\delta})$ where $k \ge 3$ and $0 < \delta \le 1$. Suppose that $\eta_{s,t}$ is a forward stochastic flow of diffeomorphisms governed by the Stratonovich equation based on $G(t)$ and $\kappa_{s,t}$, $t \in [s, T]$ is a stochastic process governed by Stratonovich's equation based on $(\eta_{s,t}{}^{-1})_* H(t)$. Then for each s $\varphi_{s,t} \equiv \eta_{s,t} \circ \kappa_{s,t}$ is governed by the Stratonovich equation based on $G + H$.*

Proof We again consider the case where M is a Euclidean space only. We shall apply the generalized Itô's formula to $\Phi(y, t) = f \circ \eta_{s,t}(y)$ and $g_t = (\kappa^1_{s,t}(x), \ldots, \kappa^d_{s,t}(x))$ where s and x are fixed. Then we have

$$\Phi(g_t, t) - \Phi(g_s, s) = \int_s^t G(\circ \mathrm{d}r) f(\eta_{s,r} \circ \kappa_{s,r}) + \int_s^t (\eta_{s,r}{}^{-1})_* H(\circ \mathrm{d}r)(f \circ \eta_{s,r})(\kappa_{s,r}).$$

Since $(\eta_{s,r}{}^{-1})_* H(r)(f \circ \eta_{s,r})(\kappa_{s,r}) = H(r) f(\eta_{s,r} \circ \kappa_{s,r})$, we get

$$f \circ \eta_{s,t} \circ \kappa_{s,t}(x) - f(x) = \int_s^t \{G(\circ \mathrm{d}r) + H(\circ \mathrm{d}r)\} f(\eta_{s,r} \circ \kappa_{s,r}(x)).$$

This proves that $\varphi_{s,t} \equiv \eta_{s,t} \circ \kappa_{s,t}$ is generated by $G + H$. □

As a special case we shall consider Itô's classical stochastic differential equations based on $F(t) = tX_0 + \sum_{l=1}^m B_t^l X_l$, namely

$$\varphi_t = x + \int_s^t X_0(\varphi_r)\,\mathrm{d}r + \sum_{l=1}^m \int_s^t X_l(\varphi_r) \circ \mathrm{d}B_r^l, \tag{23}$$

where $X_0, \ldots, X_m$ are complete C^∞-vector fields and $B_t = (B_t^1, \ldots, B_t^m)$ is a standard Brownian motion. Setting $B_t^0 \equiv t$, we often write equation (23) as

$$\varphi_t = x + \sum_{l=0}^m \int_s^t X_l(\varphi_r) \circ \mathrm{d}B_r^l. \tag{24}$$

If the Lie algebra generated by $X_0, \ldots, X_m$ is of finite dimension, the solution of the above equation defines a Brownian flow of global diffeomorphisms by Theorem 4.8.7.

In the system theory or the control theory, $B_t^1, \ldots, B_t^m$ in equation (23) are called *inputs* and the solution φ_t is called the *output*. It is an important problem in applications that we can compute the output from the input explicitly. We give two examples showing the method of calculating outputs.

Example 4.9.8 Let $X_0, \ldots, X_m$ be commutative complete vector fields. Then the Brownian flow $\varphi_{s,t}(x)$ governed by equation (23) is represented by

$$\varphi_{s,t}(x) = \exp\{(t-s)X_0\} \circ \exp\{(B_t^1 - B_s^1)X_1\} \circ \cdots \circ \exp\{(B_t^m - B_s^m)X_m\}(x), \tag{25}$$

where $\exp\{tX_l\}$ is the one parameter group of transformations generated by the vector field X_l.

We shall prove first that $\varphi_{s,t}^{(l)} \equiv \exp\{(B_t^l - B_s^l)X_l\}$ is the solution of $\mathrm{d}\varphi_t^{(l)} = X_l(\varphi_t^{(l)}) \circ \mathrm{d}B_t^l$ starting from (s, x). Set $F^{(l)}(t) = f \circ \exp\{tX_l\}$. Then by Itô's formula

$$\begin{aligned} F^{(l)}(B_t^l - B_s^l) &= F^{(l)}(0) + \int_s^t \frac{\partial F^{(l)}}{\partial r}(B_r^l - B_s^l) \circ \mathrm{d}B_r^l \\ &= F^{(l)}(0) + \int_s^t X_l f \circ \exp\{(B_r^l - B_s^l)X_l\} \circ \mathrm{d}B_r^l. \end{aligned}$$

Therefore, $\varphi_{s,t}^{(l)}$ is the solution. Then expression (25) follows from Corollary 4.9.6. □

Example 4.9.9 Consider a linear equation on $\mathbb{R}^d$:

$$\varphi_t = x + \int_s^t F\varphi_r\,\mathrm{d}r + G(B_t - B_s). \tag{26}$$

where F is a $d \times d$-matrix, G is a $d \times m$-matrix and B_t is a standard m-dimensional Brownian motion. The solution is decomposed as follows. Let $\zeta_t(x)$ be the deterministic flow governed by a linear equation $\mathrm{d}x/\mathrm{d}t = Fx$. Let $\eta_t(x)$ be the solution of equation

$$\eta_t = x + \int_s^t (\xi_{s,r}{}^{-1})_* G \, dB_r,$$

starting at x at time s. Then $\varphi_{s,t} \equiv \zeta_{s,t} \circ \eta_{s,t}$ is a solution of (26). Since $\zeta_{s,t}(x) = e^{F(t-s)}x$ we have $(\zeta_{s,t}{}^{-1})_* G = e^{-F(t-s)}G$. Therefore $\eta_{s,t}(x) = x + \int_s^t e^{-F(r-s)}G \, dB_r$. Hence the solution of equation (26) is decomposed as

$$\varphi_{s,t}(x) = e^{F(t-s)}\left(x + \int_s^t e^{-F(r-s)}G \, dB_r\right). \tag{27}$$

The complexity of expressing the solution by the inputs depends on the structure of the Lie algebra generated by the vector fields defining the equation. Let $\mathscr{L}$ be the Lie algebra generated by the vector fields $X_0, \dots, X_m$ defining equation (23). For two Lie subalgebras $\mathscr{I}$ and $\mathscr{K}$ of $\mathscr{L}$, we set $[\mathscr{I}, \mathscr{K}] = \{[X, Y] : X \in \mathscr{I}, Y \in \mathscr{K}\}$. We shall define a chain of subalgebras of $\mathscr{L}$:

$$\mathscr{G}_1 = [\mathscr{L}, \mathscr{L}], \mathscr{G}_2 = [\mathscr{G}_1, \mathscr{G}_1], \dots, \mathscr{G}_i = [\mathscr{G}_{i-1}, \mathscr{G}_{i-1}]. \tag{28}$$

The Lie algebra $\mathscr{L}$ is called *solvable* if $\mathscr{G}_i = \{0\}$ for some i. If $\mathscr{L}$ is a finite dimensional solvable Lie algebra, any element of $\mathscr{L}$ is a complete vector field. By Lie's theorem there exists a basis of $\mathscr{L}$ denoted by $\{Y_1, \dots, Y_n\}$ such that for each $i = 1, \dots, n$ the linear space $\mathscr{L}_i$ spanned by $\{Y_i, \dots, Y_n\}$ is an ideal of $\mathscr{L}$, i.e. $[\mathscr{L}, \mathscr{L}_i] \subset \mathscr{L}_i$ holds. On the other hand, consider another chain of subalgebras of $\mathscr{L}$:

$$\mathscr{G}^1 = [\mathscr{L}, \mathscr{L}], \mathscr{G}^2 = [\mathscr{L}, \mathscr{G}^1], \dots, \mathscr{G}^i = [\mathscr{L}, \mathscr{G}^{i-1}]. \tag{29}$$

If $\mathscr{G}^i = \{0\}$ for some i, the Lie algebra $\mathscr{L}$ is called *nilpotent*. A nilpotent Lie algebra is solvable, obviously. For a finite dimensional nilpotent Lie algebra, we can choose a basis $\{Y_1, \dots, Y_n\}$ such that for each $i = 1, \dots, n$ the linear space $\mathscr{L}_i$ spanned by $\{Y_i, \dots, Y_n\}$ satisfies $[\mathscr{L}, \mathscr{L}_i] \subset \mathscr{L}_{i+1}$.

Now for an arbitrary element Z of $\mathscr{L}$, the adjoint representation of Z is defined by a linear transformation $\mathrm{ad}(Z)\colon \mathscr{L} \to \mathscr{L}$ such that $\mathrm{ad}(Z)X = \mathscr{L}(Z)X = [Z, X]$ holds for any $X \in \mathscr{L}$. Suppose that $\mathscr{L}$ is a solvable Lie algebra. Let $\{Y_1, \dots, Y_n\}$ be the basis of $\mathscr{L}$ as mentioned above. Then $\mathrm{ad}(Z)Y_j$ is in $\mathscr{L}_j$ for any j so that it is written as $\sum_i c_{ij} Y_i$ where $c_{ij} = 0$ if $i < j$. We may identify $\mathrm{ad}(Z)$ with the matrix (c_{ij}). It is a lower triangular matrix of the following form.

$$\begin{pmatrix} * & & & & & & \\ & * & & 0 & & & \\ & & * & & & & \\ & & & * & & & \\ & & & & * & & \\ & * & & & & * & \\ & & & & & & * \end{pmatrix} \tag{30}$$

In particular if $\mathscr{L}$ is nilpotent, it is a nilpotent matrix, i.e. its diagonal elements are all 0. □

Theorem 4.9.10 *Assume that $X_0, \ldots, X_m$ are all complete C^∞-vector fields and generate a finite dimensional solvable Lie algebra $\mathscr{L}$. Let $\{Y_1, \ldots, Y_n\}$ be a basis of $\mathscr{L}$ as mentioned above. Then the Brownian flow $\varphi_{s,t}$, $s < t$ governed by equation (23) is represented by*

$$\varphi_{s,t}(x) = \exp\{N_t^1 Y_1\} \circ \exp\{N_t^2 Y_2\} \circ \cdots \circ \exp\{N_t^n Y_n\}(x), \tag{31}$$

where $\exp\{tY_k\}$ is the one parameter group of transformations generated by Y_k and $N_t^1, \ldots, N_t^n$ are continuous semimartingales constructed from $B_t^0 - B_s^0, \ldots, B_t^m - B_s^m$ through finite repetition of the following elementary calculations.

(i) *Linear sums and products of $B_t^0 - B_s^0, \ldots, B_t^m - B_s^m$.*
(ii) *Stratonovich integrals based on $B_t^0 - B_s^0, \ldots, B_t^m - B_s^m$.*
(iii) *Substitution of the exponential function e^x.*

Furthermore, if $\mathscr{L}$ is nilpotent $N_t^1, \ldots, N_t^n$ are constructed through (i) *and* (ii) *only.* □

Remark The algorithm for calculating $N_t^1, \ldots, N_t^n$ will be found in the proof of the theorem. It is determined by the structure constants of the Lie algebra $\mathscr{L}$ relative to the basis $\{Y_1, \ldots, Y_n\}$.

Proof Vector fields $X_0, \ldots, X_m$ are written as linear sums of $Y_1, \ldots, Y_n$, say $X_j = \sum_l a_{jl} Y_l$. Then equation (23) is written as

$$\varphi_t = x + \sum_{l=1}^n \int_s^t Y_l(\varphi_r) \circ dM_r^l, \tag{32}$$

where $M_t^l = \sum_i a_{il}(B_t^i - B_s^i)$, $l = 1, \ldots, n$.

Let $\zeta_{s,t}^{(1)}$ be the stochastic flow governed by

$$\zeta_{s,t}^{(1)}(x) = x + \int_s^t Y_1(\zeta_{s,r}^{(1)}(x)) \circ dM_r^1. \tag{33}$$

It is written as $\zeta_{s,t}^{(1)}(x) = \exp\{M_t^1 Y_1\}(x)$. Let $\eta_{s,t}^{(1)}(x)$ be the solution of the equation

$$\eta_{s,t}^{(1)}(x) = x + \sum_{l=2}^n \int_s^t (\zeta_{r,s}^{(1)})_* Y_l(\eta_{s,r}^{(1)}(x)) \circ dM_r^l, \tag{34}$$

where $\zeta_{t,s}^{(1)} = (\zeta_{s,t}^{(1)})^{-1}$. Then $\varphi_{s,t} = \zeta_{s,t}^{(1)} \circ \eta_{s,t}^{(1)}$ holds by Theorem 4.9.7. Now Itô's first formula for vector fields states

$$(\zeta_{t,s}^{(1)})_* X - X = \int_s^t (\zeta_{r,s}^{(1)})_* [Y_1, X] \circ dM_r^1.$$

(see formula (6)). We may regard $(\zeta_{t,s}^{(1)})_*$ as a linear transformation on the vector space $\mathscr{L}$. Then it satisfies

$$(\zeta_{t,s}^{(1)})_* = I + \int_s^t (\zeta_{r,s}^{(1)})_* \operatorname{ad}(Y_1) \circ dM_r^1.$$

The solution is given by $(\zeta_{t,s}^{(1)})_* = \exp\{M_t^1 \operatorname{ad}(Y_1)\}$. With the basis $\{Y_1, \dots, Y_n\}$ it is represented as a triangular matrix of the form (30). Therefore equation (34) is written as

$$\eta_{s,t}^{(1)}(x) = x + \sum_{l=2}^{n} \int_s^t \exp\{M_r^1 \operatorname{ad}(Y_1)\} Y_l(\eta_{s,r}^{(1)}(x)) \circ dM_r^l. \tag{35}$$

Using the vector notation $\hat{\mathbb{M}}_t^{(1)} = (0, M_t^2, \dots, M_t^n)$, we define an n-vector continuous semimartingale by

$$\mathbb{M}_t^{(2)} = \int_s^t \exp\{M_r^1 \operatorname{ad}(Y_1)\} \circ d\hat{\mathbb{M}}_r^{(1)}.$$

Here the integrand $\exp\{M_r^1 \operatorname{ad}(Y_1)\}$ is a triangular matrix of the form (30). Therefore the first component of the vector $\mathbb{M}_t^{(2)}$ is 0. Hence writing $\mathbb{M}_t^{(2)} = (0, M_t^{(2)2}, \dots, M_t^{(2)n})$, (35) is written as

$$\eta_{s,t}^{(1)}(x) = x + \sum_{j=2}^{n} \int_s^t Y_j(\eta_{s,r}^{(1)}(x)) \circ dM_r^{(2)j}. \tag{36}$$

We shall decompose equation (36). Let $\zeta_{s,t}^{(2)}(x)$ be the stochastic flow governed by the equation

$$\zeta_{s,t}^{(2)}(x) = x + \int_s^t Y_2(\zeta_{s,r}^{(2)}(x)) \circ dM_r^{(2)2}. \tag{37}$$

Let $\eta_{s,t}^{(2)}(x)$ be the solution of the equation

$$\eta_{s,t}^{(2)}(x) = \sum_{j=3}^{n} \int_s^t (\zeta_{r,s}^{(2)})_* Y_j(\eta_{s,r}^{(2)}(x)) \circ dM_r^{(2)j}, \tag{38}$$

starting at x at time 0. Then $\eta_{s,t}^{(1)} = \zeta_{s,t}^{(2)} \circ \eta_{s,t}^{(2)}$. We have as before that $\zeta_{s,t}^{(2)} = \exp\{M_t^{(2)2} Y_2\}$, and equation (38) is expressed as

$$\eta_{s,t}^{(2)}(x) = x + \sum_{j=3}^{n} \int_s^t Y_j(\eta_{s,r}^{(2)}(x)) \circ dM_r^{(3)j}. \tag{39}$$

Here $\mathbb{M}_t^{(3)}$ is an n-vector continuous semimartingale defined by

$$\mathbb{M}_t^{(3)} = \int_s^t \exp\{M_r^{(2)2}\,\mathrm{ad}(Y_2)\} \circ d\hat{\mathbb{M}}_r^{(2)}, \tag{40}$$

where $\hat{\mathbb{M}}_t^{(2)} = (0, 0, M_t^{(2)3}, \dots, M_t^{(2)n})$. The first and second components of $\mathbb{M}_t^{(3)}$ are 0, since the integrand of the right hand side of (40) is a triangular matrix.

Next we can decompose $\eta_{s,t}^{(2)}$ as $\zeta_{s,t}^{(3)} \circ \eta_{s,t}^{(3)}$, and repeating the above argument inductively we arrive at

$$\begin{aligned}\varphi_{s,t} &= \zeta_{s,t}^{(1)} \circ \cdots \circ \zeta_{s,t}^{(n)} \\ &= \exp\{M_t^1 Y_1\} \circ \exp\{M_t^{(2)2} Y_2\} \circ \cdots \circ \exp\{M_t^{(n)n} Y_n\}.\end{aligned} \tag{41}$$

Clearly $M_t^1, M_t^{(2)2}, \dots, M_t^{(n)n}$ are constructed from $B_t^0, \dots, B_t^m$ through (i)–(iii) of the theorem.

In the case where $\mathscr{L}$ is a nilpotent Lie algebra, matrices $\mathrm{ad}(Y_i)$ are nilpotent. Therefore any component of the matrix $\exp\{M_s^{(i)i}\,\mathrm{ad}(Y_i)\}$ is a polynomial of $M_t^{(i)i}$, not containing exponential functions. Therefore, $M_t^{(i)i}$ are constructed from $B_t^0, \dots, B_t^m$ through operations (i) and (ii) only. The proof is complete. □

Remark In the case where the Lie algebra $\mathscr{L}$ is nilpotent, the assertion of the theorem states that $N_t^1, \dots, N_t^n$ in the expression (31) are linear sums of multiple Wiener integrals of the form

$$\int_s^t \cdots \int_s^{t_{k-1}} \circ dB_{t_1}^{i_1} \circ \cdots \circ dB_{t_k}^{i_k}.$$

Supports of stochastic flows of diffeomorphisms

The stochastic flow of diffeomorphisms generated by equation (23) on the manifold M does not take all the diffeomorphisms of M in general. The possible subset of diffeomorphisms that the flow can take depends on the structure of the Lie algebra generated by vector fields $X_0, \dots, X_m$ defining the equation. In the previous theorem we have seen an explicit form of diffeomorphisms that the flow can take in the case where the Lie algebra $\mathscr{L}$ is solvable and is of finite dimension. Further Theorem 4.8.8 shows that the solution of equation (23) takes values in the Lie transformation group associated with $\mathscr{L}$ provided that $\mathscr{L}$ is a finite dimensional space. However, if the Lie algebra $\mathscr{L}$ is of infinite dimension, the Lie transformation group associated with $\mathscr{L}$ is not well defined except in the case where the underlying manifold M is compact. So we shall look at the support of diffeomorphisms from another aspect making use of Itô's formula for tensor fields. In the following we assume that the maximal solution of equation (23) defines a Brownian flow of diffeomorphisms.

Theorem 4.9.11 *Suppose that $\varphi_{s,t}$ is a Brownian flow of diffeomorphisms governed by the stochastic differential equation* (23). *Let K be a C^3-tensor field of type (p, q). Then $\varphi_{s,t}^* K = K$ a.s. if and only if $\mathscr{L}(X_l)K = 0$, $l = 0, \ldots, m$.*

Proof If $\mathscr{L}(X_l)K = 0$, $l = 0, \ldots, m$, the relation $\varphi_{s,t}^* K = K$ is clear from Corollary 4.9.4. Conversely suppose $\varphi_{s,t}^* K = K$ is satisfied. Then from (21) we have

$$\sum_{l=1}^{m} \int_s^t ((\varphi_{s,r})^* \mathscr{L}(X_l)K)(\theta^1, \ldots, \theta^p, Y_1, \ldots, Y_q)\, dB_r^l$$
$$= -\int_s^t (\varphi_{s,r})^* \left\{\frac{1}{2}\sum_{l=1}^{m} \mathscr{L}(X_l)^2 + \mathscr{L}(X_0)\right\} K(\theta^1, \ldots, \theta^p, Y_1, \ldots, Y_q)\, dr. \tag{42}$$

Since the left hand side is a continuous localmartingale and the right hand side is a process of bounded variation, both should be 0. This implies that the quadratic variation of the left hand side is 0, i.e.

$$\sum_{l=1}^{m} \int_s^t |(\varphi_{s,r})^* \mathscr{L}(X_l)K(\theta^1, \ldots, \theta^p, Y_1, \ldots, Y_q)|^2\, dr = 0$$

holds for any $\theta^1, \ldots, \theta^p \in T_x(M)^*$ and $Y_1, \ldots, Y_q \in T_x(M)$. This implies $(\varphi_{s,r})^* \mathscr{L}(X_l)K = 0$ for every $r \in [s, t]$, so that we have $\mathscr{L}(X_l)K = 0$, $l = 1, \ldots, m$. Then $\mathscr{L}(X_0)K$ should be 0, too, since the right hand side of (42) is also 0. □

As an application of the above theorem, we give here four examples, characterizing the support of diffeomorphisms.

Example 4.9.12 Let M be a Riemannian manifold with metric g. A vector field X on M is called a *Killing vector field* or an *infinitesimal motion* if

$$X(g(Y, Z)) = g([X, Y], Z) + g(Y, [X, Z])$$

is satisfied for any vector fields Y, Z. Since g is a tensor of type (0, 2), the above is equivalent to $\mathscr{L}(X)g = 0$ by formula (15). Now a diffeomorphism ψ of M is called a *motion* of M if $g(Y, Z) = g(\psi_* Y, \psi_* Z)$ (isometry) is satisfied for any vector fields Y and Z. It is equivalent to $\psi^* g = g$.

Now let $\varphi_{s,t}$ be a stochastic flow of diffeomorphisms determined by the equation (23). Then it follows from Theorem 4.9.11 that $\varphi_{s,t}$ is a motion for any $s < t$ a.s., if and only if X_j, $j = 0, \ldots, m$ are all Killing vector fields. □

Example 4.9.13 Let Ω be a positive differential form of order d (volume element). A diffeomorphism $\psi : M \to M$ is called *volume preserving* if $\psi^*\Omega = \Omega$. The solution $\varphi_{s,t}$ is volume preserving if and only if $\mathscr{L}(X_l)\Omega = 0$ for $l = 0, \ldots, m$. In particular if M is a Euclidean space, $\mathscr{L}(X_l)\Omega = 0$ is equivalent to div $X_l = 0$ (*cf.* Section 4.3).

If $X_0, \ldots, X_m$ are Killing vector fields, then the solution $\varphi_{s,t}$ is volume preserving since it is a motion. □

Example 4.9.14 Let M be a manifold where a linear connection is defined. A diffeomorphism φ is called *affine* if it maps each parallel vector field along each curve τ of M into a parallel vector field along the curve $\varphi(\tau)$. A vector field X is called an *infinitesimal affine transformation* of M if for each $x \in M$, a local one parameter group of local transformations φ_t of a neighborhood U is affine for each t. We can prove similarly to the above that $\varphi_{s,t}$ is an affine transformation for any $s < t$ a.s. if and only if $X_0, \ldots, X_m$ are all infinitesimal affine transformations. □

Example 4.9.15 Let M be a complex manifold. A map $\psi : M \to M$ is called *holomorphic* if $f \circ \psi$ is a holomorphic function on M for any holomorphic function f on M. Let J be the almost complex structure. Then a map $\psi : M \to M$ is holomorphic if and only if $J(\psi_* Y) = \psi_* JY$ holds for any vector field Y.

Let X be a real vector field. It is called *analytic* if it is written as

$$X = \sum_{i=1}^{d} X^i \frac{\partial}{\partial z_i} + \sum_{i=1}^{d} \bar{X}^i \frac{\partial}{\partial \bar{z}_i},$$

where $X^i(z)$, $i = 1, \ldots, d$ are holomorphic functions and $(z_1, \ldots, z_d)$, $z_i = x_i + \sqrt{(-1)}\, y_i$ is a holomorphic coordinate. Here $\bar{X}^i$ means the complex conjugate of X^i. Then X is analytic if and only if $J\mathscr{L}(X) = \mathscr{L}(X)J$.

Let X be a complete vector field and let φ_t be the one parameter group of transformations generated by X. Then it is known that φ_t is holomorphic for each t if and only if X is an analytic vector field. We will show the similar fact for a stochastic differential equation. □

Theorem 4.9.16 *Suppose that $\varphi_{s,t}$ is a Brownian flow of diffeomorphisms governed by the stochastic differential equation* (23). *It is holomorphic for any $s < t$ a.s. if and only if $X_0, \ldots, X_m$ are analytic vector fields.*

Proof By Theorem 4.9.1, we have

$$J(\varphi_{s,t})_* Y = JY - \sum_{l=0}^{m} \int_s^t J\mathscr{L}(X_l)(\varphi_{s,r})_* Y \circ dB_r^l \qquad (43)$$

$$(\varphi_{s,t})_* JY = JY - \sum_{l=0}^{m} \int_s^t \mathscr{L}(X_l)(\varphi_{s,r})_* JY \circ dB_r^l. \tag{44}$$

If $\varphi_{s,t}$ is holomorphic, then $(\varphi_{s,t})_* JY = J(\varphi_{s,t})_* Y$, so that we have

$$\sum_{l=0}^{m} \int_s^t (J\mathscr{L}(X_l) - \mathscr{L}(X_l)J)(\varphi_{s,r})_* Y \circ dB_r^l = 0.$$

This implies $J\mathscr{L}(X_l) = \mathscr{L}(X_l)J$, $l = 0, \ldots, m$ (see the proof of Theorem 4.9.11). Therefore X_l, $l = 0, \ldots, m$ are analytic vector fields.

Conversely suppose the X_l's are analytic. Set $\Phi_{s,t}(Y) = J((\varphi_{s,t})_* Y) - (\varphi_{s,t})_* JY$. Then (43), (44) and the relation $\mathscr{L}(X_l)J = J\mathscr{L}(X_l)$ imply

$$\Phi_{s,t}(Y) = -\sum_{l=0}^{m} \int_s^t \mathscr{L}(X_l)\Phi_{s,r}(Y) \circ dB_r^l.$$

The above linear equation has a unique solution $\Phi_{s,t}(Y)$, which should be identically 0. This proves that $\varphi_{s,t}$ is holomorphic. □

The support of the stochastic flow will be discussed further in Section 5.7 applying the approximation theorem of stochastic differential equations.

Exercise 4.9.17 Define vector fields on $\mathbb{R}^d$ by

$$X_0 = \sum_i \left(\sum_j f_{ij} x_j\right) \frac{\partial}{\partial x^i}, \qquad X_j = \sum_i g_{ij} \frac{\partial}{\partial x^i}, \qquad j = 1, \ldots, m.$$

(i) Show that the Lie algebra $\mathscr{L}$ generated by the above $X_0, \ldots, X_m$ is finite dimensional and solvable. Derive expression (27) of Example 4.9.9 from Theorem 4.9.10.

(ii) Let $\mathscr{L}(x)$ be the restriction of the Lie algebra $\mathscr{L}$ to the point x in $\mathbb{R}^d$. ($\mathscr{L}(x)$ is a subspace of the tangent space at x.) Show that $\dim \mathscr{L}(x) = d$ holds for any x if and only if the rank of the matrix $(G, FG, F^2G, \ldots, F^{d-1}G)$ is d, where $F = (f_{ij})$ and $G = (g_{ij})$ (*cf.* Exercise 3.4.8).

5

Convergence of stochastic flows

5.1 Preliminaries

Throughout this chapter we assume that a family of filtrations $\{\mathscr{F}_t^\varepsilon : 0 \le t \le T\}$ with parameter $\varepsilon > 0$ is given on a probability space and that there is defined a family of continuous $C^k(\mathbb{R}^d : \mathbb{R}^d)$-semimartingales $F_\varepsilon(x, t)$, $0 \le t \le T$ such that $F_\varepsilon(x, 0) \equiv 0$ with local characteristics $(a_\varepsilon, b_\varepsilon) = (a_\varepsilon^{ij}(x, y, t), b_\varepsilon^i(x, t))$ with parameter ε belonging to the common class $B_b^{k,\delta}$ for some $k \ge 1$ and $\delta > 0$. We have seen in Chapter 4 that for each ε the system of solutions of Itô's stochastic differential equation

$$d\varphi_t = F_\varepsilon(\varphi_t, dt) \tag{1}$$

defines a forward stochastic flow of C^k-diffeomorphism $(\varphi_\varepsilon)_{s,t}$, $0 \le s \le t \le T$. In this chapter, we will mainly consider the forward stochastic flow and its inverse. We set $\varphi_\varepsilon(x, t) = (\varphi_\varepsilon)_{0,t}(x)$. The object of this chapter is to study the weak and the strong convergences of $\{(\varphi_\varepsilon)(x, t)\}$ as $\varepsilon \to 0$ when the family of C^k-semimartingales $\{F_\varepsilon(x, t)\}$ converges to a certain C^k-Brownian motion. We wish to characterize the limiting stochastic flow by means of a stochastic differential equation. To this end, we will discuss the weak convergence of the joint laws of $\varphi_\varepsilon(t)$ and $F_\varepsilon(t)$ rather than the law of $\varphi_\varepsilon(t)$ itself.

The joint law of $(\varphi_\varepsilon(t), F_\varepsilon(t))$ is defined as follows. Let j be a non-negative integer. Let G^j be the topological group of C^j-diffeomorphisms on $\mathbb{R}^d$. Let $\hat{W}_j = C([0, T] : G^j)$ be the set of all continuous maps from $[0, T]$ into G^j. We denote each element of $\hat{W}^j$ by φ and its value at t by $\varphi(t)$. Define the metric $\mathbf{d}_j$ on $\hat{W}_j$ by

$$\mathbf{d}_j(\varphi, \psi) = \sup_{t \in [0,T]} d_j(\varphi(t), \psi(t)) \tag{2}$$

where d_j is the metric of G^j (see Section 4.1). Then $\hat{W}_j$ is a complete separable metric space. Next let $W_j = C([0, T] : C^j)$ be the set of all continuous maps from $[0, T]$ into C^j. We denote each element of W_j by F and its value at t by $F(t)$. Define the seminorms $\{\| \;\|_{j:N} : N = 1, 2, \dots\}$ by

$$\|F\|_{j:N} = \sum_{|\alpha| \le j} \sup_{t \in [0,T], |x| \le N} |D^\alpha F(x, t)|. \tag{3}$$

Then W_j is a complete separable metric space. The product space $\hat{W}_j \times W_j$ is also a complete separable metric space. In later sections, the seminorms $\{\| \ \|_{j:N}\}$ will be compared with seminorms of Sobolev type $\{\| \ \|_{j,p:N}\}$ where $1 \le p < \infty$. In such cases the seminorms $\{\| \ \|_{j:N}\}$ will be written as $\{\| \ \|_{j,\infty:N}\}$. Similarly the spaces $\hat{W}_j$ and W_j will sometimes be denoted by $\hat{W}_{j,\infty}$ and $W_{j,\infty}$ respectively.

Now the pair $(\varphi_\varepsilon(t), F_\varepsilon(t))$ can be regarded as a $\hat{W}_j \times W_j$-valued random variable for any $j \le k$. Its *law as the G^j-flow* is defined by

$$\hat{P}_\varepsilon^j(A) = P(\{\omega : (\varphi_\varepsilon(\cdot, \omega), F_\varepsilon(\cdot, \omega)) \in A\}), \quad A \in \mathscr{B}(\hat{W}_j \times W_j). \qquad (4)$$

The family of pairs $\{(\varphi_\varepsilon(t), F_\varepsilon(t))\}$ is then said to *converge weakly as G^j-flows* as $\varepsilon \to 0$ if $\{\hat{P}_\varepsilon^j\}$ converges weakly as $\varepsilon \to 0$.

The notion of the weak convergence as G^j-flows is rather strong. We shall introduced two other weak convergences which are weaker than the above convergence. The first one is the weak convergence as C^j-flows. We can define the *law* of $(\varphi_\varepsilon(t), F_\varepsilon(t))$ *as the C^j-flows* on the product space $W_j \times W_j$ by setting

$$P_\varepsilon^j(A) = P(\{\omega : (\varphi_\varepsilon(\cdot, \omega), F_\varepsilon(\cdot, \omega)) \in A\}), \quad A \in \mathscr{B}(W_j \times W_j). \qquad (5)$$

If $\{P_\varepsilon^j\}$ converges weakly as $\varepsilon \to 0$, the family of pairs $\{(\varphi_\varepsilon(t), F_\varepsilon(t))\}$ is then said to *converge weakly as C^j-flows*. Obviously if $\{(\varphi_\varepsilon(t), F_\varepsilon(t))\}$ converge weakly as G^j-flows, then they converge weakly as C^j-flows.

A weaker and more extensively studied convergence is convergence as diffusions. Let $\mathbf{x}_0 = (x_{0,1}, \ldots, x_{0,m})$ and $\mathbf{y}_0 = (y_{0,1}, \ldots, y_{0,n})$ be arbitrary fixed points in $\mathbb{R}^{md}$ and $\mathbb{R}^{nd}$, respectively. Set $\varphi_\varepsilon(\mathbf{x}_0, t) = (\varphi_\varepsilon(x_{0,1}, t), \ldots, \varphi_\varepsilon(x_{0,m}, t))$ and $F_\varepsilon(\mathbf{y}_0, t) = (F_\varepsilon(y_{0,1}, t), \ldots, F_\varepsilon(y_{0,n}, t))$. Then the pair $(\varphi_\varepsilon(\mathbf{x}_0, t), F_\varepsilon(\mathbf{y}_0, t))$ can be regarded as a continuous process (or diffusion process) with values in $\mathbb{R}^{md} \times \mathbb{R}^{nd}$. It is called an *$m + n$-point motion.* Let $V_m = C([0, T] : \mathbb{R}^{md})$ be the space of all continuous maps from $[0, T]$ into $\mathbb{R}^{md}$ equipped with the supremum norm. Let $V_{m,n} = V_m \times V_n$ be the product space. Then the law of the pair $(\varphi_\varepsilon(\mathbf{x}_0, t), F_\varepsilon(\mathbf{y}_0, t))$ is defined by

$$P_\varepsilon^{(\mathbf{x}_0, \mathbf{y}_0)}(A) = P(\{\omega : (\varphi_\varepsilon(\mathbf{x}_0, \cdot, \omega), F_\varepsilon(\mathbf{y}_0, \cdot, \omega)) \in A\}), \quad A \in \mathscr{B}(V_{m,n}). \quad (6)$$

The family of pairs $\{(\varphi_\varepsilon, F_\varepsilon)\}$ is said to *converge weakly as diffusions* as $\varepsilon \to 0$ if for each $(\mathbf{x}_0, \mathbf{y}_0) \in \mathbb{R}^{md} \times \mathbb{R}^{nd}$, $m, n = 1, 2, \ldots$, $\{P_\varepsilon^{(\mathbf{x}_0, \mathbf{y}_0)}\}$ converges weakly as $\varepsilon \to 0$.

Now if the family of laws $\{\hat{P}_\varepsilon^j\}$ or $\{P_\varepsilon^j\}$ converges weakly as $\varepsilon \to 0$, then the family of laws $\{P_\varepsilon^{(\mathbf{x}_0, \mathbf{y}_0)}\}$ converges weakly as $\varepsilon \to 0$ for every $\mathbf{x}_0$, $\mathbf{y}_0$. However, the converse assertion is not true. We need the tightness of the measures $\{\hat{P}_\varepsilon^j\}$ or $\{P_\varepsilon^j\}$.

Theorem 5.1.1 *The family of measures $\{\hat{P}_\varepsilon^j\}$ (or $\{P_\varepsilon^j\}$) converges weakly as $\varepsilon \to 0$ if and only if the following two conditions are satisfied:*

(i) *the family of measures* $\{\hat{P}_\varepsilon^j\}$ *(or* $\{P_\varepsilon^j\}$*) is tight.*
(ii) *the family of pairs* $\{(\varphi_\varepsilon, F_\varepsilon)\}$ *converges weakly as diffusions as* $\varepsilon \to 0$.
□

The proof is immediate from Theorem 1.4.5.

It is sometimes convenient to define the law of $m + n$-point motion $(\varphi_\varepsilon(\mathbf{x}_0, t), F_\varepsilon(\mathbf{y}_0, t))$ in a different way. Let $\mathscr{B}_{\mathbf{x}_0}(\hat{W}_j)$ and $\mathscr{B}_{\mathbf{y}_0}(W_j)$ be the least sub σ-fields of $\mathscr{B}(\hat{W}_j)$ and $\mathscr{B}(W_j)$ for which $\varphi(\mathbf{x}_0, t)$ and $F(\mathbf{y}_0, t)$ are measurable, respectively. Let $\mathscr{B}_{\mathbf{x}_0,\mathbf{y}_0}(\hat{W}_j \times W_j) = \mathscr{B}_{\mathbf{x}_0}(\hat{W}_j) \otimes \mathscr{B}_{\mathbf{y}_0}(W_j)$ be the product σ-field and let $Q_\varepsilon^{(\mathbf{x}_0,\mathbf{y}_0)}$ be the restriction of the measure $\hat{P}_\varepsilon^j$ to the sub σ-field $\mathscr{B}_{\mathbf{x}_0,\mathbf{y}_0}(\hat{W}_j \times W_j)$. Then $Q_\varepsilon^{(\mathbf{x}_0,\mathbf{y}_0)}$ can also be regarded as a law of $(\varphi_\varepsilon(\mathbf{x}_0, t), F_\varepsilon(\mathbf{y}_0, t))$. The merit of considering $Q_\varepsilon^{(\mathbf{x}_0,\mathbf{y}_0)}$ instead of $P_\varepsilon^{(\mathbf{x}_0,\mathbf{y}_0)}$ is that for each ε these measures $\{Q_\varepsilon^{(\mathbf{x}_0,\mathbf{y}_0)} : (\mathbf{x}_0, \mathbf{y}_0) \in \mathbb{R}^{md} \times \mathbb{R}^{nd}, m, n = 1, 2, \ldots,\}$ are a consistent family and defined on the common space $\hat{W}_j \times W_j$. They are extended uniquely to a measure Q_ε on the algebra $\mathscr{A}(\hat{W}_j \times W_j) = \bigcup_{\mathbf{x}_0,\mathbf{y}_0} \mathscr{B}_{\mathbf{x}_0,\mathbf{y}_0}(\hat{W}_j \times W_j)$ where the union runs over all $(\mathbf{x}_0, \mathbf{y}_0) \in \mathbb{R}^{md} \times \mathbb{R}^{nd}$, m, $n = 1, 2, \ldots$. The measure Q_ε coincides with the restriction of $\hat{P}_\varepsilon^j$ to $\mathscr{A}(\hat{W}_j \times W_j)$ and is called the *law as the diffusion of the pair* $(\varphi_\varepsilon, F_\varepsilon)$.

Now a family of measures $\{Q_\varepsilon^{(\mathbf{x}_0,\mathbf{y}_0)} : \varepsilon > 0\}$ is said to be *tight* or *weak convergent* if the family of the corresponding measures $\{P_\varepsilon^{(\mathbf{x}_0,\mathbf{y}_0)} : \varepsilon > 0\}$ is tight or weak convergent. Further if $\{P_\varepsilon^{(\mathbf{x}_0,\mathbf{y}_0)} : \varepsilon > 0\}$ converges weakly for any $(\mathbf{x}_0, \mathbf{y}_0)$, we say that the family of measures $\{Q_\varepsilon\}$ converges weakly as diffusions. Now suppose that the above convergence holds. Let $P_0^{(\mathbf{x}_0,\mathbf{y}_0)}$ be the weak limit of $\{P_\varepsilon^{(\mathbf{x}_0,\mathbf{y}_0)}\}$ as $\varepsilon \to 0$. Then $\{P_0^{(\mathbf{x}_0,\mathbf{y}_0)} : (\mathbf{x}_0, \mathbf{y}_0) \in \mathbb{R}^{md} \times \mathbb{R}^{nd}, m, n = 1, 2, \ldots\}$ is a consistent family of measures. Then there exists a unique measure Q_0 over the algebra $\mathscr{A}(\hat{W}_j \times W_j)$ such that its restriction to the subalgebra $\mathscr{B}_{\mathbf{x}_0,\mathbf{y}_0}(\hat{W}_j \times W_j)$ coincides with $P_0^{(\mathbf{x}_0,\mathbf{y}_0)}$ in the sense of the law. In this case we will say that $\{Q_\varepsilon\}$ *converges weakly to* Q_0 *as diffusions* as $\varepsilon \to 0$.

We shall finally define the strong convergence as stochastic flows. A family of stochastic flows $\{\varphi_\varepsilon(t)\}$ (or the pairs $\{(\varphi_\varepsilon(t), F_\varepsilon(t)\}$) is said to *converge strongly to* $\varphi(t)$ (or to $(\varphi(t), F(t))$) *as* G^j*-flows in probability* if $\mathbf{d}_j(\varphi_\varepsilon, \varphi) \to 0$ (or $|F_\varepsilon - F|_{j:N} \to 0$ for every N) in probability as $\varepsilon \to 0$, where $\mathbf{d}_j$ is the metric defined by (2) (and $|\ |_{j:N}$ are seminorms defined by (3)). Further if $E[(|\varphi_\varepsilon - \varphi|_{j:N})^p + (|\varphi_\varepsilon^{-1} - \varphi^{-1}|_{j:N})^p] \to 0$ (and $E[(|F_\varepsilon - F|_{j:N})^p] \to 0$) for any N, then it is said to *converge strongly as* G^j*-flows in* L^p. Next it is said to *converge as* C^j*-flows in probability* if $|\varphi^\varepsilon - \varphi|_{j:N} \to 0$ in probability for any N. Further if $E[(|\varphi^\varepsilon - \varphi|_{j:N})^p] \to 0$ for any N, it is said to *converge strongly as* C^j*-flows in* L^p.

The strong convergence as diffusions can be defined similarly. The family

$\{\varphi_\varepsilon(t)\}$ (or $\{\varphi_\varepsilon(t), F_\varepsilon(t)\}$) is said to *converge to* φ (or (φ, F)) *strongly as diffusions* as $\varepsilon \to 0$ if $\{\|\varphi_\varepsilon(\mathbf{x}_0) - \varphi(\mathbf{x}_0)\|\}$ (and $\{\|F_\varepsilon(\mathbf{y}_0) - F(\mathbf{y}_0)\|\}$) converges to 0 in probability (or in L^p) for any $\mathbf{x}_0$ (and $\mathbf{y}_0$), where $\| \ \|$ is the supremum norm on the space V_m (and on $V_{m,n}$).

The contents of this chapter are as follows. In Sections 5.2 and 5.3, we discuss the weak and the strong convergence of $\{(\varphi_\varepsilon(t), F_\varepsilon(t))\}$ as diffusion processes. The weak or the strong convergence of $\{F_\varepsilon(t)\}$ as $\varepsilon \to 0$ does not immediately imply the weak or the strong convergence of the associated flows $\{\varphi_\varepsilon(t)\}$. Some additional conditions will be needed on the family of local characteristics. We shall discuss two cases separately in Sections 5.2 and 5.3.

In Section 5.4 we shall discuss the weak and the strong convergence as stochastic flows. The central problem is to show the tightness of the laws. To this end, we shall obtain uniform L^p-estimates of the flows $(\varphi_\varepsilon)_{s,t}$ and their derivatives $D^\alpha(\varphi_\varepsilon)_{s,t}$.

In Section 5.5 we shall extend the convergence theorems of Sections 5.3 and 5.4 so that these can be applied to a wider class of stochastic flows. Here the method of truncation plays a very important role.

In the subsequent two sections, we shall apply these convergence theorems. In Section 5.6, various limit theorems on stochastic differential equations will be discussed. They include the following cases.

(a) Limit theorems for stochastic ordinary differential equations studied by Kesten–Papanicolaou [66], Khasminskii [68], Papanicolaou–Kohler [108] and others.

(b) Limit theorems for driven processes studied by Papanicolaou–Stroock–Varadhan [109].

In Section 5.7 we shall discuss the approximation theorems of stochastic differential equations and stochastic flows studied by Bismut [9], Dowell [27], Ikeda–Watanabe [49] and others.

5.2 Convergence as diffusions I

A convergence theorem

In this chapter if the conditional expectation of a continuous random field $X(x)$ (with respect to $\mathscr{G}$) has a continuous modification, the modification will be written as $E[X(x)|\mathscr{G}]$. The existence of such a modification can be found in Exercises 1.4.12 and 1.4.13.

Let $\{F_\varepsilon : \varepsilon > 0\}$ be a family of continuous C^k-semimartingales with local characteristics $\{(a_\varepsilon, b_\varepsilon) : \varepsilon > 0\}$ belonging to the common class $B^{k,\delta}$ for some $k \geq 1$ and $\delta > 0$. We shall introduce two assumptions on the family of local characteristics.

Condition (C.1)$_p$ *There exists a positive constant K such that*

$$E\left[\sup_x \|D_x^\alpha D_y^\beta a_\varepsilon(x, y, t)_{y=x}\|^{p/2}\right]^{2/p} \leq K, \tag{1}$$

$$E\left[\sup_x |D_x^\alpha b_\varepsilon(x, t)|^p\right]^{1/p} \leq K, \tag{2}$$

hold for any $|\alpha| \leq 1$, $|\beta| \leq 1$, t and ε. □

Condition (C.2)$_k$ *There exist deterministic functions $a(x, y, t) = (a^{ij}(x, y, t))$ and $b(x, t) = (b^i(x, t))$ belonging to the classes $\tilde{C}_b^{k,\delta}$ and $C_b^{k,\delta}$ respectively ($\delta > 0$) such that for any $s < t$ and i, j, the convergences*

$$E\left[\int_s^t a_\varepsilon^{ij}(x, y, u)\,du \,\middle|\, \mathscr{F}_s^\varepsilon\right] \to \int_s^t a^{ij}(x, y, u)\,du, \tag{3}$$

$$E\left[\int_s^t b_\varepsilon^i(x, u)\,du \,\middle|\, \mathscr{F}_s^\varepsilon\right] \to \int_s^t b^i(x, u)\,du, \tag{4}$$

hold uniformly on compact sets of spatial variables in L^1. □

Theorem 5.2.1 *Assume that the family of local characteristics $\{(a_\varepsilon, b_\varepsilon)\}$ of $\{F_\varepsilon\}$ satisfies Conditions (C.1)$_p$ and (C.2)$_k$ for some $p > 2$ and $k \geq 1$. Then the family of laws $\{Q_\varepsilon\}$ of the pairs $(\varphi_\varepsilon, F_\varepsilon)$ converges weakly as diffusions as $\varepsilon \to 0$. The limit law Q_0 can be extended uniquely to a probability measure P_0 on $\mathscr{B}(\hat{W}_k \times W_k)$. Further, the triple $(\varphi(t), F(t), P_0)$ satisfies the following properties:*

(i) *$F(x, t)$ is a C^k-Brownian motion with local characteristic (a, b),*
(ii) *$\varphi(x, t)$ is a Brownian flow of C^k-diffeomorphism generated by $F(x, t)$ in the sense of the Itô integral.* □

The proof of the theorem will be completed in three steps since it is long. In the first step we will show the tightness of the measures $\{Q_\varepsilon\}$ where Condition (C.1)$_p$ plays an important role. In Lemmas 5.2.2 and 5.2.3 we will obtain uniform L^p-estimates of the differences $F_\varepsilon(t) - F_\varepsilon(s)$ and $\varphi_\varepsilon(t) - \varphi(s)$, respectively. Then we will apply Kolmogorov's tightness criterion discussed in Chapter I.

In the second step we will characterize the arbitrary limit measure Q_0 of $\{Q_\varepsilon\}$, $\varepsilon \to 0$ by certain martingale properties: it will be shown in Lemmas 5.2.6 and 5.2.7 that with respect to Q_0, both $\varphi(x, t)$ and $F(y, t)$ are continuous semimartingales with local characteristics $(a(\varphi(x, t), \varphi(\tilde{x}, t), t)$, $b(\varphi(x, t), t))$ and $(a(y, \tilde{y}, t), b(y, t))$, respectively, which do not depend on the choice of the limit measures Q_0. This yields that $F(y, t)$ is a C^k-Brownian motion, since continuous semimartingales with deterministic characteristics are Brownian motions. There Condition $(C.2)_k$ is crucial.

In the third step we will show that these semimartingale properties determine the law Q_0 uniquely and that Q_0 can be extended to a probability measure $\hat{P}_0$ on $\mathscr{B}(\hat{W}_k \times W_k)$, which has properties (i) and (ii) of the theorem. Finally the weak convergence of $\{Q_\varepsilon\}$, $\varepsilon \to 0$ will be established.

In all the lemmas in this section, Conditions $(C.1)_p$ and $(C.2)_k$ for some $p > 2$ and $k \geq 1$ are assumed.

Uniform L^p-estimates

In the sequel p is any real number greater than 2 satisfying Condition $(C.1)_p$. We begin with the uniform L^p-estimate of $F_\varepsilon(t) - F_\varepsilon(s)$.

Lemma 5.2.2 There exists a positive constant C such that

$$E[|F_\varepsilon(y, t) - F_\varepsilon(y, s)|^p] \leq C|t - s|^{2-(2/p)} \tag{5}$$

holds for any s, t, y and ε.

Proof. We shall consider the case for $d = 1$ only. We suppress y from $F_\varepsilon(y, t)$, $a_\varepsilon(y, y, t)$, etc. and will write them as $F_\varepsilon(t)$, $a_\varepsilon(t)$, etc. Apply Itô's formula to the integrand of (5). Then we have

$$\begin{aligned}|F_\varepsilon(t) - F_\varepsilon(s)|^p = p\int_s^t b_\varepsilon(u)|F_\varepsilon(u) - F_\varepsilon(s)|^{p-1}\operatorname{sign}(u)\,du \\ + \frac{1}{2}p(p-1)\int_s^t a_\varepsilon(u)|F_\varepsilon(u) - F_\varepsilon(s)|^{p-2}\,du \\ + \text{martingale with mean } 0,\end{aligned} \tag{6}$$

where $\operatorname{sign}(u) = \operatorname{sign}(F_\varepsilon(u) - F_\varepsilon(s))$. The expectation of the first term of the right hand side is bounded by

$$pK\int_s^t E[|F_\varepsilon(u) - F_\varepsilon(s)|^p]^{1-(1/p)}\,du$$

by Hölder's inequality and inequality (2). Similarly the expectation of the second term of (6) is bounded by

$$\frac{1}{2}p(p-1)K\int_s^t E[|F_\varepsilon(u) - F_\varepsilon(s)|^p]^{1-(2/p)}\,du.$$

Therefore there exists a positive constant C_1 depending on p and K only such that

$$E[|F_\varepsilon(t) - F_\varepsilon(s)|^p] \le C_1\left\{\int_s^t E[|F_\varepsilon(u) - F_\varepsilon(s)|^p]^{1-(1/p)}\, du + \int_s^t E[|F_\varepsilon(u) - F_\varepsilon(s)|^p]^{1-(2/p)}\, du\right\}. \tag{7}$$

Using the inequality $a^{1-(1/p)} < 1 + a$ $(a > 0)$, the above yields

$$E[|F_\varepsilon(t) - F_\varepsilon(s)|^p] \le 2C_1\left\{(t-s) + \int_s^t E[|F_\varepsilon(u) - F_\varepsilon(s)|^p]\, du\right\}.$$

By Gronwall's inequality we have

$$E[|F_\varepsilon(t) - F_\varepsilon(s)|^p] \le 2C_1 e^{2(t-s)C_1}(t-s).$$

Substitute this back into (7). Then we obtain (5). The proof is complete. □

We shall next obtain a uniform L^p-estimate of the flows $\varphi_\varepsilon(t)$.

Lemma 5.2.3 *There exists a positive constant C such that*

$$E[|\varphi_\varepsilon(x, t) - \varphi_\varepsilon(x, s)|^p] \le C|t - s|^{2-(2/p)} \tag{8}$$

holds for any s, t, x and $\varepsilon > 0$.

Proof The flow $\varphi_\varepsilon(t)$ satisfies the stochastic differential equation

$$\varphi_\varepsilon(t) = \varphi_\varepsilon(s) + \int_s^t b_\varepsilon(\varphi_\varepsilon(u), u)\, du + \int_s^t Y_\varepsilon(\varphi_\varepsilon(u), du),$$

where $Y_\varepsilon(\cdot, t)$ is the martingale part of $F_\varepsilon(\cdot, t)$. Therefore we have by Itô's formula

$$|\varphi_\varepsilon(t) - \varphi_\varepsilon(s)|^p = p\int_s^t b_\varepsilon(\varphi_\varepsilon(u), u)|\varphi_\varepsilon(u) - \varphi_\varepsilon(s)|^{p-1}\, \mathrm{sign}(u)\, du + \frac{1}{2}p(p-1)\int_s^t a_\varepsilon(\varphi_\varepsilon(u), \varphi_\varepsilon(u), u)|\varphi_\varepsilon(u) - \varphi_\varepsilon(s)|^{p-2}\, du + \text{martingale with mean } 0,$$

where $\mathrm{sign}(u) = \mathrm{sign}(\varphi_\varepsilon(u) - \varphi_\varepsilon(s))$. Taking the expectation of each term of the above and applying Condition $(C.1)_p$, we obtain

$$E[|\varphi_\varepsilon(t) - \varphi_\varepsilon(s)|^p] \le pK\int_s^t E[|\varphi_\varepsilon(u) - \varphi_\varepsilon(s)|^p]^{1-(1/p)}\, du + \frac{1}{2}p(p-1)K\int_s^t E[|\varphi_\varepsilon(u) - \varphi_\varepsilon(s)|^p]^{1-(2/p)}\, du.$$

Then inequality (8) follows similarly to the proof of Lemma 5.2.2. The proof is complete. □

Now since $F_\varepsilon(y, 0) = 0$ and $\varphi_\varepsilon(x, 0) = x$ hold, from (5) and (8) we have the following:

$$E[|F_\varepsilon(y, t)|^p] \le Ct^{2-(2/p)}, \tag{9}$$

$$E[|\varphi_\varepsilon(x, t)|^p] \le 2^p(Ct^{2-(2/p)} + |x|^p). \tag{10}$$

Therefore the family of the laws $\{Q_\varepsilon\}$ is tight by Theorem 1.4.7. Consequently, for each $\mathbf{x}_0$, $\mathbf{y}_0$, there exists a sequence $\varepsilon_h \downarrow 0$ such that $\{Q_{\varepsilon_h}^{(\mathbf{x}_0, \mathbf{y}_0)}\}$ converges weakly. Let $Q_0^{(\mathbf{x}_0, \mathbf{y}_0)}$ be its limit measure. It may not be unique since it depends on the choice of a sequence $\{\varepsilon_h\}$. Our next step will be to show that both $F(t)$ and $\varphi(t)$ are continuous semimartingales with respect to $Q_0^{(\mathbf{x}_0, \mathbf{y}_0)}$ and that their local characteristics are common to any limit laws $Q_0^{(\mathbf{x}_0, \mathbf{y}_0)}$.

Semimartingale characterization of the limit measure

Before we proceed further we need two general lemmas concerning the weak convergence. In the following argument, the time s is fixed.

Let h be a bounded continuous function on $\mathbb{R}^{(m+n)ld}$. Define

$$\Phi = h(\varphi(\mathbf{x}_0, s_1), \ldots, \varphi(\mathbf{x}_0, s_l), F(\mathbf{y}_0, s_1), \ldots, F(\mathbf{y}_0, s_l)), \tag{11}$$

where $0 \le s_1 \le \cdots \le s_l \le s$. It is a bounded continuous function on $\hat{W}_j \times W_j$ where $j \le k$. Similarly

$$\Phi^\varepsilon = h(\varphi_\varepsilon(\mathbf{x}_0, s_1), \ldots, \varphi_\varepsilon(\mathbf{x}_0, s_l), F_\varepsilon(\mathbf{y}_0, s_1), \ldots, F_\varepsilon(\mathbf{y}_0, s_l)) \tag{12}$$

is a bounded random function over $(\Omega, \mathscr{F}, P)$. For any measurable functions $f(\mathbf{x}, \mathbf{y}, u)$, $g(\mathbf{x}, \mathbf{y}, u)$, $(\mathbf{x}, \mathbf{y}, u) \in \mathbb{R}^{(m+n)d} \times [0, T]$, we have

$$E[f(\varphi_\varepsilon(\mathbf{x}_0, t), F_\varepsilon(\mathbf{y}_0, t))\Phi^\varepsilon] = E_\varepsilon^{(\mathbf{x}_0, \mathbf{y}_0)}[f(\varphi(\mathbf{x}_0, t), F(\mathbf{y}_0, t))\Phi],$$

$$E\left[\left\{\int_s^t g(\varphi_\varepsilon(\mathbf{x}_0, u), F_\varepsilon(\mathbf{y}_0, u))\, du\right\}\Phi^\varepsilon\right]$$
$$= E_\varepsilon^{(\mathbf{x}_0, \mathbf{y}_0)}\left[\left\{\int_s^t g(\varphi(\mathbf{x}_0, u), F(\mathbf{y}_0, u))\, du\right\}\Phi\right],$$

where $E_\varepsilon^{(\mathbf{x}_0, \mathbf{y}_0)}$ is the expectation of the measure $Q_\varepsilon^{(\mathbf{x}_0, \mathbf{y}_0)}$, if both sides are well defined.

Now if $f(\mathbf{x}, \mathbf{y})$, $g(\mathbf{x}, \mathbf{y})$, $(\mathbf{x}, \mathbf{y}) \in \mathbb{R}^{(m+n)d}$ are bounded continuous functions then by the definition of the weak convergence, we have

$$\lim_{\varepsilon_h \to 0} E[f(\varphi_{\varepsilon_h}(\mathbf{x}_0, t), F_{\varepsilon_h}(\mathbf{y}_0, t))\Phi^{\varepsilon_h}] = E_0^{(\mathbf{x}_0, \mathbf{y}_0)}[f(\varphi(\mathbf{x}_0, t), F(\mathbf{y}_0, t))\Phi]. \tag{13}$$

$$\lim_{\varepsilon_h \to 0} E\left[\left\{\int_s^t g(\varphi_{\varepsilon_h}(\mathbf{x}_0, u), F_{\varepsilon_h}(\mathbf{y}_0, u))\,du\right\}\Phi^{\varepsilon_h}\right]$$

$$= E_0^{(\mathbf{x}_0, \mathbf{y}_0)}\left[\left\{\int_s^t g(\varphi(\mathbf{x}_0, u), F(\mathbf{y}_0, u))\,du\right\}\Phi\right]. \tag{14}$$

We will show that this is valid for some unbounded function g.

Lemma 5.2.4 *Let $f(\mathbf{x}, \mathbf{y})$, $g(\mathbf{x}, \mathbf{y})$ be continuous functions such that $|f(\mathbf{x}, \mathbf{y})| + |g(\mathbf{x}, \mathbf{y})| \le K\{1 + (|\mathbf{x}| + |\mathbf{y}|)^{p'}\}$ holds for some $K > 0$ and $p' < p$. Then we have the convergences* (13) *and* (14). □

The proof is immediate from the uniform L^p-boundedness of $\varphi_\varepsilon(x, t)$ and $F_\varepsilon(y, t)$. It is left to the reader.

The next lemma is an extension of the above lemma.

Lemma 5.2.5 *Suppose that $\{g^\varepsilon(\mathbf{x}, \mathbf{y}, t)\}$ is a family of continuous C^1-valued processes satisfying the following conditions.*

(a) *There exist positive processes $\{K_\varepsilon(t)\}$ with $\sup_\varepsilon \sup_t E[K_\varepsilon(t)^p] < \infty$ such that*

$$|D_{\mathbf{x}}^\alpha D_{\mathbf{y}}^\beta g^\varepsilon(\mathbf{x}, \mathbf{y}, t)| \le K_\varepsilon(t)(1 + |\mathbf{x}| + |\mathbf{y}|),$$

holds for all $\mathbf{x}, \mathbf{y}, t$, $|\alpha| \le 1$ and $|\beta| \le 1$. (15)

(b) *There exists a continuous deterministic function $g(\mathbf{x}, \mathbf{y}, \mathbf{t})$ with $|D_{\mathbf{x}}^\alpha D_{\mathbf{y}}^\beta g(\mathbf{x}, \mathbf{y}, t)| \le K(1 + |\mathbf{x}| + |\mathbf{y}|)^2$ for any $(\mathbf{x}, \mathbf{y}) \in \mathbb{R}^{(m+n)d}$, $|\alpha| \le 1$, $|\beta| \le 1$ such that*

$$E\left[\sup_{|\mathbf{x}|, |\mathbf{y}| \le N}\left|E\left[\int_s^t g^\varepsilon(\mathbf{x}, \mathbf{y}, u)\,du\,\middle|\,\mathscr{F}_s^\varepsilon\right] - \int_s^t g(\mathbf{x}, \mathbf{y}, u)\,du\right|\right] \underset{\varepsilon \to 0}{\longrightarrow} 0 \tag{16}$$

holds for any N and $s < t$.

Let $\varepsilon_h \downarrow 0$ be a sequence such that $\{Q_{\varepsilon_h}^{(\mathbf{x}_0, \mathbf{y}_0)}\}$ converges weakly to $Q_0^{(\mathbf{x}_0, \mathbf{y}_0)}$. Then

$$\lim_{h \to \infty} E\left[\left\{\int_s^t g^{\varepsilon_h}(\varphi_{\varepsilon_h}(\mathbf{x}_0, u), F_{\varepsilon_h}(\mathbf{y}_0, u), u)\,du\right\}\Phi^{\varepsilon_h}\right]$$

$$= E_0^{(\mathbf{x}_0, \mathbf{y}_0)}\left[\left\{\int_s^t g(\varphi(\mathbf{x}_0, u), F(\mathbf{y}_0, u), u)\,du\right\}\Phi\right] \tag{17}$$

holds for any $s < t$.

Proof We prove the lemma for the case where $m = 1$ and $n = 0$ only. The general case can be shown similarly. Observe first that by the tightness of $\{\varphi_\varepsilon(t)\}$, for any $\theta > 0$ there exists a positive constant N such that $P(\sup_t |\varphi_\varepsilon(t)| > N) < \theta/2$ holds for any ε. Associated with this N we can choose $\eta > 0$ such that

$$E\left[\sup_{|x|,|y|\le N,|x-y|<\eta} |g^\varepsilon(x,t) - g^\varepsilon(y,t)|^p\right]^{1/p} \le \theta, \qquad \text{for all } t,$$

$$\sup_{|x|,|y|\le N,|x-y|<\eta} |g(x,t) - g(y,t)| < \theta, \qquad \text{for all } t,$$

hold for any ε by (a). Further since $\{\varphi_\varepsilon(t)\}$ is tight, there exists $\zeta > 0$ such that

$$P\left(\sup_{|t-s|<\zeta} |\varphi_\varepsilon(t) - \varphi_\varepsilon(s)| < \eta\right) > 1 - \theta/2$$

holds for any ε by Ascoli–Arzela's theorem. Set

$$A_\varepsilon(N,\zeta,\eta) \equiv \left\{\omega : \sup_{|t-s|<\zeta} |\varphi_\varepsilon(t) - \varphi_\varepsilon(s)| < \eta, \quad \sup_t |\varphi_\varepsilon(t)| \le N\right\}. \tag{18}$$

Then $P(A_\varepsilon(N,\zeta,\eta)) > 1 - \theta$ for any ε.

Now let $\Delta = \{s = t_0 < t_1 < \cdots < t_l = t\}$ be a partition of $[s,t]$ such that $|\Delta| < \zeta$ and let $\Delta(u)$ be a simple function such that $\Delta(u) = t_k$ holds if $t_k < u \le t_{k+1}$. We have the inequality

$$\begin{aligned}
&\left|E\left[\left\{\int_s^t g^\varepsilon(\varphi_\varepsilon(u),u)\,du\right\}\Phi^\varepsilon\right] - E_0^{(x_0,y_0)}\left[\left\{\int_s^t g(\varphi(u),u)\,du\right\}\Phi\right]\right| \\
&\le \left|E\left[\left\{\int_s^t (g^\varepsilon(\varphi_\varepsilon(u),u) - g^\varepsilon(\varphi_\varepsilon(\Delta(u)),u))\,du\right\}\Phi^\varepsilon\right]\right| \\
&\quad + \left|E\left[\left\{\int_s^t (g^\varepsilon(\varphi_\varepsilon(\Delta(u)),u) - g(\varphi_\varepsilon(\Delta(u)),u))\,du\right\}\Phi^\varepsilon\right]\right| \\
&\quad + \left|E\left[\left\{\int_s^t g(\varphi_\varepsilon(\Delta(u)),u)\,du\right\}\Phi^\varepsilon\right] - E_0^{(x_0,y_0)}\left[\left\{\int_s^t g(\varphi(\Delta(u)),u)\,du\right\}\Phi\right]\right| \\
&\quad + \left|E_0^{(x_0,y_0)}\left[\left\{\int_s^t (g(\varphi(\Delta(u)),u) - g(\varphi(u),u))\,du\right\}\Phi\right]\right|.
\end{aligned} \tag{19}$$

We shall estimate each term of the right hand side. The first term is bounded by

$$\left\{\int_s^t E[|g^\varepsilon(\varphi_\varepsilon(u), u) - g^\varepsilon(\varphi_\varepsilon(\Delta(u)), u)| : A_\varepsilon(N, \zeta, \eta)]\, du\right\} \|h\|$$
$$+ \left\{\int_s^t E[|g^\varepsilon(\varphi_\varepsilon(u), u) - g^\varepsilon(\varphi_\varepsilon(\Delta(u)), u)| : A_\varepsilon(N, \zeta, \eta)^c]\, du\right\} \|h\|$$
$$\le \theta(t-s)\|h\| + 2 \sup_u E[K_\varepsilon(u)^p]^{1/p} E[(1+|\varphi_\varepsilon(u)|)^p]^{1/p}(t-s)\theta^{(p-2)/p}\|h\|.$$

The second term of the right hand side of (19) can be written as

$$\sum_{k=0}^{l-1} E\left[\left\{E\left[\int_{t_k}^{t_{k+1}} g^\varepsilon(\varphi_\varepsilon(t_k), u)\, du \,\middle|\, \mathscr{G}^\varepsilon_{t_k}\right] - \int_{t_k}^{t_{k+1}} g(\varphi_\varepsilon(t_k), u)\, du\right\}\Phi^\varepsilon : |\varphi_\varepsilon(t_k)| \le N\right]$$
$$+ \sum_{k=0}^{l-1} E\left[\left\{\int_{t_k}^{t_{k+1}} (g^\varepsilon(\varphi_\varepsilon(t_k), u) - g(\varphi_\varepsilon(t_k), u))\, du\right\}\Phi^\varepsilon : |\varphi_\varepsilon(t_k)| > N\right].$$

The first term of the above converges to 0 as $\varepsilon_h \to 0$ by (16). The second term is bounded by

$$E\left[\left\{\int_s^t |g^\varepsilon(\varphi_\varepsilon(\Delta(u)), u) - g(\varphi_\varepsilon(\Delta(u)), u)|\, du\right\}|\Phi^\varepsilon| : A_\varepsilon(N, \zeta, \eta)^c\right]$$
$$\le \sup_u \{E[K_\varepsilon(u)^p]^{1/p} E[(1 + |\varphi_\varepsilon(u)|)^p]^{1/p} + KE[(1 + |\varphi_\varepsilon(u)|)^p]^{2/p}\}$$
$$\times (t - s)\theta^{(p-2)/p}\|h\|,$$

where K is a positive constant such that $|g(x)| \le K(1 + |x|)^2$. Now, the third term of the right hand side of (19) converges to 0 as $\varepsilon_h \to 0$ since φ_{ε_h} converges to φ weakly. The fourth term is bounded by

$$E_0^{(x_0, y_0)}\left[\left\{\int_s^t |g(\varphi(\Delta(u)), u) - g(\varphi(u), u)|\, du\right\}\Phi : A(N, \zeta, \eta)\right]$$
$$+ E_0^{(x_0, y_0)}\left[\left\{\int_s^t |g(\varphi(\Delta(u)), u) - g(\varphi(u), u)|\, du\right\}\Phi : A(N, \zeta, \eta)^c\right]$$
$$\le \theta(t - s)\|h\| + 2K(t - s) \sup_u E_0^{(x_0, y_0)}[(1 + |\varphi(u)|)^p]^{2/p}\theta^{(p-2)/p}\|h\|.$$

Summing up these four estimations, we arrive at

$$\overline{\lim_{\varepsilon_h \to 0}} \left| E\left[\left\{\int_s^t g^{\varepsilon_h}(\varphi_{\varepsilon_h}(u), u)\, du\right\}\Phi^{\varepsilon_h}\right] - E_0^{(x_0, y_0)}\left[\left\{\int_s^t g(\varphi(u), u)\, du\right\}\Phi\right]\right|$$
$$\le (t - s)\|h\| \left(2\theta + \left\{3 \sup_{\varepsilon, u} E[K_\varepsilon(u)^p]^{1/p} E[(1 + |\varphi_\varepsilon(u)|)^p]^{1/p}\right.\right.$$
$$+ K \sup_{\varepsilon, u} E[(1 + |\varphi_\varepsilon(u)|)^p]^{2/p}$$
$$\left.\left. + 2K \sup_u E_0^{(x_0, y_0)}[(1 + |\varphi(u)|)^p]^{2/p}\right\}\theta^{(p-2)/p}\right).$$

Since θ can be chosen arbitrarily small, the above is 0. The proof is complete. □

Let $\mathscr{B}_{\mathbf{x}_0,\mathbf{y}_0}(t)$ be the least sub σ-field of $\mathscr{B}_{\mathbf{x}_0,\mathbf{y}_0}(\hat{W}_j \times W_j)$ for which $(\varphi(\mathbf{x}_0, s), F(\mathbf{y}_0, s))$, $s \le t$ are measurable. Let $\{G^{\varepsilon_h}\}$, $\varepsilon_h \downarrow 0$ and G be $\mathscr{B}_{\mathbf{x}_0,\mathbf{y}_0}(\hat{W}_j \times W_j)$-measurable functionals such that

$$E_{\varepsilon_h}^{(\mathbf{x}_0,\mathbf{y}_0)}[G^{\varepsilon_h}\Phi] \to E_0^{(\mathbf{x}_0,\mathbf{y}_0)}[G\Phi]$$

holds for any Φ of (11). We can say that $\{E_{\varepsilon_h}^{(\mathbf{x}_0,\mathbf{y}_0)}[G^{\varepsilon_h}|\mathscr{B}_{\mathbf{x}_0,\mathbf{y}_0}(s)]\}$ *converges weakly to* $E_0^{(\mathbf{x}_0,\mathbf{y}_0)}[G|\mathscr{B}_{\mathbf{x}_0,\mathbf{y}_0}(s)]$ or $\{G^{\varepsilon_h}\}$ *converges weakly to G with respect to* $\mathscr{B}_{\mathbf{x}_0,\mathbf{y}_0}(s)$*-conditional expectation*. In the following arguments, we shall denote it as

$$G^{\varepsilon_h} \to G \text{ weakly } (\mathscr{B}_{\mathbf{x}_0,\mathbf{y}_0}(s))$$

for convenience of notation.

We now discuss the semimartingale properties of $(\varphi(t), F(t))$ with respect to the measure $Q_0^{(\mathbf{x}_0,\mathbf{y}_0)}$.

Lemma 5.2.6 *The following are L^2-martingales with respect to* $(\mathscr{B}_{\mathbf{x}_0,\mathbf{y}_0}(t), Q_0^{(\mathbf{x}_0,\mathbf{y}_0)})$:

$$M(x_0, t) \equiv \varphi(x_0, t) - x_0 - \int_0^t b(\varphi(x_0, u), u)\, du, \tag{20}$$

$$Y(y_0, t) \equiv F(y_0, t) - \int_0^t b(y_0, u)\, du, \tag{21}$$

where x_0 and y_0 are any components of $\mathbf{x}_0$ and $\mathbf{y}_0$.

Proof We consider the case for $d = 1$ only. Note that $F_\varepsilon(y_0, t) - \int_0^t b_\varepsilon(y_0, u)\, du$ is a martingale adapted to $(\mathscr{F}_t^\varepsilon)$ and that Φ^ε defined by (12) is $\mathscr{F}_s^\varepsilon$-measurable. Then we have

$$E\left[\left\{F_\varepsilon(y_0, t) - F_\varepsilon(y_0, s) - \int_s^t b_\varepsilon(y_0, u)\, du\right\}\Phi^\varepsilon\right] = 0.$$

By Lemma 5.2.4, $F_\varepsilon(y_0, t) \to F(y_0, t)$ weakly $(\mathscr{B}_{\mathbf{x}_0,\mathbf{y}_0}(s))$ and by Condition (C.2)$_k$ $E[\int_s^t b_\varepsilon(y_0, u)\, du|\mathscr{F}_s^\varepsilon] \to \int_s^t b(y_0, u)\, du$ in L^1 as $\varepsilon = \varepsilon_h \to 0$. Therefore we have

$$E_0^{(\mathbf{x}_0,\mathbf{y}_0)}\left[\left\{F(y_0, t) - F(y_0, s) - \int_s^t b(y_0, u)\, du\right\}\Phi\right] = 0.$$

This proves that $Y(y_0, t)$ of (21) is a $(\mathscr{B}_{\mathbf{x}_0,\mathbf{y}_0}(s), Q_0^{(\mathbf{x}_0,\mathbf{y}_0)})$-martingale.

Observe next

$$E\left[\left\{\varphi_\varepsilon(x_0, t) - \varphi_\varepsilon(x_0, s) - \int_s^t b_\varepsilon(\varphi_\varepsilon(x_0, u), u)\,du\right\}\Phi^\varepsilon\right] = 0.$$

By Lemma 5.2.4, $\varphi_\varepsilon(x_0, t) - \varphi_\varepsilon(x_0, s) \to \varphi(x_0, t) - \varphi(x_0, s)$ weakly $(\mathscr{B}_{\mathbf{x}_0,\mathbf{y}_0}(s))$ as $\varepsilon = \varepsilon_h \to 0$. By Lemma 5.2.5, $\int_s^t b_\varepsilon(\varphi_\varepsilon(x_0, u), u)\,du \to \int_s^t b(\varphi(x_0, u), u)\,du$ weakly $(\mathscr{B}_{\mathbf{x}_0,\mathbf{y}_0}(s))$ as $\varepsilon = \varepsilon_h \to 0$. Therefore we have

$$E_0^{(\mathbf{x}_0,\mathbf{y}_0)}\left[\left\{\varphi(x_0, t) - \varphi(x_0, s) - \int_s^t b(\varphi(x_0, u), u)\,du\right\}\Phi\right] = 0.$$

This proves that $M(x_0, t)$ of (20) is a martingale. □

Lemma 5.2.7 *The following hold with respect to* $Q_0^{(\mathbf{x}_0,\mathbf{y}_0)}$:

$$\langle M^i(x_0, t), M^j(\tilde{x}_0, t)\rangle = \int_0^t a^{ij}(\varphi(x_0, u), \varphi(\tilde{x}_0, u), u)\,du, \tag{22}$$

$$\langle M^i(x_0, t), Y^j(y_0, t)\rangle = \int_0^t a^{ij}(\varphi(x_0, u), y_0, u)\,du, \tag{23}$$

$$\langle Y^i(y_0, t), Y^j(\tilde{y}_0, t)\rangle = \int_0^t a^{ij}(y_0, \tilde{y}_0, u)\,du, \tag{24}$$

where $x_0, \tilde{x}_0$ *are components of* $\mathbf{x}_0$ *and* $y_0, \tilde{y}_0$ *are components of* $\mathbf{y}_0$.

Proof We shall prove (22). (23) and (24) can be proved by a similar method. We again consider the case for $d = 1$ only. Using Itô's formula, we find that

$$\varphi_\varepsilon(x_0, t)\varphi_\varepsilon(\tilde{x}_0, t) - \varphi_\varepsilon(x_0, s)\varphi_\varepsilon(\tilde{x}_0, s) - \int_s^t \varphi_\varepsilon(x_0, u)b_\varepsilon(\varphi_\varepsilon(\tilde{x}_0, u), u)\,du$$
$$- \int_s^t \varphi_\varepsilon(\tilde{x}_0, u)b_\varepsilon(\varphi_\varepsilon(x_0, u), u)\,du - \int_s^t a_\varepsilon(\varphi_\varepsilon(x_0, u), \varphi_\varepsilon(\tilde{x}_0, u), u)\,du$$

is a martingale adapted to $(\mathscr{F}_t^\varepsilon)$ with mean 0. Let $\varepsilon_h \downarrow 0$ be a sequence such that $Q_{\varepsilon_h}^{(\mathbf{x}_0,\mathbf{y}_0)}$ converges to $Q_0^{(\mathbf{x}_0,\mathbf{y}_0)}$ weakly. By Lemmas 5.2.4 and 5.2.5, each term of the above converges weakly with respect to $\mathscr{B}_{\mathbf{x}_0,\mathbf{y}_0}(s)$-conditional expectation. Therefore

$$\varphi(x_0, t)\varphi(\tilde{x}_0, t) - \varphi(x_0, s)\varphi(\tilde{x}_0, s) - \int_s^t \varphi(x_0, u)b(\varphi(\tilde{x}_0, u), u)\,du$$
$$- \int_s^t \varphi(\tilde{x}_0, u)b(\varphi(x_0, u), u)\,du - \int_s^t a(\varphi(x_0, u), \varphi(\tilde{x}_0, u), u)\,du \tag{25}$$

is a martingale with respect to $Q_0^{(\mathbf{x}_0,\mathbf{y}_0)}$.

On the other hand, since (20) is a martingale we have by Itô's formula

$$\begin{aligned}\varphi(\mathbf{x}_0, t)\varphi(\tilde{\mathbf{x}}_0, t) - \mathbf{x}_0\tilde{\mathbf{x}}_0 &= \int_0^t \varphi(\mathbf{x}_0, u)M(\tilde{\mathbf{x}}_0, du)\\ &+ \int_0^t \varphi(\mathbf{x}_0, u)b(\varphi(\tilde{\mathbf{x}}_0, u), u)\,du\\ &+ \int_0^t \varphi(\tilde{\mathbf{x}}_0, u)M(\mathbf{x}_0, du)\\ &+ \int_0^t \varphi(\tilde{\mathbf{x}}_0, u)b(\varphi(\mathbf{x}_0, u), u)\,du\\ &+ \langle M(\mathbf{x}_0, t), M(\tilde{\mathbf{x}}_0, t)\rangle. \end{aligned} \tag{26}$$

Set $s = 0$ and substitute (26) in (25). Then we find that

$$\int_0^t \varphi(\mathbf{x}_0, u)M(\tilde{\mathbf{x}}_0, du) + \int_0^t \varphi(\tilde{\mathbf{x}}_0, u)M(\mathbf{x}_0, du) - \int_0^t a(\varphi(\mathbf{x}_0, u), \varphi(\tilde{\mathbf{x}}_0, u), u)\,du + \langle M(\mathbf{x}_0, t), M(\tilde{\mathbf{x}}_0, t)\rangle$$

is a martingale. This yields

$$\langle M(\mathbf{x}_0, t), M(\tilde{\mathbf{x}}_0, t)\rangle = \int_0^t a(\varphi(\mathbf{x}_0, u), \varphi(\tilde{\mathbf{x}}_0, u), u)\,du.$$

The proof is complete. □

Proof of Theorem 5.2.1

Although we have not yet shown the uniqueness of the law $Q_0^{(\mathbf{x}_0, \mathbf{y}_0)}$ the marginal law of $(F(y_0, t), Q_0^{(\mathbf{x}_0, \mathbf{y}_0)})$ is unique, since $F(\mathbf{y}_0, t)$ is a Brownian motion with local characteristic (a, b) by (21) and (24). We wish to show that $\varphi(t)$ is generated by $F(y, t)$ in the sense of the Itô integral. Then this will establish the uniqueness of the law of the pair $(\varphi(t), F(t))$ (see Theorem 4.2.9).

To this end, we will first show the existence of a probability measure $P^{(\mathbf{x}_0)}$ on $\mathscr{B}_{\mathbf{x}_0}(\hat{W}_k) \otimes \mathscr{B}(W_k)$ such that its restriction to the sub σ-field $\mathscr{B}_{\mathbf{x}_0}(\hat{W}_k) \otimes \mathscr{B}_{\mathbf{y}_0}(W_k)$ coincides with $Q_0^{(\mathbf{x}_0, \mathbf{y}_0)}$. Let $Q_\varepsilon^{(\mathbf{y}_0)}$ and $Q_0^{(\mathbf{y}_0)}$ be the restrictions of measures $Q_\varepsilon^{(\mathbf{x}_0, \mathbf{y}_0)}$ and $Q_0^{(\mathbf{x}_0, \mathbf{y}_0)}$ respectively to the sub σ-field $\{\phi, \hat{W}_k\} \otimes \mathscr{B}_{\mathbf{y}_0}(W_k)$. Then $Q_0^{(\mathbf{y}_0)}$ represents the law of $F(\mathbf{y}_0, t)$, so that it is unique as we have already remarked. This means that $\{Q_{\varepsilon_h}^{(\mathbf{y}_0)}\}$ converges to $Q_0^{(\mathbf{y}_0)}$ weakly as $\varepsilon_h \to 0$ for any $\mathbf{y}_0 \in \mathbb{R}^{nd}$. Consequently if the laws of $\varphi_{\tilde{\varepsilon}_h}(\mathbf{x}_0, t)$ converge weakly as $\tilde{\varepsilon}_h \downarrow 0$, then the laws of the pairs $(\varphi_{\tilde{\varepsilon}_h}(\mathbf{x}_0, t), F_{\tilde{\varepsilon}_h}(\mathbf{y}_0, t))$ always converge weakly for any $\mathbf{y}_0$. Now, given $\mathbf{x}_0$, $\mathbf{y}_0$, let $\varepsilon_h \downarrow 0$ be a sequence such that $\{Q_{\varepsilon_n}^{(\mathbf{x}_0,\mathbf{y}_0)}\}$ converges weakly to $Q_0^{(\mathbf{x}_0,\mathbf{y}_0)}$. Then for any $\tilde{\mathbf{y}}_0$, $\{Q_{\varepsilon_n}^{(\mathbf{x}_0,\tilde{\mathbf{y}}_0)}\}$ converges weakly to $Q_0^{(\mathbf{x}_0,\tilde{\mathbf{y}}_0)}$. The family of measures $\{Q_0^{(\mathbf{x}_0,\tilde{\mathbf{y}}_0)} : \tilde{\mathbf{y}}_0 \in \mathbb{R}^{nd}, n = 1, 2, \ldots\}$ is then consistent. Since (a, b) belongs to the class $B_{ub}^{k,\delta}$, there exists a unique measure $P_0^{(\mathbf{x}_0)}$ on $\mathscr{B}_{\mathbf{x}_0}(\hat{W}_k) \otimes \mathscr{B}(W_k)$ such that its restriction to $\mathscr{B}_{\mathbf{x}_0}(\hat{W}_k) \otimes \mathscr{B}_{\tilde{\mathbf{y}}_0}(W_k)$ coincides with $Q_0^{(\mathbf{x}_0,\tilde{\mathbf{y}}_0)}$ for any $\tilde{\mathbf{y}}_0$.

Now consider a stochastic integral based on $Y(y, t)$ of (21): $\tilde{M}(x_0, t) \equiv \int_s^t Y(\varphi(x_0, u), du)$ on the space $(\mathscr{B}_{\mathbf{x}_0}(\hat{W}_k) \otimes \mathscr{B}(W_k), P_0^{(\mathbf{x}_0)})$. We will prove that it coincides with $M(x_0, t)$ defined by (20). Since the local characteristic of $Y(y_0, t)$ is $(a(y_0, \tilde{y}_0, t), t)$ by (24), the local characteristic of $\tilde{M}(x_0, t)$ is $(a(\varphi(x_0, u), \varphi(\tilde{x}_0, u), u), u)$ i.e.

$$\langle \tilde{M}^i(x_0, t), \tilde{M}^j(\tilde{x}_0, t)\rangle = \int_0^t a^{ij}(\varphi(x_0, u), \varphi(\tilde{x}_0, u), u)\, du.$$

Also, $M^i(x_0, t)$ is a martingale whose quadratic variation is given by (22). Further the joint quadratic variation of $M^i(x_0, t)$ and $Y^j(y_0, t)$ is given by (23). Then we have

$$\langle M^i(x_0, t), \tilde{M}^j(x_0, t)\rangle = \int_0^t a^{ij}(\varphi(x_0, u), \varphi(x_0, u), u)\, du.$$

Therefore we have

$$\langle M^i(x_0, t) - \tilde{M}^i(x_0, t)\rangle = \langle M^i(x_0, t)\rangle - 2\langle M^i(x_0, t), \tilde{M}^i(x_0, t)\rangle + \langle \tilde{M}^i(x_0, t)\rangle = 0.$$

This proves $M^i(x_0, t) = \tilde{M}^i(x_0, t)$ a.s. for every $i = 1, \ldots, d$.

Now the relation $M(x_0, t) = \tilde{M}(x_0, t)$ can be written as

$$\varphi(x_0, t) = x_0 + \int_0^t F(\varphi(x_0, u), du). \tag{27}$$

Then the law of $(\varphi(\mathbf{x}_0, t), F(\mathbf{y}_0, t))$ is unique by Theorem 4.2.9. We have thus proved that the limit law $Q_0^{(\mathbf{x}_0,\mathbf{y}_0)}$ is unique.

Finally the uniqueness of the law $Q_0^{(\mathbf{x}_0,\mathbf{y}_0)}$ shows that the family of laws $\{Q_\varepsilon^{(\mathbf{x}_0,\mathbf{y}_0)}\}$ converges weakly to $Q_0^{(\mathbf{x}_0,\mathbf{y}_0)}$ as $\varepsilon \to 0$. Then the family of laws $\{Q_0^{(\mathbf{x}_0,\mathbf{y}_0)} : (\mathbf{x}_0, \mathbf{y}_0) \in \mathbb{R}^{md} \times \mathbb{R}^{nd}\}$ is consistent, so that it defines a law Q_0 on $\mathscr{A}(\hat{W}_k \times W_k)$. Further it can be extended to a measure P_0 on $\mathscr{B}(\hat{W}_k \times W_k)$ by Theorem 4.6.5, since (a, b) belongs to $B_b^{k,\delta}$. The latter assertion of the theorem has already been shown. □

Strong convergence

Theorem 5.2.8 *Let $\{F_\varepsilon(x, t)\}$ be a family of continuous C^1-semimartingales with local characteristics $\{(a_\varepsilon, b_\varepsilon)\}$ satisfying Conditions* (C.1)$_p$ *and* (C.2)$_k$ *for* *some $p > 2$ and $k \geq 1$. If $\{F_\varepsilon(x, t)\}$ converges strongly in probability as $\varepsilon \to 0$ for any x, then $\{F_\varepsilon(x, t)\}$ and $\{\varphi_\varepsilon(x, t)\}$ converge strongly in $L^{p'}$ as $\varepsilon \to 0$ for any $p' < p$. The limits $\hat{F}(x, t)$ and $\hat{\phi}(x, t)$ have modifications of continuous C-semimartingales. Further $\hat{\phi}$ is generated by $\hat{F}$ in the sense of the Itô integral.*

Proof Let $\hat{F}(x, t)$ be the strong limit of $\{F_\varepsilon(x, t)\}$. It is a family of continuous semimartingales with parameter x with local characteristic (a, b) as we have seen in Lemmas 5.2.6 and 5.2.7. Therefore it has a modification of C^k-Brownian motion, which we denote by the same symbol $\hat{F}(x, t)$. Let $\hat{\phi}$ be the stochastic flow generated by $\hat{F}$ in the sense of the Itô integral. Consider the 4-ple $(\varphi_\varepsilon, F_\varepsilon, \hat{\phi}, \hat{F})$. Let $(\mathbf{x}_0, \mathbf{y}_0) \in \mathbb{R}^{md} \times \mathbb{R}^{nd}$ and let $\hat{Q}_\varepsilon^{(\mathbf{x}_0, \mathbf{y}_0)}$ be the law of $(\varphi_\varepsilon(\mathbf{x}_0), F_\varepsilon(\mathbf{y}_0), \hat{\phi}(\mathbf{x}_0), \hat{F}(\mathbf{y}_0))$, which is defined on $\mathscr{B}_{\mathbf{x}_0}(\hat{W}_k) \otimes \mathscr{B}_{\mathbf{y}_0}(W_k) \otimes \mathscr{B}_{\mathbf{x}_0}(\hat{W}_k) \otimes \mathscr{B}_{\mathbf{y}_0}(W_k)$. Obviously the family of laws $\{\hat{Q}_\varepsilon^{(\mathbf{x}_0, \mathbf{y}_0)}\}$ is tight. Further it converges weakly. The proof is similar to the case of $\{Q_\varepsilon^{(\mathbf{x}_0, \mathbf{y}_0)}\}$. Let $\hat{Q}_0^{(\mathbf{x}_0, \mathbf{y}_0)}$ be the weak limit. Denote the element of $\hat{W}_k \times W_k \times \hat{W}_k \times W_k$ be $(\varphi, F, \hat{\phi}, \hat{F})$. We will show that $\varphi(x_0) = \hat{\phi}(x_0)$ and $F(y_0) = \hat{F}(y_0)$ holds a.s. $Q_0^{(\mathbf{x}_0, \mathbf{y}_0)}$. Since $F_\varepsilon(y_0) \to F(y_0)$ strongly we have the relation $F(y_0) = \hat{F}(y_0)$ a.s. $\hat{Q}_0^{(\mathbf{x}_0, \mathbf{y}_0)}$. Now since φ and F are related by Itô's stochastic differential equation by Theorem 5.2.1, we get $\varphi(x_0) = \hat{\phi}(x_0)$ a.s. $\hat{Q}_0^{(\mathbf{x}_0, \mathbf{y}_0)}$ by the pathwise uniqueness of the solution of the stochastic differential equation.

Now since $\{\hat{Q}_\varepsilon^{(\mathbf{x}_0, \mathbf{y}_0)}\}$ converges to $\hat{Q}_0^{(\mathbf{x}_0, \mathbf{y}_0)}$ weakly and $\sup_t |\varphi_\varepsilon(x_0, t)|$ is uniformly L^p-bounded, $\{E_\varepsilon^{(\mathbf{x}_0, \mathbf{y}_0)}[f(\varphi(\mathbf{x}_0), \hat{\phi}(\mathbf{x}_0))]\}$ converges to $E_0^{(\mathbf{x}_0, \mathbf{y}_0)}[f(\varphi(\mathbf{x}_0), \hat{\phi}(\mathbf{x}_0))]$ for any continuous function f on $V_m \times V_n$ such that $|f(\varphi, \hat{\phi})| \leq C(1 + \|\varphi\| + \|\hat{\phi}\|)^{p'}$, where $\|\varphi\| = \sup_t |\varphi(t)|$ and $p' < p$ (*cf.* Lemma 5.2.4). Setting $f(\varphi, \hat{\phi}) = \|\varphi - \hat{\phi}\|^{p'}$, we obtain

$$E[\|\varphi_\varepsilon(\mathbf{x}_0) - \hat{\phi}(\mathbf{x}_0)\|^{p'}] = E_\varepsilon^{(\mathbf{x}_0, \mathbf{y}_0)}[\|\varphi(x_0) - \hat{\phi}(x_0)\|^{p'}]$$

$$\underset{\varepsilon \to 0}{\longrightarrow} E_0^{(\mathbf{x}_0, \mathbf{y}_0)}[\|\varphi(x_0) - \hat{\phi}(x_0)\|^{p'}] = 0.$$

Therefore $\{\varphi_\varepsilon(x_0)\}$ converges to $\hat{\phi}(x_0)$ strongly in $L^{p'}$ if $p' < p$. □

Corollary 5.2.9 *Assume Conditions* (C.1)$_p$ *and* (C.2)$_k$ *for $p > 2$ and $k \geq 1$ with $a = 0$. Then both $\{F_\varepsilon(x, t)\}$ and $\{\varphi_\varepsilon(x, t)\}$ converge strongly in $L^{p'}$ for $p' < p$ as $\varepsilon \to 0$.*

Proof Let $Q_0^{(\mathbf{x}_0, \mathbf{y}_0)}$ be the weak limit of $\{Q_\varepsilon^{(\mathbf{x}_0, \mathbf{y}_0)}\}$. Then $F(y_0, t)$ is equal to $\int_0^t b(y_0, u)\, du$ a.s. with respect to $Q_0^{(\mathbf{x}_0, \mathbf{y}_0)}$. This means that $Q_0^{(\mathbf{x}_0, \mathbf{y}_0)}$ is concentrated at the single curve $(\varphi^0(t), \int_0^t b(y_0, u)\, du)$ where $\varphi^0(t)$ is determined by the ordinary differential equation $d\varphi^0(t)/dt = b(\varphi^0(t), t)$ with the initial condition $\varphi^0(0) = x_0$. Then we have

$$E\left[\left\|F_\varepsilon(y_0)-\int b(y_0,u)\,du\right\|^{p'}\right]$$
$$=E_\varepsilon^{(x_0,y_0)}\left[\left\|F(y_0)-\int b(y_0,u)\,du\right\|^{p'}\right]$$
$$\xrightarrow[\varepsilon\to 0]{} E_0^{(x_0,y_0)}\left[\left\|F(y_0)-\int b(y_0,u)\,du\right\|^{p'}\right]=0$$

for any $p'<p$. Therefore $F_\varepsilon(y_0,t)$ converges to $\int_0^t b(y_0,u)\,du$ strongly in $L^{p'}$. The proof is complete. □

Exercise 5.2.10 *(Averaging problem. Khasminskii* [68]) Consider a family of stochastic ordinary differential equations with parameter $\varepsilon>0$:

$$\frac{dx}{dt}=f\left(x,\frac{t}{\varepsilon},\omega\right).$$

Suppose that there exists a positive constant such that $|D^\alpha f(x,t)|\le C$ holds for any $x,y\in\mathbb{R}^d$, $t>0$ and $|\alpha|\le 1$ and further there exists a deterministic function $\bar f(x)$ such that

$$\sup_{t_0} E\left[\left|\frac{1}{T}\int_{t_0}^{t_0+T} D^\alpha f(x,u)\,du-D^\alpha\bar f(x)\right|\right]\xrightarrow[T\to 0]{}0$$

holds for any x and $|\alpha|\le 1$, Let $(\varphi_\varepsilon)_t(x_0)$ be the solution of the stochastic ordinary differential equation starting at x_0 at time 0. Show that $(\varphi_\varepsilon)_t(x_0)$ converges to the deterministic curve $(\varphi_0)_t(x_0)$ uniformly on compact sets and it satisfies the equation $dx/dt=\bar f(x,t)$.

5.3 Convergence as diffusions II

Statements of theorems

In the previous section we studied the weak convergence of the pairs of the stochastic flows φ_ε and their Itô's infinitesimal generators F_ε as $\varepsilon\to 0$. A basic assumption was that both of the martingale parts Y_ε and the absolute continuous parts $\int b_\varepsilon(\cdot,u)\,du$ of F_ε converge to a Brownian motion Y with mean 0 (Gaussian martingale) and $\int b(\cdot,u)\,du$, respectively. In this section we shall study the case where the processes $\int b_\varepsilon(\cdot,u)\,du$ do not converge to a process like $\int b(\cdot,u)\,du$ but converge to another Brownian motion which is independent of Y.

We shall formulate the problem precisely. We assume that b_ε is decomposed to the sum of two measurable random fields $\bar b_\varepsilon(x,t)$ and $\tilde b_\varepsilon(x,t)$, k-times differentiable in x, such that $\bar b_\varepsilon(x,t)$ is $(\mathscr{F}_t^\varepsilon)$-adapted and $\tilde b_\varepsilon(x,t)$ is

$(\mathscr{F}_0^\varepsilon)$-adapted for any t. Further, the family of triples $\{(a_\varepsilon, \bar{b}_\varepsilon, \tilde{b}_\varepsilon)\}$ satisfies the following properties.

(a) The family of pairs $\{(a_\varepsilon, \bar{b}_\varepsilon)\}$ satisfies Conditions $(C.1)_p$ and $(C.2)_k$ for some $p > 2$ and $k \geq 1$. We denote the limit of $E[\int_s^t \bar{b}_\varepsilon(x, u)\, du | \mathscr{F}_s^\varepsilon]$ by $\int_s^t \bar{b}(x, u)\, du$.

(b) Set

$$\mathscr{G}_s^\varepsilon \equiv \sigma(Y_\varepsilon(x, r), \bar{b}_\varepsilon(x, r), \tilde{b}_\varepsilon(x, r) : x \in \mathbb{R}^d, r \leq s), \tag{1}$$

and

$$K_\varepsilon^{\alpha\beta}(x, y, s, t) \equiv \left| E\left[\int_s^t D_x^\alpha \tilde{b}_\varepsilon(x, u)\, du \,\middle|\, \mathscr{G}_s^\varepsilon\right]\right| (1 + |D_y^\beta \tilde{b}_\varepsilon(y, s)| + |D_y^\beta \bar{b}_\varepsilon(y, s)|), \tag{2}$$

$$L_\varepsilon^{\alpha\beta}(x, y, s, t) \equiv \left| E\left[\int_s^t D_x^\alpha \tilde{b}_\varepsilon(x, u)\, du \,\middle|\, \mathscr{G}_s^\varepsilon\right]\right| \|D_x^\beta D_y^\beta a_\varepsilon(x, y, s)\|. \tag{3}$$

These random fields have modifications which are continuous in x, y.

Condition $(C.3)_p$ *There exists a positive constant* K *such that*

$$E\left[\sup_{x,y} K_\varepsilon^{\alpha\beta}(x, y, s, t)^p\right]^{1/p} \leq K, \qquad \text{for } |\alpha| \leq 2, |\beta| \leq 1, \tag{4}$$

$$E\left[\sup_{x,y} L_\varepsilon^{\alpha\beta}(x, y, s, t)^p\right]^{1/p} \leq K, \qquad \text{for } |\alpha| \leq 3, |\beta| \leq 1. \tag{5}$$

hold for any s, t *and* ε. □

Further, set

$$A_\varepsilon^{ij}(x, y, t, s) = E\left[\int_s^t \tilde{b}_\varepsilon^i(x, u)\, du \,\middle|\, \mathscr{G}_s^\varepsilon\right] b_\varepsilon^j(y, s), \tag{6}$$

$$c_\varepsilon^i(x, t, s) = \sum_j \frac{\partial A_\varepsilon^{ij}}{\partial x^j}(x, y, t, s)|_{y=x}. \tag{7}$$

Condition $(C.4)_k$ *There exist deterministic functions* $(A^{ij}(x, y, t), c^i(x, t))$ *belonging to the class* $\tilde{C}_b^{k,\delta}$ *and* $C_b^{k,\delta}$ *respectively for some* $\delta > 0$ *satisfying the following properties: for each* $s < t$ *and* i, j, *the convergences*

$$E\left[\int_s^t A_\varepsilon^{ij}(x, y, t, u)\, du \,\middle|\, \mathscr{G}_s^\varepsilon\right] \to \int_s^t A^{ij}(x, y, u)\, du, \tag{8}$$

$$E\left[\int_s^t c_\varepsilon^i(x, t, u)\, du \,\middle|\, \mathscr{G}_s^\varepsilon\right] \to \int_s^t c^i(x, u)\, du, \tag{9}$$

$$\left|E\left[\int_s^t D^\alpha \tilde{b}_\varepsilon(x,u)\,du\,\middle|\,\mathscr{G}_s^\varepsilon\right]\right|(1+\|a_\varepsilon(y,y,s)\|+|\bar{b}_\varepsilon(y,s)|)\to 0, \quad \text{for } |\alpha|\le 2 \tag{10}$$

hold uniformly on compact sets of spatial variables in L^1. □

Now let Q_ε be the law of $(\varphi_\varepsilon, F_\varepsilon)$ as diffusion defined on the algebra $\mathscr{A}(\hat{W}_j \times W_j)$. The weak convergence as diffusions is stated as follows.

Theorem 5.3.1 *Assume that the family of local characteristics* $\{(a_\varepsilon, \bar{b}_\varepsilon, \tilde{b}_\varepsilon)\}$ *of* $\{F_\varepsilon\}$ *satisfies Conditions* $(C.1)_p$, $(C.2)_k$, $(C.3)_p$, $(C.4)_k$ *for some* $p > 3$ *and* $k \ge 1$. *Then the family of laws* $\{Q_\varepsilon\}$ *converges weakly as diffusions as* $\varepsilon \to 0$. *The limit law* Q_0 *can be extended uniquely to a probability measure* P_0 *on* $\mathscr{B}(\hat{W}_k \times W_k)$. *Further, the triple* $(\varphi(t), F(t), P_0)$ *satisfies these properties:*

(i) $F(y,t)$ *is a* C^k*-valued Brownian motion with local characteristic* $(a + \bar{A}, \bar{b})$, *where*

$$\bar{A}^{ij}(x,y,t) = A^{ij}(x,y,t) + A^{ji}(y,x,t). \tag{11}$$

(ii) $\varphi(x,t)$ *is a Brownian flow of* C^k*-diffeomorphisms generated by* $F(x,t) + \int_0^t c(x,u)\,du$ *in the sense of the Itô integral. Its infinitesimal mean is* $\bar{b} + c$ *and its infinitesimal covariance is* $a + \bar{A}$. □

In the case where $a_\varepsilon \equiv 0$, F_ε is given by $\int b_\varepsilon(\cdot, u)\,du$. Then equation (1) can be regarded as a stochastic ordinary differential equation

$$\frac{d\varphi_t}{dt} = \bar{b}_\varepsilon(\varphi_t, t) + \tilde{b}_\varepsilon(\varphi_t, t). \tag{12}$$

The assumptions of the theorem can be relaxed slightly in this case. In fact we have the following.

Theorem 5.3.1′ *Assume that the family* $\{\bar{b}_\varepsilon, \tilde{b}_\varepsilon\}$ *of coefficients of stochastic ordinary differential equations* (12) *satisfy Conditions* $(C.1)_p$, $(C.2)_k$, $(C.3)_p$ *and* $(C.4)_k$ *for some* $p > 2$ *and* $k \ge 1$. *Then the assertion of Theorem* 5.3.1 *is valid for* $a \equiv 0$. □

Before we proceed further, we will make a preliminary observation about the assumptions of the theorem and what follows from them, since they look complicated. Let Y_ε be the martingale part of F_ε. Its quadratic variation is given by $\int a_\varepsilon(x,y,u)\,du$. It converges to a deterministic function $\int a(x,y,u)\,du$ in the sense of Condition $(C.2)_k$. It was shown in Section 5.2 that $\{Y_\varepsilon\}$ then converges weakly to a Brownian motion with mean 0 and covariance $\int a(x, y, u)\,du$. Next, consider the bounded variation part. It is decomposed to the sum of the following two: $\bar{B}_\varepsilon \equiv \int \bar{b}_\varepsilon(\cdot, u)\,du$, $\tilde{B}_\varepsilon \equiv \int \tilde{b}_\varepsilon(\cdot, u)\,du$. Then $\{\bar{B}_\varepsilon\}$ converges weakly to $\bar{B} = \int \bar{b}(\cdot, r)\,dr$ as $\varepsilon \to 0$ as we have seen in the previous section. Concerning $\{\tilde{B}_\varepsilon\}$, we have

$$E[\tilde{B}_\varepsilon(x, t) - \tilde{B}_\varepsilon(x, s) | \mathscr{G}_s^\varepsilon] \to 0, \tag{13}$$

as $\varepsilon \to 0$ by (10). Also,

$$\begin{aligned} &E[(\tilde{B}_\varepsilon^i(x, t) - \tilde{B}_\varepsilon^i(x, s))(\tilde{B}_\varepsilon^j(y, t) - \tilde{B}_\varepsilon^j(y, s)) | \mathscr{G}_s^\varepsilon] \\ &\quad = E\left[\int_s^t \{\tilde{A}_\varepsilon^{ij}(x, y, t, u) + \tilde{A}_\varepsilon^{ji}(y, x, t, u)\}\,du \,\Big|\, \mathscr{G}_s^\varepsilon\right] \end{aligned} \tag{14}$$

where

$$\tilde{A}_\varepsilon^{ij}(x, y, t, s) = E\left[\int_s^t \tilde{b}_\varepsilon^i(x, u)\,du \,\Big|\, \mathscr{G}_s^\varepsilon\right] \tilde{b}_\varepsilon^j(y, s). \tag{15}$$

Observe that convergences (8) and (10) are equivalent to (10) and (16):

$$E\left[\int_s^t \tilde{A}_\varepsilon^{ij}(x, y, t, u)\,du \,\Big|\, \mathscr{G}_s^\varepsilon\right] \to \int_s^t A^{ij}(x, y, u)\,du. \tag{16}$$

Then (14) converges to $\int_s^t \bar{A}^{ij}(x, y, u)\,du$. Therefore $\{\tilde{B}^\varepsilon(x, t)\}$ will converge weakly as $\varepsilon \to 0$ to a martingale with deterministic quadratic variation $\int_0^t \bar{A}^{ij}(x, y, u)\,du$. This shows that the weak limit is a Brownian motion. Consequently, $\{F_\varepsilon(x, t)\}$ converges weakly to a Brownian motion $F(x, t)$ with mean $\int_0^t \bar{b}(x, u)\,du$ and covariance $\int_0^t (a + \bar{A})(x, y, u)\,du$. However, the existence of the limit (8) is not sufficient for the existence of the limit of flows $\{\varphi_\varepsilon(t)\}$. Convergence (9) will be needed for this problem. Further the weak limit $\varphi(t)$ is not generated by $F(x, t)$. A correction term $c(x, t)$ is needed for the generator of the limit diffusion process.

Now condition (C.3)$_p$ is needed for the tightness of $\{Q_\varepsilon\}$. Note that the $\{\tilde{B}_\varepsilon(x, t)\}$ satisfy (13) but their time derivatives $\{\tilde{b}_\varepsilon(x, t)\}$ diverge (or converge to a white noise, loosely speaking). Condition (C.3)$_p$ indicates that the products of these two are balanced and are uniformly L^p-bounded.

The arguments of the proofs of Theorem's 5.3.1 and 5.3.1′ are quite close to those of Theorem 5.2.1. However, the required computations are more complicated since we have to deal with the term $\tilde{b}_\varepsilon(x, t)$. The proof will be divided into two steps. In the first step, we will prove the tightness of measures $\{Q_\varepsilon\}$ by giving the uniform L^p-estimates of $\{F_\varepsilon(t) - F_\varepsilon(s)\}$ and $\{\varphi_\varepsilon(t) - \varphi_\varepsilon(s)\}$ (see Lemmas 5.3.2 and 5.3.3). In the second step, we will characterize any limit measure of $\{Q_\varepsilon\}$ by its semimartingale properties.

This will complete the proof of the theorem similarly to the proof of Theorem 5.2.1.

Uniform L^p-estimates

Lemma 5.3.2 Assume Conditions (C.1)$_p$ *and* (C.3)$_p$ *for some* $p > 2$. *There exists a positive constant C such that*

$$E[|F_\varepsilon(x,t) - F_\varepsilon(x,s)|^p] \le C|t-s|^{2-(2/p)} \tag{17}$$

holds for any s, t, x and ε.

Proof We shall only consider the case where $d = 1$. Let $\bar{F}_\varepsilon(x,t)$ be the sum of $Y_\varepsilon(x,t)$ and $\int_0^t \bar{b}_\varepsilon(x,u)\,du$. Then the assertion of the lemma is valid for $\bar{F}_\varepsilon(x,t)$ as we have seen in Lemma 5.2.2. We shall consider the term $\int_0^t \tilde{b}_\varepsilon(x,\tau)\,d\tau$. We suppress x for simplicity. Since $(|x|^p)' = p\,\mathrm{sign}(x)\cdot|x|^{p-1}$ and $(\mathrm{sign}(x)\cdot|x|^{p-1})' = (p-1)|x|^{p-2}$, we have by integration by parts

$$\left|\int_s^t \tilde{b}_\varepsilon(\tau)\,d\tau\right|^p = p(p-1)\int_s^t\left(\int_\sigma^t \tilde{b}_\varepsilon(\tau)\,d\tau\right)\tilde{b}_\varepsilon(\sigma)\left|\int_s^\sigma \tilde{b}_\varepsilon(u)\,du\right|^{p-2} d\sigma.$$

Therefore we have

$$E\left[\left|\int_s^t \tilde{b}_\varepsilon(\tau)\,d\tau\right|^p\right] = p(p-1)E\left[\int_s^t \tilde{A}_\varepsilon(t,\sigma)\left|\int_s^\sigma \tilde{b}_\varepsilon(u)\,du\right|^{p-2} d\sigma\right].$$

By Condition (C.3)$_p$, the above is bounded by

$$p(p-1)K\int_s^t d\sigma\, E\left[\left|\int_s^\sigma \tilde{b}_\varepsilon(u)\,du\right|^p\right]^{1-(2/p)}.$$

Therefore there exists a positive constant $C = C(p, K)$ such that the above is bounded by $C|t-s|^{2-2/p}$ for any s, t similarly to the proof of Lemma 5.2.2. The proof is complete. □

Lemma 5.3.3

(i) *Assume Conditions* (C.1)$_p$ *and* (C.3)$_p$ *for some* $p > 3$. *Then there exists a positive constant C such that*

$$E[|\varphi_\varepsilon(x,t) - \varphi_\varepsilon(x,s)|^p] \le C|t-s|^{2-(3/p)} \tag{18}$$

holds for any s, t, x and ε.

(ii) *Assume* $a_\varepsilon \equiv 0$ *and Conditions* (C.1)$_p$, (C.3)$_p$ *for some* $p > 2$. *Then there exists a positive constant C such that*

$$E[|\varphi_\varepsilon(x,t) - \varphi_\varepsilon(x,s)|^p] \le C|t-s|^{2-(2/p)} \tag{19}$$

holds for any s, t, x and *ε.*

Proof We shall only prove the case where $d = 1$. We first consider case (i). Note that

$$\varphi_\varepsilon(t) - \varphi_\varepsilon(s) = \int_s^t \bar{b}_\varepsilon(\varphi_\varepsilon(u), u)\, du + \int_s^t \tilde{b}_\varepsilon(\varphi_\varepsilon(u), u)\, du + \int_s^t Y_\varepsilon(\varphi_\varepsilon(u), du).$$

By Itô's formula, we have

$$\begin{aligned}|\varphi_\varepsilon(t) - \varphi_\varepsilon(s)|^p = p & \int_s^t \bar{b}_\varepsilon(u)|\varphi_\varepsilon(u) - \varphi_\varepsilon(s)|^{p-1} \operatorname{sign}(u)\, du \\ &+ p\int_s^t \tilde{b}_\varepsilon(u)|\varphi_\varepsilon(u) - \varphi_\varepsilon(s)|^{p-1} \operatorname{sign}(u)\, du \\ &+ p\int_s^t |\varphi_\varepsilon(u) - \varphi_\varepsilon(s)|^{p-1} \operatorname{sign}(u)\, Y_\varepsilon(\varphi_\varepsilon(u), du) \\ &+ \frac{1}{2}p(p-1)\int_s^t a_\varepsilon(u)|\varphi_\varepsilon(u) - \varphi_\varepsilon(s)|^{p-2}\, du,\end{aligned}$$

where $\bar{b}_\varepsilon(u) = \bar{b}_\varepsilon(\varphi_\varepsilon(u), u)$, $a_\varepsilon(u) = a_\varepsilon(\varphi_\varepsilon(u), \varphi_\varepsilon(u), u)$ etc. and $\operatorname{sign}(u) = \operatorname{sign}(\varphi_\varepsilon(u) - \varphi_\varepsilon(s))$. Denote the ith term of the right hand side by I_i^ε, $i = 1, \dots, 4$. We shall estimate the expectations of the I_i^ε.

For the estimates of $E[I_i^\varepsilon]$, $i = 1, 4$ apply Hölder's inequality and Condition $(C.1)_p$. Then similarly to the proof of Lemma 5.2.3, we have

$$|E[I_1^\varepsilon]| \le pK \int_s^t E[|\varphi_\varepsilon(u) - \varphi_\varepsilon(s)|^p]^{1-(1/p)}\, du,$$

$$|E[I_4^\varepsilon]| \le \frac{1}{2}p(p-1)K \int_s^t E[|\varphi_\varepsilon(u) - \varphi_\varepsilon(s)|^p]^{1-(2/p)}\, du.$$

Since Y^ε is a martingale, we have $E[I_3^\varepsilon] = 0$.

We will calculate I_2^ε. Since $\tilde{b}(x, u)$ is $\mathscr{F}_0^\varepsilon$-measurable for each fixed u, we can apply Itô's formula for the composed process

$$\tilde{b}_\varepsilon(\varphi_\varepsilon(\sigma), u)|\varphi_\varepsilon(\sigma) - \varphi_\varepsilon(s)|^{p-1} \operatorname{sign}(\sigma), \qquad \sigma \in [s, u].$$

The integrand of I_2^ε is then represented by

$$\begin{aligned}&\tilde{b}_\varepsilon(\varphi_s(u), u)|\varphi_\varepsilon(u) - \varphi_\varepsilon(s)|^{p-1} \operatorname{sign}(u) \\ &= \int_s^u \left\{\frac{\partial \tilde{b}_\varepsilon}{\partial x}(\varphi_\varepsilon(\sigma), u) b_\varepsilon(\sigma) + \frac{1}{2}\frac{\partial^2 \tilde{b}_\varepsilon}{\partial x^2}(\varphi_\varepsilon(\sigma), u) a_\varepsilon(\sigma)\right\}|\varphi_\varepsilon(\sigma) - \varphi_\varepsilon(s)|^{p-1} \operatorname{sign}(\sigma)\, d\sigma \\ &\quad + (p-1)\int_s^u \left\{\tilde{b}_\varepsilon(\varphi_\varepsilon(\sigma), u) b_\varepsilon(\sigma) + \frac{\partial \tilde{b}_\varepsilon}{\partial x}(\varphi_\varepsilon(\sigma), u) a_\varepsilon(\sigma)\right\}|\varphi_\varepsilon(\sigma) - \varphi_\varepsilon(s)|^{p-2}\, d\sigma \\ &\quad + \frac{1}{2}(p-1)(p-2)\int_s^u \tilde{b}_\varepsilon(\varphi_\varepsilon(\sigma), u) a_\varepsilon(\sigma)|\varphi_\varepsilon(\sigma) - \varphi_\varepsilon(s)|^{p-3} \operatorname{sign}(\sigma)\, d\sigma \\ &\quad + \text{martingale with mean } 0.\end{aligned}$$

Integrate each term of the above with respect to u on the interval $[s, t]$ and then take the expectation. Setting

$$e_\varepsilon(x, y, t, s) \equiv E\left[\int_s^t \tilde{b}_\varepsilon(x, u)\, du \,\middle|\, \mathscr{G}_s^\varepsilon\right] a_\varepsilon(y, y, s), \tag{20}$$

$E[I_2^\varepsilon]$ is written as

$$p\int_s^t E\left[\left\{c_\varepsilon(t, \sigma) + \frac{1}{2}\frac{\partial^2 e_\varepsilon}{\partial x^2}(t, \sigma)\right\}|\varphi_\varepsilon(\sigma) - \varphi_\varepsilon(s)|^{p-1}\operatorname{sign}(\sigma)\right] d\sigma$$

$$+ p(p-1)\int_s^t E\left[\left\{A_\varepsilon(t, \sigma) + \frac{\partial e_\varepsilon}{\partial x}(t, \sigma)\right\}|\varphi_\varepsilon(\sigma) - \varphi_\varepsilon(s)|^{p-2}\right] d\sigma$$

$$+ \frac{1}{2}p(p-1)(p-2)\int_s^t E[e_\varepsilon(t, \sigma)|\varphi_\varepsilon(\sigma) - \varphi_\varepsilon(s)|^{p-3}\operatorname{sign}(\sigma)]\, d\sigma, \tag{21}$$

where $c_\varepsilon(t, \sigma) = c_\varepsilon(\varphi_\varepsilon(\sigma), t, \sigma)$, $(\partial e_\varepsilon/\partial x)(t, \sigma) = (\partial e_\varepsilon/\partial x)(\varphi_\varepsilon(\sigma), \varphi_\varepsilon(\sigma), t, \sigma)$ etc. Apply Hölder's inequality and Condition $(C.3)_p$. Then there exists a positive constant $C_1 = C_1(p, K)$ such that

$$|E[I_2^\varepsilon]| \le C_1\int_s^t \Big\{E[|\varphi_\varepsilon(\sigma) - \varphi_\varepsilon(s)|^p]^{1-(1/p)}$$

$$+ E[|\varphi_\varepsilon(\sigma) - \varphi_\varepsilon(s)|^p]^{1-(2/p)} + E[|\varphi_\varepsilon(\sigma) - \varphi_\varepsilon(s)|^p]^{1-(3/p)}\Big\}\, d\sigma. \tag{22}$$

We now sum these estimates for $E[I_i^\varepsilon]$, $i = 1, \ldots, 4$. Then we find that $E[|\varphi_\varepsilon(t) - \varphi_\varepsilon(s)|^p]$ is bounded by the same quantity as the right hand side of (22) with a different constant C_1'. This implies inequality (18), by a similar argument to that in Lemma 5.2.2.

Now if $a^\varepsilon \equiv 0$, the last term of (21) is 0. Then the term $E[|\varphi_\varepsilon(t) - \varphi_\varepsilon(s)|^p]^{1-(3/p)}$ in (22) vanishes and we get (19). The proof is complete. □

Characterization of the limit measure

Let $(\mathbf{x}_0, \mathbf{y}_0)$ be an arbitrary point in $\mathbb{R}^{md} \times \mathbb{R}^{nd}$. Let $(\varphi_\varepsilon(\mathbf{x}_0), F_\varepsilon(\mathbf{y}_0))$ be the $m + n$-point motions and let $Q_\varepsilon^{(\mathbf{x}_0, \mathbf{y}_0)}$ be their laws. By the above two lemmas, the family of measures $\{Q_\varepsilon^{(\mathbf{x}_0, \mathbf{y}_0)}\}$ is tight. Then there exists a sequence $\varepsilon_h \downarrow 0$ such that $\{Q_{\varepsilon_h}^{(\mathbf{x}_0, \mathbf{y}_0)}, h = 1, 2, \ldots\}$ converges weakly. Denote the limit by $Q_0^{(\mathbf{x}_0, \mathbf{y}_0)}$. We shall characterize the limit measure by its semimartingale properties. The following Lemmas 5.3.4 and 5.3.5 correspond to Lemmas 5.2.6 and 5.2.7. However, note that $\varphi(\mathbf{x}_0)$ has an additional drift term $\int_0^t c(\varphi(\mathbf{x}_0, u), u)\, du$ compared to Lemma 5.2.6.

Lemma 5.3.4 *The following are L^2-martingales with respect to the measure $Q_0^{(\mathbf{x}_0, \mathbf{y}_0)}$.*

$$Y(y_0, t) \equiv F(y_0, t) - \int_0^t \bar{b}(y_0, u)\, du, \tag{23}$$

$$M(x_0, t) \equiv \varphi(x_0, t) - x_0 - \int_0^t \{\bar{b}(\varphi(x_0, u), u) + c(\varphi(x_0, u), u)\}\, du, \tag{24}$$

where y_0 and x_0 are components of $\mathbf{y}_0$ and $\mathbf{x}_0$, respectively.

Proof The martingale property of $Y(y_0, t)$ can be proved similarly to the proof of Lemma 5.2.6. We shall prove the martingale property of $M(x_0, t)$. Take a sequence $\varepsilon_h \downarrow 0$ such that $\{Q_{\varepsilon_h}^{(\mathbf{x}_0, \mathbf{y}_0)}\}$ converges to $Q_0^{(\mathbf{x}_0, \mathbf{y}_0)}$ weakly. For each ε, the equality

$$E\left[\left\{\varphi_\varepsilon(t) - \varphi_\varepsilon(s) - \int_s^t \bar{b}_\varepsilon(\varphi_\varepsilon(u), u)\, du - \int_s^t \tilde{b}_\varepsilon(\varphi_\varepsilon(u), u)\, du\right\} \Phi^\varepsilon\right] = 0$$

holds where Φ^ε is defined in (12) of Section 5.2. Now $\varphi_\varepsilon(t) - \varphi_\varepsilon(s)$ converges to $\varphi(t) - \varphi(s)$ weakly $(\mathscr{B}_{\mathbf{x}_0, \mathbf{y}_0}(s))$ as $\varepsilon = \varepsilon_h \to 0$. By Conditions $(\mathrm{C}.1)_p$, $(\mathrm{C}.2)_k$ and Lemma 5.2.5, $\{\int_s^t \bar{b}_\varepsilon(\varphi_\varepsilon(u), u)\, du\}$ converges to $\int_s^t \bar{b}(\varphi(u), u)\, du$ weakly $(\mathscr{B}_{\mathbf{x}_0, \mathbf{y}_0}(s))$ as $\varepsilon = \varepsilon_h \to 0$. We shall consider the term $\int_s^t \tilde{b}_\varepsilon(\varphi_\varepsilon(u), u)\, du$. By Itô's formula, we have

$$\begin{aligned}
\int_s^t & \tilde{b}_\varepsilon(\varphi_\varepsilon(u), u)\, du \\
&= \int_s^t \tilde{b}_\varepsilon(\varphi_\varepsilon(s), u)\, du + \int_s^t \left\{\int_s^u \frac{\partial \tilde{b}_\varepsilon}{\partial x}(\varphi_\varepsilon(\sigma), u) b_\varepsilon(\varphi_\varepsilon(\sigma), \sigma)\, d\sigma\right\} du \\
&\quad + \frac{1}{2}\int_s^t \left\{\int_s^u \frac{\partial^2 \tilde{b}_\varepsilon}{\partial x^2}(\varphi_\varepsilon(\sigma), u) a_\varepsilon(\varphi_\varepsilon(\sigma), \varphi_\varepsilon(\sigma), \sigma)\, d\sigma\right\} du \\
&\quad + (\mathscr{G}_t^\varepsilon)\text{-martingale with mean } 0.
\end{aligned}$$

Denote the ith term of the right hand side by J_i^ε, $i = 1, \ldots, 4$. By Condition $(\mathrm{C}.4)_k$ $J_1^{\varepsilon_h} \to 0$ weakly $(\mathscr{B}_{\mathbf{x}_0, \mathbf{y}_0}(s))$. Also

$$E[J_2^\varepsilon | \mathscr{G}_s^\varepsilon] = E\left[\int_s^t c_\varepsilon(\varphi_\varepsilon(\sigma), t, \sigma)\, d\sigma \,\middle|\, \mathscr{G}_s^\varepsilon\right] \to \int_s^t c(\varphi(\sigma), \sigma)\, d\sigma$$

weakly $(\mathscr{B}_{\mathbf{x}_0, \mathbf{y}_0}(s))$ as $\varepsilon = \varepsilon_h \to 0$, since c_ε satisfies the conditions of Lemma 5.2.5. Also, from Condition $(\mathrm{C}.3)_p$ and (10) we get

$$E[J_3^\varepsilon | \mathscr{G}_s^\varepsilon] = E\left[\int_s^t \frac{\partial^2 e_\varepsilon}{\partial x^2}(\varphi_\varepsilon(\sigma), \varphi_\varepsilon(\sigma), t, \sigma)\, d\sigma \,\middle|\, \mathscr{G}_s^\varepsilon\right] \to 0$$

in L^1 as $\varepsilon = \varepsilon_h \to 0$. Finally we have $E[J_4^\varepsilon | \mathscr{G}_s^\varepsilon] = 0$ since J_4^ε is a martingale. We have thus shown that $\{\int_s^t \tilde{b}_\varepsilon(\varphi_\varepsilon(\sigma), \sigma)\, d\sigma\}$ converges to $\int_s^t c(\varphi(\sigma), \sigma)\, d\sigma$

weakly ($\mathscr{B}_{\mathbf{x}_0,\mathbf{y}_0}(s)$) as $\varepsilon = \varepsilon_h \to 0$. Therefore we have

$$E_0^{(\mathbf{x}_0,\mathbf{y}_0)}\left[\left\{\varphi(t) - \varphi(s) - \int_s^t \bar{b}(\varphi(u), u)\,du - \int_s^t c(\varphi(u), u)\,du\right\}\Phi\right] = 0.$$

This proves that (24) is a martingale with respect to $Q_0^{(\mathbf{x}_0,\mathbf{y}_0)}$. The proof is complete. □

Lemma 5.3.5 *With respect to the measure $Q_0^{(\mathbf{x}_0,\mathbf{y}_0)}$ we have*

$$\langle M^i(x_0, t), M^j(\tilde{x}_0, t)\rangle = \int_0^t (a^{ij} + \bar{A}^{ij})(\varphi(x_0, u), \varphi(\tilde{x}_0, u), u)\,du, \quad (25)$$

$$\langle M^i(x_0, t), Y^j(y_0, t)\rangle = \int_0^t (a^{ij} + \bar{A}^{ij})(\varphi(x_0, u), y_0, u)\,du, \quad (26)$$

$$\langle Y^i(y_0, t), Y^j(\tilde{y}_0, t)\rangle = \int_0^t (a^{ij} + \bar{A}^{ij})(y_0, \tilde{y}_0, u)\,du, \quad (27)$$

where x_0, $\tilde{x}_0$ are components of $\mathbf{x}_0$, and y_0, $\tilde{y}_0$ are components of $\mathbf{y}_0$.

Proof We will prove the first assertion (25). The others can be proved similarly. Again we only consider the case where $d = 1$. By Itô's formula,

$$\varphi_\varepsilon(x_0, t)\varphi_\varepsilon(\tilde{x}_0, t) - \varphi_\varepsilon(x_0, s)\varphi_\varepsilon(\tilde{x}_0, s) - \int_s^t \varphi_\varepsilon(x_0, u)b_\varepsilon(\varphi_\varepsilon(\tilde{x}_0, u), u)\,du$$
$$- \int_s^t \varphi_\varepsilon(\tilde{x}_0, u)b_\varepsilon(\varphi_\varepsilon(x_0, u), u)\,du - \int_s^t a_\varepsilon(\varphi_\varepsilon(x_0, u), \varphi_\varepsilon(\tilde{x}_0, u), u)\,du$$

is a $(\mathscr{G}_t^\varepsilon)$-martingale with mean 0. The first and fourth terms converge weakly with respect to $\mathscr{B}_{\mathbf{x}_0,\mathbf{y}_0}(s)$-conditional expectations as $\varepsilon = \varepsilon_h \to 0$. We have also

$$\int_s^t \varphi_\varepsilon(x_0, u)\bar{b}_\varepsilon(\varphi_\varepsilon(\tilde{x}_0, u), u)\,du \to \int_s^t \varphi(x_0, u)\bar{b}(\varphi(\tilde{x}_0, u), u)\,du$$

as $\varepsilon = \varepsilon_h \to 0$. We shall prove the following weak convergence ($\mathscr{B}_{\mathbf{x}_0,\mathbf{y}_0}(s)$).

$$\int_s^t \varphi_\varepsilon(x_0, u)\tilde{b}_\varepsilon(\varphi_\varepsilon(\tilde{x}_0, u), u)\,du$$
$$\to \int_s^t \varphi(x_0, u)c(\varphi(\tilde{x}_0, u), u)\,du + \int_s^t A(\varphi(x_0, u), \varphi(\tilde{x}_0, u), u)\,du \quad (28)$$

as $\varepsilon = \varepsilon_h \to 0$. By Itô's formula, the integrand is written as

$$\varphi_\varepsilon(x_0, u)\tilde{b}_\varepsilon(\varphi_\varepsilon(\tilde{x}_0, u), u)$$

$$= \varphi_\varepsilon(x_0, s)\tilde{b}_\varepsilon(\varphi_\varepsilon(\tilde{x}_0, s), u) + \int_s^u \varphi_\varepsilon(x_0, \sigma)\frac{\partial \tilde{b}_\varepsilon}{\partial x}(\varphi_\varepsilon(\tilde{x}_0, \sigma), u) b_\varepsilon(\varphi_\varepsilon(\tilde{x}_0, \sigma), \sigma)\, d\sigma$$

$$+ \int_s^u b_\varepsilon(\varphi_\varepsilon(x_0, \sigma), \sigma)\tilde{b}_\varepsilon(\varphi_\varepsilon(\tilde{x}_0, \sigma), u)\, d\sigma$$

$$+ \int_s^u \frac{\partial \tilde{b}_\varepsilon}{\partial x}(\varphi_\varepsilon(\tilde{x}_0, \sigma), u) a_\varepsilon(\varphi_\varepsilon(x_0, \sigma), \varphi_\varepsilon(\tilde{x}_0, \sigma), \sigma)\, d\sigma$$

$$+ \frac{1}{2}\int_s^u \varphi_\varepsilon(x_0, \sigma)\frac{\partial^2 \tilde{b}_\varepsilon}{\partial x^2}(\varphi(\tilde{x}_0, \sigma), u) a_\varepsilon(\varphi_\varepsilon(\tilde{x}_0, \sigma), \varphi_\varepsilon(\tilde{x}_0, \sigma), \sigma)\, d\sigma$$

$$+ (\mathscr{G}_t^\varepsilon)\text{-martingale with mean } 0.$$

Denote each term of the right hand side by $I_i^\varepsilon(u)$, $i = 1, \ldots, 6$. Then, by assumption (10) we have $\int_s^t I_1^\varepsilon(u)\, du \to 0$ weakly $(\mathscr{B}_{x_0, y_0}(s))$ as $\varepsilon = \varepsilon_h \to 0$. By assumption (9) and Lemma 5.2.5

$$E\left[\int_s^t I_2^\varepsilon(u)\, du \middle| \mathscr{G}_s^\varepsilon\right] = E\left[\int_s^t \varphi_\varepsilon(x_0, \sigma) c_\varepsilon(\varphi_\varepsilon(\tilde{x}_0, \sigma), t, \sigma)\, d\sigma \middle| \mathscr{G}_s^\varepsilon\right]$$

$$\to E_0^{(x_0, y_0)}\left[\int_s^t \varphi(x_0, \sigma) c(\varphi(\tilde{x}_0, \sigma), \sigma)\, d\sigma \middle| \mathscr{B}_{x_0, y_0}(s)\right]$$

weakly as $\varepsilon = \varepsilon_h \to 0$. Similarly by assumption (8) and Lemma 5.2.5 we have

$$\int_s^t I_3^\varepsilon(u)\, du \to \int_s^t A(\varphi(x_0, \sigma), \varphi(\tilde{x}_0, \sigma), \sigma)\, d\sigma$$

weakly $(\mathscr{B}_{x_0, y_0}(s))$ as $\varepsilon = \varepsilon_h \to 0$. Similarly for $i = 4, 5$ we have $\int_0^t I_i^\varepsilon(u)\, du \to 0$ weakly $(\mathscr{B}_{x_0, y_0}(s))$ as $\varepsilon = \varepsilon_h \to 0$ by assumption (10). Also $\int_0^t I_6^\varepsilon(u)\, du$ converges to 0 weakly $(\mathscr{B}_{x_0, y_0}(s))$ as $\varepsilon = \varepsilon_h \to 0$. Summing these computations we obtain (28).

We have thus shown that

$$\varphi(x_0, t)\varphi(\tilde{x}_0, t) - \int_0^t \varphi(x_0, \sigma)(\bar{b} + c)(\varphi(\tilde{x}_0, \sigma), \sigma)\, d\sigma$$

$$- \int_0^t \varphi(\tilde{x}_0, \sigma)(\bar{b} + c)(\varphi(x_0, \sigma), \sigma)\, d\sigma$$

$$- \int_0^t (a + \bar{A})(\varphi(x_0, \sigma), \varphi(\tilde{x}_0, \sigma), \sigma)\, d\sigma \tag{29}$$

is a martingale with respect to $Q_0^{(x_0, y_0)}$. Then noting that (24) is a martingale we can show similarly to the proof of Lemma 5.2.7 that the joint quadratic variation of $M(x_0, t)$ and $M(\tilde{x}_0, t)$ is given by the last term of (29). The proof of (25) is complete. □

We can now complete the proof of Theorem 5.3.1 similarly to that of Theorem 5.2.1 using Lemmas 5.3.2–5.3.5. The details are left to the reader.

Theorem 5.3.6 *Let $\{F_\varepsilon(x, t)\}$ be a family of continuous semimartingales with local characteristics $\{(a_\varepsilon, b_\varepsilon)\}$ satisfying Conditions* $(C.1)_p$, $(C.2)_k$, $(C.3)_p$ *and* $(C.4)_k$ *for some $p > 3$ and $k \geq 1$. If $\{F_\varepsilon(x, t)\}$ converges strongly in probability for every x, then $\{\varphi_\varepsilon(x, t)\}$ converges strongly in $L^{p'}$ for every x for any $p' < p$.* □

The proof can be carried out similarly to the proof of Theorem 5.2.8. The details are omitted.

5.4 Convergence as stochastic flows

Convergence as C^j-flows

In the previous two sections we discussed the weak convergences as diffusions of the pairs of stochastic flows $\varphi_\varepsilon(t)$ and their Itô's random infinitesimal generators $F_\varepsilon(t)$. In this section we wish to discuss the weak convergence of $\{(\varphi_\varepsilon(t), F_\varepsilon(t))\}$ as stochastic flows. The topology required for the convergence as stochastic flows is much stronger than that as diffusions. Thus we will have to assume stronger conditions than those introduced in the previous section.

We begin by introducing two classes of stochastic processes. Let $\{K_\varepsilon(t)\}$ be a family of real processes. It is called *uniformly L^p-bounded* if

$$\sup_\varepsilon \sup_t E[|K_\varepsilon(t)|^p] < \infty \tag{1}$$

is satisfied. It is called *uniformly exponentially bounded in mean* if it is uniformly L^p-bounded for any $p > 1$ and satisfies

$$\sup_\varepsilon E\left[\exp \lambda \int_0^T K_\varepsilon(s)\, ds\right] < \infty \tag{2}$$

for any $\lambda > 0$. Families of random fields $\{A_\varepsilon(x, y, s)\}$, $\{A_\varepsilon(x, y, t, s)\}$ etc. are also called *uniformly exponentially bounded in mean* if there exists $\{K_\varepsilon(s)\}$ mentioned above such that

$$\sup_{x,y} |A_\varepsilon(x, y, s)| \leq K_\varepsilon(s), \qquad \sup_{x,y,t} |A_\varepsilon(x, y, s, t)| \leq K_\varepsilon(s)$$

are satisfied.

Throughout this section we will assume that a family of filtrations $\{\mathscr{F}_{s,t}^{\varepsilon} : 0 \le s \le t \le T\}, \varepsilon > 0$ of sub σ-fields of $\mathscr{F}$ is given. We set $\mathscr{F}_{0,t}^{\varepsilon} = \mathscr{F}_t^{\varepsilon}$. Now suppose that for every $\varepsilon > 0$ we are given a forward–backward C^k-semimartingale $F^{\varepsilon}(x, t)$ adapted to $(\mathscr{F}_{s,t}^{\varepsilon})$ with the same forward and backward local characteristic $(a_{\varepsilon}, b_{\varepsilon}, t)$ belonging to the class $B^{k+2,\delta}$. We assume that b_{ε} is decomposed as $b_{\varepsilon} = \bar{b}_{\varepsilon} + \tilde{b}_{\varepsilon}$ where $\tilde{b}_{\varepsilon}$ is $\mathscr{F}_0^{\varepsilon}$-measurable and $k + 3$-times continuously differentiable with respect to x. We will define $K_{\varepsilon}^{\alpha\beta}$, $L_{\varepsilon}^{\alpha\beta}$, A_{ε} and c_{ε} by equations (2)–(7) of Section 5.3 using this $\tilde{b}_{\varepsilon}$.

Condition (C.1)$_{k,\infty}$ *The families of random fields $\{D_x^{\alpha} D_y^{\beta} a_{\varepsilon}(x, y, s) : \varepsilon > 0\}$, $|\alpha| \le k$, $|\beta| \le k$ and $\{D_x^{\alpha} \bar{b}_{\varepsilon}(x, s) : \varepsilon > 0\}$, $|\alpha| \le k$ are uniformly exponentially bounded in mean.* □

Condition (C.3)$_{k,\infty}$ *The families of random fields $\{K_{\varepsilon}^{\alpha\beta}(x, y, s, t) : \varepsilon > 0\}$, $|\alpha| \le k + 1$, $|\beta| \le k$ and $\{L_{\varepsilon}^{\alpha\beta}(x, y, s, t) : \varepsilon > 0\}$, $|\alpha| \le k + 2$, $|\beta| \le k$ are all uniformly exponentially bounded in mean.* □

We first establish the weak convergence as C^{k-1}-flows.

Theorem 5.4.1 *Assume that the family of local characteristics $\{(a_{\varepsilon}, \bar{b}_{\varepsilon}, \tilde{b}_{\varepsilon})\}$ of $\{F_{\varepsilon}\}$ satisfies Conditions* (C.1)$_{k,\infty}$, (C.2)$_k$, (C.3)$_{k,\infty}$ *and* (C.4)$_k$ *for some $k \ge 1$. Then $\{(\varphi_{\varepsilon}(t), F_{\varepsilon}(t))\}$ converges weakly as C^{k-1}-flows as $\varepsilon \to 0$, i.e. the laws P_{ε}^{k-1} of $(\varphi_{\varepsilon}(t), F_{\varepsilon}(t))$ on the space $W_{k-1} \times W_{k-1}$ converge weakly as $\varepsilon \to 0$. Furthermore, the limit law P_0^{k-1} satisfies properties* (i) *and* (ii) *of Theorem* 5.3.1. □

In the proof of the theorem, a fundamental role will be played by the uniform L^p-estimates of the derivatives $D^{\alpha}\varphi_{\varepsilon}(t)$. They are stated as follows.

Theorem 5.4.2 *Let $\{(\varphi_{\varepsilon})_{s,t}\}$ be a family of stochastic flows of C^k-diffeomorphisms generated by $\{F_{\varepsilon}\}$, respectively. Assume that the family of local characteristic $\{(a_{\varepsilon}, \bar{b}_{\varepsilon}, \tilde{b}_{\varepsilon})\}$ of $\{F_{\varepsilon}\}$ satisfies Conditions* (C.1)$_{k,\infty}$ *and* (C.3)$_{k,\infty}$. *Then for each $p > 3$ and $N > 0$, there exists a positive constant C such that for every ε and $|\alpha| \le k$*

$$E[|D^{\alpha}(\varphi_{\varepsilon})_{s,t}(x) - D^{\alpha}(\varphi_{\varepsilon})_{s,t'}(x)|^p] \le C|t - t'|^{2-(3/p)} \tag{3}$$

holds for any $s \le t$, $s \le t'$ and $|x| \le N$. □

The proof of Theorem 5.4.2 will be given later since it is long.

In the sequel we will give the proof of Theorem 5.4.1 by applying Theorem 5.4.2. For this we need some Sobolev spaces. Let j be a non-negative integer and let $1 < p < \infty$. Let B_N denote the ball in $\mathbb{R}^d$ with center the origin and radius N. Let $f : \mathbb{R}^d \to \mathbb{R}^d$ be a function such that $D^\alpha f \in L^p(B_N)$ for all α with $|\alpha| \le j$, where the derivatives are in the sense of distributions. For such a function f, we define seminorms $\{\| \ \|_{j,p:N} : N = 1, 2, \dots\}$ by

$$\|f\|_{j,p:N} = \left(\sum_{|\alpha| \le j} \int_{B_N} |D^\alpha f(x)|^p \, dx \right)^{1/p}. \tag{4}$$

Set

$$H_{j,p}^{\text{loc}} = \{f : \mathbb{R}^d \to \mathbb{R}^d, \ \|f\|_{j,p:N} < \infty \text{ for any } N\}. \tag{5}$$

Then the seminorms $\{\| \ \|_{j,p:N} : N = 1, 2, \dots\}$ make $H_{j,p}^{\text{loc}}$ a real separable semi-reflexive Frechet space. By Sobolev's imbedding theorem, we have $H_{j+1,p}^{\text{loc}} \subset C^j \subset H_{j,p}^{\text{loc}}$ if $p > d$. Furthermore the imbedding $i : H_{j+1,p}^{\text{loc}} \subset C^j$ is a compact operator by the Rellich–Kondrachov theorem (see Adams [1]).

Let $W_{j,p} = C([0, T] : H_{j,p}^{\text{loc}})$. For $\varphi \in W_{j,p}$ define

$$|\varphi|_{j,p:N} = \sup_{t \in [0,T]} \|\varphi(t)\|_{j,p:N}, \qquad N = 1, 2, \dots \tag{6}$$

With this family of seminorms, $W_{j,p}$ is a separable Frechet space. Denote the product space $W_{j,p} \times W_{j,p}$ by $W_{j,p}^2$. We have $W_{j+1,p}^2 \subset W_{j,\infty}^2 \subset W_{j,p}^2$ if $p > d$ by Sobolev's imbedding theorem, where $W_{j,\infty}$ is the space W_j introduced in Section 5.1.

Proof of Theorem 5.4.1 Define the law of $(\varphi_\varepsilon, F_\varepsilon)$ by

$$P_\varepsilon^{k,p}(A) \equiv P(\{\omega : (\varphi_\varepsilon(\omega), F_\varepsilon(\omega)) \in A\}), \qquad A \in \mathscr{B}(W_{k,p}^2). \tag{7}$$

We shall prove that the family of measures $\{P_\varepsilon^{k,p} : \varepsilon > 0\}$ satisfies Kolmogorov's criteria, (15) and (16) of Theorem 1.4.7. Set $s = 0$ in the inequality (3) and integrate both sides by the Lebesgue measure over the ball B_N and sum for all α such that $|\alpha| \le k$. Then we have by Fubini's theorem

$$E[(\|\varphi_\varepsilon(t) - \varphi_\varepsilon(t')\|_{k,p:N})^p] \le C'|t - t'|^{2-(3/p)}. \tag{8}$$

Further, as in Lemma 5.3.2, we have

$$E[|D^\alpha F_\varepsilon(y, t) - D^\alpha F_\varepsilon(y, t')|^p] \le C|t - t'|^{2-(2/p)}, \qquad \text{for } |\alpha| \le k \tag{9}$$

and any $t, t', |x| \le N$ and ε. This yields

$$E[(\|F_\varepsilon(t) - F_\varepsilon(t')\|_{k,p:N})^p] \le C'|t - t'|^{2-(2/p)}. \tag{10}$$

Then condition (15) of Theorem 1.4.7 is satisfied. Condition (16) of Theorem 1.4.7 is obvious from (8) and (10) since $D^\alpha \varphi_\varepsilon(0) = \text{constant}$ and $D^\alpha F_\varepsilon(0) = 0$. Therefore the family of measures $\{P_\varepsilon^{k,p}\}$ is right with respect to the semi-weak topology of $W_{k,p}^2$. Now since $i : H_{k,p}^{\text{loc}} \subset C^{k-1}$ is a compact operator, the family of measures $\{\mathrm{P}_\varepsilon^{k-1}\}$ on $W_{k-1,\infty}^2$ is tight with respect to the strong topology of $W_{k-1,\infty}^2$.

Now assume Conditions (C.2)$_k$ and (C.4)$_k$ hold. Then the family of measures $\{\mathrm{P}_\varepsilon^{k-1}\}$ converges weakly and the limit measure P_0^{k-1} satisfies (i) and (ii) of Theorem 5.3.1, since $\{(\varphi_\varepsilon, F_\varepsilon)\}$ converges weakly as diffusions. We have thus proved Theorem 5.4.1. □

As an application of Theorem 5.4.1, we have the strong L^p-convergence of stochastic flows. We will now assume that the family of continuous C^k-semimartingales $\{F_\varepsilon(t)\}$ is defined on the common probability space $(\Omega, \mathscr{F}, P)$ and is adapted to the common filtration $\{\mathscr{F}_t : t > 0\}$.

Theorem 5.4.3 *Assume that the family of local characteristics* $\{(a_\varepsilon, \bar{b}_\varepsilon, \tilde{b}_\varepsilon)\}$ *of* $\{F_\varepsilon\}$ *satisfies Conditions* (C.1)$_{k,\infty}$, (C.2)$_k$, (C.3)$_{k,\infty}$ *and* (C.4)$_k$ *for some* $k \geq 1$. *If* $\{F_\varepsilon(x, t)\}$ *converges strongly in probability for each* x, *the family of pairs* $\{(F_\varepsilon, \varphi_\varepsilon)\}$ *converges strongly as* C^{k-1}*-flows in* L^p *for any* $p > 1$. □

The proof can be carried out similarly to that of Theorem 5.2.8. It is left to the reader.

Uniform L^p-estimate of stochastic flows

The proof of Theorem 5.4.2 is not as simple as in the case of $|\alpha| = 0$ (Lemmas 5.3.2 and 5.3.3). In fact, the Jacobian matrix $\partial(\varphi_\varepsilon)_{s,t} = (\partial(\varphi_\varepsilon)_{s,t}^i / \partial x^j)$ satisfies Itô's linear stochastic differential equation

$$\partial(\varphi_\varepsilon)_{s,t} = I + \int_s^t \partial F_\varepsilon((\varphi_\varepsilon)_{s,r}, \circ dr) \partial(\varphi_\varepsilon)_{s,r}, \tag{11}$$

where $\partial F_\varepsilon(x, t) = ((\partial F_\varepsilon^i / \partial x^j)(x, t))$. Then the local characteristic of $\partial F_\varepsilon(x, t)y$ is not a bounded function of y a.s., so that Conditions (C.1)$_p$ and (C.3)$_p$ are not satisfied. Thus we can not apply Lemma 5.3.3 to $\partial(\varphi_\varepsilon)_{s,t}$. Instead we will try to obtain the uniform L^p-estimate for solutions of certain systems of stochastic differential equations, which are enlargements of the stochastic differential equations based on $F_\varepsilon(x, t)$. A typical example of such enlarged systems will be the equations for the pairs $((\varphi_\varepsilon)_{s,t}, \partial(\varphi_\varepsilon)_{s,t})$ described above.

Let $G_{0,\varepsilon}(x, t)$, $x \in \mathbb{R}^d$, $t \in [0, T]$ be a continuous forward–backward semimartingale with values in $C^k(\mathbb{R}^d : \mathbb{R}^d)$ adapted to $(\mathscr{F}_{s,t}^\varepsilon)$ with the same forward and backward local characteristic belonging to the class $B^{2,\delta}$ where $k \geq 3$ and $0 < \delta \leq 1$, and $\varepsilon > 0$ denotes a parameter. We assume that it is written as

$$G_{0,\varepsilon}(x, t) = \begin{pmatrix} G_{0,\varepsilon}^1(x^1, t) \\ G_{0,\varepsilon}^2(x^1, x^2, t) \end{pmatrix}$$

where $x = (x^1, x^2)$, $x^1 \in \mathbb{R}^{d_1}$, $x^2 \in \mathbb{R}^{d_2}$, $d = d_1 + d_2$ and $G_{0,\varepsilon}^1$ and $G_{0,\varepsilon}^2$ are $C^k(\mathbb{R}^{d_1} : \mathbb{R}^{d_1})$-valued and $C^k(\mathbb{R}^d : \mathbb{R}^{d_2})$-valued semimartingales, respectively. We denote the local characteristic of $G_{0,\varepsilon}^1$ by $(a_{0,\varepsilon}, b_{0,\varepsilon})$. Let $G_{1,\varepsilon}(x^1, t)$ and $G_{2,\varepsilon}(x, t)$ be forward–backward semimartingales adapted to $(\mathscr{F}_{s,t}^\varepsilon)$ with values in $C^k(\mathbb{R}^{d_1} : \mathbb{R}^{d'} \otimes \mathbb{R}^{d'})$ and $C^k(\mathbb{R}^d : \mathbb{R}^{d'})$, and with the same forward and backward local characteristics $(a_{1,\varepsilon}, b_{1,\varepsilon})$, $(a_{2,\varepsilon}, b_{2,\varepsilon})$ belonging to the class $B^{k,\delta}$, respectively. As before $b_{i,\varepsilon} (i = 0, 1, 2)$ are assumed to be decomposed as $b_{i,\varepsilon} = \bar{b}_{i,\varepsilon} + \tilde{b}_{i,\varepsilon}$ where $\bar{b}_{i,\varepsilon}$ are $(\mathscr{F}_{0,t}^\varepsilon)$-adapted and $\tilde{b}_{i,\varepsilon}$ are $(\mathscr{F}_{0,0}^\varepsilon)$-adapted and $\tilde{b}_{i,\varepsilon}$ is 3-times continuously differentiable. We set

$$\mathscr{G}_t^\varepsilon = \sigma(G_{i,\varepsilon}(\cdot, u), \bar{b}_{i,\varepsilon}(\cdot, u), \tilde{b}_{i,\varepsilon}(\cdot, u) : 0 \le u \le t, i = 0, 1, 2).$$

For the family of local characteristics $\{(a_{i,\varepsilon}, b_{i,\varepsilon}) : \varepsilon > 0\}$ we assume the following condition.

Condition (C.5) *There exists a non-negative integer m such that the following family of random fields are uniformly exponentially bounded in mean:*

$$a_{i,\varepsilon}(1 + |x|)^{-m(i-1)^+}(1 + |x'|)^{-m(i-1)^+},$$

$$\bar{b}_{i,\varepsilon}(1 + |x|)^{-m(i-1)^+},$$

$$K_{ij,\varepsilon}^\alpha(x, x', s, t)(1 + |x|)^{-m(i-1)^+}(1 + |x'|)^{-m(i-1)^+}, \qquad |\alpha| \le 1,$$

$$L_{ij,\varepsilon}^\alpha(x, x', s, t)(1 + |x|)^{-m(i-1)^+}(1 + |x'|)^{-m(i-1)^+}, \qquad |\alpha| \le 2,$$

where $i, j = 0, 1, 2$, $(i - 1)^+ = (i - 1) \vee 0$. The $K_{ij,\varepsilon}^\alpha$ and $L_{ij,\varepsilon}^\alpha$ are continuous random fields defined by

$$K_{ij,\varepsilon}^\alpha(x, x', s, t) = \left| \int_s^t E[D_x^\alpha \tilde{b}_{i,\varepsilon}(x, r) | \mathscr{G}_s^\varepsilon] \, dr \right| (1 + |\bar{b}_{j,\varepsilon}(x', s)| + |\tilde{b}_{j,\varepsilon}(x', s)|), \quad (12)$$

$$L_{ij,\varepsilon}^\alpha(x, x', s, t) = \left| \int_s^t E[D_x^\alpha \tilde{b}_{i,\varepsilon}(x, r) | \mathscr{G}_s^\varepsilon] \, dr \right| \| a_{j,\varepsilon}(x', x', s) \|. \quad (13)$$

□

Now let $(\varphi_\varepsilon)_{s,t} = ((\varphi_\varepsilon^1)_{s,t}, (\varphi_\varepsilon^2)_{s,t}))$ be the stochastic flow determined by

$$(\varphi_\varepsilon^1)_{s,t}(x^1) = x^1 + \int_s^t G_{0,\varepsilon}^1((\varphi_\varepsilon^1)_{s,r}(x^1), dr),$$
$$(\varphi_\varepsilon^2)_{s,t}(x) = x^2 + \int_s^t G_{0,\varepsilon}^2((\varphi_\varepsilon)_{s,r}(x), dr). \quad (14)$$

Consider Itô's stochastic differential equation:

$$\eta_t = y + \int_s^t G_{1,\varepsilon}((\varphi_\varepsilon^1)_{s,r}, dr)\eta_r + \int_s^t G_{2,\varepsilon}((\varphi_\varepsilon)_{s,r}(x), dr). \tag{15}$$

For any s and (x, y), it has a unique solution, denoted by $(\eta_\varepsilon)_{s,t}(x, y)$. Then $(\varphi_\varepsilon, \eta_\varepsilon)$ defines a stochastic flow of C^k-diffeomorphisms for each ε. We will obtain the uniform L^p-estimates for $(\varphi_\varepsilon, \eta_\varepsilon)$.

Lemma 5.4.4 *Assume Condition* (C.5). *Then for each $p > 3$ and $N > 0$ there exists a positive constant C such that for every ε*

$$E[|(\varphi_\varepsilon^1)_{s,t}(x)|^p] \le C, \tag{16}$$

$$E[|(\varphi_\varepsilon^1)_{s,t}(x) - (\varphi_\varepsilon^1)_{s,t'}(x)|^p] \le C|t - t'|^{2-(3/p)} \tag{17}$$

holds for any s, t, t', $|x| \le N$. Assume further that inequalities (16) *and* (17) *are valid for $(\varphi_\varepsilon^2)_{s,t}$, too. Then there exists a positive constant C' such that for every ε*

$$E[|(\eta_\varepsilon)_{s,t}(x, y)|^p] \le C', \tag{18}$$

$$E[|(\eta_\varepsilon)_{s,t}(x, y) - (\eta_\varepsilon)_{s,t'}(x, y)|^p] \le C'|t - t'|^{2-(3/p)} \tag{19}$$

holds for any s, t, t', $|x|$, $|y| \le N$. □

The proof of the lemma will be given later since it is long. Here we will complete the proof of Theorem 5.4.2 using the above lemma.

Proof of Theorem 5.4.2 The case $|\alpha| = 0$ is immediate from (17) by setting $\varphi_\varepsilon = \varphi_\varepsilon^1$. We shall consider the case where $|\alpha| = 1$. Let φ_ε be the stochastic flow generated by F_ε. Set $G_{0,\varepsilon}^1 = F_\varepsilon$, $G_{0,\varepsilon}^2 = G_{2,\varepsilon} = 0$ and $G_{1,\varepsilon}(x, t) = ((\partial F_\varepsilon^i/\partial x^j)(x, t))_{i,j=1,\dots,d}$. Then $(\eta_\varepsilon)_{s,t}(x) = ((\partial(\varphi_\varepsilon)_{s,t}^i/\partial x^j)(x))$ satisfies Itô's stochastic differential equations

$$(\eta_\varepsilon)_{s,t}(x) = I + \int_s^t G_{1,\varepsilon}((\varphi_\varepsilon)_{s,r}(x), dr)(\eta_\varepsilon)_{s,r}(x), \tag{20}$$

where I is the identity matrix. Their local characteristics satisfy Condition (C.5) because of the assumption of the theorem that Conditions $(\mathrm{C}.1)_{k,\infty}$ and $(\mathrm{C}.3)_{k,\infty}$ hold. Regarding φ_ε as φ_ε^1 and the identity flow ($\varphi_{s,t}(x) = x$) as φ_ε^2 in Lemma 5.4.4, we have inequality (19) for $(\eta_\varepsilon)_{s,t} = D^\alpha(\varphi_\varepsilon)_{s,t}$ ($|\alpha| = 1$). This proves the assertion in the case where $|\alpha| = 1$.

We next consider the case where $|\alpha| = 2$. Set $x^1 = x$, $x^2 = (x_{ij})$, $y = (y_{ijk})$, $i, j, k = 1, \dots, d$ and define

$$G_{1,\varepsilon}(x^1, t)y + G_{2,\varepsilon}(x^1, x^2, t)$$
$$= \left(\sum_l \frac{\partial F_\varepsilon^i}{\partial x^l}(x^1, t)y_{ljk} + \sum_{m,l} \frac{\partial^2 F_\varepsilon^i}{\partial x^m \partial x^l}(x^1, t)x_{lj}x_{mk}\right).$$

Then the derivatives $(\zeta_\varepsilon)_{s,t} = (\partial^2(\varphi_\varepsilon)^i_{s,t}/\partial x^k \partial x^j)$, $i, j, k = 1, \ldots, d$, satisfy

$$(\zeta_\varepsilon)_{s,t} = \int_s^t G_{1,\varepsilon}((\varphi_\varepsilon)_{s,r}, dr)(\zeta_\varepsilon)_{s,r}$$

$$+ \int_s^t G_{2,\varepsilon}((\varphi_\varepsilon)_{s,r}, \partial(\varphi_\varepsilon)_{s,r}/\partial x, dr). \tag{21}$$

We can show that the local characteristics of $(G^1_{0,\varepsilon}, G_{1,\varepsilon}, G_{2,\varepsilon})$ satisfy Condition (C.5) with $m = 2$, regarding F_ε as $G^1_{0,\varepsilon}$ and $(\partial F_\varepsilon/\partial x)$ as $G^2_{0,\varepsilon}$. Furthermore, both $\varphi^1_\varepsilon \equiv \varphi_\varepsilon$ and $\varphi^2_\varepsilon \equiv (\partial\varphi_\varepsilon/\partial x)$ satisfy estimates (16) and (17) as we have shown above. Consequently $\zeta_\varepsilon(t)$ satisfies (19). We can now complete the proof of the theorem by induction on the degree of the derivatives in the same way. □

Proof of Lemma 5.4.4

Inequalities (16) and (17) follow from Lemma 5.3.3. In fact Lemma 5.3.3 is valid not only for $\varphi_\varepsilon(x, t) \equiv (\varphi_\varepsilon)_{0,t}(x)$ but for any $(\varphi_\varepsilon)_{s,t}(x)$.

For inequalities (18) and (19), we will only prove the case where $d = d' = 1$. For simplicity, we often suppress ε and x, y from $(\varphi_\varepsilon)_{s,t}(x)$, $(\eta_\varepsilon)_{s,t}(x, y)$ etc. and write them as $\varphi_{s,t}$ and $\eta_{s,t}$. The martingale parts of $G_i = G_{i,\varepsilon}$ are denoted by Y_i and the joint local characteristics of Y_i and Y_j are denoted by a_{ij}. Further $b_i(\varphi_{s,r}, r)$ and $a_{ij}(\varphi_{s,r}, \varphi_{s,r}, r)$ etc. are denoted by $b_i(r)$ and $a_{ij}(r)$ etc.

The proof of (18) is most complicated. We shall first obtain a linear stochastic integral inequality regarding $|\eta_{s,t}|^{2p}$ as an unknown. Since $\eta_{s,t}$ satisfies (15), by Itô's formula we have

$$|\eta_{s,t}|^{2p} - |y|^{2p} = 2p \int_s^t b_1(r)|\eta_{s,r}|^{2p}\, dr$$

$$+ 2p \int_s^t b_2(r)\eta_{s,r}|\eta_{s,r}|^{2p-2}\, dr$$

$$+ p(2p-1)\int_s^t \{a_1(r)\eta^2_{s,r} + a_{12}(r)\eta_{s,r} + a_2(r)\}|\eta_{s,r}|^{2p-2}\, dr$$

$$+ 2p \int_s^t Y_1(\varphi_{s,r}, dr)|\eta_{s,r}|^{2p}$$

$$+ 2p \int_s^t Y_2(\varphi_{s,r}, dr)\eta_{s,r}|\eta_{s,r}|^{2p-2}. \tag{22}$$

The first term of the right hand side is

$$2p \int_s^t \bar{b}_1(\varphi_{s,r}, r)|\eta_{s,r}|^{2p}\, dr + 2p \int_s^t \tilde{b}_1(\varphi_{s,r}, r)|\eta_{s,r}|^{2p}\, dr.$$

We shall rewrite the last term. Applying Itô's formula to $\tilde{b}_1(\varphi_{s,r}, r)|\eta_{s,r}|^{2p}$ and then integrating it over the interval $[s, t]$, we obtain

$$\int_s^t \tilde{b}_1(\varphi_{s,r}, r)|\eta_{s,r}|^{2p}\, dr$$

$$= \left(\int_s^t \tilde{b}_1(x, r)\, dr\right)|y|^{2p} + \int_s^t \left\{\left(\int_u^t \frac{\partial \tilde{b}_1}{\partial x}(\varphi_{s,u}, r)\, dr\right) b_0(u)\right.$$

$$\left. + \frac{1}{2}\left(\int_u^t \frac{\partial^2 \tilde{b}_1}{\partial x^2}(\varphi_{s,u}, r)\, dr\right) a_0(u)\right\} |\eta_{s,u}|^{2p}\, du$$

$$+ 2p \int_s^t \left(\int_u^t \tilde{b}_1(\varphi_{s,u}, r)\, dr\right)\left(\{b_1(u)\eta_{s,u} + b_2(u)\}\eta_{s,u}|\eta_{s,u}|^{2p-2}\right.$$

$$\left. + \left(p - \frac{1}{2}\right)\{a_1(u)\eta_{s,u}^2 + a_{12}(u)\eta_{s,u} + a_2(u)\}|\eta_{s,u}|^{2p-2}\right) du$$

$$+ 2p \int_s^t \left(\int_u^t \frac{\partial \tilde{b}_1}{\partial x}(\varphi_{s,u}, r)\, dr\right)\eta_{s,u}|\eta_{s,u}|^{2p-2}\{a_{01}(u)\eta_{s,u} + a_{02}(u)\}\, du$$

$$+ M_t - M_s, \tag{23}$$

where M_t is the term involving the stochastic integrals based on $Y_i(x, t)$, $i = 0, 1, 2$. We can express the second term of the right hand side of (22) in the similar way. Summing these expressions, we arrive at the following equality:

$$|\eta_{s,t}|^{2p} - |y|^{2p}$$

$$= 2p\left\{\int_s^t \tilde{b}_1(x, r)\, dr + \left(\int_s^t \tilde{b}_2(x, r)\, dr\right)\frac{y}{|y|^2}\right\}|y|^{2p}$$

$$+ \sum_{l=0}^{3} \int_s^t \Psi_l(t, u, \varphi_{s,u})|\eta_{s,u}|^{2p-l}\, \mathrm{sign}(u)^l(1 + |\varphi_{s,u}|)^{ml}\, du$$

$$+ \sum_{l=0}^{3} \int_s^t \Phi_l(r, \varphi_{s,r})|\eta_{s,r}|^{2p-l}\, \mathrm{sign}(r)^l(1 + |\varphi_{s,r}|)^{ml}\, dr$$

$$+ \sum_{i=0}^{2} \int_s^t \Xi_i(t, u, \varphi_{s,u}, \eta_{s,u})\, Y_i(\varphi_{s,u}, du). \tag{24}$$

Here $\Psi_l = \Psi_{l,\varepsilon}$, $l = 0, 1, 2, 3$ are sums of the terms involving integrals such as $\{\int_u^t (\partial\tilde{b}_i/\partial x)(x, r)\, dr\} b_j(x, u)(1 + |x|)^{-n(i,j)m}$, where $n(i, j) = (i - 1)^+ + (j - 1)^+$. The family $\{\Psi_{l,\varepsilon} : \varepsilon > 0\}$ is not uniformly exponentially bounded in mean, but the family of conditional expectations $\{\hat{\Psi}_{l,\varepsilon} : \varepsilon > 0\}$ ($\hat{\Psi}_{l,\varepsilon}(t, u, x) \equiv E[\Psi_{l,\varepsilon}(t, u, x)|\mathscr{G}_u^\varepsilon]$) are uniformly exponentially bounded in mean by Condition (C.5). We set $\mathrm{sign}(u) = \mathrm{sign}(\eta_{s,u})$. The random fields $\Phi_l = \Phi_{l,\varepsilon}$, $l = 0, 1, 2, 3$ are sums of the terms such as $\bar{b}_i(1 + |x|)^{-n(i)m}$ and $a_{ij}(1 + |x|)^{-n(i,j)m}$ etc. where $n(i) = (i - 1)^+$. The family $\{\Phi_{l,\varepsilon} : \varepsilon > 0\}$ is uniformly exponentially bounded in mean. The random fields $\Xi_i = \Xi_{i,\varepsilon}(t, u, x, y)$, $i = 0, 1, 2$ are $(\mathscr{F}_u^\varepsilon)$-adapted. Now set

$$Z(x, y, s, t) = 2p\left\{\int_s^t \tilde{b}_1(x, r)\, dr + \left(\int_s^t \tilde{b}_2(x, r)\, dr\right)\frac{y}{|y|^2}\right\}, \tag{25}$$

$$\begin{aligned} Y_{t_0}(x, y, s, t) = &\sum_{i=0}^{2} \int_s^t \Xi_i(t_0, u, \varphi_{s,u}(x), \eta_{s,u}(x, y))\, Y_i(\varphi_{s,u}(x), du) \\ &+ \sum_{l=0}^{3} \int_s^t (\Psi_l - \hat{\Psi}_l)(t_0, u, \varphi_{s,u}(x))|\eta_{s,u}(x, y)|^{2p-l} \\ &\times \operatorname{sign}(u)^l (1 + |\varphi_{s,u}(x)|)^{ml}\, du, \end{aligned} \tag{26}$$

(ε is suppressed). Taking modifications, we may assume that these random fields are continuous in (x, y, s, t) a.s. Then (24) is written as

$$\begin{aligned} |\eta_{s,t}|^{2p} = {}&|y|^{2p}(1 + Z(\cdot, \cdot, s, t)) \\ &+ \sum_{l=0}^{3} \int_s^t (\hat{\Psi}_l(t, r, \varphi_{s,r}) + \Phi_l(r, \varphi_{s,r}))|\eta_{s,r}|^{2p-l} \operatorname{sign}(r)^l (1 + |\varphi_{s,r}|)^{ml}\, dr \\ &+ Y_t(\cdot, \cdot, s, t). \end{aligned} \tag{27}$$

There exists a family of positive processes $\{K_\varepsilon(t)\}$ satisfying (2) for any λ such that for each ε the random fields $\hat{\Psi}_{l,\varepsilon} + \Phi_{l,\varepsilon}$, $l = 0, 1, 2, 3$ are bounded by K_ε. Then using the inequality $ab \le a^{p'}/p' + b^{q'}/q'$ where $p', q' > 1$ and $(p')^{-1} + (q')^{-1} = 1$, from (27) we obtain the inequality

$$\begin{aligned} |\eta_{s,t}(x, y)|^{2p} \le |y|^{2p}(1 + |Z(x, y, s, t)|) + \int_s^t K(r)|\eta_{s,r}(x, y)|^{2p}\, dr \\ + \int_s^t K(r)(1 + |\varphi_{s,r}(x)|)^{2mp}\, dr + Y_t(x, y, s, t). \end{aligned} \tag{28}$$

(ε is suppressed from K_ε). The above can be regarded as a linear integral inequality for $|\eta_{s,t}(x, y)|^{2p}$.

The inequality (28) can not be solved directly. We shall transform it to a backward inequality by substituting $x = \varphi_{t,s}(x') = \varphi_{s,t}^{-1}(x')$ and $y = \eta_{t,s}(x', y') = \eta_{s,t}^{-1}(x', y')$ into (28). Set

$$\alpha_{t,s}(x', y') = |\eta_{t,s}(x', y')|^{2p}(1 + |Z(\varphi_{t,s}(x'), \eta_{t,s}(x', y'), s, t)|).$$

Then we get

$$|y'|^{2p} \le \alpha_{t,s}(x', y') + \int_s^t K(r)\alpha_{t,r}(x', y')\, dr$$

$$+ \int_s^t K(r)(1 + |\varphi_{t,r}(x')|)^{2mp}\, dr + Y_t(\varphi_{t,s}(x'), \eta_{t,s}(x', y'), s, t). \tag{29}$$

The last term is a backward semimartingale with respect to s by the generalized Itô's formula (Theorem 3.3.2). Inequality (29) can be regarded as a backward linear inequality for $\alpha_{t,s}(x', y')$ with time variable s, where t, x', y' are fixed. Then $\alpha_{t,s}(x', y')$ satisfies

$$\alpha_{t,s}(x', y') \ge |y'|^{2p} \exp\left\{-\int_s^t K(r)\, dr\right\}$$

$$- \int_s^t \exp\left\{-\int_s^r K(u)\, du\right\} K(r)(1 + |\varphi_{t,r}(x')|)^{2mp}\, dr$$

$$- \int_s^t \exp\left\{-\int_s^r K(u)\, du\right\} Y_t(\varphi_{t,r}(x'), \eta_{t,r}(x', y'), \circ\hat{d}r, t), \tag{30}$$

using the backward Stratonovich integral. Indeed denote the right hand side of (30) by Z_s. It satisfies the backward linear equation

$$|y'|^{2p} = Z_s + \int_s^t K(r)Z_r\, dr + \int_s^t K(r)(1 + |\varphi_{t,r}(x')|)^{2mp}\, dr$$

$$+ Y_t(\varphi_{t,s}(x'), \eta_{t,s}(x', y'), s, t).$$

Therefore, $W_s \equiv Z_s - \alpha_{t,s}(x', y')$ satisfies $W_s \le \int_s^t K(r) W_r\, dr$ for any $s < t$. This implies $W_s \le 0$ by Gronwall's inequality.

Now substitute $x' = \varphi_{s,t}(x)$, $y' = \eta_{s,t}(x, y)$ into (30). Then we obtain

$$|y|^{2p}(1 + |Z(x, y, s, t)|)$$

$$\ge |\eta_{s,t}(x, y)|^{2p} \exp\left\{-\int_s^t K(r)\, dr\right\}$$

$$- \int_s^t \exp\left\{-\int_s^r K(u)\, du\right\} K(r)(1 + |\varphi_{s,r}(x)|)^{2mp}\, dr$$

$$- \int_s^t \exp\left\{-\int_s^r K(u)\, du\right\} Y_t(x, y, s, \circ dr), \tag{31}$$

where the last term is the forward Stratonovich integral. It is equal to the Itô integral since the integrand is a process of bounded variation. Therefore its expectation is 0. Then from (31) we obtain

$$E\left[|\eta_{s,t}(x,y)|^{2p}\exp\left\{-\int_s^t K(r)\,dr\right\}\right]$$
$$\leq |y|^{2p}E[(1+|Z(x,y,s,t)|]$$
$$+E\left[\int_s^t \exp\left\{-\int_s^r K(u)\,du\right\}K(r)(1+|\varphi_{s,r}(x)|)^{2mp}\,dr\right]. \tag{32}$$

Since

$$E[|Z(x,y,s,t)|^2]\leq 16p^2E\left[\int_s^t\left(\int_u^t \tilde{b}_1(x,r)\,dr\right)\tilde{b}_1(x,u)\,du\right]$$
$$+16p^2|y|^{-2}E\left[\int_s^t\left(\int_u^t \tilde{b}_2(x,r)\,dr\right)\tilde{b}_2(x,u)\,du\right],$$

the first term of the right hand side of (32) is uniformly bounded on compact sets. The second term is also uniformly bounded on compact sets since $\{(\varphi_\varepsilon)_{s,t}(x)\}$ is uniformly L^p-bounded on compact sets for any p. Now observe the inequality

$$E[|\eta_{s,t}(x,y)|^p]$$
$$\leq E\left[|\eta_{s,t}(x,y)|^{2p}\exp\left\{-\int_s^t K(r)\,dr\right\}\right]^{1/2}E\left[\exp\left\{\int_s^t K(r)\,dr\right\}\right]^{1/2}. \tag{33}$$

Then we have estimate (18).

We shall finally prove (19). Following the method used to obtain (24), we have for $s<t'<t$,

$$|\eta_{s,t}-\eta_{s,t'}|^p=\sum_{l=1}^3\int_{t'}^t \Psi_l^{(2)}(t,r)|\eta_{s,r}-\eta_{s,t'}|^{p-l}\operatorname{sign}(r)^l\,dr$$
$$+Y_t^{(2)}(\cdot,\cdot,t',t). \tag{34}$$

Here $\Psi_l^{(2)}=\Psi_{l,\varepsilon}^{(2)}$, $l=1,2,3$ are written as

$$\Psi_l^{(2)}(t,r,x,y)=\sum_{k=0}^{l}\Psi_{kl}(t,r,\varphi_{s,r}(x))(1+|\varphi_{s,r}(x)|)^{km}\eta_{s,r}(x,y)^{l-k},$$

where $\Psi_{kl}(t,r,x)$ are suitable linear combinations of the functionals $\Phi_l(r,x)$ and $\Psi_{l'}(t,r,x)$ which appeared in (24). The sign(r) is sign$(\eta_{s,r}-\eta_{s,t'})$ and $Y_{t_0}^{(2)}$ is defined similarly to (26) and its expectation is 0. Denote the conditional expectations of $\Psi_{kl}(t,r)$ and $\Psi_l^{(2)}(t,r)$ with respect to $\mathscr{G}_r^\varepsilon$ by $\hat{\Psi}_{k,l}(t,r)$ and $\hat{\Psi}_l^{(2)}(t,r)$, respectively. Since $\{\hat{\Psi}_{kl,\varepsilon}(t,r,(\varphi_\varepsilon)_{s,r}(x))\}$, $\{(\varphi_\varepsilon)_{s,r}(x)\}$ and

$\{(\eta_\varepsilon)_{s,r}(x, y)\}$ are uniformly L_p-bounded on compact sets for any $p > 1$, $\{\hat{\Psi}^{(2)}_{l,\varepsilon}(t, r, x, y) : \varepsilon > 0\}$ is also uniformly L^p-bounded on compact sets for any $p > 1$. Now take the expectations of each term of (34) and apply Hölder's inequality. Then

$$E[|\eta_{s,t} - \eta_{s,t'}|^p] \le \sum_{l=1}^{3} \int_{t'}^{t} E[|\hat{\Psi}^{(2)}_l(t, r)|^{p/l}]^{l/p} E[|\eta_{s,r} - \eta_{s,t'}|^p]^{1-(l/p)}\, dr.$$

Therefore for any N there exists a positive constant C such that for every ε

$$E[|\eta_{s,t}(x, y) - \eta_{s,t'}(x, y)|^p] \le C \sum_{l=1}^{3} \int_{t'}^{t} E[|\eta_{s,r}(x, y) - \eta_{s,t'}(x, y)|^p]^{1-(l/p)}\, dr \tag{35}$$

holds for any $|x|, |y| \le N$ and s. This implies (19) as in the case of Lemma 5.2.2. □

Convergence as G^j-flows

We will now discuss the weak convergence of $(\varphi_\varepsilon(t), F_\varepsilon(t))$ as G^j-stochastic flows. We will assume that $F_\varepsilon(x, t)$ is a continuous forward–backward C^k-semimartingale such that $F^\varepsilon(x, t) - F^\varepsilon(x, s)$ is adapted to the filtrations $\{\mathscr{F}^\varepsilon_{s,t} : 0 \le s \le t \le T\}$. We have seen in Section 4.4 that the inverse map $(\varphi_\varepsilon)_{t,s} = (\varphi_\varepsilon)^{-1}_{s,t}$ satisfies the backward Itô's stochastic differential equation

$$(\varphi_\varepsilon)_{t,s}(x) = x - \int_s^t F_\varepsilon((\varphi_\varepsilon)_{t,r}, \hat{d}r) + 2 \int_s^t \bar{c}_\varepsilon((\varphi_\varepsilon)_{t,r}, r)\, dr, \tag{36}$$

where

$$\bar{c}^i_\varepsilon(x, t) = \frac{1}{2} \sum_j \frac{\partial a^{ij}_\varepsilon}{\partial x^j}(x, y, t)|_{y=x}. \tag{37}$$

Note that $a_\varepsilon(x, y, t), b_\varepsilon(x, t), \bar{c}_\varepsilon(x, t)$ are $\mathscr{F}^\varepsilon_{t-,t+} \equiv \bigcap_{\delta>0} \mathscr{F}_{t-\delta,t+\delta}$ measurable. We will assume that b_ε is decomposed as $\bar{b}_\varepsilon + \tilde{b}_\varepsilon$, such that the triple $\{(a_\varepsilon, \bar{b}_\varepsilon, \tilde{b}_\varepsilon)\}$ satisfies the following.

Condition $(C.3)^{\wedge}_{k,\infty}$ *$\tilde{b}_\varepsilon$ is $\bigcap_t \mathscr{F}^\varepsilon_{t-,t+}$-measurable. Further the pairs $\{(a_\varepsilon, \tilde{b}_\varepsilon)\}$ satisfy Condition* $(C.3)_{k,\infty}$ *to the backward direction.* □

Theorem 5.4.5 *Assume that the family of local characteristics $\{(a_\varepsilon, \bar{b}_\varepsilon, \tilde{b}_\varepsilon)\}$ of $\{F_\varepsilon\}$ satisfies Conditions* $(C.1)_{k,\infty}$, $(C.2)_k$, $(C.3)_{k,\infty}$, $(C.3)^{\wedge}_{k,\infty}$ *and* $(C.4)_k$ *for some $k \ge 2$. Then $\{(\varphi_\varepsilon(t), F_\varepsilon(t))\}$ converges weakly as G^{k-2}-flows. The limit measure satisfies* (i) *and* (ii) *of Theorem* 5.3.1. □

For the proof of the theorem, it is sufficient to show the tightness of $\{(\varphi_\varepsilon, F_\varepsilon)\}$ on $\hat{W}_{k-2} \times W_{k-2}$. We have already shown the tightness of F_ε. In the following we will prove the tightness of $\{\varphi_\varepsilon\}$. We denote the law of φ_ε on $\hat{W}_j$ by the same character $\hat{P}_\varepsilon^j$.

Given a φ of $\hat{W}_j$ we denote by φ^{-1} an element of $\hat{W}_j$ defined by $\varphi^{-1}(t) = \varphi(t)^{-1}$. By the definition of the metric of $\hat{W}_j$ it is clear that a subset K of $\hat{W}_j$ is relatively compact if there exist two relatively compact subsets K_1, K_2 of W_j such that $K = (K_1 \cap \hat{W}_j) \cap \{\varphi \in \hat{W}_j : \varphi^{-1} \in K_2\}$. We have seen in the proof of Theorem 5.4.1 that for any $\delta > 0$ there exists a compact subset K_1 of W_{k-2} such that $\hat{P}_\varepsilon^{k-2}(K_1 \cap \hat{W}_{k-2}) > 1 - \delta/2$ for all $\varepsilon > 0$. Therefore it is sufficient to show the existence of a relatively compact subset K_2 of W_{k-2} such that $\hat{P}_\varepsilon^{k-2}(\{\varphi \in \hat{W}_{k-1} : \varphi^{-1} \in K_2\}) > 1 - \delta/2$ for all $\varepsilon > 0$. Thus the problem is reduced to showing the tightness of the laws of the inverse flow on the space W_{k-2}.

The tightness of the inverse flow $(\varphi_\varepsilon)\hat{}_{t,s} \equiv (\varphi_\varepsilon)_{s,t}^{-1}$ on the space W_{k-2} follows from the following theorem.

Theorem 5.4.6 *Assume that the family of local characteristics $\{(a_\varepsilon, \bar{b}_\varepsilon, \tilde{b}_\varepsilon)\}$ of $\{F_\varepsilon\}$ satisfies Conditions $(\mathrm{C.1})_{k,\infty}$, $(\mathrm{C.3})_{k,\infty}$ and $(\mathrm{C.3})\hat{}_{k,\infty}$ for some $k \geq 2$. Then for each $p > 3$ and $N > 0$, there exists a positive constant C such that for every ε and $|\alpha| \leq k - 1$*

$$E[|D^\alpha(\varphi_\varepsilon)\hat{}_{t,s}(x) - D^\alpha(\varphi_\varepsilon)\hat{}_{t',s}(x)|^p] \leq C|t - t'|^{2-(3/p)} \tag{38}$$

holds for any $s < t$, $s < t'$ and $|x| \leq N$.

Proof We will only give the proof of (38) in the case where $\alpha = 0$. The case $1 \leq |\alpha| \leq k - 1$ can be proved similarly. The inverse flow $(\varphi_\varepsilon)\hat{}_{t,s}$ satisfies the backward stochastic differential equation (36) and the local characteristic of $F_\varepsilon + \int 2\bar{c}_\varepsilon(r)\,dr$ satisfies Condition $(\mathrm{C.3})\hat{}_{k,\infty}$. Then we can apply Theorem 5.4.2 to the backward direction and we find that $\{D^\alpha(\varphi_\varepsilon)\hat{}_{t,s}(x)\}$ are uniformly L^p-bounded on compact set for any α with $|\alpha| \leq k$.

Now the inverse flow $(\varphi_\varepsilon)\hat{}_{t,s}$ satisfies also a forward Stratonovich stochastic differential equation:

$$(\varphi_\varepsilon)\hat{}_{t,s}(x) = x + \int_s^t \partial(\varphi_\varepsilon)\hat{}_{r,s}(x)\mathring{F}(x, \circ dr) \tag{39}$$

by Theorem 4.4.2, where $\mathring{F}_\varepsilon \equiv F_\varepsilon + \int \bar{c}_\varepsilon\,dr$. Therefore if $s < t' < t$ we have

$$|(\varphi_\varepsilon)\hat{}_{t,s} - (\varphi_\varepsilon)\hat{}_{t',s}|^p$$

$$= p\int_{t'}^t |(\varphi_\varepsilon)\hat{}_{r,s} - (\varphi_\varepsilon)\hat{}_{t',s}|^{p-1}\,\mathrm{sign}(r)\partial(\varphi_\varepsilon)\hat{}_{r,s}(x)\mathring{F}_\varepsilon(x, \circ dr),$$

where $\mathrm{sign}(r) = \mathrm{sign}((\varphi_\varepsilon)\hat{}_{r,s} - (\varphi_\varepsilon)\hat{}_{t',s})$. Then similarly to the proof of Lemma 5.4.4 the right hand side of the above is written as

$$\sum_{l=1}^{3} \int_{t'}^{t} \Psi_{l,\varepsilon}^{(3)}(t, r, (\varphi_\varepsilon)_{r,s}, \partial(\varphi_\varepsilon)_{r,s}) |(\varphi_\varepsilon)_{r,s} - (\varphi_\varepsilon)_{t',s}|^{p-l} \operatorname{sign}(r)^l \, dr$$

$$+ Y_{t,\varepsilon}^{(3)}(\cdot, t', t).$$

Here we denote by $\hat{\Psi}_{l,\varepsilon}^{(3)}(t, r)$ the conditional expectation of $\Psi_{l,\varepsilon}^{(3)}(t, r)$ with respect to $\mathscr{G}_r^\varepsilon$, $\{\hat{\Psi}_{l,\varepsilon}^{(3)}(t, r, (\varphi_\varepsilon)_{r,s}, \partial(\varphi_\varepsilon)_{r,s}) : \varepsilon > 0\}$ is uniformly L_p-bounded on compact sets for any $p > 1$. The $Y_{l,\varepsilon}^{(3)}(\cdot, t', t)$ is of mean 0. Then we have an inequality similar to (35) and hence for any N there exists a constant C such that for every ε

$$E[|(\varphi_\varepsilon)_{t,s}(x) - (\varphi_\varepsilon)_{t',s}(x)|^p] \leq C|t - t'|^{2-(3/p)}$$

holds for any $|x| \leq N$ and ε. This proves (38) with $\alpha = 0$. □

Combining Theorems 5.4.2 and 5.4.6, we obtain a more general uniform L^p-estimate.

Theorem 5.4.7 *Assume the same condition as in Theorem 5.4.6. Then for each $p > 3$ and $N > 0$, there exists a positive constant C such that for every ε and $|\alpha| \leq k - 1$*

$$E[|D^\alpha(\varphi_\varepsilon)_{s,t}(x) - D^\alpha(\varphi_\varepsilon)_{s',t'}(x)|^p] \leq C\{|t - t'|^{2-(3/p)} + |s - s'|^{2-(3/p)}\} \quad (40)$$

holds for any $s < t$, $s' < t'$ and $|x| \leq N$. Further the inverse flows $(\varphi_\varepsilon)_{t,s}$ also satisfy (40).

Proof Consider first the case where $t = t'$. Then the flow $(\varphi_\varepsilon)_{s,t}$ satisfies a backward equation with respect to the variable s similar to (39). Then Theorem 5.4.6 can be applied for $(\varphi_\varepsilon)_{s,t}$ (to the backward direction) and we obtain

$$E[|D^\alpha(\varphi_\varepsilon)_{s,t}(x) - D^\alpha(\varphi_\varepsilon)_{s',t}(x)|^p] \leq C|s - s'|^{2-(3/p)}.$$

This and (38) imply (40) immediately. The latter assertion of the theorem can be shown in exactly the same way. □

Finally we state the strong convergence as G^{k-2}-flows.

Theorem 5.4.8 *Assume that $\{(a_\varepsilon, \bar{b}_\varepsilon, \tilde{b}_\varepsilon)\}$ satisfies Conditions $(C.1)_{k,\infty}$, $(C.2)_k$, $(C.3)_{k,\infty}$, $(C.3)_{\hat{k},\infty}$ and $(C.4)_k$ for some $k \geq 2$. If $\{F_\varepsilon(x, t)\}$ converges strongly in probability for each x, the family of pairs $\{(F_\varepsilon, \varphi_\varepsilon)\}$ converges strongly as G^{k-2}-flows in L^p for any $p > 1$.* □

Exercise 5.4.9 *(Averaging problem)* In Exercise 5.2.10, assume further that $|D^\alpha f(x, t)| \le C$ holds for any α with $|\alpha| \le k$, where k is a positive integer greater than 2. Show that the stochastic flows $\{\varphi_\varepsilon(x, t)\}$ converge to the deterministic flow $\varphi_0(x, t)$ in the following sense:

$$E\left[\sup_{t\in[0,T],|x|\le N} |D_x^\alpha \varphi_\varepsilon(x, t) - D_x^\alpha \varphi_0(x, t)|^p\right] \to 0, \qquad \text{for } |\alpha| \le k-1$$

as $\varepsilon \to 0$.

5.5 Extensions of convergence theorems

Extensions of convergence theorems as diffusions

Let $\{F_\varepsilon(t)\}$ be a family of continuous C^k-semimartingales adapted to $(\mathscr{F}_t^\varepsilon)$ with local characteristics $(a_\varepsilon, b_\varepsilon, t)$ belonging to the class $B^{k,\delta}$, $\delta > 0$ and let $\varphi_\varepsilon(t) \equiv (\varphi_\varepsilon)_{0,t}$ be the stochastic flow generated by $F_\varepsilon(t)$ in the sense of the Itô integral. In Section 5.2 and 5.3 we studied the weak and strong convergences of the pairs $(\varphi_\varepsilon(t), F_\varepsilon(t))$ as diffusions, assuming Conditions $(C.1)_p$–$(C.4)_k$ for the family of local characteristics $(a_\varepsilon, b_\varepsilon, t)$. Conditions $(C.1)_p$ and $(C.3)_p$ required the uniform boundedness of the local characteristics. However in some applications these conditions are too restrictive. In this section we will replace them by localized Conditions $(C.1)_p^{\text{loc}}$ and $(C.3)_p^{\text{loc}}$ to be stated later and under these relaxed conditions we will discuss the weak and strong convergence of $(\varphi_\varepsilon, F_\varepsilon)$ as diffusions. A basic idea is the use of the 'truncated process' as the intermediate.

Let us introduce an assumption on the local characteristics $(a_\varepsilon, b_\varepsilon, t)$.

Condition $(C.1)_p^{\text{loc}}$ *For each $N > 0$ there exists a positive constant K such that*

$$E\left[\sup_{|x|,|y|\le N} \|D_x^\alpha D_y^\beta a_\varepsilon(x, y, t)\|^{p/2}\right]^{2/p} \le K, \qquad \textit{for } |\alpha|, |\beta| \le 1, \tag{1}$$

$$E\left[\sup_{|x|\le N} |D_x^\alpha b_\varepsilon(x, t)|^p\right]^{1/p} \le K, \qquad \textit{for } |\alpha| \le 1, \tag{2}$$

holds for any t and ε. □

We will first extend the results of Section 5.2. Note that Condition $(C.1)_p^{\text{loc}}$ is a localization of Condition $(C.1)_p$ of Section 5.2. It will be shown that Theorem 5.2.1 is valid under the above localized condition.

Theorem 5.5.1 *Assume that the family of local characteristics $\{(a_\varepsilon, b_\varepsilon)\}$ of $\{F_\varepsilon\}$ satisfies Conditions $(C.1)_p^{\text{loc}}$ and $(C.2)_k$ for some $p > 2$ and $k \ge 1$. Then*

$\{(\varphi_\varepsilon, F_\varepsilon)\}$ converges weakly as diffusions as $\varepsilon \to 0$. The limit measure satisfies properties (i) *and* (ii) *of Theorem* 5.2.1. □

Corollary 5.5.2 *Assume that the family of local characteristics $\{(a_\varepsilon, b_\varepsilon)\}$ of $\{F_\varepsilon\}$ satisfies Condition* $(C.1)_p^{\mathrm{loc}}$ *and one of Conditions* $(C.2)'_k$ *or* $(C.2)''_k$ *for some $p > 2$ and $k \geq 1$. Then $\{(\varphi_\varepsilon, F_\varepsilon)\}$ converges strongly as diffusions.* □

Next suppose that the local characteristic $b_\varepsilon(x, t)$ is decomposed as $b_\varepsilon(x, t) = \bar{b}_\varepsilon(x, t) + \tilde{b}_\varepsilon(x, t)$ where $\tilde{b}_\varepsilon(x, t)$ is $\mathscr{F}_0^\varepsilon$-measurable and $k + 2$-times differentiable with respect to x. Random functions $K_\varepsilon^{\alpha\beta}$, $L_\varepsilon^{\alpha\beta}$, A_ε, c_ε, are defined from $\tilde{b}_\varepsilon$ in exactly the same way as in Section 5.3. We shall relax Condition $(C.3)_p$ to the following Condition $(C.3)_p^{\mathrm{loc}}$ with $k = 1$.

Condition $(C.3)_p^{\mathrm{loc}}$ *For each $N > 0$ there exists a positive constant K such that*

$$E\left[\sup_{|x|,|y|\leq N} |K_\varepsilon^{\alpha\beta}(x, y, t, s)|^p\right]^{1/p} \leq K, \qquad \text{for } |\alpha| \leq 2, |\beta| \leq 1, \tag{3}$$

$$E\left[\sup_{|x|,|y|\leq N} |L_\varepsilon^{\alpha\beta}(x, y, t, s)|^p\right]^{1/p} \leq K, \qquad \text{for } |\alpha| \leq 3, |\beta| \leq 1, \tag{4}$$

hold for any s, t and ε. □

We can show that the convergence theorems in Section 5.3 are valid under the above assumption.

Theorem 5.5.3 *Assume that the family of local characteristics $\{(a_\varepsilon, \bar{b}_\varepsilon, \tilde{b}_\varepsilon)\}$ of $\{F_\varepsilon\}$ satisfies Conditions* $(C.1)_p^{\mathrm{loc}}$, $(C.2)_k$, $(C.3)_p^{\mathrm{loc}}$ *and* $(C.4)_k$ *for some $p > 3$ and $k \geq 1$. Then $\{(\varphi_\varepsilon, F_\varepsilon)\}$ converges weakly as diffusions as $\varepsilon \to 0$. Further, the limit measure satisfies* (i) *and* (ii) *of Theorem* 5.3.1. □

Corollary 5.5.4 *Assume further that $\{F_\varepsilon\}$ converges strongly as $\varepsilon \to 0$, then $\{\varphi_\varepsilon\}$ converges strongly as $\varepsilon \to 0$.* □

We shall give the proof of Theorem 5.5.3 only, since the proof of Theorem 5.5.1 is similar.

Under Conditions $(C.1)_p^{\mathrm{loc}}$ and $(C.3)_p^{\mathrm{loc}}$, we can not expect to get the uniform L^p-boundedness of φ_ε as in Lemma 5.3.3. However, as we will see soon, the family of the truncated processes $\{\varphi_\varepsilon^N\}$ has the property of the uniform L^p-boundedness for any N. This fact will help us to show the weak convergence of $\{\varphi_\varepsilon\}$ itself.

Given a positive integer N let $\psi_N(x)$ be a C^∞-function on $\mathbb{R}^d$ such that

$\psi_N(x) = 1$ if $|x| \le N$, $= 0$ if $|x| > N + 1$ and $0 \le \psi_N \le 1$ if $N \le |x| \le N + 1$. Set $F_\varepsilon^N(x, t) = F_\varepsilon(x, t)\psi_N(x)$. Let $\varphi_\varepsilon^N(t)$ be the stochastic flow generated by $F_\varepsilon^N(x, t)$ in the sense of the Itô integral. Let $P_{\varepsilon,N}^{(\mathbf{x}_0,\mathbf{y}_0)}$ be the law of the truncated $n + m$-point motion $(\varphi_\varepsilon^N(\mathbf{x}_0, t), F_\varepsilon^N(\mathbf{y}_0, t))$. Then for each N, $\{P_{\varepsilon,N}^{(\mathbf{x}_0,\mathbf{y}_0)}\}$ converges weakly as $\varepsilon \to 0$ by Theorem 5.3.1 since the local characteristics satisfy Conditions $(C.1)_p$, $(C.2)_k$, $(C.3)_p$ and $(C.4)_k$. Let $P_{0,N}^{(\mathbf{x}_0,\mathbf{y}_0)}$ be its limiting measure. By Theorem 5.3.1 there exists a probability measure $P_{0,N}$ on $(\hat{W}_k \times W_k, \mathscr{B}(\hat{W}_k \times W_k))$ with the following properties:

(i) $F(t)$ is a C^k-Brownian motion with local characteristic $(a(x, y, t)\psi_N(x)\psi_N(y), b(x, t)\psi_N(x))$,

(ii) φ_t is a Brownian flow of C^k-diffeomorphisms generated by $F(x, t) + \int_0^t \tilde{c}(x, r)\psi_N(x)\,dr$ in the sense of the Itô integral where $\tilde{c}(x, r) = c(x, r) + \sum_j a^{\cdot j}(x, x, r)\dfrac{\partial^4 N}{\partial x^j}(x)$. Here $F(t)$ and $\varphi(t)$ are the second and the first components of elements of $\hat{W}_k \times W_k$, respectively.

Now let P_0 be a probability measure on $(\hat{W}_k \times W_k, \mathscr{B}(\hat{W}_k \times W_k))$ satisfying (i) and (ii) of Theorem 5.3.1. Let $P_0^{(\mathbf{x}_0,\mathbf{y}_0)}$ be the law of $n + m$-point motion $(\varphi(\mathbf{x}_0, t), F(\mathbf{y}_0, t))$ with respect to P_0 defined on $V_{m,n}$. We wish to show that $\{P_\varepsilon^{(\mathbf{x}_0,\mathbf{y}_0)}\}$ converges weakly to $P_0^{(\mathbf{x}_0,\mathbf{y}_0)}$ as $\varepsilon \to 0$. To this end, we shall compare two measures $P_0^{(\mathbf{x}_0,\mathbf{y}_0)}$ and $P_{0,N}^{(\mathbf{x}_0,\mathbf{y}_0)}$. Note that the local characteristics with respect to $P_0^{(\mathbf{x}_0,\mathbf{y}_0)}$ and $P_{0,N}^{(\mathbf{x}_0,\mathbf{y}_0)}$ coincide if $|x| \le N$ and $|y| \le N$. Then the two measures $P_0^{(\mathbf{x}_0,\mathbf{y}_0)}$ and $P_{0,N}^{(\mathbf{x}_0,\mathbf{y}_0)}$ coincide with each other up to the time that $\varphi(t)$ leaves the ball B_N, i.e. we have

$$P_0^{(\mathbf{x}_0,\mathbf{y}_0)}(A \cap \{\varphi : \|\varphi\| < N\}) = P_{0,N}^{(\mathbf{x}_0,\mathbf{y}_0)}(A \cap \{\varphi : \|\varphi\| < N\})$$

for any Borel set A where $\|\varphi\| = \sup_t |\varphi(t)|$, provided $|\mathbf{x}_0|, |\mathbf{y}_0| < N$.

We shall now prove that $\{P_\varepsilon^{(\mathbf{x}_0,\mathbf{y}_0)}\}$ converges to $P_0^{(\mathbf{x}_0,\mathbf{y}_0)}$ as $\varepsilon \to 0$ for each $\mathbf{x}_0, \mathbf{y}_0$. It is sufficient to show that $\overline{\lim}_{\varepsilon\to 0} P_\varepsilon^{(\mathbf{x}_0,\mathbf{y}_0)}(S) \le P_0^{(\mathbf{x}_0,\mathbf{y}_0)}(S)$ holds for any closed set S of $V_{m,n}$. Now for any $\delta > 0$ choose N such that $P_0^{(\mathbf{x}_0,\mathbf{y}_0)}(G_N) > 1 - \delta$, where $G_N = \{\varphi : \|\varphi\| < N\}$. It is an open subset of $V_{m,n}$. Then we have $P_{0,N}^{(\mathbf{x}_0,\mathbf{y}_0)}(G_N) = P_0^{(\mathbf{x}_0,\mathbf{y}_0)}(G_N) > 1 - \delta$. Since $\{P_{\varepsilon,N}^{(\mathbf{x}_0,\mathbf{y}_0)}\}$ converges weakly as $\varepsilon \to 0$, we have $\underline{\lim}_{\varepsilon\to 0} P_{\varepsilon,N}^{(\mathbf{x}_0,\mathbf{y}_0)}(G_N) \ge P_{0,N}^{(\mathbf{x}_0,\mathbf{y}_0)}(G_N)$ (Theorem 1.1.3). Therefore, there exists $\varepsilon_0 = \varepsilon_0(N)$ such that $P_{\varepsilon,N}^{(\mathbf{x}_0,\mathbf{y}_0)}(G_N) \ge 1 - 2\delta$ holds for all $\varepsilon \le \varepsilon_0$. Using the set G_N, we have for any closed set S,

$$\begin{aligned} P_\varepsilon^{(\mathbf{x}_0,\mathbf{y}_0)}(S) &= P_\varepsilon^{(\mathbf{x}_0,\mathbf{y}_0)}(S \cap G_N) + P_\varepsilon^{(\mathbf{x}_0,\mathbf{y}_0)}(S \cap G_N^c) \\ &\le P_{\varepsilon,N}^{(\mathbf{x}_0,\mathbf{y}_0)}(S \cap G_N) + 2\delta, \end{aligned}$$

if $\varepsilon \le \varepsilon_0$. Therefore,

$$\begin{aligned} \overline{\lim_{\varepsilon\to 0}}\, P_\varepsilon^{(\mathbf{x}_0,\mathbf{y}_0)}(S) &\le \overline{\lim_{\varepsilon\to 0}}\, P_{\varepsilon,N}^{(\mathbf{x}_0,\mathbf{y}_0)}(S \cap \bar{G}_N) + 2\delta \\ &\le P_{0,N}^{(\mathbf{x}_0,\mathbf{y}_0)}(S \cap \bar{G}_N) + 2\delta \le P_0^{(\mathbf{x}_0,\mathbf{y}_0)}(S) + 2\delta. \end{aligned}$$

Since δ is arbitrary, we get $\overline{\lim}_{\varepsilon\to 0} P_\varepsilon^{(x_0,y_0)}(S) \le P_0^{(x_0,y_0)}(S)$. The proof of Theorem 5.5.3 is complete. □

Extensions of convergence theorems as stochastic flows

We shall next extend a theorem on the convergence as stochastic flows which was discussed in Section 5.4. We wish to replace Conditions $(C.1)_{k,\infty}$ and $(C.3)_{k,\infty}$ in Theorem 5.4.1 with the localized conditions introduced below.

Families of random fields $\{A_\varepsilon(x, y, s)\}$ or $\{A_\varepsilon(x, y, t, s)\}$ are called uniformly exponentially bounded in mean on compact sets, if for any $N > 0$, there exists a family of processes $\{K_\varepsilon^N(s)\}$ uniformly exponentially bounded in mean such that

$$\sup_{|x|,|y|\le N} |A_\varepsilon(x, y, s)| \le K_\varepsilon^N(s) \quad \text{or} \quad \sup_{|x|,|y|\le N,t} |A_\varepsilon(x, y, t, s)| \le K_\varepsilon^N(s)$$

holds for any ε and s.

Now let $\{F_\varepsilon\}$ be a family of forward–backward semimartingales with common forward and backward local characteristics $\{(a_\varepsilon, b_\varepsilon, t)\}$ belonging to the class B^{k+2}.

Condition $(C.1)_{k,\infty}^{\text{loc}}$ *The family of random fields* $\{D_x^\alpha D_y^\beta a_\varepsilon(x, y, s) : \varepsilon > 0\}$, $|\alpha| \le k, |\beta| \le k$ *and* $\{D_x^\alpha \bar{b}_\varepsilon(x, s) : \varepsilon > 0\}$, $|\alpha| \le k$ *are uniformly exponentially bounded in mean on compact sets.* □

Condition $(C.3)_{k,\infty}^{\text{loc}}$ *The family of random fields* $\{K_\varepsilon^{\alpha,\beta}(x, y, s, t) : \varepsilon > 0\}$, $|\alpha| \le k + 1, |\beta| \le k$ *and* $\{L_\varepsilon^{\alpha,\beta}(x, y, s, t) : \varepsilon > 0\}$, $|\alpha| \le k + 2, |\beta| \le k$ *are uniformly exponentially bounded in mean on compact sets.* □

Theorem 5.5.5 *Assume that the family of local characteristics* $\{(a_\varepsilon, \bar{b}_\varepsilon, \tilde{b}_\varepsilon)\}$ *of* $\{F_\varepsilon\}$ *satisfies Conditions* $(C.1)_{k,\infty}^{\text{loc}}$, $(C.2)_k$, $(C.3)_{k,\infty}^{\text{loc}}$ *and* $(C.4)_k$. *Then* $\{(\varphi_\varepsilon, F_\varepsilon)\}$ *converges weakly as* C^{k-1}*-flows as* $\varepsilon \to 0$. *Further the limit law satisfies* (i) *and* (ii) *of Theorem* 5.3.1. □

Corollary 5.5.6 *Assume further that* $\{F_\varepsilon(t)\}$ *converges strongly. Then* $\{\varphi_\varepsilon(t)\}$ *converges strongly as* C^{k-1}*-flows.* □

Proof of Theorem 5.5.5 We shall consider the truncated flow $\varphi_\varepsilon^N(t)$ generated by $F_\varepsilon^N(x, t) \equiv F(x, t)\psi_N(x)$ as before. Let $P_{\varepsilon,N}^{k-1}$ be the law of the pair $(\varphi_\varepsilon^N, F_\varepsilon^N)$ defined on $W_{k-1} \times W_{k-1}$. The local characteristics of F_ε^N, $\varepsilon > 0$ satisfy Conditions $(C.1)_{k,\infty}$, $(C.2)_k$, $(C.3)_{k,\infty}$ and $(C.4)_k$ for each N. Therefore $\{P_{\varepsilon,N}^{k-1} : \varepsilon > 0\}$ converges weakly as $\varepsilon \to 0$ by Theorem 5.4.1. Now let P_ε^{k-1} be the law of $(\varphi_\varepsilon, F_\varepsilon)$ defined on $W_{k-1} \times W_{k-1}$. Then similarly to the proof of Theorem 5.5.3 we can show that $\{P_\varepsilon^{k-1} : \varepsilon > 0\}$ converges weakly as $\varepsilon \to 0$. □

Finally we discuss the weak convergence as G^j-flows. We denote by $\hat{K}_\varepsilon^{\alpha\beta}$ and $\hat{L}_\varepsilon^{\alpha\beta}$ the random fields of types (2) and (3) of Section 5.3, defined in the backward direction. We introduce a further condition.

Condition (C.3)$\hat{}_{k,\infty}^{\text{loc}}$ *The family of random fields* $\{\hat{K}_\varepsilon^{\alpha,\beta} : \varepsilon > 0\}, |\alpha| \le k+1, |\beta| \le k$ *and* $\{\hat{L}_\varepsilon^{\alpha,\beta} : \varepsilon > 0\}, |\alpha| \le k+2, |\beta| \le k$ *are uniformly exponentially bounded in mean on compact sets.* □

Theorem 5.5.7 *Assume further that* $\{(a_\varepsilon, \bar{b}_\varepsilon, \tilde{b}_\varepsilon)\}$ *satisfies Condition* (C.3)$\hat{}_{k,\infty}^{\text{loc}}$ *in Theorem 5.5.5. Then* $\{(\varphi_\varepsilon, F_\varepsilon)\}$ *converges weakly as* G^{k-2}*-flows. The limit satisfies properties* (i) *and* (ii) *of Theorem 5.3.1.* □

The proof is left to the reader. It is obvious that we can prove the strong convergence as G^{k-2}-flows similarly.

5.6 Some limit theorems for stochastic differential equations

Central limit theorem for stochastic flows. I Mixing case

Let $f(x, t), g(x, t), x \in \mathbb{R}^d, t \in [0, \infty)$ be $\mathbb{R}^d$-valued continuous random fields, smooth in x such that their first derivatives are bounded in (x, t) a.s. Let $Y(x, t)$ be an $\mathbb{R}^d$-valued continuous forward–backward C^∞-semimartingale with the same local characteristic $(\tilde{a}(x, y, t), 0, t)$. Consider an Itô's stochastic differential equation with parameter $\varepsilon > 0$

$$\varphi_t = x + \varepsilon \int_0^t f(\varphi_r, r)\, dr + \varepsilon^2 \int_0^t g(\varphi_r, r)\, dr + \varepsilon \int_0^t Y(\varphi_r, dr). \tag{1}$$

Let $(\varphi_\varepsilon)_{0,t}$ be the associated stochastic flow. Then $\{(\varphi_\varepsilon)_{0,t}\}$ converges to a trivial flow $\varphi_{0,t}(x) \equiv x$ as ε tend to 0 since the right hand side of the above converges to 0. However, $\psi_\varepsilon(t) = (\varphi_\varepsilon)_{0,t/\varepsilon}$ satisfies

$$\psi_\varepsilon(t) = x + \int_0^t f\left(\psi_\varepsilon(r), \frac{r}{\varepsilon}\right) dr + \varepsilon \int_0^t g\left(\psi_\varepsilon(r), \frac{r}{\varepsilon}\right) dr + \varepsilon^{1/2} \int_0^t Z_\varepsilon(\psi_\varepsilon(r), dr), \tag{2}$$

where $Z_\varepsilon(x, t) = \varepsilon^{-1/2} Y\left(x, \frac{t}{\varepsilon}\right)$. Note that the second and third terms of the right hand side of (2) are negligible compared with the first term when ε is small. Then under an ergodic assumption on the random field $f(x, t)$, $\{\psi_\varepsilon(t)\}$ converges strongly to a deterministic flow ψ_t and it satisfies a differential equation $dx/dt = \bar{f}(x)$ where $\bar{f}(x) \equiv \lim_{T\to\infty} T^{-1}\int_0^T f(x, t)\,dt$ (see Exercise 5.2.10).

In this section we shall consider the case $\bar{f} = 0$. Then the limit flow ψ_t is again a trivial flow, so we shall change the time scale and consider $\tilde{\psi}_\varepsilon(t) = (\varphi_\varepsilon)_{0,t/\varepsilon^2}$. Then the flow is generated by

$$\tilde{F}_\varepsilon(x, t) = \frac{1}{\varepsilon}\int_0^t f\left(x, \frac{r}{\varepsilon^2}\right) dr + \int_0^t g\left(x, \frac{r}{\varepsilon^2}\right) dr + \int_0^t \tilde{Z}_\varepsilon(x, dr) \tag{3}$$

in the sense of the Itô integral, where

$$\tilde{Z}_\varepsilon(x, t) = \frac{1}{\varepsilon} Y\left(x, \frac{t}{\varepsilon^2}\right). \tag{4}$$

We shall study the weak convergence of $\{(\tilde{\psi}_\varepsilon(t), \tilde{F}_\varepsilon(t))\}$ as $\varepsilon \to 0$.

Such a limit theorem has been studied extensively in the case where $Y \equiv 0$, i.e. in the case of the stochastic ordinary differential equation. See Kesten–Papanicolaou [66], Khasminskii [68], Papanicolaou–Kohler [108]. In these works, the weak convergences as diffusions are studied. Here we will discuss the weak convergence both as diffusions and as stochastic flows including the case $Y \neq 0$.

We shall introduce some conditions on the local characteristics.

***Condition* $(C.6)_p^{\mathrm{loc}}$** *For each $N > 0$,*

$$\sup_t E\left[\sup_{|x|,|y|\le N} \|D_x^\alpha D_y^\beta a(x, y, t)\|^{2p}\right] < \infty, \qquad |\alpha| \le 1, \quad |\beta| \le 1,$$

$$\sup_t E\left[\sup_{|x|\le N} |D_x^\alpha f(x, t)|^{2p}\right] < \infty, \qquad |\alpha| \le 3,$$

$$\sup_t E\left[\sup_{|x|\le N} |D_x^\alpha g(x, t)|^{2p}\right] < \infty. \qquad |\alpha| \le 1. \quad \square$$

***Condition* $(C.6)_{k,\infty}^{\mathrm{loc}}$** *The triple (a, f, g) satisfies Condition $(C.6)_p^{\mathrm{loc}}$ for any p. Further for each $N > 0$ and $\lambda > 0$,*

$$\sup_t E\left[\exp\left\{\lambda \sup_{|x|,|y|\le N} \|D_x^\alpha D_y^\beta a(x, y, t)\|^2\right\}\right] < \infty, \quad |\alpha| \le k, \ |\beta| \le k,$$

$$\sup_t E\left[\exp\left\{\lambda \sup_{|x|\le N} |D^\alpha f(x, t)|^2\right\}\right] < \infty, \quad |\alpha| \le k + 2,$$

$$\sup_t E\left[\exp\left\{\lambda \sup_{|x|\le N} |D^\alpha g(x, t)|^2\right\}\right] < \infty, \quad |\alpha| \le k. \quad \square$$

Condition (C.7)$_k$ *There exist continuous functions $\bar{a} = (\bar{a}^{ij}(x, y, t))$, $A = (A^{ij}(x, y, t))$, $b = (b^i(x, t))$, $c = (c^i(x, t))$, $i, j = 1, \ldots, d$ which are k-times continuously differentiable with respect to x, y, the first derivatives being bounded. These satisfy the following properties:*

$$\left|\bar{a}^{ij}(x, y, t) - \frac{1}{\varepsilon}\int_t^{t+\varepsilon} E\left[a^{ij}\left(x, y, \frac{r}{\varepsilon^2}\right)\right] dr\right| \to 0, \quad (5)$$

$$\left|A^{ij}(x, y, t) - \frac{1}{\varepsilon^3}\int_t^{t+\varepsilon} d\tau \int_t^{\tau} d\sigma\, E\left[f^i\left(x, \frac{\sigma}{\varepsilon^2}\right) f^j\left(y, \frac{\tau}{\varepsilon^2}\right)\right]\right| \to 0, \quad (6)$$

$$\left|b^i(x, t) - \frac{1}{\varepsilon}\int_t^{t+\varepsilon} E\left[g^i\left(x, \frac{r}{\varepsilon^2}\right)\right] dr\right| \to 0, \quad (7)$$

$$\left|c^i(x, y, t) - \frac{1}{\varepsilon^3}\int_t^{t+\varepsilon} d\tau \int_t^{\tau} d\sigma \sum_j E\left[f^j\left(x, \frac{\sigma}{\varepsilon^2}\right) \frac{\partial f^i}{\partial x^j}\left(x, \frac{\tau}{\varepsilon^2}\right)\right]\right| \to 0, \quad (8)$$

uniformly on compact sets as ε tends to 0. □

Set for $s < t$,

$$\mathscr{G}_{s,t} \equiv \sigma(f(\cdot, u), g(\cdot, u), Y(\cdot, u) - Y(\cdot, v) : s \le u, v \le t).$$

Associated with $\mathscr{G}_{s,t}$ we introduce the *uniform mixing rate* ρ defined as

$$\rho(t) = \sup_{s \ge 0} \sup_{A \in \mathscr{G}_{s+t,\infty}, B \in \mathscr{G}_{0,s}} |P(A|B) - P(A)|. \quad (9)$$

We introduce a condition on the mixing rate.

Condition (C.8)$_p$ *$\rho(t)$ decreases to 0 as $t \uparrow \infty$ and satisfies*

$$\int_0^{\infty} \rho(s)^{(1/2)(1-(1/p))}\, ds < \infty. \quad (10)$$

Theorem 5.6.1 *Assume that the coefficients of equation* (1) *satisfy Conditions* (C.6)$_p^{\text{loc}}$, (C.7)$_k$ *and* (C.8)$_p$ *for some $k \ge 1$ and $p > 3$. Then the families of pairs $\{(\tilde{\psi}_\varepsilon(t), \tilde{F}_\varepsilon(t))\}$ converge weakly as diffusions as $\varepsilon \to 0$. Assume that the coefficients of equation* (1) *satisfy Conditions* (C.6)$_{k,\infty}^{\text{loc}}$, (C.7)$_k$ *and* (C.8)$_p$ *for some $k \ge 2$. Then they converge as G^{k-2}-flows. In both cases, the limit $(\tilde{\psi}, \tilde{F}, \tilde{P}_0)$ satisfies the following properties.*

(i) *$\tilde{F}(t)$ is a C^k-Brownian motion with local characteristic $(\bar{a} + \bar{A}, b)$ where $\bar{A}^{ij}(x, y, t) \equiv A^{ij}(x, y, t) + A^{ji}(y, x, t)$.*

(ii) *$\tilde{\psi}(t)$ is a C^k-Brownian flow generated by $\tilde{F}(x, t) + \int_0^t c(x, r)\, dr$ in the sense of the Itô integral.* □

Before we proceed to the proof, we prove two mixing lemmas. Let $p > 1$. We denote by q the conjugate of p, i.e. $p^{-1} + q^{-1} = 1$. The supremum norm of continuous functions on the set $\{x \in \mathbb{R}^d : |x| \le N\}$ is denoted by $\| \ \|_N$.

Lemma 5.6.2
(i) *Let X be a $\mathscr{G}_{t,\infty}$-measurable random variable such the $E[X] = 0$ and $E[|X|^p] < \infty$. Then for each $s < t$ we have*

$$E[|E[X|\mathscr{G}_{0,s}]|^p]^{1/p} \le 2\rho(t-s)^{1/q} E[|X|^p]^{1/p}. \tag{11}$$

(ii) *Let $X(x), x \in \mathbb{R}^d$ be a continuous random field, $\mathscr{G}_{t,\infty}$-measurable for each X, such that $E[X(x)] = 0$ for any x and $E[\|X\|_N^p] < \infty$ for any N. Suppose that $E[X(x)|\mathscr{G}_{0,s}]$ has a modification of a continuous random field. Then for each $s < t$ and $N > 0$ we have*

$$E[\|E[X(\cdot)|\mathscr{G}_{0,s}]\|_N^p]^{1/p} \le 2\rho(t-s)^{1/q} E[\|X\|_N^p]^{1/p}. \tag{12}$$

Proof Suppose that X is a discrete valued random variable of the form $X = \sum_i c_i \chi(A_i)$, where c_i are constants, A_i are elements of $\mathscr{G}_{t,\infty}$ and $\chi(A_i)$ are the indicator functions of the sets A_i. Then by Hölder's inequality

$$\begin{aligned} &|E[X|\mathscr{G}_{0,s}] - E[X]| \\ &= \left|\sum_i c_i(P(A_i|\mathscr{G}_{0,s}) - P(A_i))\right| \\ &\le \left\{\sum_i |c_i|^p |P(A_i|\mathscr{G}_{0,s}) - P(A_i)|\right\}^{1/p} \left\{\sum_i |P(A_i|\mathscr{G}_{0,s}) - P(A_i)|\right\}^{1/q}. \end{aligned} \tag{13}$$

Further, we have

$$\sum_i |c_i|^p |P(A_i|\mathscr{G}_{0,s}) - P(A_i)| \le E[|X|^p|\mathscr{G}_{0,s}] + E[|X|^p].$$

Let $P(\cdot|\mathscr{G}_{0,s}) - P(\cdot) = \mu_1 - \mu_2$ be the Hahn decomposition of the signed measure. Then we have $\mu_i(\Omega) \le \rho(t-s)$ a.s. for each $i = 1, 2$. Therefore

$$\sum_i |P(A_i|\mathscr{G}_{0,s}) - P(A_i)| \le 2\rho(t-s).$$

Then (13) implies

$$|E[X|\mathscr{G}_{0,s}] - E[X]|^p \le \{E[|X|^p|\mathscr{G}_{0,s}] + E[|X|^p]\}(2\rho(t-s))^{p/q}. \tag{14}$$

The above is valid for any X of $L_p(\Omega)$. Now take the expectations of both sides of the above to get (11).

Now let $X(x), x \in \mathbb{R}^d$ be a continuous random field satisfying the conditions of the lemma. From (14) we have

$$\|E[X(\cdot)|\mathscr{G}_{0,s}]\|_N^p \le \{E[\|X\|_N^p|\mathscr{G}_{0,s}] + E[\|X\|_N^p]\}(2\rho(t-s))^{p/q}. \tag{15}$$

Take the expectations of both sides of the above. Then we get (12).

Lemma 5.6.3

(i) *Let $s < t < u$ and X, Y be $\mathscr{G}_{u,u}$-measurable and $\mathscr{G}_{t,t}$-measurable random variables respectively such that $E[|X|^{2p}] < \infty$, $E[X] = 0$, $E[|Y|^{2p}] < \infty$, $E[Y] = 0$. Then we have*

$$E[|E[XY|\mathscr{G}_{0,s}] - E[XY]|^p]^{1/p} \le 4\rho(u-t)^{1/2q}\rho(t-s)^{1/2q}E[|X|^{2p}]^{1/2p}E[|Y|^{2p}]^{1/2p}. \quad (16)$$

(ii) *Let $s < t < u$ and $X(x)$, $Y(x)$, $x \in \mathbb{R}^d$ be continuous random fields, $\mathscr{F}_{u,u}$-measurable and $\mathscr{G}_{t,t}$-measurable respectively, such that $E[\|X\|_N^{2p}] < \infty$, $E[X(x)] = 0$, $E[\|Y\|_N^{2p}] < \infty$, $E[Y(y)] = 0$ hold for any x, y and $N > 0$. Suppose that $E[X(x)Y(y)|\mathscr{G}_{0,s}]$ has a modification of a continuous random field. Then for any N we have*

$$E\left[\sup_{|x|,|y|\le N} |E[X(x)Y(y)|\mathscr{G}_{0,s}] - E[X(x)Y(y)]|^p\right]^{1/p} \le 4\rho(u-t)^{1/2q}\rho(t-s)^{1/2q}E[\|X\|_N^{2p}]^{1/2p}E[\|Y\|_N^{2p}]^{1/2p}. \quad (17)$$

Proof We have by (14)

$$|E[X|\mathscr{G}_{0,t}]Y|^p \le \{E[|XY|^p|\mathscr{G}_{0,t}] + E[|X|^p]|Y|^p\}(2\rho(u-t))^{p/q}. \quad (18)$$

Therefore

$$E[|E[XY|\mathscr{G}_{0,s}]|^p]^{1/p} \le E[|E[XY|\mathscr{G}_{0,t}]|^p]^{1/p} \le 2\rho(u-t)^{1/q}E[|X|^{2p}]^{1/2p}E[|Y|^{2p}]^{1/2p}.$$

Then $|E[XY]|$ is also bounded by the same bound. Therefore we have

$$E[|E[XY|\mathscr{G}_{0,s}] - E[XY]|^p]^{1/p} \le 4\rho(u-t)^{1/q}E[|X|^{2p}]^{1/2p}E[|Y|^{2p}]^{1/2p}. \quad (19)$$

Further since XY is $\mathscr{G}_{t,\infty}$-measurable, we have by the previous lemma

$$E[|E[XY|\mathscr{G}_{0,s}] - E[XY]|^p]^{1/p} \le 2\rho(t-s)^{1/q}E[|X|^{2p}]^{1/2p}E[|Y|^{2p}]^{1/2p}. \quad (20)$$

The two inequalities (19) and (20) imply inequality (16).

We will next prove (17). We have from (18)

$$\sup_{|x|,|y|\le N} |E[X(x)Y(y)|\mathscr{G}_{0,t}]|^p \le \left\{E\left[\sup_{|x|,|y|\le N} |X(x)Y(y)|^p \middle| \mathscr{G}_{0,t}\right] + E[\|X\|_N^p]\,\|Y\|_N^p\right\}(2\rho(u-t))^{p/q}. \quad (21)$$

Therefore similarly to case (i), we obtain

$$E\left[\sup_{|x|,|y|\le N}|E[X(x)Y(y)|\mathscr{G}_{0,s}]-E[X(x)Y(y)]|^p\right]^{1/p}$$
$$\le 4\rho(u-t)^{1/q}E[\|X\|_N^{2p}]^{1/2p}E[\|Y\|_N^{2p}]^{1/2p}. \tag{22}$$

Further since $X(x)Y(y)$ is $\mathscr{G}_{t,\infty}$-measurable we have from the previous lemma

$$E\left[\sup_{|x|,|y|\le N}|E[X(x)Y(y)|\mathscr{G}_{0,s}]-E[X(x)Y(y)]|^p\right]^{1/p}$$
$$\le 2\rho(t-s)^{1/q}E[\|X\|_N^{2p}]^{1/2p}E[\|Y\|_N^{2p}]^{1/2p}. \tag{23}$$

Inequalities (22) and (23) yield (17). The proof is complete. □

Proof of Theorem 5.6.1 We first assume Conditions $(C.6)_p^{loc}$, $(C.7)_k$ and $(C.8)_p$ and prove the weak convergence as diffusions. It is sufficient to check Conditions $(C.1)_p^{loc}$, $(C.2)_k$, $(C.3)_p^{loc}$ and $(C.4)_k$ by setting $a_\varepsilon(x, y, t) = a(x, y, t/\varepsilon^2)$, $\bar{b}_\varepsilon(x, t) = g(x, t/\varepsilon^2)$, $\tilde{b}_\varepsilon(x, t) = \varepsilon^{-1}f(x, t/\varepsilon^2)$ and $\mathscr{F}_t^\varepsilon = \mathscr{G}_{0,t/\varepsilon^2}$. Condition $(C.1)_p^{loc}$ is immediate from Condition $(C.6)_p^{loc}$. We shall show that Condition $(C.2)_k$ holds. Set $\bar{a}_\varepsilon(x, y, t) = E[a(x, y, t/\varepsilon^2)]$. Then by Condition $(C.7)_k$, we have

$$\int_r^{r+\varepsilon}\bar{a}_\varepsilon^{ij}(x, y, u)\,du = \varepsilon\bar{a}^{ij}(x, y, r) + o(\varepsilon, x, y, r)$$

where $o(\varepsilon, x, y, r)/\varepsilon$ converge to 0 as $\varepsilon \to 0$ uniformly on compact sets with respect to x, y, r. Let $s = t_0 < t_1 < \cdots < t_n = t$ be a partition such that $t_h - t_{h-1} = \varepsilon$. Then

$$\left|\int_s^t \bar{a}_\varepsilon^{ij}(x, y, u)\,du - \int_s^t \bar{a}^{ij}(x, y, u)\,du\right|$$
$$\le (t-s)\sup_h \sup_{t_h<u\le t_{h+1}}|\bar{a}^{ij}(x, y, t_h) - \bar{a}^{ij}(x, y, u)|$$
$$+\left(\frac{t-s}{\varepsilon}+1\right)\sup_h o(\varepsilon, x, y, t_h).$$

This converges to 0 uniformly in x, y on compact sets as $\varepsilon \to 0$. Set $\tilde{a}_\varepsilon^{ij}(x, y, t) = a_\varepsilon^{ij}(x, y, t) - \bar{a}_\varepsilon^{ij}(x, y, t)$. Let q be the conjugate of $2p$. Lemma 5.6.2 yields

$$E\left[\sup_{|x|,|y|\le N}\left|\int_s^t E[\tilde{a}_\varepsilon^{ij}(x, y, u)|\mathscr{F}_s^\varepsilon]\,du\right|^{2p}\right]^{1/2p}$$
$$\le \int_s^t E\left[\sup_{|x|,|y|\le N}|E[\tilde{a}_\varepsilon^{ij}(x, y, u)|\mathscr{F}_s^\varepsilon]|^{2p}\right]^{1/2p}du$$
$$\le 2\left\{\int_s^t \rho\left(\frac{u-s}{\varepsilon^2}\right)^{1/q}du\right\}\sup_u E\left[\sup_{|x|,|y|\le N}|\tilde{a}_\varepsilon^{ij}(x, y, u)|^{2p}\right]^{1/2p}. \tag{24}$$

Since

$$\int_s^t \rho\left(\frac{u-s}{\varepsilon^2}\right)^{1/q} du \le \varepsilon^2 \int_0^\infty \rho(u)^{1/q}\, du \tag{25}$$

and the last integral is finite by Condition (C.8)$_p$, the last member of (24) converges to 0 as $\varepsilon \to 0$. These two convergences yield that for any $s < t$,

$$\int_s^t E[a_\varepsilon^{ij}(x, y, u)|\mathscr{F}_s^\varepsilon]\, du \to \int_s^t \bar{a}^{ij}(x, y, u)\, du$$

uniformly on compact sets in L^p. Similarly we can show

$$\int_s^t E\left[g^i\left(x, \frac{r}{\varepsilon^2}\right)\middle|\mathscr{F}_s^\varepsilon\right] dr \to \int_s^t b^i(x, r)\, dr$$

uniformly in x on compact sets. Therefore Condition (C.2)$_k$ is verified.

We will next prove Condition (C.3)$_p^{\text{loc}}$. It is sufficient to prove that

$$\sup_{|x|,|y|\le N} \frac{1}{\varepsilon^2}\left|\int_s^t E\left[D^\alpha f^i\left(x, \frac{u}{\varepsilon^2}\right)\middle|\mathscr{F}_s^\varepsilon\right] du\right|\left|D^\beta f^j\left(y, \frac{s}{\varepsilon^2}\right)\right|, \quad |\alpha| \le 1, \quad |\beta| \le 1, \tag{26}$$

are uniformly L^p-bounded, or

$$K_\varepsilon(s) \equiv \frac{1}{\varepsilon^2}\int_s^T \left\| E\left[D^\alpha f^i\left(\frac{u}{\varepsilon^2}\right)\middle|\mathscr{F}_s^\varepsilon\right]\right\|_N du, \qquad |\alpha| \le 1, \tag{27}$$

are uniformly L^{2p}-bounded. Let q be the conjugate of $2p$. Then by (15) we have

$$\begin{aligned} K_\varepsilon(s) \le{}& \frac{1}{\varepsilon^2}\left(\int_0^T 2^{1/q}\rho\left(\frac{u-s}{\varepsilon^2}\right)^{1/q} du\right) \sup_{u,\varepsilon} E\left[\left\|D^\alpha f^i\left(\frac{u}{\varepsilon^2}\right)\right\|_N^{2p}\right]^{1/2p} \\ &+ \frac{1}{\varepsilon^2}\int_0^T 2^{1/q}\rho\left(\frac{u-s}{\varepsilon^2}\right)^{1/q} E\left[\left\|D^\alpha f^i\left(\frac{u}{\varepsilon^2}\right)\right\|_N^{2p}\middle|\mathscr{F}_s^\varepsilon\right]^{1/2p} du. \end{aligned} \tag{28}$$

The first term is uniformly bounded with respect to ε and s, since (25) is satisfied. Then the second term is uniformly L^{2p}-bounded with respect to ε and s for the same reason. Therefore (27) is uniformly L^{2p}-bounded.

In order to prove Condition (C.4)$_k$ (8), consider

$$K_\varepsilon^{ij}(\tau, \sigma, x, y) = \frac{1}{\varepsilon^2} f^i\left(x, \frac{\tau}{\varepsilon^2}\right) f^j\left(y, \frac{\sigma}{\varepsilon^2}\right).$$

Set $\bar{K}_\varepsilon^{ij} = E[K_\varepsilon^{ij}]$, $\tilde{K}_\varepsilon^{ij} = K_\varepsilon^{ij} - \bar{K}_\varepsilon^{ij}$. Then by Condition (C.6)$_k$ we have

$$\int_t^{t+\varepsilon} d\tau \int_t^\tau d\sigma\, \bar{K}_\varepsilon^{ij}(\tau, \sigma, x, y) = \varepsilon A^{ij}(x, y, t) + o(\varepsilon, x, y, t)$$

where $o(\varepsilon, x, y, t)/\varepsilon$ converge to 0 uniformly in (x, y, t) on compact sets as $\varepsilon \to 0$. Further we can prove that $\{\int_s^t K_\varepsilon^{ij}(\tau, \sigma, x, y)\, d\sigma : \varepsilon > 0\}$ is uniformly L^p-bounded in (τ, x, y) on compact sets similarly to the case of random fields (26). Then $\{\int_t^{t+\varepsilon}(\int_s^t \bar{K}_\varepsilon^{ij}(\tau, \sigma, x, y)\, d\sigma)\, d\tau : \varepsilon > 0\}$ converges to 0 as $\varepsilon \to 0$ uniformly in (x, y, t) on compact sets. Therefore we have

$$\int_t^{t+\varepsilon} d\tau \int_s^\tau d\sigma\, \bar{K}_\varepsilon^{ij}(\tau, \sigma, x, y) = \varepsilon A^{ij}(x, y, t) + o(\varepsilon, x, y, t).$$

This proves

$$\lim_{\varepsilon \to 0} \int_s^t d\tau \int_s^\tau d\sigma\, \bar{K}_\varepsilon^{ij}(\tau, \sigma, x, y) = \int_s^t A^{ij}(x, y, \tau)\, d\tau$$

uniformly in (x, y) on compact sets as before.

On the other hand, we have by Lemma 5.6.3,

$$E\left[\left|\int_s^t d\tau \int_s^\tau d\sigma \sup_{|x|,|y| \le N} E[\tilde{K}_\varepsilon^{ij}(\tau, \sigma, x, y)|\mathscr{F}_s^\varepsilon]\right|^p\right]^{1/p}$$
$$\le 4\left(\int_s^t d\tau \int_s^\tau d\sigma\, \rho\left(\frac{\tau - \sigma}{\varepsilon^2}\right)^{(\frac{1}{2}-\frac{1}{p})} \rho\left(\frac{\sigma - s}{\varepsilon^2}\right)^{(\frac{1}{2}-\frac{1}{p})}\right)$$
$$\times \frac{1}{\varepsilon^2}\left(\sup_{t,l} E[\|f^l(t)\|_N^{2p}]^{1/p}\right). \tag{29}$$

Since

$$\frac{1}{\varepsilon^2}\int_s^t d\tau \int_s^\tau d\sigma\, \rho\left(\frac{\tau - \sigma}{\varepsilon^2}\right)^{(\frac{1}{2}-\frac{1}{p})} \rho\left(\frac{\sigma - s}{\varepsilon^2}\right)^{(\frac{1}{2}-\frac{1}{p})}$$
$$\le \varepsilon^2\left(\int_0^\infty \rho(u)^{(\frac{1}{2}-\frac{1}{p})}\, du\right)^2,$$

(29) converges to 0 as $\varepsilon \to 0$. These two computations yield

$$\lim_{\varepsilon \to 0} E\left[\int_s^t d\tau \int_s^\tau d\sigma\, K_\varepsilon^{ij}(\tau, \sigma, x, y)\middle|\mathscr{F}_s^\varepsilon\right] = \int_s^t A^{ij}(x, y, \tau)\, d\tau$$

uniformly in (x, y) on compact sets in L^p. This proves property (8) of Condition $(C.4)_k$.

Property (9) of Condition $(C.4)_k$ can be proved similarly. Property (10) of Condition $(C.4)_k$ follows from

$$E\left[\left\|E\left[\int_s^t \frac{1}{\varepsilon} D^\alpha f\left(\frac{u}{\varepsilon^2}\right) du\middle|\mathscr{G}_s^\varepsilon\right]\right\|_N^{2p}\right]^{1/2p}$$
$$\le \frac{2}{\varepsilon}\left(\int_s^t \rho\left(\frac{u - s}{\varepsilon^2}\right)^{(1-\frac{1}{p})} du\right)\left(\sup_t E[\|D^\alpha f(t)\|_N^{2p}]^{1/2p}\right) \underset{\varepsilon \to 0}{\longrightarrow} 0.$$

Therefore Condition $(C.4)_k$ is satisfied.

We will next prove the weak convergence as G^{k-2}-flows, assuming Conditions $(C.6)_{k,\infty}^{loc}$, $(C.7)_k$ and $(C.8)_\infty$. It is sufficient to prove Conditions $(C.1)_{k,\infty}^{loc}$, $(C.3)_{k,\infty}^{loc}$ and $(C.3)_{k,\infty}^{\wedge\, loc}$. Condition $(C.1)_{k,\infty}^{loc}$ is immediate from Condition $(C.6)_{k,\infty}^{loc}$. Indeed, since $\exp \lambda p$ is a convex function, we have

$$\sup_\varepsilon E\left[\exp\left\{\lambda \int_0^T \sup_{|x|,|y|\le N}\left|D_x^\alpha D_y^\beta a\left(x, y, \frac{u}{\varepsilon^2}\right)\right| du\right\}\right]$$

$$\le \sup_\varepsilon \frac{1}{T}\int_0^T E\left[\exp\left\{\lambda T \sup_{|x|,|y|\le N}\left|D_x^\alpha D_y^\beta a\left(x, y, \frac{u}{\varepsilon^2}\right)\right|\right\}\right] du < \infty.$$

We will next prove Condition $(C.3)_{k,\infty}^{loc}$. Set $2p = q = 2$ in (28). Then there exist positive constants c_1, c_2 such that

$$K_\varepsilon(s)^2 \le c_1 + c_2\int_0^{T/\varepsilon^2} \rho(v)^{1/2} E[\|D^\alpha f^i(v)\|_N^2 | \mathscr{F}_s^\varepsilon]\, dv.$$

Since $\exp \lambda x$ is a convex function we have by Jensen's inequality (Theorem 1.1.6),

$$E[\exp\{\lambda K_\varepsilon(s)^2\}] \le \exp(\lambda c_1)a^{-1}\int_0^{T/\varepsilon^2} \rho(v)^{1/2} E[\exp\{\lambda c_2 a \|D^\alpha f^i(v)\|_N^2\}]\, dv$$

where $a = \int_0^\infty \rho(v)^{1/2}\, dv$. Therefore $K_\varepsilon(s)^2$ is uniformly exponentially bounded in mean. Hence the first inequality of Condition $(C.3)_{k,\infty}^{loc}$ is established. The others can be shown similarly. □

Central limit theorem for stochastic flows. II Markov case

Let us again consider the limiting behavior of $\{(\tilde{\psi}_\varepsilon(t), \tilde{F}_\varepsilon(t))\}$ as $\varepsilon \to 0$, where $\tilde{F}_\varepsilon(t)$ is defined by (3) and $\tilde{\psi}_\varepsilon(t)$ is the stochastic flow generated by $\tilde{F}_\varepsilon(t)$. In the sequel we shall discuss the weak convergence of $\{(\tilde{\psi}_\varepsilon(t), \tilde{F}_\varepsilon(t))\}$ when $\tilde{F}_\varepsilon(t)$ is given by a functional of a Markov process $z(t)$.

Let $z(t)$ be a Feller process with state space S, a locally compact complete separable metric space. We assume it satisfies Condition (A) of Section 1.3 and is recurrent in the sense of Harris with invariant probability Λ.

Let $f(x, z)$, $g(x, z)$ and $a(x, y, z)$, $x, y \in \mathbb{R}^d$, $z \in S$ be $\mathbb{R}^d$-valued deterministic functions infinitely continuously differentiable in x and y, whose derivatives are bounded and continuous in (x, z) and (x, y, z) respectively. We assume that the random fields $f(x, t)$, $g(x, t)$ of equation (1) and the local characteristic $a(x, y, t)$ of $Y(x, t)$ are given by

$$f(x, t) = f(x, z(t)),\ g(x, t) = g(x, z(t)),\ a(x, y, t) = a(x, y, z(t)).$$

Instead of the mixing property and the existence of the infinitesimal quantities, we assume the following:

Condition (C.9) *The family of functions* $\{f(x,\cdot): x \in \mathbb{R}^d\}$ *is supported by a compact subset of S. Further,*

$$\int_S f(x,z)\Lambda(dz) = 0, \qquad \textit{for any } x. \quad \square$$

Let $\{P_t(z,\cdot)\}$ and $\{T_t\}$ be the transition probability and the semigroup of the Feller process z_t, respectively. Set

$$T_t f(x,z) = \int_S f(x,z')P_t(z,dz').$$

Then by Theorem 1.3.15, $\int_0^t T_s f(x,z)\,ds$ converges boundedly as $t \to \infty$. We denote the limit by $\hat{f}(x,z)$, i.e.

$$\hat{f}(x,z) = \lim_{t\uparrow\infty} \int_0^t T_s f(x,z)\,ds.$$

Further let $W(z,\cdot)$ be a recurrent potential kernel of $\{T_t\}$. Then we have

$$\hat{f}(x,z) = \int f(x,z')W(z,dz') - w(x),$$

where w is the integral of the first term of the right hand side with respect to z by the invariant measure Λ. Therefore $\hat{f}(x,z)$ is also a bounded C^∞-function of x with bounded derivatives.

Theorem 5.6.4 *Assume that the coefficients of equation* (1) *satisfy Condition* (C.9). *Then* $\{(\tilde{\psi}_\varepsilon(t), \tilde{F}_\varepsilon(t))\}$ *converges weakly as* C^∞*-flows. The limit* $(\tilde{\psi}(t), \tilde{F}(t), P_0)$ *satisfies the following properties.*

(i) $\tilde{F}(t)$ *is a* C^∞*-Brownian motion with local characteristic* $(\bar{a}, \bar{b})$, *where*

$$\bar{a}^{ij}(x,y) = \int_S a^{ij}(x,y,z)\Lambda(dz) + \int_S \{\hat{f}^i(x,z)f^j(y,z) + \hat{f}^j(y,z)f^i(x,z)\}\Lambda(dz),$$

$$\bar{b}(x) = \int_S g(x,z)\Lambda(dz).$$

(ii) $\tilde{\psi}(t)$ *is a* C^∞*-Brownian flow generated by* $\tilde{F}(x,t) + \int_0^t c(x,r)\,dr$ *in the sense of the Itô integral, where*

$$c^j(x) = \sum_{i=1}^d \int_S \frac{\partial \hat{f}^j}{\partial x^i}(x,z)f^i(x,z)\Lambda(dz).$$

Proof Set $\mathscr{G}_s^\varepsilon = \sigma\left(Y(x, s), z(s) : 0 \le s \le \frac{t}{\varepsilon^2}\right)$. We will show that Conditions (C.2)$_k$ and (C.4)$_k$ hold for any k. From the Markov property and the ergodic property of $z(t)$, we have

$$E\left[\int_s^t D_x^\alpha g\left(x, z\left(\frac{u}{\varepsilon^2}\right)\right) du \,|\, \mathscr{G}_s^\varepsilon\right]$$

$$= \varepsilon^2 \int_0^{(t-s)/\varepsilon^2} T_u D_x^\alpha g(x, z(s/\varepsilon^2))\, du \to (t-s) \int_S D_x^\alpha g(x, z)\Lambda(dz) \quad \text{a.s.}$$

for each x and $s < t$ (see Theorem 1.3.12). We have further

$$E\left[\int_s^t D_x^\alpha D_y^\beta a\left(x, y, z\left(\frac{u}{\varepsilon^2}\right)\right) du \,\middle|\, \mathscr{G}_s^\varepsilon\right] \to (t-s) \int_S D_x^\alpha D_y^\beta a(x, y, z)\Lambda(dz) \text{ a.s.}$$

$$\frac{1}{\varepsilon^2} E\left[\int_s^t D_x^\alpha f^i\left(x, z\left(\frac{\tau}{\varepsilon^2}\right)\right) d\tau \int_s^\tau D_y^\beta f^j\left(y, z\left(\frac{\sigma}{\varepsilon^2}\right)\right) d\sigma \,\middle|\, \mathscr{G}_s^\varepsilon\right]$$

$$= \varepsilon^2 E\left[\int_{s/\varepsilon^2}^{t/\varepsilon^2} D_y^\beta f^j(y, z(\sigma))\, d\sigma \int_\sigma^{t/\varepsilon^2} D_x^\alpha f^i(x, z(\tau))\, d\tau \,\middle|\, \mathscr{G}_s^\varepsilon\right]$$

$$= \varepsilon^2 \int_0^{(t-s)/\varepsilon^2} \left(\int_S P_\sigma\left(z\left(\frac{s}{\varepsilon^2}\right), dz\right) D_y^\beta f^j(y, z)\right.$$

$$\left. \times \int_0^{t/\varepsilon^2 - \sigma} T_\tau(D_x^\alpha f^i)(x, z)\, d\tau\right) d\sigma$$

$$\underset{\varepsilon \to 0}{\longrightarrow} (t-s) \int_S D_y^\beta f^j(y, z)\widehat{D_x^\alpha f^i}(x, z)\Lambda(dz) \quad \text{a.s.}$$

for any x, y and $s < t$. The above convergences occur uniformly in x or in (x, y) on compact sets by Sobolev's inequality. We have similarly,

$$\sum_j \frac{1}{\varepsilon^2} E\left[\int_s^t \frac{\partial f^i}{\partial x^j}\left(x, z\left(\frac{\tau}{\varepsilon^2}\right)\right) d\tau \int_s^\tau f^j\left(x, z\left(\frac{\sigma}{\varepsilon^2}\right)\right) d\sigma \,\middle|\, \mathscr{G}_s^\varepsilon\right]$$

$$\underset{\varepsilon \to 0}{\longrightarrow} \sum_j (t-s) \int_S \frac{\partial \hat{f}^i}{\partial x^j}(x, z) f^j(x, z)\Lambda(dz)$$

uniformly in x on compact sets for any $s < t$.

We next show that Conditions (C.1)$_{k,\infty}$ and (C.3)$_{k,\infty}$ hold. Condition (C.1)$_{k,\infty}$ is obvious since a and g and their derivatives are bounded. We have for any α and β

$$\left|\frac{1}{\varepsilon}E\left[\int_s^t D^\alpha f^i\left(x, z\left(\frac{r}{\varepsilon^2}\right)\right) \mathrm{d}r \,\middle|\, \mathscr{G}_s^\varepsilon\right]\right|$$

$$\times\left(1+\left|\frac{1}{\varepsilon}D^\beta f^j\left(y, z\left(\frac{s}{\varepsilon^2}\right)\right)\right|+\left|D^\beta g^j\left(y, z\left(\frac{s}{\varepsilon^2}\right)\right)\right|\right)$$

$$=\left|\left(\int_0^{(t-s)/\varepsilon^2} T_r(D^\alpha f^i)\left(x, z\left(\frac{s}{\varepsilon^2}\right)\right) \mathrm{d}r\right)\right|$$

$$\times\left(\varepsilon+\left|D^\beta f^j\left(y, z\left(\frac{s}{\varepsilon^2}\right)\right)\right|+\varepsilon\left|D^\beta g^j\left(y, z\left(\frac{s}{\varepsilon^2}\right)\right)\right|\right)$$

which is bounded by a constant not depending on $s, t, x, \varepsilon, \omega$ by Condition (C.9). The proof is complete. □

Remark Let μ be a measure on S appearing in Condition (A) in Section 1.3. Assume that the invariant measure Λ has a strictly positive continuous density function $\varphi(x)$ with respect to μ. Then $\hat{P}_t(y, A) = (\int \Lambda(\mathrm{d}x)P_t(x, y))/\varphi(y)$ is a transition probability with the invariant measure Λ. If the initial distribution of $z(t)$ is identical to the invariant measure, then $z(t)$ is a stationary Markov process. The time inverse $z(T-t) = \hat{z}(t), 0 \le t \le T$ is also a Markov process with transition probability $\hat{P}_t(z, \cdot)$. The potential theory of Section 1.3 is valid to $\{\hat{P}_t\}$. Then Condition $(\mathrm{C}.3)^{\wedge}_{k,\infty}$ is verified. Therefore $(\varphi_\varepsilon(t), F_\varepsilon(t))$ converge weakly as G^k-flows as $\varepsilon \to 0$ for any k.

Exercise 5.6.5 (*Papanicolaou–Stroock–Varadhan* [109]). Consider the following system of Itô's stochastic differential equations:

$$x_\varepsilon(t) = x + \frac{1}{\varepsilon}\int_0^t f(x_\varepsilon(u), z_\varepsilon(u))\,\mathrm{d}u + \int_0^t g(x_\varepsilon(u), z_\varepsilon(u))\,\mathrm{d}u$$

$$+\sum_{l=1}^n \int_0^t \sigma_{\cdot l}(x_\varepsilon(u), z_\varepsilon(u))\,\mathrm{d}\beta^l(u), \tag{30}$$

$$z_\varepsilon(t) = z + \frac{1}{\varepsilon^2}\int_0^t \tilde{f}(z_\varepsilon(u))\,\mathrm{d}u + \frac{1}{\varepsilon}\sum_{l=1}^m \int_0^t \tilde{\sigma}_{\cdot l}(z_\varepsilon(u))\,\mathrm{d}\tilde{\beta}^l(u),$$

where $(\beta^1(t), \ldots, \beta^n(t))$ and $(\tilde{\beta}^1(t), \ldots, \tilde{\beta}^m(t))$ are independent Brownian motions. The coefficients of the equations are all C_b^∞-functions and $\tilde{\sigma}\tilde{\sigma}^t$ is non-degenerate. Suppose that $z_1(t)$ has an invariant probability Λ. Suppose further $\int f(x, z)\Lambda(\mathrm{d}z) = 0$. Let $\varphi_\varepsilon^z(x, t)$ be the stochastic flow determined by equation (30) above. Show that $\{\varphi_\varepsilon^z(x, t)\}$ converges weakly as the G^k-flow.

5.7 Approximations of stochastic differential equations, supports of stochastic flows

Preliminaries

In this section we will apply the convergence theorems obtained in Sections 2–5 to get the approximation theorems for stochastic differential equations. A classical Stratonovich's stochastic differential equation is written as

$$\varphi_t = x + \int_s^t f_0(\varphi_u, u)\,du + \sum_{l=1}^n \int_s^t f_l(\varphi_u, u) \circ dB^l(u), \tag{1}$$

where $B(t) = (B^1(t), \ldots, B^n(t))$ is an n-dimensional Brownian motion and $f_l(x, t)$, $l = 1, \ldots, n$ (and $f_0(x, t)$) are continuous functions of the class $C^{k+1,\delta}$ ($C^{k,\delta}$, respectively) for some $k \geq 2$ and $\delta > 0$. Loosely it is written as

$$\frac{d\varphi_t}{dt} = f_0(\varphi_t, t) + \sum_{l=1}^n f_l(\varphi_t, t) b^l(t), \tag{2}$$

where $b^l(t) = dB^l(t)/dt$, $l = 1, \ldots, n$ are so called white noise processes: the processes $b^l(t)$ are independent at any different times. The term $\sum_{l=1}^n f_l(x, t) b^l(t)$ could be regarded as a random noise or a random force disturbing the dynamical system given by $d\varphi_t/dt = f_0(\varphi_t, t)$. However, the derivative of the Brownian motion does not exist at any time a.s., since the sample paths $B^l(t, \omega)$ are not of bounded variation at any time interval a.s. Thus equation (2) can not have a definite meaning. It can be defined rigorously through Itô's or Stratonovich's stochastic differential equations.

In a physical system the random noises or forces are not exactly white noises, since the noises must be more or less dependent on the past history. Further, the noise process should be a piecewise smooth function of the time. Thus a physical random equation could be written in the form

$$\frac{d\varphi_t}{dt} = f_0(\varphi_t, t) + \sum_{l=1}^n f_l(\varphi_t, t) v^l(t), \tag{3}$$

where $v(t) = (v^1(t), \ldots, v^n(t))$ is a piecewise smooth noise process which may be close to the white noise process. Equation (1) can be considered as an idealized equation of the physical one.

The problem that we will study in this section is as follows. Let $\{v_\varepsilon(t) = (v_\varepsilon^1(t), \ldots, v_\varepsilon^n(t))\}$ be a family of piecewise smooth processes such that it converges to a white noise or more precisely $\{B_\varepsilon(t) \equiv \int_0^t v_\varepsilon(r)\,dr\}$ converges to a Brownian motion $B(t)$. Let $(\varphi_\varepsilon)_t(x) \equiv \varphi_\varepsilon(x, t)$ be the solution of the stochastic ordinary differential equation

$$\frac{dx}{dt} = f_0(x, t) + \sum_{l=1}^n f_l(x, t) v_\varepsilon^l(t), \tag{4}$$

starting from x at time 0. It is a stochastic flow of C^k-diffeomorphisms. Then does the flow $\varphi_\varepsilon(t)$ converge as $\varepsilon \to 0$ to a Brownian flow φ_t determined by the stochastic differential equation (1)? The answer is not always affirmative: an additional term may appear in the right hand side in some cases. Such an example was found by Ikeda–Nakao–Yamato [48]. The details will be stated in Theorem 5.7.1.

We will first study the approximation problem in the framework of the weak convergence. The result is almost a direct consequence of the weak convergence theorems in Sections 5.4 and 5.5. Then we will study the strong approximation problem.

As an application of the approximation theorem, we will discuss the support of the Brownian flow governed by Stratonovich's stochastic differential equation (1). For this it is convenient to regard $v(t) = (v_1(t), \dots, v_n(t))$ in equation (3) as a deterministic piecewise smooth function. It is often called a *control function*. Denote by $\mathscr{V}$ the set of all control functions. Then for each v of $\mathscr{V}$, the solution of equation (3) defines a deterministic flow of C^k-diffeomorphisms, which we denote by $\varphi^v(t) \equiv \varphi^v_{0,t}$. Then the collection of the flows $\{\varphi^v : v \in \mathscr{V}\}$ can be regarded as a subset of $\hat{W}_j = C([0, T] : G^j)$ or $W_j = C([0, T] : C^j)$ if $j \le k$. Now let $\varphi(t)$ be the Brownian flow determined by Stratonovich's equation (1). Then its law can be defined on $\hat{W}_j$ and W_j for any $j \le k$, which we denote by $\hat{P}^j$ and P^j, respectively. The support of the measure is by definition the smallest closed subset F of $\hat{W}_j$ (or W_j) such that $\hat{P}^j(F) = 1$ (or $P^j(F) = 1$ respectively). The support of $\hat{P}^j$ is called the *support of the G^j-flow* and the support of P^j is the *support of the C^j-flow*. We will show in Theorem 5.7.6 that the support of the C^{k-1}-flow is the closure of $\{\varphi^v : v \in \mathscr{V}\}$ in the space W_{k-1}. Therefore we can determine the support of the Brownian flow through the deterministic flows associated with the control functions v. This fact will provide an alternative proof of the support theorems and examples stated in Section 4.9.

A weak approximation theorem

Let us first introduce some conditions for the processes $\{v_\varepsilon(t) : \varepsilon > 0\}$ in equation (3). Set

$$\mathscr{G}^\varepsilon_{s,t} \equiv \sigma(v_\varepsilon(u) : s \le u \le t). \tag{5}$$

Condition (C.10)$_\infty$ *The family of processes $\{K_\varepsilon(s)\}$ defined by*

$$K_\varepsilon(s) = \sum_{l,m=1}^{n} \left(\int_s^T \left| E\left[v^l_\varepsilon(r) | \mathscr{G}^\varepsilon_{0,s} \right] \right| dr \right) (1 + |v^m_\varepsilon(s)|), \qquad \varepsilon > 0 \tag{6}$$

is uniformly L^p-bounded for any $p > 1$ and uniformly exponentially bounded in mean. □

Condition (C.10)$^{\wedge}_{\infty}$ *The family of processes $\{\hat{k}_\varepsilon(s)\}$ defined by*

$$\hat{k}_\varepsilon(t) = \sum_{l,m=1}^{n} \left(\int_0^t \left| E\left[v_\varepsilon^l(r) | \mathscr{G}_{t,T}^\varepsilon \right] \right| dr \right) (1 + |v_\varepsilon^m(t)|), \qquad \varepsilon > 0 \tag{7}$$

is uniformly L^p-bounded for any $p > 1$ and is uniformly exponentially bounded in mean. □

Condition (C.11) *For each $s < t$, the convergence*

$$\int_s^t \left| E\left[v_\varepsilon^l(u) | \mathscr{G}_{0,\varepsilon}^\varepsilon \right] \right| du \xrightarrow[\varepsilon \to 0]{} 0 \tag{8}$$

holds in L^2. Further, there exist deterministic bounded measurable functions $v^{lm}(t)$ such that for any $s < t$

$$E\left[\int_s^t v_\varepsilon^l(\tau)\, d\tau \int_s^\tau v_\varepsilon^m(\sigma)\, d\sigma \middle| \mathscr{G}_{0,s}^\varepsilon \right] \xrightarrow[\varepsilon \to 0]{} \int_s^t v^{lm}(u)\, du \tag{9}$$

holds in L^1. □

The convergences (8) and (9) indicate that $\{B_\varepsilon(t)\}$ converges weakly to a continuous martingale with quadratic variation $\int_s^t (v^{lm}(u) + v^{ml}(u))\, du$. Therefore the limit $B(t)$ should be a Brownian motion with mean 0 and covariance equal to the quadratic variation. Condition (C.10)$_\infty$ ensures the tightness of the laws $\{B_\varepsilon(t) : \varepsilon > 0\}$. Since $v_\varepsilon^l(s)$ should converge to a white noise, its moment should diverge. Condition (C.10)$_\infty$ indicates that the rate of the divergence of $v_\varepsilon(s)$ and the convergence (8) should be balanced.

In order to state the weak convergence as C^j-flows or as G^j-flows, it is convenient to define the law of the pair $(\varphi_\varepsilon(t), B_\varepsilon(t))$ on the space $W_j \times V_n$ or $\hat{W}_j \times V_n$ where $W_j = C([0, T] : C^j)$, $\hat{W} = C([0, T] : G^j)$ and $V_n = C([0, T] : \mathbb{R}^n)$. Set the law on $W_j \times V_n$ and $\hat{W}_j \times V_n$ by P_ε^j and $\hat{P}_\varepsilon^j$ respectively. Then the family of pairs $\{\varphi_\varepsilon(t), B_\varepsilon(t)\}$ is said to *converge weakly as C^j-flows* (or *G^j-flows*) if $\{P_\varepsilon^j\}$ (or $\{\hat{P}_\varepsilon^j\}$) converges weakly.

Theorem 5.7.1 *Assume that the coefficients $f_l(x, t)$, $l = 1, \dots, n$ (and $f_0(x, t)$) of equation (4) are functions of the class $C_b^{k+1,\delta}$ ($C_b^{k,\delta}$, respectively) for some $k \geq 2$ and $\delta > 0$. Assume further that $\{v_\varepsilon(t)\}$ in equation (4) satisfies Conditions (C.10)$_\infty$ (and (C.10)$^{\wedge}_{\infty}$) and (C.11). Then the families of pairs $\{(\varphi_\varepsilon(t), B_\varepsilon(t))\}$ converge weakly as C^{k-1}-flows (and G^{k-2}-flows). Furthermore, the limit measure P_0^{k-1} (and $\hat{P}_0^{k-2}$) satisfies the following properties.*

(i) *$B(t)$ is an n-dimensional Brownian motion with mean 0 and covariance $(\int_0^t (v^{lm}(u) + v^{ml}(u))\,du)$.*
(ii) *$\varphi(t)$ and $B(t)$ are related by the Stratonovich's equation:*

$$\varphi(t) = x + \sum_l \int_0^t f_l(\varphi(u), u) \circ dB^l(u) + \int_0^t f_0(\varphi(u), u)\,du$$
$$+ \sum_{1 \le l \le m \le n} \int_0^t s_{l,m}(u)[f_l, f_m](\varphi(u), u)\,du, \tag{10}$$

where $s_{l,m}(t) = 2^{-1}(v^{l,m}(t) - v^{m,l}(t))$ and

$$[f_l, f_m]^j(x, t) = \sum_i f_l^i(x, t)\frac{\partial f_m^j}{\partial x^i}(x, t) - \sum_i f_m^i(x, t)\frac{\partial f_l^j}{\partial x^i}(x, t). \tag{11}$$

Before we proceed to the proof of the theorem, we give a lemma.

Lemma 5.7.2 *Assume Conditions* (C.10)$_\infty$ *and* (C.11). *Then for every continuous function $f(x, t)$ and $g(x, t)$ on $\mathbb{R}^d \times [0, T]$, the convergence*

$$E\left[\int_s^t f(x, \tau)v_\varepsilon^i(\tau)\,d\tau \int_s^\tau g(y, \sigma)v_\varepsilon^j(\sigma)\,d\sigma \,\middle|\, \mathscr{G}_{0,s}^\varepsilon\right]$$
$$\xrightarrow[\varepsilon \to 0]{} \int_s^t f(x, u)g(y, u)v^{ij}(u)\,du \tag{12}$$

holds uniformly in (x, y) on compact sets in L^1 for any $s < t$.

Proof Set $\mathscr{G}_s^\varepsilon = \mathscr{G}_{0,s}^\varepsilon$ for simplicity. By Condition (C.10)$_\infty$, there exists a positive constant K such that

$$E\left[\left|E\left[\int_s^t \tilde{f}(\tau)v_\varepsilon^l(\tau)\,d\tau \int_s^\tau \tilde{g}(\sigma)v_\varepsilon^m(\sigma)\,d\sigma \,\middle|\, \mathscr{G}_s^\varepsilon\right]\right|^p\right]$$
$$= E\left[\left|E\left[\int_s^t \left(\int_\sigma^t \tilde{f}(\tau)E[v_\varepsilon^l(\tau)|\mathscr{G}_\sigma^\varepsilon]\,d\tau\right)\tilde{g}(\sigma)v_\varepsilon^m(\sigma)\,d\sigma \,\middle|\, \mathscr{G}_s^\varepsilon\right]\right|^p\right]$$
$$\le K\{(t - s)\|\tilde{f}\|\,\|\tilde{g}\|\}^p \tag{13}$$

holds for any bounded measurable functions $\tilde{f}$, $\tilde{g}$ on $[0, T]$ where $\|\ \|$ is the supremum norm. Therefore it is sufficient to show the lemma in the case where f, g are step functions. We assume that there exists a partition $s = t_0 < \cdots < t_l = t$ such that $f(x, t) = f(x, t_h)$, $g(y, t) = g(y, t_h)$ holds if $t \in (t_h, t_{h+1}]$. Then

$$E\left[\int_s^t f(x,\tau)v_\varepsilon^l(\tau)\,d\tau\int_s^\tau g(y,\sigma)v_\varepsilon^m(\sigma)\,d\sigma\,\middle|\,\mathscr{G}_s^\varepsilon\right]$$
$$=\sum_h E\left[f(x,t_h)g(y,t_h)\int_{t_h}^{t_{h+1}}v_\varepsilon^l(\tau)\,d\tau\int_{t_h}^{\tau}v_\varepsilon^m(\sigma)\,d\sigma\,\middle|\,\mathscr{G}_s^\varepsilon\right]$$
$$+\sum_h E\left[f(x,t_h)\int_{t_h}^{t_{h+1}}v_\varepsilon^l(\tau)\,d\tau\int_s^{t_h}g(y,\sigma)v_\varepsilon^m(\sigma)\,d\sigma\,\middle|\,\mathscr{G}_s^\varepsilon\right].$$

By Condition (C.11), the first term converges to

$$\sum_h f(x,t_h)g(y,t_h)\int_{t_h}^{t_{h+1}}v^{lm}(u)\,du=\int_s^t f(x,u)g(y,u)v^{lm}(u)\,du$$

uniformly in x, y on compact sets in L^1 as $\varepsilon\to 0$. Concerning the second term, we have

$$E\left[\left|\sum_h f(x,t_h)E\left[\int_{t_h}^{t_{h+1}}E[v_\varepsilon^l(\tau)|\mathscr{G}_{t_h}^\varepsilon]\,d\tau\int_s^{t_h}g(y,\sigma)v_\varepsilon^m(\sigma)\,d\sigma\,\middle|\,\mathscr{G}_s^\varepsilon\right]\right|\right]$$
$$\le\sum_h|f(x,t_h)|E\left[\left|\int_{t_h}^{t_{h+1}}E[v_\varepsilon^l(\tau)|\mathscr{G}_{t_h}^\varepsilon]\,d\tau\right|^2\right]^{1/2}$$
$$\times E\left[\left|E\left[\int_s^{t_h}g(y,\sigma)v_\varepsilon^m(\sigma)\,d\sigma\,\middle|\,\mathscr{G}_s^\varepsilon\right]\right|^2\right]^{1/2}.$$

Expectations $\{E[|\int_{t_h}^{t_{h+1}}\dots|^2]:\varepsilon>0\}$ converge to 0 by (8), and further $\{E[|\int_s^{t_{h+1}}\dots|^2]:\varepsilon>0\}$ is bounded by (13). Therefore the above sum converges to 0 uniformly on compact sets as $\varepsilon\to 0$. The assertion is thus verified for step functions $f(x,t)$ and $g(x,t)$. □

Proof of Theorem 5.7.1 Set

$$F_\varepsilon(x,t)=\int_0^t f_0(x,u)\,du+\sum_{l=1}^n\int_0^t f_l(x,u)v_\varepsilon^l(u)\,du.$$

We shall consider the weak convergence of the triples $\{(\varphi_\varepsilon(t),F_\varepsilon(t),B_\varepsilon(t))\}$ instead of the pairs $\{(\varphi_\varepsilon(t),B_\varepsilon(t))\}$. The law of the triple $(\varphi_\varepsilon(t),F_\varepsilon(t),B_\varepsilon(t))$ is defined on the space $W_{k-1}\times W_{k-1}\times V_n$ or $\hat{W}_{k-2}\times W_{k-2}\times V_n$. We denote it by the same symbol P_ε^{k-1} or $\hat{P}_\varepsilon^{k-2}$. If Conditions $(C.3)_{k,\infty}^{loc}$ (and $(C.3)_{k,\infty}^{\wedge\,loc}$) and $(C.4)_k$ are satisfied, then we can show that the family of laws $\{P_\varepsilon^{k-1}\}$ (and $\{\hat{P}_\varepsilon^{k-2}\}$) converges weakly as $\varepsilon\to 0$ in exactly the same way as in Theorem 5.5.5 (or Theorem 5.5.7).

Now, set $\tilde{b}_\varepsilon(x,r)=\sum_{l=1}^n f_l(x,r)v_\varepsilon^l(r)$ and $\bar{b}_\varepsilon(x,r)=f_0(x,r)$. Let $K_\varepsilon^{\alpha\beta}$ be random fields defined by (2) of Section 5.3. Then

$$K_\varepsilon^{\alpha\beta}(x, y, t, s) \le \sum_{l,m} \left\{ \int_s^t |D_x^\alpha f_l(x, r)| |E[v_\varepsilon^l(r)|\mathcal{G}_s^\varepsilon]| \, dr \right\}$$

$$\times (1 + |D_y^\beta f_m(y, s)| |v_\varepsilon^m(s)| + |D_y^\beta f_0(y, s)|).$$

Then for each $N > 0$ $\sup_{|x|, |y| \le N, t} K_\varepsilon^{\alpha,\beta}(x, y, t, s)$ is bounded by a constant c times $K_\varepsilon(s)$ of (6). Therefore Condition $(C.3)_{k,\infty}^{\text{loc}}$ is satisfied. Condition $(C.3)_{k,\infty}^{\wedge\,\text{loc}}$ can be shown in the same way if Condition $(C.10)_\infty^\wedge$ is satisfied.

We shall next examine Condition $(C.4)_k$. Observe the inequality

$$\sup_{|x| \le N} \left| E\left[\int_s^t D^\alpha \tilde{b}_\varepsilon(x, r) \, dr \middle| \mathcal{G}_s^\varepsilon \right] \right|$$

$$\le \sum_l \int_s^t \sup_{|x| \le N} |D^\alpha f_l(x, r)| |E[v_\varepsilon^l(r)|\mathcal{G}_s^\varepsilon]| \, dr.$$

The right hand side converges to 0 in L^1 by Condition (C.11). Let $\tilde{A}_\varepsilon^{ij}$ be defined by (15) of Section 5.3. Then since

$$E\left[\int_s^t \tilde{A}_\varepsilon^{ij}(x, y, t, r) \, dr \middle| \mathcal{G}_s^\varepsilon \right]$$

$$= \sum_{l,m} E\left[\int_s^t f_l^i(x, \tau) v_\varepsilon^l(\tau) \, d\tau \int_s^\tau f_m^j(y, \sigma) v_\varepsilon^m(\sigma) \, d\sigma \middle| \mathcal{G}_s^\varepsilon \right],$$

it converges to

$$\sum_{l,m} \int_s^t f_l^i(x, r) f_m^j(y, r) v^{ij}(r) \, dr$$

uniformly in (x, y) on compact sets in L^1 for any $s < t$ by Lemma 5.7.2. This proves (16) in Section 5.3 or equivalently (8) in Condition $(C.4)_k$. The convergence (9) in Condition $(C.4)_k$ can be proved similarly. The convergence (10) is obvious from Condition (C.11).

Let $(\varphi(t), F(t), B(t), P_0^{k-1})$ be the weak limit of $\{P_\varepsilon^{k-1}\}$ as $\varepsilon \to 0$. Then the pair $(\varphi(t), F(t))$ satisfies the properties stated in Theorem 5.3.1, i.e.

(i) $F(x, t)$ is a C^k-Brownian motion with local characteristic given by

$$a^{ij}(x, y, t) = \sum_{l,m} f_l^i(x, t) f_m^j(y, t) \bar{v}^{lm}(t), \qquad \bar{b}^i(x, t) = f_0^i(x, t)$$

where $\bar{v}^{lm}(t) = v^{lm}(t) + v^{ml}(t)$.

(ii) $\varphi(t)$ is a Brownian flow generated by $F(x, t) + \int_0^t c(x, r) \, dr$ in the sense of the Itô integral, where

$$c(x, t) = \sum_{l,m,i} \frac{\partial f_l}{\partial x^i}(x, t) f_m^i(x, t) v^{lm}(t).$$

We wish to show that $(B(t), P_0^{k-1})$ has the following property:

(iii) $B(t)$ is a Brownian motion with mean 0 and covariance $\int_0^t \bar{v}^{ij}(r)\, dr$. Furthermore, $F(x, t)$ is represented as

$$F(x, t) = \sum_{l=1}^{n} \int_0^t f_l(x, u)\, dB^l(u) + \int_0^t f_0(x, u)\, du. \tag{14}$$

Similarly to Lemmas 5.3.4 and 5.3.5 we can show that $B(t)$ is a continuous martingale such that

$$\langle B^l(t), B^m(t)\rangle = \int_0^t \bar{v}^{lm}(u)\, du,$$

$$\langle Y^i(x, t), B^m(t)\rangle = \sum_l \int_0^t f_l^i(x, u)\bar{v}^{lm}(u)\, du,$$

where $Y(x, t) = F(x, t) - \int_0^t f_0(x, u)\, du$. Set $\tilde{Y}(x, t) \equiv \sum_l \int_0^t f_l(x, u)\, dB^l(u)$. Then we have for every i, j

$$\langle \tilde{Y}^i(x, t), \tilde{Y}^j(x, t)\rangle = \langle Y^i(x, t), \tilde{Y}^j(x, t)\rangle = \int_0^t a^{ij}(x, x, u)\, du.$$

Further $\langle Y^i(x, t), Y^j(x, t)\rangle$ is also equal to the above by property (i). Therefore we have $\langle Y^i(x, t) - \tilde{Y}^i(x, t)\rangle = 0$ for any x and t. This yields $Y = \tilde{Y}$, proving (14).

We have thus shown that $\varphi(x, t)$ satisfies an Itô's equation:

$$\varphi(x, t) = x + \sum_l \int_0^t f_l(\varphi(x, u), u)\, dB^l(u) + \int_0^t (f_0 + c)(\varphi(x, u), u)\, du.$$

We will rewrite the above equation using the Stratonovich integral. We have

$$\begin{aligned} &\int_0^t f_l(\varphi(x, u), u) \circ dB^l(u) \\ &= \int_0^t f_l(\varphi(x, u), u)\, dB^l(u) \\ &\quad + \frac{1}{2}\sum_{l,m} \int_0^t \frac{\partial f_l}{\partial x^i}(\varphi(x, u), u) f_m^i(\varphi(x, u), u)\bar{v}^{lm}(u)\, du \end{aligned}$$

(see Section 3.4). Furthermore, since

$$c(x, u) - \frac{1}{2} \sum_{l,m,i} \frac{\partial f_l}{\partial x^i}(x, u) f_m^i(x, u) \bar{v}^{lm}(u)$$

$$= \frac{1}{2} \sum_{1 \le l \le m \le n} (v^{lm}(u) - v^{ml}(u)) [f_l, f_m](x, u),$$

$\varphi(t)$ satisfies equation (10). The proof is complete. □

Strong approximation theorems

We will now study the strong approximation problem of the solution of a stochastic differential equation.

Theorem 5.7.3 *Assume the same conditions as in Theorem* 5.7.1. *If* $\{B_\varepsilon(t)\}$ *converges to* $B(t)$ *strongly as* $\varepsilon \to 0$, *then* $\{\varphi_\varepsilon(t)\}$ *converges strongly to a Brownian flow* $\varphi(t)$ *as* C^{k-1}*-flows in* L^p *for any* $p > 1$. *The limit* $\varphi(t)$ *satisfies the stochastic differential equation* (10).

Assume further that $\{v_\varepsilon(t)\}$ *satisfies Condition* (C.10)$_\infty^\wedge$. *Then* $\{\varphi_\varepsilon(t)\}$ *converges strongly as* G^{k-2}*-flows in* L^p *for any* $p > 1$. □

The proof can be carried out similarly to the proof of Theorem 5.2.8, using Theorem 5.7.1. Details are left to the reader.

Remark The above theorem gives us an alternative method of proving that the solution of the stochastic differential equation (1) defines a stochastic flow of C^{k-2}-diffeomorphisms, not relying on the arguments of Sections 4.5 and 4.6. Indeed it is obvious that the solutions $(\varphi^\varepsilon)_t$ of the stochastic ordinary differential equation (4) define stochastic flows of C^k-diffeomorphisms. Then Theorem 5.7.3 tells us that $\{(\varphi_\varepsilon)_t : \varepsilon > 0\}$ converges strongly as G^{k-2}-flows. Then the limit φ_t must be a stochastic flow governed by equation (1) in the case where $s_{l,m}(t) = 0$. See further Bismut [9] and Ikeda–Watanabe [49].

Example 5.7.4 (*Polygonal approximations*) Let $B(t) = (B^1(t), \ldots, B^n(t))$ be an n-dimensional standard Brownian motion. Set for $l = 1, \ldots, n$.

$$v_\varepsilon^l(t) = \frac{1}{\varepsilon} \Delta_h^\varepsilon B^l, \qquad \text{if } \varepsilon h \le t \le \varepsilon(h+1), \tag{15}$$

where $\Delta_h^\varepsilon B^l = B^l(\varepsilon(h+1)) - B^l(\varepsilon h)$. Then for each ε, $B_\varepsilon^l(t) \equiv \int_0^t v_\varepsilon^l(u)\,du$ is a polygonal function of t and converges to $B^l(t)$ uniformly in t as $\varepsilon \to 0$.

We will show that this approximation satisfies Condition (C.10)$_p$. Since $v_\varepsilon^l(t)$ and $v_\varepsilon^l(s)$ are independent if $(t - s) > \varepsilon$ we have

$$\int_s^t |E[v_\varepsilon^l(u)|\mathscr{G}_s^\varepsilon]|\, du \le |\Delta_h^\varepsilon B^l| \qquad \text{if } \varepsilon h \le s \le \varepsilon(h+1). \tag{16}$$

The variance of $\Delta_h^\varepsilon B^l$ is ε and its pth moment is $c\varepsilon^{p/2}$ since it is a Gaussian random variable. Therefore the pth moment of $K_\varepsilon(s)$ of (6) is bounded by a constant independent of ε.

Further we have

$$E\left[\exp\left\{\lambda \int_0^T ds\left(\int_s^T |E[v_\varepsilon^l(u)|\mathscr{G}_s^\varepsilon]|\, du\right)|v_\varepsilon^j(s)|\right\}\right]$$
$$\le E\left[\exp\left\{\lambda \sum_{h=0}^{[T/\varepsilon]} |\Delta_h^\varepsilon B^l||\Delta_h^\varepsilon B^j|\right\}\right]$$
$$\le E\left[\exp\left\{\lambda \sum_{h=0}^{[T/\varepsilon]} |\Delta_h^\varepsilon B^l|^2\right\}\right]^{1/2} E\left[\exp\left\{\lambda \sum_{h=0}^{[T/\varepsilon]} |\Delta_h^\varepsilon B^j|^2\right\}\right]^{1/2}.$$

Since $\lambda^{1/2}\Delta_h^\varepsilon B^l$ is a Gaussian random variable with mean 0 and covariance $\lambda\varepsilon$, we have

$$E[\exp(\lambda|\Delta_h^\varepsilon B^l|^2)] = (2\pi\lambda\varepsilon)^{-1/2}\int_{-\infty}^{\infty} \exp(-x^2/2\lambda\varepsilon)\exp(x^2)\, dx$$
$$= (1 - 2\lambda\varepsilon)^{-1/2}$$

if $2\lambda\varepsilon < 1$. Therefore we have

$$E\left[\exp\left\{\lambda \sum_{h=0}^{[T/\varepsilon]} |\Delta_h^\varepsilon B^l|^2\right\}\right] = \prod_{h=1}^{[T/\varepsilon]} (1 - 2\lambda\varepsilon)^{-1/2}.$$

This converges to $e^{\lambda T}$ as $\varepsilon \to 0$. Then the uniform exponential boundedness in mean of $K_\varepsilon(t)$ of (6) follows.

The convergence (8) is obvious from (16). The convergence (9) is easily verified with $v_{ij} = \delta_{ij}$. Therefore Theorem 5.7.3 is valid for any polygonal approximation. The limit $\varphi(t)$ satisfies Stratonovich's stochastic differential equation (10) with $s_{l,m} = 0$ since $v^{ij} = v^{ji} = 2^{-1}\delta_{ij}$. □

Supports of stochastic flows

Let $\varphi^v(t)$ be a deterministic flow determined by an ordinary differential equation (3), where v is an element of $\mathscr{V}$. We will first establish a lemma.

Lemma 5.7.5 *For any $\theta, N > 0$, we have*

$$P\left(\|\varphi - \varphi^v\|_{j:N} < \theta \,\Big|\, \sup_t |B(t) - V(t)| < \delta\right) \to 1 \tag{17}$$

as $\delta \to 0$, where $V(t) = \int_0^t v(s)\,ds$ and $P(A|B)$ is the conditional probability of event A given event B.

Proof Let M be a positive number such that $M/2 \geq |\varphi^v|_{j:N} + \theta$. Let $\varphi^M(t)$ be a truncated process such that $|\varphi^M|_{j:N} \leq M$ a.s. and $\varphi(t) = \varphi^M(t)$ holds for $t < \tau_M$, where

$$\tau_M = \inf\left\{t \in [0, T] : \sum_{|\alpha| \leq j} \sup_{|x| \leq N} |D^\alpha \varphi(x, t)| \geq \frac{M}{2}\right\}.$$

Then $\varphi(t) = \varphi^M(t)$ holds for all $t \in [0, T]$ if $\varphi^M(t)$ satisfies $\|\varphi^M - \varphi^v\|_{j:N} < \theta$. Therefore, instead of (17) it is sufficient to prove

$$P\left(|\varphi^M - \varphi^v|_{j:N} < \theta \,\middle|\, \sup_t |B(t) - V(t)| < \delta\right) \to 1 \tag{18}$$

as $\delta \to 0$. By a proof similar to that in Stroock–Varadhan [117], we can prove that for each x with $|x| \leq N$,

$$P\left(\sup_t |D^\beta \varphi^M(x, t) - D^\beta \varphi^v(x, t)| > \theta \,\middle|\, \sup_t |B(t) - V(t)| < \delta\right) \to 0$$

as $\delta \to 0$. Therefore we have

$$E\left[\sup_t |D^\beta \varphi^M(x, t) - D^\beta \varphi^v(x, t)|^p \,\middle|\, \sup_t |B(t) - V(t)| < \delta\right] \to 0$$

as $\delta \to 0$. Integrate the above with respect to x over the ball B_N and sum over β such that $|\beta| \leq j + 1$. Then we obtain

$$E\left[(|\varphi^M - \varphi^v|_{j+1,p:N})^p \,\middle|\, \sup_t |B(t) - V(t)| < \delta\right] \to 0.$$

Sobolev's inequality then yields

$$E\left[(|\varphi^M - \varphi^v|_{j:N})^p \,\middle|\, \sup_t |B(t) - V(t)| < \delta\right] \to 0$$

as $\delta \to 0$ provided $p > d$. This implies (18).

Theorem 5.7.6 *Assume that the coefficients $f_l(x, t)$, $l = 1, \ldots, n$ (and $f_0(x, t)$) of the Stratonovich equation (1) are of the class $C_b^{k+1,\delta}$ ($C_b^{k,\delta}$, respectively) for some $k \geq 2$ and $\delta > 0$. Let $\varphi_{s,t}$ be the Brownian flow determined by equation (1). Then the support of $\varphi(t) \equiv \varphi_{0,t}$ as the C^{k-1}-flow is equal to the closure of $\{\varphi_t^v : v \in \mathscr{V}\}$ in the space W_{k-1}. Further the support as the G^{k-2}-flow is included in the closure of $\{\varphi_t^v : v \in \mathscr{V}\}$ in $\hat{W}_{k-2}$.*

Proof We shall use the polygonal approximation stated in Example 5.7.4. Let $\varphi_\varepsilon(t) \equiv (\varphi_\varepsilon)_{0,t}$ be the stochastic flow determined by equation (4) where $v_\varepsilon(t)$ is given by (15). Then $\{\varphi_\varepsilon : \varepsilon > 0\}$ converges to φ strongly in the sense of Theorem 5.7.3. Let P_ε^{k-1} be the law of φ_ε on the space W_{k-1}. Then obviously we have $P_\varepsilon^j(F) = 1$ where F is the closure of $\{\varphi^v : v \in \mathscr{V}\}$ in W_{k-1}. Since $\{P_\varepsilon^{k-1} : \varepsilon > 0\}$ converges weakly to P^{k-1}, we have $P^{k-1}(F) = 1$ (Theorem 1.1.3). Therefore the support of P^{k-1} is included in F. A similar argument is valid for the support of $\hat{P}^{k-2}$, proving the second assertion of the theorem.

We shall next prove that the support of P^{k-1} includes F. Lemma 5.7.5 states that for any $N, \theta > 0$ there exists $\delta > 0$ such that

$$P\left(\|\varphi - \varphi^v\|_{k-1:N} < \theta, \sup_t |B(t) - V(t)| < \delta\right)$$

$$\geq \frac{1}{2} P\left(\sup_t |B(t) - V(t)| > \delta\right) > 0.$$

This shows that φ_t^v belongs to the support of the measure P^{k-1}. The proof is complete. □

Exercise 5.7.7. (*McShane's approximations. Ikeda–Watanabe* [49]) Let $B(t) = (B^1(t), B^2(t))$ be a 2-dimensional Brownian motion. Let $\xi_1(t)$ and $\xi_2(t)$ be smooth functions of t on $[0, 1]$. Define $v_\varepsilon^l(t)$, $l = 1, 2$ for $\varepsilon h \leq t \leq \varepsilon(h + 1)$ by

$$v_\varepsilon^l(t) = \begin{cases} \dfrac{1}{\varepsilon}\dot{\xi}_l((t - h\varepsilon)/\varepsilon)\Delta_h^\varepsilon B^l & \text{if } \Delta_h^\varepsilon B^1 \Delta_h^\varepsilon B^2 \geq 0, \\[2ex] \dfrac{1}{\varepsilon}\dot{\xi}_{3-l}((t - h\varepsilon)/\varepsilon)\Delta_h^\varepsilon B^l & \text{if } \Delta_h^\varepsilon B^1 \Delta_h^\varepsilon B^2 < 0. \end{cases}$$

Show that the solution $(\varphi_\varepsilon)_t$ of equation (4) converges strongly in the sense of Theorem 5.7.3. Show further

$$s_{lm} = \frac{1}{\pi}\int_0^t \{\xi_l(s)\dot{\xi}_m(s) - \xi_m(s)\dot{\xi}_l(s)\}\, ds.$$

Exercise 5.7.8 (*Mollifier approximation. Malliavin* [93], *Shu* [114]). Let $\rho(t) \geq 0$ be a smooth function with support included in $[0, 1]$ such that $\int_0^1 \rho(s)\, ds = 1$. Set $\rho_\varepsilon(t) = \varepsilon^{-1}\rho(t/\varepsilon)$ and

$$v_\varepsilon^l(t) = -\int_0^\infty \dot{\rho}_\varepsilon(s - t)B^l(s)\, ds = \int_0^\infty \rho_\varepsilon(s - t)\, dB^l(s),$$

where $B(t) = (B^1(t), \ldots, B^n(t))$ is a standard Brownian motion. Show that Theorem 5.7.3 is satisfied.

Exercise 5.7.9 *(Approximation by Ornstein–Uhlenbeck processes, cf. Dowell [27])* Let $v_\varepsilon(t) = (v_\varepsilon^1(t), \ldots, v_\varepsilon^n(t))$ be an Ornstein–Uhlenbeck process determined by the linear stochastic differential equation

$$v_\varepsilon(t) = x - \frac{1}{\varepsilon}\int_0^t v_\varepsilon(r)\, dr + \frac{1}{\varepsilon}B(t)$$

where $B(t)$ is a standard n-dimensional Brownian motion and x is a Gaussian random variable with mean 0 and covariance $\left(\frac{1}{2\varepsilon}\delta_{lm}\right)_{l,m=1,\ldots,n}$, independent of $B(t)$. Show that $B_\varepsilon(t)$ converges to $B(t)$ strongly. Show further that Theorem 5.7.3 is satisfied.

Exercise 5.7.10 Check Examples 4.9.12–4.9.15 using the support theorem of the stochastic flow.

6

Stochastic partial differential equations

6.1 First order stochastic partial differential equations

Preliminaries

In this chapter we shall study initial value problems of a certain class of stochastic partial differential equations of the first and second order. We will not discuss a wide class of equations applying analytic methods developed in the theory of partial differential equations, but will restrict our attention to equations that can be handled by purely probabilistic methods. In these discussions the theory of stochastic flows will play a central role.

Let us first recall the theory of the deterministic partial differential equation of the first order. Consider the initial value problem of a nonlinear equation of the first order given in the canonical form:

$$\frac{\partial u}{\partial t} = F(x, u, u_x, t), \qquad u|_{t=0} = f, \tag{1}$$

where $F(x, u, p, t)$ is a smooth function of $(x, u, p, t) \in \mathbb{R}^d \times \mathbb{R}^1 \times \mathbb{R}^d \times [0, T]$ and $u_x = (\partial u/\partial x^1, \ldots, \partial u/\partial x^d)$. It is known that the problem of integrating equation (1) can be reduced to the characteristic system of ordinary differential equations

$$\frac{dx^i}{dt} = -F_{p^i}, \qquad \frac{du}{dt} = -\sum_i p^i F_{p^i} + F, \qquad \frac{dp^i}{dt} = F_{x^i} + F_u p^i, \tag{2}$$

where $F_{x^i} = \partial F/\partial x^i$, $F_{p^i} = \partial F/\partial p^i$. Indeed, let $(\varphi_t(x, u, p), \eta_t(x, u, p), \chi_t(x, u, p))$ be the solution of the above equation starting from (x, u, p) at $t = 0$. Then the local solution of equation (1) is represented near $t = 0$ by means of the solution of the associated characteristic equation in the following form

$$u(x, t) = \bar{\eta}_t(\bar{\varphi}_t^{-1}(x)), \tag{3}$$

where $\bar{\varphi}_t(x) = \varphi_t(x, f(x), \partial f(x))$, $\bar{\eta}_t(x) = \eta_t(x, f(x), \partial f(x))$ and $\bar{\varphi}_t^{-1}$ is the inverse map of the map $\bar{\varphi}_t : \mathbb{R}^d \to \mathbb{R}^d$. (See Courant–Hilbert [22].)

In this section we shall study the initial value problem of the stochastic partial differential equation of the first order given by

$$u(x, t) - f(x) = \int_0^t F(x, u, u_x, \circ \mathrm{d}s). \tag{4}$$

Here $F(x, u, p, t)$, $(x, u, p) \in \mathbb{R}^d \times \mathbb{R}^1 \times \mathbb{R}^d$, $t \in [0, T]$ is a continuous forward $C^{k,\delta}$-semimartingale with local characteristic belonging to the class $(B^{k+1,\delta}, B^{k,\delta})$ where $k \geq 4$ and $\delta > 0$. The differential $F(\ldots, \circ \mathrm{d}t)$ means the Stratonovich differential.

The equation may not have a global solution in general though semilinear and linear equations can have global solutions. We will define a local solution. A local random field $u(x, t)$, $x \in \mathbb{R}^d$, $t \in [0, T(x))$ with values in $\mathbb{R}^1$ is called a *local solution* of equation (1) with the initial value $f(x)$, if the following conditions are satisfied.

(i) $T(x, \omega)$ is an accessible, lower semicontinuous stopping time less than or equal to T.

(ii) $u(x, t)$, $0 \leq t < T(x)$ is a local $C^{1,\varepsilon}$-semimartingale for some $\varepsilon > 0$ and satisfies

$$u(x, t) = f(x) + \int_0^t F(x, u(x, r), u_x(x, r), \circ \mathrm{d}r) \tag{5}$$

for all (x, t) such that $t < T(x)$ a.s.

If there exists a solution of equation (5) such that the terminal time $T(x, \omega)$ is equal to T for all x a.s., it is called a *global solution* of equation (5).

We will solve the above nonlinear equation (5) by introducing the characteristic system of stochastic differential equations. Then in the latter half of this section we shall apply the result to quasi-linear and linear stochastic partial differential equations.

In this section we shall make use of the Stratonovich integral only. It enables us to deal with the stochastic analysis for irregular functionals of time variables (semimartingales for example) in the same way as the deterministic analysis for regular (smooth) functionals of time variables, due to the generalized Itô's formula for the Stratonovich's differential (Theorem 3.3.2).

Throughout this chapter we will often use abreviated notations on stochastic calculus. For example, a stochastic differential equation based on a continuous C-valued semimartingale $F(x, t)$ is written as

$$\mathrm{d}\varphi_t = F(\varphi_t, \circ \mathrm{d}t),$$

and the generalized Itô's formula II (Theorem 3.3.2) is written as

$$\circ \mathrm{d}F(g_t, t) = F(g_t, \circ \mathrm{d}t) + \sum \frac{\partial F^i}{\partial x^i}(g_t, t) \circ \mathrm{d}g_t^i,$$

instead of formula (9) in Section 3.3.

Stochastic characteristic system

As usual we will use the notations $F_x = (\partial F/\partial x^1, \ldots, \partial F/\partial x^d)$ and $F_p = (\partial F/\partial p^1, \ldots, \partial F/\partial p^d)$. For two vectors p,q, $p \cdot q$ denotes the inner product.

The *stochastic characteristic system associated with* (5) is defined by a system of Stratonovich's stochastic differential equations of the form

$$\begin{aligned} \mathrm{d}\varphi_t &= -F_p(\varphi_t, \eta_t, \chi_t, \circ \mathrm{d}t), \\ \mathrm{d}\eta_t &= F(\varphi_t, \eta_t, \chi_t, \circ \mathrm{d}t) - \chi_t \cdot F_p(\varphi_t, \eta_t, \chi_t, \circ \mathrm{d}t), \qquad (6) \\ \mathrm{d}\chi_t &= F_x(\varphi_t, \eta_t, \chi_t, \circ \mathrm{d}t) + F_u(\varphi_t, \eta_t, \chi_t, \circ \mathrm{d}t)\chi_t. \end{aligned}$$

Let $(\varphi_t(x, u, p), \eta_t(x, u, p), \chi_t(x, u, p))$, $t \in [0, T(x, u, p))$ be the maximal solution of the above equation starting at (x, u, p) at time 0, where $T(x, u, p)$ is the explosion time. It has a modification of a local $C^{k-1,\varepsilon}$-semimartingale for every $\varepsilon < \delta$. Furthermore, it defines a forward stochastic flow of local C^{k-1}-diffeomorphisms by Theorem 4.7.3.

Now let $f(x)$ be a function of the class $C^{k,\delta}$ corresponding to the initial value of equation (5). We define a local $C^{k-1,\varepsilon}$-semimartingale $(\bar{\varphi}_t(x), \bar{\eta}_t(x), \bar{\chi}_t(x))$, $t \in [0, \bar{T}(x))$ where $\varepsilon < \delta^2$ and $\bar{T}(x) = T(x, f(x), f_x(x))$ by

$$\begin{aligned} \bar{\varphi}_t(x) &= \varphi_t(x, f(x), f_x(x)), \\ \bar{\eta}_t(x) &= \eta_t(x, f(x), f_x(x)), \qquad (7) \\ \bar{\chi}_t(x) &= \chi_t(x, f(x), f_x(x)). \end{aligned}$$

The triple $(\bar{\varphi}_t(x), \bar{\eta}_t(x), \bar{\chi}_t(x))$ is called the *stochastic characteristic curve of equation* (5). We shall study the process $\bar{\varphi}_t(x)$ in detail in the sequel. Processes $\bar{\eta}_t(x)$ and $\bar{\chi}_t(x)$ will be discussed later.

The map $\bar{\varphi}_t(\cdot, \omega): \{x | \bar{T}(x, \omega) > t\}$ into $\mathbb{R}^d$, is not a diffeomorphism in general, since the Jacobian matrix $\partial\bar{\varphi}_t(x)$ can be singular at some t less than $\bar{T}(x)$. Define $\tau(x) = \inf\{t > 0 : \det \partial\bar{\varphi}_t(x) = 0\} \wedge \bar{T}(x)$. It is an accessible, lower semicontinuous stopping time. Further, we have $\lim_{t \uparrow \tau(x)} \det \partial\bar{\varphi}_t(x) = 0$ if $\tau(x) < \bar{T}(x)$. We will show that $\bar{\varphi}_t$ defines a diffeomorphism if we restrict the map $\bar{\varphi}_t$ to the domain $\{\tau > t\}$. It is convenient to introduce the *adjoint stopping time* of $\tau(x)$ in the following manner: $\sigma(y) = \inf\{t \in [0, \bar{T}(x)) : y \in \bar{\varphi}_t(\{\tau > t\})\}$ $(= \bar{T}(x)$ if $\{\ldots\} = \emptyset)$, where $\bar{\varphi}_t(\{\tau > t\})$ is the range of the set $\{x | \tau(x) > t\}$ by the map $\bar{\varphi}_t$.

Lemma 6.1.1

(i) *The map $\bar{\varphi}_t$ from the domain $\{\tau > t\}$ into $\mathbb{R}^d$ is a C^{k-1}-diffeomorphism for every t a.s.*

(ii) *The inverse $\bar{\psi}_t(y) \equiv \bar{\varphi}_t^{-1}(y)$, $t < \sigma(y)$ is a continuous local C^{k-1}-process and a local $C^{k-2,\varepsilon}$-semimartingale for some $\varepsilon > 0$ and satisfies*

$$d\bar{\psi}_t(y) = \partial\bar{\varphi}_t^{-1}(\bar{\psi}_t(y))F_p(y, \bar{\eta}_t \circ \bar{\psi}_t(y), \bar{\chi}_t \circ \bar{\psi}_t(y), \circ dt), \tag{8}$$

where $\partial\bar{\varphi}_t^{-1}(y)$ is the inverse matrix of the Jacobian matrix $\partial\bar{\varphi}_t(y)$.

(iii) *$\sigma(y)$ is an accessible, lower semicontinuous stopping time such that if $\sigma(y) < T$*

$$\lim_{t \uparrow \sigma(y)} |\det \partial\bar{\psi}_t(y)| = \infty \quad \text{or} \quad \lim_{t \uparrow \sigma(y)} \bar{\psi}_t(y) \in \{x \mid \bar{T}(x) > \sigma(y)\},$$

where $\partial\bar{\psi}_t$ is the Jacobian matrix of the map $\bar{\psi}_t$.

Proof Consider the Stratonovich's equation based on $G(x, t)$ where

$$G(x, t) = \int_0^t \partial\bar{\varphi}_s^{-1}(x)F_p(\bar{\varphi}_s(x), \bar{\eta}_s(x), \bar{\chi}_s(x), \circ ds). \tag{9}$$

Since it is a continuous local $C^{k-1,\varepsilon'}$-semimartingale with local characteristic of the class $B^{k-1,\varepsilon'}$ for some $\varepsilon' > 0$, for each $y \in \mathbb{R}^d$ the equation has a unique solution $\bar{\psi}_t(y)$, $t \in [0, \hat{\sigma}(y))$ such that $\bar{\psi}_0(y) = y$ and $\bar{\psi}_t(y) \in \{x \mid \tau(x) > t\}$ for any $t \in [0, \hat{\sigma}(y))$, where $\hat{\sigma}(y)$ is the explosion time of the process $\bar{\psi}_t(y)$. The solution $\bar{\psi}_t(y)$ is a continuous local $C^{k-2,\varepsilon}$-semimartingale with local characteristic of the class $B^{k-2,\varepsilon}$ for any $\varepsilon < \varepsilon'$ (*cf.* Theorem 4.7.3 and its proof). Further, $\hat{\sigma}(y)$ is an accessible and lower semicontinuous stopping time. We have

$$\lim_{t \uparrow \hat{\sigma}(y)} |\det \partial\bar{\varphi}_t(\bar{\psi}_t(y))| = 0 \quad \text{or} \quad \lim_{t \uparrow \hat{\sigma}(y)} \bar{\psi}_t(y) \in \{x \mid \bar{T}(x) > \hat{\sigma}(y)\}. \tag{10}$$

We shall prove $\bar{\varphi}_t \circ \bar{\psi}_t(y) = y$ for $t < \hat{\sigma}(y)$. Note that $\bar{\varphi}_t(x)$ satisfies

$$\bar{\varphi}_t(x) = x - \int_0^t F_p(\bar{\varphi}_s(x), \bar{\eta}_s(x), \bar{\chi}_s(x), \circ ds). \tag{11}$$

It is a continuous local $C^{k-1,\varepsilon}$-semimartingale with local characteristic belonging to the class $B^{k-1,\varepsilon}$ for every $\varepsilon < \delta^2$. Then we can apply the generalized Itô's formula (Theorem 3.3.2) to the above $\bar{\varphi}(x, t) = \bar{\varphi}_t(x)$ and $\bar{\psi}_t$, since $\bar{\varphi}$ is a C^3-semimartingale. Then,

$$\bar{\varphi}_t(\bar{\psi}_t) - \bar{\varphi}_0(\bar{\psi}_0) = \int_0^t \bar{\varphi}(\bar{\psi}_r, \circ dr) + \sum_r \int_0^t \frac{\partial\bar{\varphi}_r}{\partial x^i}(\bar{\psi}_r) \circ d\bar{\psi}_r^i.$$

The first term of the right hand side is equal to

$$-\int_0^t F_p(\bar{\varphi}_r \circ \bar{\psi}_r, \bar{\eta}_r \circ \bar{\psi}_r, \bar{\chi}_r \circ \bar{\psi}_r, \circ \mathrm{d}r)$$

(see Theorem 3.3.4). By (9), the second term is

$$\int_0^t \partial\bar{\varphi}_r(\bar{\psi}_r)(\partial\bar{\varphi}_r)^{-1}(\bar{\psi}_r)F_p(\bar{\varphi}_r \circ \bar{\psi}_r, \bar{\eta}_r \circ \bar{\psi}_r, \bar{\chi}_r \circ \bar{\psi}_r, \circ \mathrm{d}r).$$

Therefore we have $\bar{\varphi}_t(\bar{\psi}_t) - \bar{\varphi}_0(\bar{\psi}_0) = 0$ or $\bar{\varphi}_t(\bar{\psi}_t(y)) = y$. Since the Jacobian matrix $\partial\bar{\varphi}_t(\bar{\psi}_t(y))$ is nonsingular at $t < \hat{\sigma}(y)$, the implicit function theorem shows that $\bar{\psi}_t(x)$ is a continuous C^{k-1}-process.

Define now

$$\hat{\tau}(x) = \inf\{t \in [0, \tau(x)) : \bar{\varphi}_t(x) \in \{\hat{\sigma} > t\} \text{ or } |\det \partial\bar{\psi}_t(\bar{\varphi}_t(x))| = \infty\},$$
$$(= \tau(x) \text{ if } \{\dots\} = \varnothing).$$

We shall prove $\bar{\psi}_t \circ \bar{\varphi}_t(x) = x$ if $t < \hat{\tau}(x)$. Since $\bar{\varphi}_t(\bar{\psi}_t(y)) = y$, we see $\partial\bar{\varphi}_t(\bar{\psi}_t(y))\partial\bar{\psi}_t(y) =$ identity matrix. Therefore, equation (11) is written as

$$\mathrm{d}\bar{\varphi}_t = -\partial\bar{\psi}_t^{-1}(\bar{\varphi}_t)\partial\bar{\varphi}_t^{-1}(\bar{\psi}_t \circ \bar{\varphi}_t)F_p(\bar{\varphi}_t, \bar{\eta}_t, \bar{\chi}_t, \circ \mathrm{d}t).$$

We can apply the generalized Itô's formula to $\bar{\psi}_t$ and $\bar{\varphi}_t$. Then $\bar{\psi}_t \circ \bar{\varphi}_t$ satisfies the following

$$\begin{aligned} \mathrm{d}(\bar{\psi}_t \circ \bar{\varphi}_t) &= \mathrm{d}\bar{\psi}_t(\bar{\varphi}_t) + \sum_i \frac{\partial\bar{\psi}_t}{\partial x^i}(\bar{\varphi}_t) \circ \mathrm{d}\bar{\varphi}_t^i \\ &= (\partial\bar{\varphi}_t)^{-1}(\bar{\psi}_t \circ \bar{\varphi}_t)F_p(\bar{\varphi}_t, \bar{\eta}_t \circ \bar{\psi}_t \circ \bar{\varphi}_t, \bar{\chi}_t \circ \bar{\psi}_t \circ \bar{\varphi}_t, \circ \mathrm{d}t) \\ &\quad - (\partial\bar{\varphi}_t)^{-1}(\bar{\psi}_t \circ \bar{\varphi}_t)F_p(\bar{\varphi}_t, \bar{\eta}_t, \bar{\chi}_t, \circ \mathrm{d}t). \end{aligned}$$

It can be regarded as an equation for $\bar{\psi}_t \circ \bar{\varphi}_t$. The equation has a unique solution. The solution satisfies $\bar{\psi}_t \circ \bar{\varphi}_t(x) = x$ for all $t < \hat{\tau}(x)$.

We shall next prove $\tau(x) = \hat{\tau}(x)$ a.s. If $\hat{\tau}(x)$ coincides with $\inf\{t \in [0, \tau(x)) : \bar{\varphi}_t(x) \in \{\hat{\sigma} > t\}\}$ $(= \tau(x)$ if $\{\dots\} = \varnothing)$, then from (10) we have

$$\lim_{t \uparrow \hat{\tau}(x)} |\det \partial\bar{\varphi}_t(\bar{\psi}_t \circ \bar{\varphi}_t(x))| = 0 \qquad \text{if } \hat{\tau}(x) < \bar{T}(x).$$

This implies $\tau(x) = \hat{\tau}(x)$. On the other hand, if $\hat{\tau}(x)$ is equal to $\inf\{t > 0 : |\det \partial\bar{\psi}_t(\bar{\varphi}_t(x))| = \infty\} \wedge \tau(x)$, we have $\lim_{t\uparrow\hat{\tau}(x)} \det \partial\bar{\varphi}_t(x) = 0$ if $\hat{\tau}(x) < \bar{T}(x)$, since $\partial\bar{\psi}_t(\bar{\varphi}_t(x))\partial\bar{\varphi}_t(x) =$ identity. This implies $\hat{\tau}(x) = \tau(x)$.

The one to one property of the map $\bar{\varphi}_t$ on the domain $\{\tau > t\}$ is immediate. Indeed, suppose $\bar{\varphi}_t(x) = \bar{\varphi}_t(x')$ holds for $x, x' \in \{\tau > t\}$. Then, since $\bar{\psi}_t \circ \bar{\varphi}_t(x) = x$ holds on $\{\tau > t\}$, we get $x = x'$. Now since the Jacobian matrix $\partial\bar{\varphi}_t(x)$ is nonsingular, $\bar{\varphi}_t$ is a local C^{k-1}-diffeomorphism by the implicit function theorem. Consequently $\bar{\varphi}_t$ defines a C^{k-1}-diffeomorphism from $\{\tau > t\}$ into $\mathbb{R}^d$. We have thus proved (i) of the lemma. The relation $\bar{\psi}_t \circ \bar{\varphi}_t = \bar{\varphi}_t \circ \bar{\psi}_t$ shows that $\bar{\psi}_t$ is equal to the inverse of $\bar{\varphi}_t$.

Now consider the stopping time $\hat{\sigma}$. If $\hat{\sigma}(y) < T$

$$\lim_{t \uparrow \hat{\sigma}(y)} |\det \partial\bar{\psi}_t(y)| = \infty \quad \text{or} \quad \lim_{t \uparrow \hat{\sigma}(y)} \bar{\psi}_t(y) \in \{x \,|\, \bar{T}(x) > \sigma(y)\}.$$

In fact, $\partial\bar{\varphi}_t(\bar{\psi}_t(x))\partial\bar{\psi}_t(x)$ is the identity matrix and (10) is satisfied for $\partial\bar{\varphi}_t(\bar{\psi}_t(x))$. Therefore it only remains to prove $\sigma = \hat{\sigma}$. We have $\{\hat{\sigma} > t\} \supset \bar{\varphi}_t(\{\tau > t\})$ since $\bar{\psi}_t(y) = \bar{\varphi}_t^{-1}(y)$ holds for any $y \in \bar{\varphi}_t(\{\tau > t\})$. From the definition of $\bar{\psi}_t$, we have $\bar{\psi}_t(\{\hat{\sigma} > t\}) \subset \{\tau > t\}$, so that we have $\bar{\varphi}_t \circ \bar{\psi}_t(\{\hat{\sigma} > t\}) \subset \bar{\varphi}_t(\{\tau > t\})$. Since $\bar{\varphi}_t \circ \bar{\psi}_t(x) = x$ holds on $\{\hat{\sigma} > t\}$, we get $\{\hat{\sigma} > t\} \subset \bar{\varphi}_t(\{\tau > t\})$. Therefore we have $\{\hat{\sigma} > t\} = \bar{\varphi}_t(\{\tau > t\})$ for any t a.s. This proves $\hat{\sigma}(y) = \sigma(y)$ for any y a.s. The proof is complete. □

Existence and uniqueness of solutions

We will prove the existence and uniqueness of solutions of nonlinear stochastic partial differential equations of the first order, using stochastic characteristic curves of the previous subsection. We first establish the existence of the solution.

Theorem 6.1.2 *Assume that F of equation (5) is a continuous $C^{k,\alpha}$-semimartingale with local characteristic belonging to the class $(B^{k+1,\delta}, B^{k,\delta})$ for some $k \geq 4$ and $\delta > 0$ and f is a function of $C^{k,\delta}$. Let $(\bar{\varphi}_t, \bar{\eta}_t, \bar{\chi}_t)$ be the stochastic characteristic curve of equation (5). Then $u(x, t)$ defined by*

$$u(x, t) \equiv \bar{\eta}_t(\bar{\varphi}_t^{-1}(x)), \; t \in [0, \sigma(x)) \tag{12}$$

is a local solution of equation (5). Further it is a continuous local $C^{k-1,\varepsilon}$-semimartingale for some $\varepsilon > 0$. □

Before the proof, we need a lemma.

Lemma 6.1.3 *Set $\bar{\psi}_t = \bar{\varphi}_t^{-1}$. The following relations hold*

$$\frac{\partial}{\partial x^i}(\bar{\eta} \circ \bar{\psi}_t) = \bar{\chi}_t^i \circ \bar{\psi}_t, \qquad i = 1, \ldots, d. \tag{13}$$

Proof We first claim:

$$\frac{\partial \bar{\eta}_t}{\partial x^i} = \bar{\chi}_t \cdot \frac{\partial \bar{\varphi}_t}{\partial x^i}. \tag{14}$$

Set $\theta_t^i = \partial\bar{\eta}_t/\partial x^i - \bar{\chi}_t \cdot (\partial\bar{\varphi}_t/\partial x^i)$. We wish to show $\theta_t^i = 0$, $i = 1, \ldots, d$. For this, we shall obtain a stochastic differential equation governing θ_t^i. Observe the stochastic characteristic equation of $\bar{\eta}_t$. We can change the order of $\partial/\partial x^i$ and the stochastic integrals by Theorem 3.3.4. Then we have

$$\begin{aligned}\frac{\partial \bar{\eta}_t}{\partial x^i} - \frac{\partial f}{\partial x^i} &= \int_0^t \frac{\partial \bar{\varphi}_s}{\partial x^i} \cdot F_x(\bar{\varphi}_s, \bar{\eta}_s, \bar{\chi}_s, \circ ds) + \int_0^t \frac{\partial \bar{\eta}_s}{\partial x^i} F_u(\bar{\varphi}_s, \bar{\eta}_s, \bar{\chi}_s, \circ ds) \\ &+ \int_0^t \frac{\partial \bar{\chi}_s}{\partial x^i} \cdot F_p(\bar{\varphi}_s, \bar{\eta}_s, \bar{\chi}_s, \circ ds) - \int_0^t \frac{\partial \bar{\chi}_s}{\partial x^i} \cdot F_p(\bar{\varphi}_s, \bar{\eta}_s, \bar{\chi}_s, \circ ds) \\ &- \int_0^t \bar{\chi}_s \cdot \frac{\partial}{\partial x^i}(F_p(\bar{\varphi}_s, \bar{\eta}_s, \bar{\chi}_s, \circ ds)).\end{aligned}$$

We have further

$$\begin{aligned}\bar{\chi}_t \cdot \frac{\partial \bar{\varphi}_t}{\partial x^i} - \frac{\partial f}{\partial x^i} &= \int_0^t \bar{\chi}_s \cdot \circ d \frac{\partial \bar{\varphi}_s}{\partial x^i} + \int_0^t \frac{\partial \bar{\varphi}_s}{\partial x^i} \cdot \circ d\bar{\chi}_s \\ &= -\int_0^t \bar{\chi}_s \cdot \frac{\partial}{\partial x^i}(F_p(\bar{\varphi}_s, \bar{\eta}_s, \bar{\chi}_s, \circ ds)) \\ &+ \int_0^t \frac{\partial \bar{\varphi}_s}{\partial x^i} \cdot F_x(\bar{\varphi}_s, \bar{\eta}_s, \bar{\chi}_s, \circ ds) \\ &+ \int_0^t \frac{\partial \bar{\varphi}_s}{\partial x^i} \cdot \bar{\chi}_s F_u(\bar{\varphi}_s, \bar{\eta}_s, \bar{\chi}_s, \circ ds)\end{aligned}$$

by the same theorem. Therefore θ_t^i satisfies the following linear stochastic differential equation

$$\theta_t^i = \int_0^t \theta_s^i F_u(\bar{\varphi}_s, \bar{\eta}_s, \bar{\chi}_s, \circ ds).$$

The above equation has a unique solution $\theta_t^i \equiv 0$. This proves (14).

We now proceed to the proof of the lemma. Note that

$$\partial(\bar{\eta}_t \circ \bar{\psi}_t) = \partial \bar{\eta}_t(\bar{\psi}_t) \partial \bar{\psi}_t = \partial \bar{\eta}_t(\bar{\psi}_t) \partial \bar{\varphi}_t(\bar{\psi}_t)^{-1}.$$

Using (14), the right hand side of the above equals

$$\bar{\chi}_t(\bar{\psi}_t) \partial \bar{\varphi}_t(\bar{\psi}_t) \partial \bar{\varphi}_t(\bar{\psi}_t)^{-1} = \bar{\chi}_t(\bar{\psi}_t).$$

Therefore (13) is established. □

Proof of Theorem 6.1.2 Set $u(x, t) = \bar{\eta}_t(\bar{\psi}_t(x))$, $t < \sigma(x)$. It is a continuous local C^{k-1}-process, since $\bar{\eta}_t$ and $\bar{\psi}_t$ are continuous C^{k-1}-processes. Further

$u_{x^i}(x, t)$ is a local $C^{k-2,\varepsilon}$-semimartingale for some ε by Lemma 6.1.3. Therefore, $u(x, t)$ is a continuous local $C^{k-1,\varepsilon}$-semimartingale. We shall apply the generalized Itô's formula (Theorem 3.3.2). Then we get

$$\mathrm{d}(\bar{\eta}_t \circ \bar{\psi}_t) = \mathrm{d}\bar{\eta}_t(\bar{\psi}_t) + \sum_i \frac{\partial \bar{\eta}_t}{\partial x^i}(\bar{\psi}_t) \circ \mathrm{d}\bar{\psi}_t^i.$$

The first term of the right hand side is

$$F(\cdot, \bar{\eta}_t \circ \bar{\psi}_t, \bar{\chi}_t \circ \bar{\psi}_t, \circ \mathrm{d}t) - \bar{\chi}_t \circ \bar{\psi}_t \cdot F_p(\cdot, \bar{\eta}_t \circ \bar{\psi}_t, \bar{\chi}_t \circ \bar{\psi}_t, \circ \mathrm{d}t)$$

by (6). The second term equals

$$\partial \bar{\eta}_t(\bar{\psi}_t) \partial \bar{\varphi}_t(\bar{\psi}_t)^{-1} \cdot F_p(\cdot, \bar{\eta}_t \circ \bar{\psi}_t, \bar{\chi}_t \circ \bar{\psi}_t, \circ \mathrm{d}t)$$

by (8). Note the relation $\partial \eta_t = \bar{\chi}_t \cdot \partial \bar{\varphi}_t$ of (13). Then we get

$$u(\cdot, \circ \mathrm{d}t) = F(\cdot, \bar{\eta}_t \circ \bar{\psi}_t, \bar{\chi}_t \circ \bar{\psi}_t, \circ \mathrm{d}t).$$

Since $u = \bar{\eta}_t \circ \bar{\psi}_t$ and $u_{x^i} = \bar{\chi}_t^i \circ \bar{\psi}_t$ hold by Lemma 6.1.3, u is the solution of equation (5). The proof is complete. □

We will next consider the uniqueness of the solution.

Theorem 6.1.4 *Assume the same conditions for F and f as in Theorem* 6.1.2. *Let $u(x, t)$, $t \in [0, T(x))$ be an arbitrary local solution of equation* (5) *such that it is a continuous local $C^{4,\varepsilon}$-semimartingale for some $\varepsilon > 0$. Then it is represented as $u(x, t) = \bar{\eta}_t \circ \bar{\psi}_t(x)$ for $t \in [0, T(x) \wedge \sigma(x))$.*

Proof Let $u(x, t)$ be a solution of equation (5) satisfying the conditions of the theorem. We shall prove $u(\bar{\varphi}_t, t) = \bar{\eta}_t$ holds for $t < T(x) \wedge \sigma(x)$. The local characteristic of $u(x, t)$ belongs to $B^{3,\varepsilon}$ since it satisfies equation (5). Then by the generalized Itô's formula we have

$$\begin{aligned} d_t\{u(\bar{\varphi}_t, t)\} &= u(\bar{\varphi}_t, \circ \mathrm{d}t) + \sum_i \frac{\partial u}{\partial x^i}(\bar{\varphi}_t, t) \circ \mathrm{d}\bar{\varphi}_t^i \\ &= F(\bar{\varphi}_t, u(\bar{\varphi}_t, t), u_x(\bar{\varphi}_t, t), \circ \mathrm{d}t) - u_x(\bar{\varphi}_t, t) \cdot F_p(\bar{\varphi}_t, \bar{\eta}_t, \bar{\chi}_t, \circ \mathrm{d}t) \end{aligned} \tag{15}$$

by (6). Therefore from (15) and (6) we have

$$\begin{aligned} d_t(u(\bar{\varphi}_t, t) - \bar{\eta}_t) &= F(\bar{\varphi}_t, u(\bar{\varphi}_t, t), u_x(\bar{\varphi}_t, t), \circ \mathrm{d}t) - F(\bar{\varphi}_t, \bar{\eta}_t, \bar{\chi}_t, \circ \mathrm{d}t) \\ &\quad - (u_x(\bar{\varphi}_t, t) - \bar{\chi}_t) \cdot F_p(\bar{\varphi}_t, \bar{\eta}_t, \bar{\chi}_t, \circ \mathrm{d}t). \end{aligned} \tag{16}$$

Similarly we can apply the generalized Itô's formula to $\partial u(x, t)$ and $\bar{\varphi}_t$. We have

$$\begin{aligned}
&\mathrm{d}(\partial u(\bar{\varphi}_t, t) - \bar{\chi}_t) \\
&\quad = F_x(\bar{\varphi}_t, u(\bar{\varphi}_t, t), u_x(\bar{\varphi}_t, t), \circ \mathrm{d}t) - F_x(\bar{\varphi}_t, \bar{\eta}_t, \bar{\chi}_t, \circ \mathrm{d}t) \\
&\qquad + F_u(\bar{\varphi}_t, u(\bar{\varphi}_t, t), u_x(\bar{\varphi}_t, t), \circ \mathrm{d}t) u_x(\bar{\varphi}_t, t) - F_u(\bar{\varphi}_t, \bar{\eta}_t, \bar{\chi}_t, \circ \mathrm{d}t)\bar{\chi}_t \\
&\qquad + \{F_p(\bar{\varphi}_t, u(\bar{\varphi}_t, t), u_x(\bar{\varphi}_t, t), \circ \mathrm{d}t) - F_p(\bar{\varphi}_t, \bar{\eta}_t, \bar{\chi}_t, \circ \mathrm{d}t)\} u_{xx}(\varphi_t, t). \quad (17)
\end{aligned}$$

We can regard the above two equations (16) and (17) as forming a system of stochastic differential equations for $u(\bar{\varphi}_t, t) - \bar{\eta}_t$ and $u_x(\bar{\varphi}_t, t) - \bar{\chi}_t$. The initial values are 0 since $u(\bar{\varphi}_0, 0) = \eta_0 = f$ and $u_x(\bar{\varphi}_0, 0) = \chi_0 = f_x$. Further, the local characteristics of the semimartingales appearing on the right hand sides of (16) and (17) belong to $B^{2,\varepsilon}$. Therefore the system has a unique solution, $u(\bar{\varphi}_t, t) - \bar{\eta}_t = 0$ and $u_x(\bar{\varphi}_t, t) - \bar{\chi}_t = 0$. The proof is complete. □

Summing up Theorems 6.1.2 and 6.1.4, we have the following.

Theorem 6.1.5 *Assume the same conditions as in Theorem* 6.1.2. *Then equation* (1) *has a unique local solution such that it is a continuous local $C^{k-1,\varepsilon}$-semimartingale for some* $\varepsilon > 0$. □

Quasi-linear and semi-linear equations

A first order stochastic partial differential equation

$$u(x, t) = f(x) + \sum_{i=1}^{d} \int_0^t F^i(x, u(x, r), \circ \mathrm{d}r) \frac{\partial u}{\partial x^i}(x, r) + \int_0^t F^{d+1}(x, u(x, r), \circ \mathrm{d}r) \quad (18)$$

is called a *quasi-linear equation.* Here $F^i(x, u, t)$, $i = 1, \ldots, d+1$, are continuous $C^{k,\delta}$-semimartingales with local characteristics belonging to the class $(B_b^{k+1,\delta}, B_b^{k,\delta})$ for some $k \geq 3$ and $\delta > 0$. It is of course a special case of the nonlinear equation (1) obtained by setting

$$F(x, u, p, t) = F(x, u, t) \cdot p + F^{d+1}(x, u, t),$$

where $F(x, u, t) = (F^1(x, u, t), \ldots, F^d(x, u, t))$. However, we can verify the existence and the uniqueness of the solution of the quasi-linear equation under a condition weaker than that of the nonlinear equation we have been discussing until now.

First observe the relations

$$F_p(x, u, p, t) = F(x, u, t),$$

$$F(x, u, p, t) - F_p(x, u, p, t) \cdot p = F^{d+1}(x, u, t),$$

which do not contain the variable p. Therefore in the stochastic characteristic equation for the quasi-linear equation, (φ_t, η_t) satisfies a closed system of a stochastic differential equation

$$\begin{aligned} d\varphi_t &= -F(\varphi_t, \eta_t, \circ dt), \\ d\eta_t &= F^{d+1}(\varphi_t, \eta_t, \circ dt) \qquad (\chi_t \text{ is not involved}). \end{aligned} \tag{19}$$

Then the solution defines a global stochastic flow of C^k-diffeomorphisms $(\varphi_t(x, u), \eta_t(x, u))$, $t \in [0, T]$ since the local characteristic of F^i belongs to the class $(B_b^{k+1,\delta}, B_b^{k,\delta})$, $k \geq 2$. Let f be a $C^{k,\delta}$-function. Set $\bar{\varphi}_t(x) = \varphi_t(x, f(x))$, $\bar{\eta}_t(x) = \eta_t(x, f(x))$ and let $\bar{\psi}_t(x)$, $t < \sigma(x)$ be the inverse of $\bar{\varphi}_t(x)$. Then $u(x, t) = \bar{\eta}_t(\bar{\psi}_t(x))$, $t \in [0, \sigma(x))$ is a local solution of the quasi-linear equation. Further, it is a continuous local $C^{k,\varepsilon}$-semimartingale for some $\varepsilon > 0$ by Lemma 6.1.3.

Conversely let $u(x, t)$ be a local solution such that it is a continuous local C^3-semimartingale. Then similarly to the proof of Theorem 6.1.4, the generalized Itô's formula implies

$$\begin{aligned} d\{u(\bar{\varphi}_t, t) - \bar{\eta}_t\} = \sum_i \{F^i(\bar{\varphi}_t, u(\bar{\varphi}_t, t), \circ dt) - F^i(\bar{\varphi}_t, \bar{\eta}_t, \circ dt)\} \frac{\partial u}{\partial x^i} \\ + F^{d+1}(\bar{\varphi}_t, u(\bar{\varphi}_t, t), \circ dt) - F^{d+1}(\bar{\varphi}_t, \bar{\eta}_t, \circ dt). \end{aligned}$$

We can regard this as an equation for $u(\bar{\varphi}_t, t) - \bar{\eta}_t$. Its initial value is 0 since $u(\bar{\varphi}_0, 0) = \bar{\eta}_0 = f$. Note that the local characteristics of $(F^1, \ldots, F^{d+1})$ belong to the class $(B_b^{k+1,\delta}, B_b^k)$ for $k \geq 3$. Then the above equation has a unique solution, i.e. $u(\bar{\varphi}_t, t) = \bar{\eta}_t$. Therefore we have the following.

Theorem 6.1.6 *Assume that $(F^1, \ldots, F^d)$ of the quasi-linear equation* (18) *are continuous $C^{k,\delta}$-semimartingales with local characteristics belonging to the class $(B_b^{k+1,\delta}, B_b^{k,\delta})$ for $k \geq 3$ and $\delta > 0$. Then for any f of $C^{k,\delta}$, the quasi-linear equation has a unique local solution such that it is a continuous local $C^{k,\varepsilon}$-semimartingale for some $\varepsilon > 0$.* □

We will study the asymptotic behavior of the solution as t tends to $\sigma(x)$.

Theorem 6.1.7 *Assume that the initial function $f(x)$ satisfies* $\lim_{x\to\infty} |f(x)| = \infty$. *Then the solution $u(x, t) = \bar{\eta}_t \circ \bar{\psi}_t(x)$, $t \in [0, \sigma(x))$ satisfies*

$$\lim_{t\uparrow\sigma(x)} |u(x, t)| = \infty \quad \textit{or} \quad \lim_{t\uparrow\sigma(x)} |u_x(x, t)| = \infty \quad \textit{if} \quad \sigma(x) < T. \tag{20}$$

Proof Observe the equation for $\bar{\psi}_t$:

$$d\bar{\psi}_t(x) = (\partial\bar{\varphi}_t)^{-1}(\bar{\psi}_t(x))F(x, \bar{\eta}_t \circ \bar{\psi}_t(x), \circ dt).$$

Since $\varphi_t(x)$ is strictly conservative, the terminal time $\sigma(x)$ of $\bar{\psi}_t(x)$ satisfies

$$\text{(i)} \qquad \lim_{t\uparrow\sigma(x)} \bar{\psi}_t(x) = \infty \quad \text{or}$$
$$\text{(ii)} \qquad \lim_{t\uparrow\sigma(x)} |\det \partial\bar{\psi}_t(x)| = \infty \tag{21}$$

if $\sigma(x) < T$ by Lemma 6.1.1.

We first consider case (i). We have $\lim_{|x|\to\infty} |\bar{\eta}_t \circ \bar{\psi}_t(x)| = \infty$ if $\sigma(x) < T$. Therefore we have $\lim_{t\uparrow\sigma(x)} |\bar{\eta}_t \circ \bar{\psi}_t(x)| = \infty$ if $\sigma(x) < T$. Let us next consider the case that (ii) holds but (i) does not hold. Set $\partial\bar{\eta}_t = (\partial\bar{\eta}_t/\partial x^1, \ldots, \partial\bar{\eta}_t/\partial x^d)$. Changing the order of the derivative ∂ and the stochastic integral in equation (19), we get the following linear equation for the $d \times (d+1)$-matrix $(\partial\bar{\varphi}_t, \partial\bar{\eta}_t)^t$ (transpose):

$$d\partial\bar{\varphi}_t = -\partial F(\bar{\varphi}_t, \bar{\eta}_t, \circ dt)\partial\bar{\varphi}_t - F_u(\bar{\varphi}_t, \bar{\eta}_t, \circ dt)\partial\bar{\eta}_t$$
$$d\partial\bar{\eta}_t = F_x^{d+1}(\bar{\varphi}_t, \bar{\eta}_t, \circ dt)\partial\bar{\varphi}_t + F_u^{d+1}(\bar{\varphi}_t, \bar{\eta}_t, \circ dt)\partial\bar{\eta}_t.$$

The adjoint $(d+1) \times d$-matrix equation (A_t, B_t) is defined by

$$dA_t = A_t\partial F(\bar{\varphi}_t, \bar{\eta}_t, \circ dt) + B_t F_u(\bar{\varphi}_t, \bar{\eta}_t, \circ dt)$$
$$dB_t = -A_t F_x^{d+1}(\bar{\varphi}_t, \bar{\eta}_t, \circ dt) - B_t F_u^{d+1}(\bar{\varphi}_t, \bar{\eta}_t, \circ dt).$$

By Itô's formula $d\{(A_t, B_t)(\partial\bar{\varphi}_t, \partial\bar{\eta}_t)^t\} = 0$. Taking the initial condition of (A_t, B_t) as $(I, 0)$ where I is the $d \times d$-unit matrix, we get $(A_t, B_t)(\partial\bar{\varphi}_t, \partial\bar{\eta}_t)^t = I$. This proves that the rank of the matrix $(\partial\bar{\varphi}_t(x), \partial\bar{\eta}_t(x))$ is d for any t a.s. The same property is valid for $\lim_{t\uparrow\sigma(y)} (\partial\bar{\varphi}_t(\bar{\psi}_t(y)), \partial\bar{\eta}_t(\bar{\psi}_t(y)))$. Therefore,

$$\varliminf_{t\uparrow\sigma(y)} \{|\det \partial\bar{\varphi}_t(\bar{\psi}_t(y))| + |\partial\bar{\eta}_t(\bar{\psi}_t(y))|\} > 0 \quad \text{if} \quad \sigma(y) < T.$$

Since $\lim_{t\uparrow\sigma(y)} |\det \partial\bar{\psi}_t(y)| = \infty$, we have $\lim_{t\uparrow\sigma(y)} \det \partial\bar{\varphi}_t(\bar{\psi}_t(y)) = 0$. This implies $\varliminf_{t\uparrow\sigma(y)} |\partial\bar{\eta}_t(\bar{\psi}_t(y))| > 0$. Consequently we have

$$\lim_{t\uparrow\sigma(y)} |u_x(y, t)| = \lim_{t\uparrow\sigma(y)} |\partial\bar{\eta}_t(\bar{\psi}_t(y))\partial\bar{\psi}_t(y)| = \infty.$$

The proof is complete. □

A quasi-linear equation (18) is called *semi-linear* if $F^i(x, u, t)$, $i = 1, \ldots, d$ of the equation do not depend on u. Then the characteristic equation of the semi-linear equation takes the form

$$d\varphi_t = -F(\varphi_t, \circ dt), \qquad d\eta_t = F^{d+1}(\varphi_t, \eta_t, \circ dt). \tag{22}$$

Assume that the local characteristic of F belongs to $(B_b^{k+1,\delta}, B_b^{k,\delta})$ for some $k \geq 3$ and $\delta > 0$. Then the solution $\varphi_t(x)$ defines a global flow of C^k-diffeomorphisms. Since $\bar{\varphi}_t(x) = \varphi_t(x)$, Jacobian matrix $\partial\bar{\varphi}_t(x)$ is always non-singular and the inverse $\bar{\psi}_t(x)$ is defined for all t, x. Therefore (12) defines a global solution of the semi-linear equation. Thus we have the following.

Theorem 6.1.8 *Assume that* $(F^1, \ldots, F^d)$ *of the semi-linear equation are continuous* $C^{k,\delta}$*-semimartingales with local characteristics belonging to* $(B_b^{k+1,\delta}, B_b^{k,\delta})$ *for some* $k \geq 3$ *and* $\delta > 0$. *Then the semi-linear equation has a unique global solution such that it is a continuous local* $C^{k,\varepsilon}$*-semimartingale for some* $\varepsilon > 0$. □

Linear equations

A first order stochastic partial differential equation

$$u(x, t) = f(x) + \sum_{i=1}^{d} \int_0^t F^i(x, \circ ds) \frac{\partial u}{\partial x^i}(x, s) + \int_0^t F^{d+1}(x, \circ ds) u(x, s) + F^{d+2}(x, t) \tag{23}$$

is called a *linear equation.* It is a special case of the semi-linear equation. We shall integrate the equation making use of the stochastic flow generated by the characteristic system of equations associated with equation (23).

We will assume that the coefficients $F^1, \ldots, F^{d+2}$ of the equation satisfy the following Condition $(D.1)_{k,\delta}$, where k is an integer greater than or equal to 3 and δ is a positive number less than or equal to 1.

Condition $(D.1)_{k,\delta}$ $(F^1, \ldots, F^{d+2})$ *is a continuous* C^k*-semimartingale with local characteristic belonging to the class* $(B_b^{k+1,\delta}, B_b^{k,\delta})$. □

Also we often assume a stronger condition $(D.1)_{k,\delta}^u$.

Condition $(D.1)_{k,\delta}^u$ $(F^1, \ldots, F^{d+2})$ *is a continuous* C^k*-semimartingale with local characteristic belonging to the class* $(B_{ub}^{k+1,\delta}, B_{ub}^{k,\delta})$ *such that* $A_t \equiv t$. □

The associated characteristic system of the Stratonovich's stochastic differential equation is given by

$$\varphi_{s,t}(x) = x - \int_s^t F(\varphi_{s,r}(x), \circ dr) \tag{24}$$

where $F(x, t) = (F^1(x, t), \ldots, F^d(x, t))$ and

$$\eta_{s,t}(x, u) = u + \int_s^t \eta_{s,r}(x, u) F^{d+1}(\varphi_{s,r}(x), \circ dr) + \int_s^t F^{d+2}(\varphi_{s,r}(x), \circ dr). \tag{25}$$

The solution defines a global stochastic flow of C^k-diffeomorphisms. We denote $(\varphi_{0,t}(x), \eta_{0,t}(x,u))$ by $(\varphi_t(x), \eta_t(x,u))$. Set $\bar\eta_t(x) = \eta_t(x, f(x))$ and $u(x,t) = \bar\eta_t(\psi_t(x))$ where $\psi_t = \varphi_t^{-1}$. Then it is the unique global solution of equation (23) by Theorems 6.1.2 and 6.1.8. Now since $\eta_t(x,u)$ satisfies the linear equation (25), $\bar\eta_t(y)$ is calculated by

$$\bar\eta_t(y) = \exp\left\{\int_0^t F^{d+1}(\varphi_s(y), \circ ds)\right\}$$
$$\times\left[f(y) + \int_0^t \exp\left\{-\int_0^s F^{d+1}(\varphi_r(y), \circ dr)\right\} F^{d+2}(\varphi_s(y), \circ ds)\right]. \tag{26}$$

Therefore the solution $u(x,t)$ of equation (23) is represented by

$$u(x,t)$$
$$= \exp\left\{\int_0^t F^{d+1}(\varphi_s(y), \circ ds)|_{y=\psi_t(x)}\right\}$$
$$\times\left[f(\psi_t(x)) + \int_0^t \exp\left\{-\int_0^s F^{d+1}(\varphi_r(y), \circ dr)\right\} F^{d+2}(\varphi_s(y), \circ ds)|_{y=\psi_t(x)}\right]. \tag{27}$$

Now assume that $(F^1, \ldots, F^{d+2})$ is a forward–backward semimartingale, satisfying Condition $(D.1)_{k,\delta}$ for the backward direction. Then the inverse $\varphi_{t,s} = \varphi_{s,t}^{-1}$ satisfies the backward Stratonovich's equation

$$\varphi_{t,s}(x) = x + \int_s^t F(\varphi_{t,r}(x), \circ \hat{d}r), \qquad s < t \tag{28}$$

by Theorem 4.4.4. Note that $\psi_t(x) = \varphi_{t,0}(x)$. We have

$$\exp\left\{\int_0^t F^{d+1}(\varphi_s(y), \circ ds)\right\}$$
$$\times \int_0^t \exp\left\{-\int_0^s F^{d+1}(\varphi_r(y), \circ dr)\right\} F^{d+2}(\varphi_s(y), \circ ds)\Big|_{y=\varphi_{t,0}(x)}$$
$$= \int_0^t \exp\left\{\int_s^t F^{d+1}(\varphi_{t,r}(x), \circ \hat{d}r)\right\} F^{d+2}(\varphi_{t,s}(x), \circ \hat{d}s)$$

(see the proof of Theorem 4.2.10). Then (27) is represented by

$$u(x,t) = \exp\left\{\int_0^t F^{d+1}(\varphi_{t,s}(x), \circ \hat{d}s)\right\} f(\varphi_{t,0}(x))$$
$$+ \int_0^t \exp\left\{\int_s^t F^{d+1}(\varphi_{t,r}(x), \circ \hat{d}r)\right\} F^{d+2}(\varphi_{t,s}(x), \circ \hat{d}s). \tag{29}$$

Consequently, we have the following.

Theorem 6.1.9 *Assume that $(F^1, \ldots, F^d)$ of linear equation (23) satisfies Condition $(\mathrm{D}.1)_{k,\delta}$ for some $k \geq 3$ and $\delta > 0$. If the initial function is of $C^{k,\delta}$, the linear equation has a unique global solution which is a continuous $C^{k,\varepsilon}$-semimartingale for some $\varepsilon > 0$. It is represented by (27). Assume further that $(F^1, \ldots, F^{d+2})$ is a forward–backward semimartingale satisfying Condition $(\mathrm{D}.1)_{k,\delta}$ for the backward direction. Then the solution is represented by (29).* □

There is an alternative method of proving that (29) is a solution of equation (23). Let $(\varphi_{t,s}(x), \xi_{t,s}(x, u), \zeta_{t,s}(x, u, v))$, $s \leq t$ be a backward stochastic flow of diffeomorphisms determined by the system of backward stochastic differential equations (28), (30) and (31) described below.

$$\xi_{t,s}(x, u) = u + \int_s^t \xi_{t,r}(x, u) F^{d+1}(\varphi_{t,r}(x), \circ \hat{d}r) \tag{30}$$

$$= u \exp\left\{\int_s^t F^{d+1}(\varphi_{t,r}(x), \circ \hat{d}r)\right\}$$

$$\zeta_{t,s}(x, u, v) = v + \int_s^t \xi_{t,r}(x, u) F^{d+2}(\varphi_{t,r}(x), \circ \hat{d}r). \tag{31}$$

Define

$$u(x, u, v, t) = \xi_{t,0}(x, u) f(\varphi_{t,0}(x)) + \zeta_{t,0}(x, u, v). \tag{32}$$

We will apply Itô's first formula to the variable t of the backward flow by setting $\tilde{f}(\tilde{u}, \tilde{v}, \tilde{x}) = \tilde{u} f(\tilde{x}) + \tilde{v}$ and $\tilde{u} = \xi_{t,0}(x, u)$, $\tilde{x} = \varphi_{t,0}(x)$ and $\tilde{v} = \zeta_{t,0}(x, u, v)$. Then Theorem 4.4.5 implies

$$u(x, u, v, t) = f(x) + \sum_{i=1}^d \int_0^t F^i(x, \circ dr) \frac{\partial u}{\partial x^i}(x, u, v, r)$$

$$+ u \int_0^t F^{d+1}(x, \circ dr) \frac{\partial u}{\partial u}(x, u, v, r)$$

$$+ u \int_0^t F^{d+2}(x, \circ dr) \frac{\partial u}{\partial v}(x, u, v, r).$$

Note that

$$\frac{\partial u}{\partial u}(x, u, v, r) = u(x, 1, 0, r), \qquad \frac{\partial u}{\partial v}(x, u, v, r) = 1.$$

Then we get

$$u(x, 1, 0, t) = f(x) + \sum_{i=1}^d \int_0^t F^i(x, \circ dr) \frac{\partial u}{\partial x^i}(x, 1, 0, r)$$

$$+ \int_0^t F^{d+1}(x, \circ dr) u(x, 1, 0, r) + \int_0^t F^{d+2}(x, \circ dr).$$

This proves that $u(x, t) \equiv u(x, 1, 0, t)$ is a solution of equation (23).

We shall study the integrability and the growth property of the solution of equation (23) as $|x| \to \infty$. The results will be applied in the next section for proving the existence and the uniqueness of the solution of a second order stochastic partial differential equation. A continuous C^k-valued process $u(x, t)$ is called *slowly increasing* if there exists a positive integer m such that

$$\lim_{|x|\to\infty} \sup_t \frac{|u(x, t)|}{(1 + |x|)^m} = 0 \qquad \text{a.s. } P$$

is satisfied. The process $u(x, t)$ is called *rapidly decreasing* if for any positive integer n,

$$\lim_{|x|\to\infty} \sup_t |u(x, t)|(1 + |x|)^n = 0 \qquad \text{a.s. } P$$

is satisfied.

Theorem 6.1.10 *Assume that $(F^1, \ldots, F^{d+2})$ of linear equation (23) is a forward–backward semimartingale satisfying Condition $(\mathrm{D.1})_{k,\delta}$, $k \geq 3$, $\delta > 0$ both for the forward and the backward direction. Assume that the local characteristic of F^{d+1} is uniformly bounded for the backward direction. Then if the initial function is slowly increasing, the solution is also slowly increasing. Further, assume Condition $(\mathrm{D.1})^u_{k,\delta}$ in place of Condition $(\mathrm{D.1})_{k,\delta}$ for the backward direction. Then*

$$E\left[\sup_{|x|\leq N} |D^\alpha u(x, t)|^p\right] < \infty$$

holds for any $|\alpha| \leq k$ and $N, p > 1$.

Proof The theorem can be reduced to the similar integrabilities and growth properties of the backward stochastic flow $(\varphi_{t,s}(x), \xi_{t,s}(x, 1), \zeta_{t,s}(x, 1, 0))$ determined by equations (28), (30) and (31). Indeed, assume first Condition $(\mathrm{D.1})^u_{k,\delta}$. Then for any $0 < \varepsilon < 1$, we have $\lim_{|x|\to\infty} \sup_{s,t} |\varphi_{t,s}(x)|/(1 + |x|)^{1+\varepsilon} = 0$ (see Exercise 4.5.9). We can show a similar property for $\xi_{t,s}(x, 1)$ and $\zeta_{t,s}(x, 1, 0)$. Since $u(x, t)$ is represented by (32), it is slowly increasing. The case for Condition $(\mathrm{D.1})_{k,\delta}$ can be reduced to the above case by changing the scale of time.

The second assertion of the theorem is also reduced to the corresponding assertion for the flow $(\varphi_{t,s}, \xi_{t,s}, \zeta_{t,s})$. In fact, $\sup_{|x|\leq N} |D^\alpha \varphi_{t,s}(x)|^p$ is integrable for any N and p by Corollary 4.6.7. Similar properties are valid for ξ and ζ. Then the assertion holds. □

We shall next consider the case where the initial function is rapidly decreasing.

Theorem 6.1.11 *Assume the same conditions as in Theorem 6.1.10. Let $u(x, t)$ be the solution of equation (23) where $F^{d+2} \equiv 0$. If the initial function is rapidly decreasing, then $u(x, t)$ is also rapidly decreasing. Further if Condition $(D.1)^u_{k,\delta}$, $k \geq 3$, $\delta > 0$ is satisfied for the backward direction, then we have for any positive integer n*

$$E\left[\left[\int_{\mathbf{R}^d} (1 + |x|)^n |u(x, t)| dx\right]\right] < \infty, \qquad \textit{for all } t \in [0, T]. \tag{33}$$

Proof Note that the solution $u(x, t)$ is represented by

$$u(x, t) = \xi_{t,0}(x, 1) f(\varphi_{t,0}(x)) \tag{34}$$

by (32). For any $0 < \varepsilon < 1$, $|\xi_{t,0}(x, 1)|(1 + |x|)^{-(1+\varepsilon)}$ converges to 0 uniformly in t a.s. as $|x| \to \infty$. Exercise 4.5.9). Further, since f is rapidly decreasing for any $p \geq 2$ there exists a positive constant C such that $|f(x)| \leq C(1 + |x|)^{-p}$. Then we have

$$|f(\varphi_{t,0}(x))|(1 + |x|)^{p-1} \leq C(1 + |\varphi_{t,0}(x)|)^{-p}(1 + |x|)^{p-1} \to 0$$

uniformly in t as $|x| \to \infty$ (Exercise 4.5.10). Consequently for $n = p - \varepsilon - 2$, $|u(x, t)|(1 + |x|)^n$ converges to 0 uniformly in t a.s. as $|x| \to \infty$.

Now assume further Condition $(D.1)^u_{k,\delta}$, $k \geq 3$, $\delta > 0$ for the backward direction. Then the above consideration yields that $\sup_{x,t} |u(x, t)|(1 + |x|)^n$ has finite moments of any order (Exercises 4.5.9 and 4.5.10). Then $\int_{\mathbf{R}^d} |u(x, t)|(1 + |x|)^{n-d-1} dx$ has finite moments of any order. Thus property (33) is verified. The proof is complete. □

6.2 Second order stochastic partial differential equations

Preliminaries

In this section we shall study the initial value problem of the second order linear partial differential equation with random coefficients:

$$u(x, t) = f(x) + \int_0^t L_s u(x, s)\, ds + \sum_{i=1}^d \int_0^t F^i(x, \circ ds) \frac{\partial u}{\partial x^i}(x, s)$$
$$+ \int_0^t F^{d+1}(x, \circ ds) u(x, s). \tag{1}$$

Here L_s is an elliptic operator of the form

$$L_s u = \frac{1}{2} \sum_{i,j} a^{ij}(x, s) \frac{\partial^2 u}{\partial x^i \partial x^j} + \sum_i b^i(x, s) \frac{\partial u}{\partial x^i} + d(x, s) u, \tag{2}$$

and the random field $(F^1, \ldots, F^{d+1})$ is a continuous forward–backward

$C^{k,\delta}$-semimartingale satisfying Condition $(D.1)_{k,\delta}$ for some $k \geq 3$ and $\delta > 0$ both for the forward and backward direction. The differentials $F^i(x, \circ ds)$ are the Stratonovich differentials. A continuous C^2-semimartingale $u(x, t)$ is then called a *solution with the initial value f* if it satisfies equation (1) for any x and t a.s.

In the sequel we assume that the coefficients of the operator L_s satisfy the following Condition $(D.2)_{k,\delta}$ for some $k \geq 3$ and $\delta > 0$.

Condition $(D.2)_{k,\delta}$ *There exists a non-negative and symmetric continuous function $a^{ij}(x, y, s)$ belonging to the class $C_{ub}^{k+1,\delta}$ such that $a^{ij}(x, s) = a^{ij}(x, x, s)$. The function b^i is continuous in (x, s) belonging to the class $C_u^{k,\delta}$. The function $d(x, s)$ is continuous in (x, s) belonging to the class $C^{k,\delta}$. Further a^{ij} is bounded and $d/(1 + |x|)$ is bounded from the above.* □

Before we proceed we remark that any function of $C_b^{k,\delta}$ is not bounded but of linear growth. We shall rewrite equation (1) using the Itô integral. Let (A^{ij}, B^i, t) be the local characteristic of $(F^1, \ldots, F^{d+1})$. Define a second order operator $\tilde{L}_t$ by

$$\tilde{L}_t u = \frac{1}{2} \sum_{i,j=1}^{d} A^{ij}(x, x, t) \frac{\partial^2 u}{\partial x^i \partial x^j} + \sum_{i=1}^{d} \left\{ A^{i,d+1}(x, x, t) + \frac{1}{2} C^i(x, t) \right\} \frac{\partial u}{\partial x^i} + \frac{1}{2} \{ D(x, t) + A^{d+1,d+1}(x, x, t) \} u, \tag{3}$$

where

$$C^j(x, t) = \sum_{i=1}^{d} \frac{\partial A^{ij}}{\partial y^i}(x, y, t)|_{y=x},$$

$$D(x, t) = \sum_{i=1}^{d} \frac{\partial A^{i,d+1}}{\partial y^i}(x, y, t)|_{y=x}.$$

Then using the Itô integral, equation (1) can be written as

$$u(x, t) = f(x) + \int_0^t (L_s + \tilde{L}_s) u(x, s)\, ds + \sum_{i=1}^{d} \int_0^t F^i(x, ds) \frac{\partial u}{\partial x^i}(x, s) + \int_0^t F^{d+1}(x, ds) u(x, s). \tag{4}$$

Indeed, the Stratonovich integrals are represented by

$$\int_0^t F^i(x, \circ ds)\frac{\partial u}{\partial x^i} = \int_0^t F^i(x, ds)\frac{\partial u}{\partial x^i} + \frac{1}{2}\left\langle F^i(x, t), \frac{\partial u}{\partial x^i}\right\rangle$$
$$= \int_0^t F^i(x, ds)\frac{\partial u}{\partial x^i} + \frac{1}{2}\left(\int_0^t \sum_{j=1}^d A^{ij}(x, x, s)\frac{\partial^2 u}{\partial x^i \partial x^j} ds\right.$$
$$+ \int_0^t \left\{\sum_{j=1}^d \frac{\partial A^{ij}}{\partial y^i}(x, y, s)\bigg|_{y=x}\frac{\partial u}{\partial x^j} + A^{i,d+1}(x, x, s)\frac{\partial u}{\partial x^i}\right\} ds$$
$$\left. + \int_0^t \frac{\partial A^{i,d+1}}{\partial y^i}(x, y, s)|_{y=x} u \, ds\right).$$

We have similarly

$$\int_0^t F^{d+1}(x, \circ ds)u = \int_0^t F^{d+1}(x, ds)u$$
$$+ \frac{1}{2}\left\{\int_0^t \left(\sum_{j=1}^d A^{d+1,j}(x, x, s)\frac{\partial u}{\partial x^j}\right) ds\right.$$
$$\left. + \int_0^t A^{d+1,d+1}(x, x, s)u \, ds\right\}.$$

Substituting these relations in (1), we obtain equation (4).

Conversely let $u(x, t)$ be a continuous C^2-process satisfying the equation represented by the Itô integrals:

$$u(\cdot, t) = f + \int_0^t A_s u(s) \, ds + \sum_{i=1}^d \int_0^t F^i(\cdot, ds)\frac{\partial u}{\partial x^i}$$
$$+ \int_0^t F^{d+1}(\cdot, ds)u(s). \qquad (5)$$

If $u(x, t)$ is a C^2-semimartingale, it is represented by (1), replacing L_s by $A_s - \tilde{L}_s$ where $\tilde{L}_s$ is defined by (3).

It should be noted that the principal part (second order part) of the operator L_t in equation (1) and that of the operator $A_t = L_t + \tilde{L}_t$ in equation (4) are different provided that the local characteristic $A^{ij}(x, y, t)$ of $F(x, t)$ is not identically 0. Conversely, solutions of the equations with the common principal part and the common random first order part can have different properties if one is written by the Stratonovich integral and the other is written by the Itô integral. Equation (5) represented by the Itô integrals looks like a second order equation even in the case $A_t = \tilde{L}_t$ though the same equation represented by the Stratonovich integrals is just a first order equation. Furthermore if the coefficient $a^{ij}(x, t)$ of the operator L_t is less than $A^{ij}(x, x, t)$, equation (5) does not seem to have any solution, since the second order part of the same equation represented by the Stratonovich integrals is no longer non-negative definite. Thus the stochastic partial differential equation represented by the Stratonovich integrals can have more qualitative properties than that represented by the Itô integral.

For this reason, we shall discuss the equation represented by the Stratonovich integrals.

Existence and uniqueness of solutions

We will study the existence and the uniqueness problem under Conditions $(D.1)_{k,\delta}$ and $(D.2)_{k,\delta}$ for some $k \geq 3$ and $\delta > 1$. We can assume that the coefficients b^i in the operator L_t are identically 0 since the first order term in L_t can be shifted to the random first order term of the equation.

Note that the right hand side of equation (1) consists of the first order part with random coefficients and the deterministic second order part $\int_0^t L_s u \, ds$. We can regard the first order part as a perturbation term adjoined to the second order part. In what follows we show that the existence and uniqueness of the solution of equation (1) can be reduced to the existence and uniqueness of a certain deterministic second order equation, which is a modification of the equation $\partial u/\partial t = L_t u$ affected by the perturbation term.

We first consider the first order part. Let $w_t(f)$ be the solution of the first order equation (1) where $L_t \equiv 0$. Then by Theorem 6.1.9 it is represented by

$$w_t(f)(x) = \xi_{t,0}(x) f(\varphi_{t,0}(x)) \tag{6}$$

where $\varphi_{t,0}$ is the inverse of the stochastic flow $\varphi_{0,t}$ generated by $-(F^1, \dots, F^d)$ in the sense of the Stratonovich integral. $\xi_{t,0}(x)$ is an abbreviation of $\xi_{t,0}(x, 1)$ in the previous section, namely,

$$\xi_{t,0}(x) = \exp\left\{\int_0^t F^{d+1}(\varphi_{t,s}(x), \circ \hat{d}s)\right\}. \tag{7}$$

We may consider that for almost all ω, w_t is a linear map on $C^{3,\varepsilon}(\mathbb{R}^d : \mathbb{R})$ for some $\varepsilon \in [0, \delta]$. It is one to one and onto. The inverse map is given by

$$w_t^{-1}(f)(x) = \xi_{t,0}(\varphi_{0,t}(x))^{-1} f(\varphi_{0,t}(x)) = \xi_{0,t}(x) f(\varphi_{0,t}(x)), \tag{8}$$

where

$$\xi_{0,t}(x) = \exp\left\{\int_0^t F^{d+1}(\varphi_{0,s}(x), \circ ds)\right\}. \tag{9}$$

Lemma 6.2.1 *Define the operator L_t^w by $w_t^{-1}L_t w_t$. Then it is a second order differential operator represented by*

$$L_t^w = \frac{1}{2}\sum_{i,j} a_w^{ij}(x,t)\frac{\partial^2}{\partial x^i \partial x^j} + \sum_i b_w^i(x,t)\frac{\partial}{\partial x^i} + d_w(x,t). \tag{10}$$

Here a_w^{ij}, b_w^i, d_w are smooth functions with random parameter ω defined by

$$a_w^{ij}(x,t) = \sum_{k,l} a^{kl}(\varphi_{0,t}(x),t)\partial_k(\varphi_{t,0}^i)(\varphi_{0,t}(x))\partial_l(\varphi_{t,0}^j)(\varphi_{0,t}(x)),$$

$$\begin{aligned} b_w^i(x,t) &= \frac{1}{2}\sum_{k,l} a^{kl}(\varphi_{0,t}(x),t)\{\partial_k\partial_l(\varphi_{t,0}^i)(\varphi_{0,t}(x)) \\ &\quad + \xi_{t,0}(\varphi_{0,t}(x))^{-1}\partial_l(\xi_{t,0})(\varphi_{0,t}(x))\partial_k(\varphi_{t,0}^i)(\varphi_{0,t}(x))\}, \end{aligned} \tag{11}$$

$$d_w(x,t) = \sum_{k,l} a^{kl}(\varphi_{0,t}(x),t)\xi_{t,0}(\varphi_{0,t}(x))^{-1}\partial_k\partial_l(\xi_{t,0})(\varphi_{0,t}(x)) + d(x,t),$$

where $\partial_k = \partial/\partial x^k$. □

The proof is immediate from a direct computation.

Note that the above coefficients a_w, b_w, d_w are not bounded, but have the following growth properties by Conditions $(D.1)_{k,\delta}^w$ and $(D.2)_{k,\delta}$ for $k \geq 3$ and $\delta > 0$. For any $0 < \varepsilon < 1$ there exists a positive constant G depending on ω such that

$$|a_w^{ij}(x,t)| \leq G(1+|x|)^\varepsilon, \tag{12}$$

$$|b_w^i(x,t)| \leq G(1+|x|)^\varepsilon, \tag{13}$$

$$d_w(x,t) \leq G(1+|x|). \tag{14}$$

Indeed if $1 \leq |\alpha| \leq k$ we have for any $\varepsilon > 0$

$$\lim_{|x|\to\infty} \frac{|D^\alpha \varphi_{t,0}(\varphi_{0,t}(x))|}{(1+|x|)^\varepsilon} = 0 \qquad \text{uniformly in } t \quad \text{a.s.}$$

(see Exercise 4.6.9). We can show similarly

$$\lim_{|x|\to\infty} \frac{\left|\int_0^t D^\alpha(F^{d+1}\circ\varphi_{t,s})(y,\circ\hat{d}s)\right|_{y=\varphi_{0,t}(x)}}{(1+|x|)^\varepsilon} = 0 \qquad \text{uniformly in } t.$$

Note that

$$\xi_{t,0}(y)^{-1}D^\alpha(\xi_{t,0})(y) = \int_0^t D^\alpha(F^{d+1}\circ\varphi_{t,s})(y,\circ\hat{d}s).$$

Then direct estimations of a^w, b^w, d^w given by (11) lead to (12), (13), and (14).

We borrow a theorem from the theory of partial differential equations. Let β be a positive number. A function $f(x)$ is said to be *of exponential growth with exponent* β or simply a function *with exponent* β if $f(x)\exp\{-c|x|^\beta\}$ is a bounded function for any positive constant c.

Theorem 6.2.2 *Let $f(x)$ be a function with an exponent less than 2 belonging to C^3. Then the parabolic equation with the initial condition*

$$\frac{\partial u}{\partial s} = L_s^w u, \qquad \lim_{s\downarrow 0} u(\cdot, s) = f \tag{15}$$

has a solution $u(x, t)$ with an exponent less than 2. Further it is the unique solution among those with exponents less than 2.

Proof We shall follow Friedman [33]. Let f be a function with exponent β. Let 2γ be a positive number greater than $1 \vee \beta$ and less than 2. Introduce the function

$$H(x, t) = \exp\left\{-\frac{(1 + k|x|^2)^\gamma}{1 - \mu t}\right\},$$

where k, μ are positive constants to be fixed later. Define

$$\tilde{L}_t^w u = H(x, t)\left(-\frac{\partial}{\partial t} + L_t^w\right)H(x, t)^{-1}u. \tag{16}$$

Then we have

$$\tilde{L}_t^w = -\frac{\partial}{\partial t} + \frac{1}{2}\sum_{i,j} a_w^{ij}\frac{\partial^2}{\partial x^i \partial x^j} + \sum_i \tilde{b}_w^i \frac{\partial}{\partial x^i} + \tilde{d}_w \tag{17}$$

where

$$\begin{aligned}\tilde{d}_w = d_w &+ 2k^2\gamma^2\left\{\frac{(1 + k|x|^2)^{2(\gamma-1)}}{(1 - \mu t)^2} + \frac{\gamma - 1}{\gamma}\frac{(1 + k|x|^2)^{\gamma-2}}{1 - \mu t}\right\}\left\{\sum_{i,j} a_w^{ij}x^i x^j\right\} \\ &+ \frac{k\gamma(1 + k|x|^2)^{\gamma-1}}{1 - \mu t}\left\{\sum_i a_w^{ii}\right\} + \frac{2k\gamma(1 + k|x|^2)^{\gamma-1}}{1 - \mu t}\left\{\sum_i b_w^i x^i\right\} \\ &- \frac{\mu(1 + k|x|^2)^\gamma}{(1 - \mu t)^2}.\end{aligned} \tag{18}$$

Since a_w^{ij}, b_w^i, d_w satisfy inequalities (12)–(14) for ε such that $\varepsilon + 2\gamma < 2$, we can choose positive constants k, μ such that $\tilde{d}_w(x, t) \le 0$ for any $x \in \mathbb{R}^d$ and $t \in [0, T]$. Since $fH(\cdot, 0)$ is a bounded function, the equation

$$\frac{\partial}{\partial t}u = \tilde{L}_t^w u, \qquad u|_{t=0} = fH(\cdot, 0) \tag{19}$$

has a unique bounded solution (see Stroock–Varadhan [119], Theorem

3.1.1). Then $u(x, t)H(x, t)$ is the unique solution of equation (15) with an exponent less than 2. The proof is complete. □

We now discuss the relation between equations (1) and (15).

Lemma 6.2.3 *Let $u(x, t)$ be a continuous C^3-process and C^2-semimartingale. It is a solution of equation (1) if and only if $u'(x, t) \equiv w_t^{-1}(u(x, t))$ is a solution of equation (15).*

Proof Let $u(x, t)$ be a solution of (1) such that it is a continuous C^3-process and a C^2-semimartingale. Set $u'(x, t) = w_t^{-1}(u(x, t))$. It is again a continuous C^3-process and C^2-semimartingale. Write $\varphi_{0,t} = (\varphi_{0,t}^1, \dots, \varphi_{0,t}^d)$. Then by the generalized Itô's formula, we have

$$\begin{aligned}
u'(x, \circ dt) &= d_t\{u(\varphi_{0,t}(x), t)\xi_{0,t}(x)^{-1}\} \\
&= \left\{u(\varphi_{0,t}(x), \circ dt) + \sum_i \frac{\partial u}{\partial x^i}(\varphi_{0,t}(x), t) \circ d\varphi_{0,t}^i(x)\right\}\xi_{0,t}(x)^{-1} \\
&\quad - u(\varphi_{0,t}(x), t)F^{d+1}(\varphi_{0,t}(x), \circ dt)\xi_{0,t}(x)^{-1} \\
&= L_t u(\varphi_{0,t}(x), t)\xi_{0,t}(x)^{-1}\, dt \\
&= w_t^{-1}(L_t w_t(u'))(x)\, dt. \qquad (20)
\end{aligned}$$

Therefore $u'(x, t)$ satisfies (15).

Conversely if $u'(x, t)$ is a solution of equation (15), we can show similarly that $u(x, t) \equiv w_t(u'(x, t))$ is a solution of equation (1). □

Theorem 6.2.4 *Assume that in equation (1) the operator L_t satisfies Condition $(D.2)_{k,\delta}$ and $(F^1, \dots, F^d)$ satisfies Condition $(D.1)_{k,\delta}$ for some $k \geq 3$ and $\delta > 0$. Let f be a function with an exponent less than 2. Then equation (1) has a unique solution with exponents less than 2.*

Proof We first show the existence of the solution. Let $u'(x, t)$ be a solution of equation (15) with an exponent less than 2. Set $u(x, t) = w_t(u'(x, t))$. Then it is a solution of the equation with an exponent less than 2 by Lemma 6.2.3. Conversely let $u_1(x, t)$ and $u_2(x, t)$ be two solutions of equation (1) with exponents less than 2. Set $u_i'(x, t) = w_t^{-1}(u_i(x, t))$, $i = 1, 2$. Then both are solutions of equation (15) with exponents less than 2. Therefore $u_1' = u_2'$ holds by Lemma 6.2.3. This shows $u_1 = u_2$. The proof is complete. □

Probabilistic representation of solutions

We shall now construct a solution of equation (1) by a probabilistic method under a slightly stronger assumption than in Theorem 6.2.4. Let $(W, \mathscr{B}, Q)$

be another probability space where a $C^{k,\gamma}$-valued Brownian motion $X(x, t)$ with the local characteristic $a^{ij}(x, y, t)$ and $m(x, t) \equiv b(x, t) - c(x, t)$ is given, where $c(x, t)$ is the correction term based on (a^{ij}). On the product probability space $(\Omega \times W, \mathscr{F} \otimes \mathscr{B}, P \times Q)$ we consider a first order stochastic partial differential equation:

$$v(\cdot, t) = f + \sum_{i=1}^{d} \int_0^t \{X^i(\cdot, \circ ds) + F^i(\cdot, \circ ds)\} \frac{\partial v}{\partial x^i}$$
$$+ \int_0^t \{d(\cdot, s)\, ds + F^{d+1}(\cdot, \circ ds)\} v. \tag{21}$$

By Theorem 6.1.9, the equation has a unique global solution such that it is a continuous $C^{k,\varepsilon}$-semimartingale. Further it is represented by

$$v(x, t) = \bar{\xi}_{t,0}(x) f(\bar{\varphi}_{t,0}(x)), \tag{22}$$

where $\bar{\varphi}_{t,0}$ is the inverse of the flow $\bar{\varphi}_{0,t}$ generated by $-(X^1 + F^1, \ldots, X^d + F^d)$ in the sense of the Stratonovich integral and

$$\bar{\xi}_{t,0}(x) = \exp\left\{\int_0^t d(\bar{\varphi}_{t,s}(x), s)\, ds + \int_0^t F^{d+1}(\bar{\varphi}_{t,s}(x), \circ \hat{d}s)\right\}. \tag{23}$$

Theorem 6.2.5 *Assume that in equation* (1) *the operator L_t satisfies Condition* $(\mathrm{D.2})_{k,\delta}$ *and $(F^1, \ldots, F^d)$ satisfies Condition* $(\mathrm{D.1})_{k,\delta}$ *for some $k \geq 3$ and $\delta > 0$. Assume further that the backward local characteristics of F^{d+1} and $d(x, t)$ are uniformly bounded. If the function f is slowly increasing, the partial expectation*

$$u(x, t) = E_Q[\bar{\xi}_{t,0}(x) f(\bar{\varphi}_{t,0}(x))] \tag{24}$$

is well defined, slowly increasing and is the unique solution of equation (1). *Further if the function f is rapidly decreasing, the above is rapidly decreasing.*

Proof Rewrite the term involving $X^i(x, \circ ds)$ in (21) using the Itô integral. Then noting that X^i and F^i are independent, we have

$$v(x, t) = f(x) + \int_0^t L_s v(x, s)\, ds + \sum_{i=1}^{d} \int_0^t Y^i(x, ds) \frac{\partial v}{\partial x^i}$$
$$+ \sum_{i=1}^{d} \int_0^t F^i(x, \circ ds) \frac{\partial v}{\partial x^i} + \int_0^t F^{d+1}(x, \circ ds) v, \tag{25}$$

where $Y(x, t)$ is the martingale part of the semimartingale $X(x, t)$.

Now, take the expectation of each term of (25) with respect to Q. By Theorem 6.1.10, $\sup_{|x| \leq N} |D^\alpha v(x, t)|$ $(|\alpha| \leq k)$ have moments of any order with respect to $P \times Q$. Therefore they have moments of any order with respect to Q a.s. P. Consequently, we can change the order of the integration by Q and derivatives, stochastic integrals, etc. In fact for each fixed x and t, we have the following relations.

$$E_Q\left[\int_0^t L_s v(x, s)\, ds\right] = \int_0^t L_s E_Q[v(x, s)]\, ds \qquad \text{a.s. } P, \tag{26}$$

$$E_Q\left[\int_0^t Y^i(x, ds)\frac{\partial v}{\partial x^i}\right] = 0 \qquad \text{a.s. } P, \tag{27}$$

$$E_Q\left[\int_0^t F^i(x, \circ ds)\frac{\partial v}{\partial x^i}\right] = \int_0^t F^i(x, ds)\frac{\partial}{\partial x^i}E_Q[v] \qquad \text{a.s. } P, \tag{28}$$

$$E_Q\left[\int_0^t F^{d+1}(x, \circ ds)v\right] = \int_0^t F^{d+1}(x, \circ ds)E_Q[v] \qquad \text{a.s. } P, \tag{29}$$

which will be shown shortly. Therefore, for each fixed x and t, the partial expectation $u(x, t) = E_Q[v(x, t)]$ satisfies equation (1).

If the initial function is slowly increasing, the solution $v(x, t)$ of equation (21) is slowly increasing and there exists a positive integer m such that $\sup_{x,t}|v(x, t)|(1 + |x|)^{-m}$ has finite moments of any order with respect to $P \times Q$ (see Theorem 6.1.10 and Exercise 4.5.9). Then the partial expectation $E_Q[v(x, t)]$ is slowly increasing. The second assertion can be proved similarly. □

Lemma 6.2.6 *Equalities* (26)–(29) *hold.*

Proof The first equality (26) is straightforward from the integrability of $\sup_{|\alpha|\le N}|D^\alpha v(x, t)|$. We will prove (27). Note that for a fixed x and t.

$$\int_0^t Y^i(x, ds)\frac{\partial v}{\partial x^i} = \lim_{|\Delta|\to 0}\sum_{k=0}^{n-1}(Y^i(x, t_{k+1}) - Y^i(x, t_k))\frac{\partial v}{\partial x^i}(t_k)$$

in $L^2(P \times Q)$ where $\Delta = \{0 = t_0 < \cdots < t_n = t\}$. Then we have

$$E_Q\left[\int_0^t Y^i(x, ds)\frac{\partial v}{\partial x^i}\right] = \lim_{|\Delta|\to 0} E_Q\left[\sum_{k=0}^{n-1}(Y^i(x, t_{k+1}) - Y^i(x, t_k))\frac{\partial v}{\partial x^i}(t_k)\right]$$

in $L^2(P)$. The right hand side is 0 since

$$\begin{aligned} &E_Q\left[(Y^i(x, t_{k+1}) - Y^i(x, t_k))\frac{\partial v}{\partial x^i}(t_k)\right] \\ &\quad = E_Q\left[E_Q[(Y^i(x, t_{k+1}) - Y^i(x, t_k))|\mathscr{B}_{t_k}]\frac{\partial v}{\partial x^i}(t_k)\right] \\ &\quad = 0, \end{aligned}$$

where $\mathscr{B}_{t_k} = \sigma(X(\cdot, s) : s \le t_k)$. For the proof of (28), observe the relation

$$\int_0^t F_i(x, \circ ds)\frac{\partial v}{\partial x^i}$$

$$= \lim_{|\Delta|\to 0} \sum_{k=0}^{n-1} (F_i(x, t_{k+1}) - F_i(x, t_k))\frac{1}{2}\left(\frac{\partial v}{\partial x^i}(t_{k+1}) + \frac{\partial v}{\partial x^i}(t_k)\right)$$

in $L^2(P \times Q)$. Then we have

$$E_Q\left[\int_0^t F_i(x, \circ ds)\frac{\partial v}{\partial x^i}\right]$$

$$= \lim_{|\Delta|\to 0} \frac{1}{2}\sum_{k=0}^{n-1} (F_i(x, t_{k+1}) - F_i(x, t_k))\left(\frac{\partial u}{\partial x^i}(t_{k+1}) + \frac{\partial u}{\partial x^i}(t_k)\right).$$

The right hand side converges to the right hand side of (28).

Equality (29) can be shown similarly. The proof is complete. □

Remark If $F^1, \dots, F^{d+1}$ are identically 0 in equation (1), the equation is a second order (deterministic) partial differential equation

$$\frac{\partial u}{\partial t} = L_t u, \qquad u|_{t=0} = f. \tag{30}$$

The solution is then represented by

$$u(x, t) = E_Q\left[\exp\left\{\int_0^t d(\bar{\varphi}_{t,s}(x), s)\, ds\right\} f(\bar{\varphi}_{t,0}(x))\right] \tag{31}$$

where $\bar{\varphi}_{t,s}$ is the inverse of the stochastic flow $\bar{\varphi}_{s,t}$ generated by $-X(x, t)$ in the sense of the Stratonovich integral.

The use of the inverse flow provides an interesting probabilistic interpretation for diffusion or heat equations. We consider the case $c = d = 0$. Then $u(x, t) = E[f(\bar{\varphi}_{t,0}(x))]$ is a solution to a heat equation with the initial condition $u(x, 0) = f(x)$. Now a particle starting from $\bar{\varphi}_{t,0}(x)$ at time 0 and moving along the trajectory $\bar{\varphi}_{0,s}(\bar{\varphi}_{t,0}(x))$, $0 \le s \le t$ will arrive at x at time t since $\bar{\varphi}_{0,t}(\bar{\varphi}_{t,0}(x)) = x$. Hence $f(\bar{\varphi}_{t,0}(x))$ can be interpreted as the temperature at the state x at time t, which is carried through the above trajectories from the point $\bar{\varphi}_{t,0}(x)$. Then by the law of large numbers, its expectation $u(x, t)$ indicates the temperature at the state x at time t.

Adjoint equation

We shall next study a stochastic partial differential equation for a measure valued process, which can be considered as an adjoint of equation (1). Some notation is necessary.

Let L_t be a second order linear partial differential operator defined by (2) satisfying Condition $(D.2)_{k,\delta}$ for some $k \geq 3$ and $\delta > 0$. Let $X_t^1, \ldots, X_t^n$ be continuous semimartingales. For simplicity we assume that $\langle X^i, X^j \rangle = \delta_{ij} t$ holds for any $i, j = 1, \ldots, n$. Let $f_j^i(x, t)$, $h_j(x, t)$ be continuous functions belonging to the class C_b^k. Define the first order differential operators by

$$D_j(t)f = \sum_i f_j^i(\cdot, t)\frac{\partial}{\partial x^i}, \qquad M_j(t)f = D_j(t)f + h_j(\cdot, t)f. \tag{32}$$

We shall consider a stochastic partial differential equation for a continuous process ρ_t with values in $\mathscr{M}(\mathbb{R}^d)$, where $\mathscr{M}(\mathbb{R}^d)$ is the space of Borel measures on $\mathbb{R}^d$:

$$\rho_t(f) = \rho_0(f) + \int_0^t \rho_s(L_s f)\,\mathrm{d}s + \sum_j \int_0^t \rho_s(M_j(s)f) \circ \mathrm{d}X_s^j, \tag{33}$$

where $\rho_t(f) \equiv \int_{\mathbb{R}^d} \rho_t(\mathrm{d}x) f(x)$ and f is a smooth function with compact support. Rewrite the last term of the above using the Itô integral. The equation is equivalent to the following Itô's stochastic differential equation:

$$\rho_t(f) = \rho_0(f) + \int_0^t \rho_s((L_s + \tilde{L}_s)f)\,\mathrm{d}s + \sum_j \int_0^t \rho_s(M_j(s)f)\,\mathrm{d}X_s^j \tag{34}$$

where

$$\tilde{L}_s f = \frac{1}{2}\sum_j D_j(s)^2 f + \frac{1}{2}\sum_j h_j(s)D_j(s)f + \left(\frac{1}{2}\sum_j M_j(s)h_j(s)\right)f. \tag{35}$$

Further if $\rho_t(\mathrm{d}x)$ has a smooth density function $v(x, t)$, it satisfies the following stochastic partial differential equation:

$$\begin{aligned} v(x, t) = v(x, 0) &+ \int_0^t L_s^* v(x, s)\,\mathrm{d}s - \sum_i \int_0^t F^i(x, \circ\mathrm{d}s)\frac{\partial v}{\partial x^i} \\ &- \int_0^t \operatorname{div} F(x, \circ\mathrm{d}s)v(x, s) + \int_0^t F^{d+1}(x, \circ\mathrm{d}s)v(x, s), \end{aligned} \tag{36}$$

where L_s^* is the formal adjoint operator of L_s and the F^i are given by

$$F^i(x, t) = \sum_{k=1}^n \int_0^t f_k^i(x, s)\,\mathrm{d}X_s^k, \qquad F^{d+1}(x, t) = \sum_{k=1}^n \int_0^t h_k(x, s)\,\mathrm{d}X_s^k. \tag{37}$$

We shall now study the initial value problem of equation (33). Our approach to this problem can be regarded as an adjoint to the discussion for equation (1). We shall consider the first order random part of equation (33) as a perturbation term. Let $\hat{w}_t(f)$ be the solution of

$$\hat{w}_t(f)(x) = f(x) + \sum_j \int_0^t \hat{w}_r(M_j(s)f) \circ \mathrm{d}X_s^j. \tag{38}$$

Define the random operator $\hat{L}_t^w$ by $\hat{w}_t L_t \hat{w}_t^{-1}$. It is a second order operator represented as in Lemma 6.2.1.

Lemma 6.2.7 *Let ρ_t be a continuous $\mathscr{M}(\mathbb{R}^d)$-valued process. It is a solution of equation* (33) *if and only if $\rho'_t(f) \equiv \rho_t(\hat{w}_t^{-1}(f))$ satisfies*

$$\rho'_t(f_t) = \rho'_0(f_0) + \int_0^t \rho'_s\left(\left(\frac{\partial}{\partial s} + \hat{L}_s^w\right) f_s\right) \mathrm{d}s \tag{39}$$

for any smooth function $f_t(x)$ with compact support.

Proof Let $\{\psi_\varepsilon(x) : \varepsilon > 0\}$ be a family of bounded non-negative C^∞-functions such that $\int \psi_\varepsilon(x)\,\mathrm{d}x = 1$ for any ε and as $\varepsilon \to 0$ $\int \psi_\varepsilon(x - y) f(y)\,\mathrm{d}y$ converges to $f(x)$ for any bounded continuous function f. Define

$$\rho_t^\varepsilon(y) = \int \psi_\varepsilon(y - z)\rho_t(\mathrm{d}z).$$

Then it satisfies

$$\rho_t^\varepsilon(y) = \rho_0^\varepsilon(y) + \int_0^t \hat{L}_s \rho_s^\varepsilon(y)\,\mathrm{d}s - \sum_j \int_0^t D_j(s)\rho_s^\varepsilon(y) \circ \mathrm{d}X_s^j$$
$$+ \int_0^t \rho_s^\varepsilon(d)(y)\,\mathrm{d}s + \sum_j \int_0^t \rho_s^\varepsilon(h_j)(y) \circ \mathrm{d}X_s^j,$$

where

$$\hat{L}_s = \frac{1}{2}\sum_{i,j} a^{ij}(x, x, t)\frac{\partial^2}{\partial x^i \partial x^j} - \sum_i b^i(x, t)\frac{\partial}{\partial x^i}$$

and $\rho_s^\varepsilon(d)(y) = \int \psi_\varepsilon(y - z) d(z) \rho_s(\mathrm{d}z)$ etc. Set $\hat{w}_t^{-1}(x) = \hat{w}_t^{-1}(f_t)(x)$. It satisfies

$$\hat{w}_t^{-1} = \hat{w}_0^{-1} - \sum_j \int_0^t M_j(s)\hat{w}_s^{-1} \circ \mathrm{d}X_s^j + \int_0^t \hat{w}_s^{-1}\left(\frac{\partial}{\partial s} f_s\right) \mathrm{d}s.$$

Then we have by Itô's formula

$$\rho_t^\varepsilon \hat{w}_t^{-1} = \rho_0^\varepsilon \hat{w}_0^{-1} + \left(\int_0^t \hat{w}_s^{-1}\hat{L}_s \rho_s^\varepsilon\,\mathrm{d}s + \int_0^t \hat{w}_s^{-1}\rho_s^\varepsilon(d)\,\mathrm{d}s\right)$$
$$+ \left(-\sum_j \int_0^t \hat{w}_s^{-1} D_j(s)\rho_s^\varepsilon \circ \mathrm{d}X_s^j + \sum_j \int_0^t \hat{w}_s^{-1}\rho_s^\varepsilon(h_j) \circ \mathrm{d}X_s^j\right)$$
$$- \sum_j \int_0^t (M_j(s)\hat{w}_s^{-1})\rho_s^\varepsilon \circ \mathrm{d}X_s^j + \int_0^t \hat{w}_s^{-1}\left(\frac{\partial}{\partial s} f_s\right)\rho_s^\varepsilon\,\mathrm{d}s.$$

Integrate each term of the above by $\mathrm{d}y$ over $\mathbb{R}^d$ and then let ε tend to 0.

Then we obtain

$$\rho_t(\hat{w}_t^{-1}) = \rho_0(\hat{w}_0^{-1}) + \int_0^t \rho_s(L_s\hat{w}_s^{-1})\,\mathrm{ds} + \int_0^t \rho_s\hat{w}_s^{-1}\left(\frac{\partial}{\partial s}f_s\right)\mathrm{ds}.$$

This proves the lemma. □

Theorem 6.2.8 *Assume that in equation* (33) *the operator L_t satisfies Condition* $(\mathrm{D}.2)_{k,\delta}$ *for some $k \geq 3$ and $\delta > 0$ and the coefficients of the operator $M_j(t)$, $j = 1, \ldots, n$ belong to the class $C_b^{k,\delta}$. If the initial measure ρ_0 has a compact support a.s. equation* (33) *has a unique solution.*

Proof We first consider the existence of the solution. Let ρ_t' be a solution of equation

$$\frac{\partial \rho_t'(f)}{\partial t} = \rho_t'(\hat{L}_t^w f), \qquad \rho_0' = \rho_0. \tag{40}$$

Then $\rho_t(f)(x) \equiv \rho_t'(\hat{w}_t(f))$ is a solution of equation (33) by Lemma 6.2.7. Suppose conversely that ρ_t and $\tilde{\rho}_t$ are solutions of equation (33) with the same initial value ρ_0 which is of compact support a.s. Set $\rho_t'(f) = \rho_t(\hat{w}_t^{-1}(f))$ and $\tilde{\rho}_t'(f) = \tilde{\rho}_t(\hat{w}_t^{-1}(f))$. Then both satisfy (39). Now let f_t be a solution of the following backward equation with a terminal condition at time t,

$$\frac{\partial}{\partial r}f_r = -\hat{L}_r^w f_r, \qquad f_{r|r=t} = f. \tag{41}$$

Then Lemma 6.2.7 shows that both $\rho_t'(f)$ and $\tilde{\rho}_t'(f)$ coincide with $\rho_0(f_0)$ for any f. This implies $\rho_t = \tilde{\rho}_t$ and the uniqueness follows. The proof is complete. □

The measure valued solution ρ_t has a smooth density function if the coefficients of equation (1) are smooth and the initial distribution $\rho_0(\mathrm{d}x)$ has a smooth density.

Corollary 6.2.9 *Assume that in equation* (33) *the operator L_t satisfies Condition* $(\mathrm{D}.2)_{k,\delta}$ *for some $k \geq 3$ and $\delta > 0$ and coefficients of operators $M_j(t), j = 1, \ldots, n$ belong to the class $C_b^{k,\delta}$. Assume further that $h_j, j = 1, \ldots, n$ and d are uniformly bounded. If the initial distribution $\rho_0(\mathrm{d}x)$ has a C^{k-1}-density function $v(x, 0)$ which is rapidly decreasing, then the solution $\rho_t(\mathrm{d}x)$ also has a rapidly decreasing density $v(x, t)$ of such that it is a continuous C^{k-1}-process and C^{k-2}-semimartingale. Further it satisfies equation* (36).

Proof Consider equation (36) with the initial function $v(x, 0)$. Then by Theorem 6.2.4, it has a unique solution $v(x, t)$ which is a continuous C^{k-1}-process and C^{k-2}-semimartingale. Further it is rapidly decreasing by Theorem 6.2.5. Set $\rho_t(f) = \int v(x, t)f(x)\,dx$. Then it clearly satisfies equation (33) and has finite moments of any order a.s. The proof is complete. □

Remark The existence of the smooth density of $\rho_t(dx)$ has been shown without assuming the existence of the smooth density of the initial distribution $\rho_0(dx)$. See Bismut–Michel [11], Kusuoka–Stroock [85]. In these works, the Malliavin calculus is a basic tool for solving the problem. A fundamental assumption for the existence of the smooth density is that the operator L_t is non-degenerate in a certain sense, which is not needed for Corollary 6.2.9. The non-degeneracy is stated in connection with Hörmander's hypoellipticity. For the details, see the papers cited above. An example will be given in Exercise 6.3.6. □

Now we shall construct the solution of equation (33) by a probabilistic method. We will use the same notation as before. Let $(W, \mathscr{B}, Q)$ be another probability space where a $C^{k,\gamma}$-valued Brownian motion $X(x, t)$ with local characteristic $(a^{ij}(x, y, t), m^i(x, t))$ is given. Let $\tilde{\varphi}_{s,t}$ be the stochastic flow generated by $(X^1 + F^1, \ldots, X^d + F^d)$ in the sense of the Stratonovich integral on the product space $(\Omega \times W, \mathscr{F} \otimes \mathscr{B}, P \times Q)$. Set

$$\tilde{\xi}_{0,t}(x) = \exp\left\{\int_0^t d(\tilde{\varphi}_{0,r}(x), r)\,dr + \int_0^t F^{d+1}(\tilde{\varphi}_{0,r}(x), \circ dr)\right\}. \tag{42}$$

We show that its expectation $E_{P\times Q}[\tilde{\xi}_{0,t}(x)]$ is bounded in x under a condition to be stated soon. Let $e(x, t)$ be the coefficient of the 0-th order term of the operator $L_t + \tilde{L}_t$. It is written as

$$e(x, t) \equiv \frac{1}{2}\sum_j M_j(t)h_j(x, t) + d(x, t), \tag{43}$$

where d is the 0-th order term of the operator L_t. Then, using the Itô integral, $\tilde{\xi}_{0,t}(x)$ is written as

$$\exp\left\{\sum_j \int_0^t h_j(\tilde{\varphi}_{0,r}(x), r)\,dX_r^j - \frac{1}{2}\sum_j \int_0^t h_j(\tilde{\varphi}_{0,r}(x), r)^2\,dr\right\}$$
$$\times \exp\left\{\int_0^t e(\tilde{\varphi}_{0,r}(x), r)\,dr\right\}. \tag{44}$$

Now the expectation of the first member of the above with respect to the measure $P \times Q$ is less than or equal to 1 for any t and x since it is a positive localmartingale for any x. The last member of (44) is bounded if (43) is bounded from above. Therefore the expectation of $\tilde{\xi}_{0,t}(x)$ with respect to $P \times Q$ is bounded in x.

Let ρ_0 be an $\mathscr{M}(\mathbb{R}^d)$-valued random variable such that $\rho_0(\mathbb{R}^d) < \infty$ a.s. Then for any bounded continuous function f,

$$\rho_t(f) = \int_{\mathbb{R}^d} E_Q[\tilde{\xi}_{0,t}(x) f(\tilde{\varphi}_{0,t}(x))] \rho_0(dx), \tag{45}$$

is well defined. It is a real continuous semimartingale. We claim that it is a solution of equation (33).

Theorem 6.2.10 *Assume that in equation* (33) *the operator L_t satisfies Condition* (D.2)$_{k,\delta}$ *and* $(F^1, \ldots, F^d)$ *satisfies Condition* (D.1)$_{k,\delta}$ *for some $k \geq 3$ and $\delta > 0$. Assume further that* (43) *is bounded from above. Then* (45) *is a solution of equation* (33) *with the initial value ρ_0.*

Proof Set

$$\tilde{v}(x, t) = \tilde{\xi}_{0,t}(x) f(\tilde{\varphi}_{0,t}(x)), \tag{46}$$

where f is a smooth function with compact support. Then Itô's formula yields

$$\begin{aligned} d\tilde{v}(x, t) = & \Big\{ L_t f(\tilde{\varphi}_{0,t}(x))\, dt + \sum_i Y^i(\tilde{\varphi}_{0,t}(x), dt) \frac{\partial f}{\partial x^i}(\tilde{\varphi}_{0,t}(x)) \\ & + \sum_i F^i(\tilde{\varphi}_{0,t}(x), \circ dt) \frac{\partial f}{\partial x^i}(\tilde{\varphi}_{0,t}(x)) \Big\} \tilde{\xi}_{0,t}(x) \\ & + F^{d+1}(\tilde{\varphi}_{0,t}(x), \circ dt) \tilde{v}(x, t). \end{aligned} \tag{47}$$

Take the partial expectation of each term of the above by the measure Q. Since $Y^i(x, t)$ is a martingale with respect to the measure Q, we have

$$E_Q\left[\int_0^t Y^i(\tilde{\varphi}_{0,s}, ds) \frac{\partial f}{\partial x^i}(\tilde{\varphi}_{0,s}) \tilde{\xi}_{0,s}\right] = 0 \qquad \text{a.s.}$$

Therefore $\rho_t(f)$ defined by (45) satisfies (33). The proof is complete. □

6.3 Applications to nonlinear filtering theory

Filtering problem

Let θ_t be a continuous stochastic process on $\mathbb{R}^d$ governed by a Stratonovich's stochastic differential equation

$$\theta_t = \theta_0 + \int_0^t X(\theta_s, \circ ds) + \sum_{j=1}^n \int_0^t f_j(\theta_s, s) \circ dN_s^j. \tag{1}$$

Here $X(\cdot, t)$ is a $C^k(\mathbb{R}^d : \mathbb{R}^d)$-valued Brownian motion with local characteristic belonging to $(B_{ub}^{k+1,\delta}, B_{ub}^{k,\delta})$ and the $f_j(x, t)$ are bounded $\mathbb{R}^d$-valued functions of $C_b^{k,\delta}$ where $k \geq 3$ and $\delta > 0$. Further $N_t = (N_t^1, \ldots, N_t^n)$ is an n-dimensional standard Brownian motion. These $X(\cdot, t)$, θ_0 and N_t are assumed to be independent. The process θ_t is often called a *system process* or a *signal process*. Suppose that we wish to observe the state θ_t at each time t, but we cannot observe it directly. The data concerning θ_t is provided by the process

$$Y_t = \int_0^t h(\theta_s, s)\, ds + N_t, \tag{2}$$

where $h(x, t) = (h_1(x, t), \ldots, h_n(x, t))$ is continuous in (x, t) and belongs to the class $C_b^{k,\delta}$. The problem we are concerned with is to get the estimation of the state θ_t for every t through the observed data $\{Y_s : s \leq t\}$.

The observed data can be provided by the σ-field generated by Y_t, i.e.

$$\mathscr{G}_t = \sigma(Y_s : s \leq t). \tag{3}$$

The least square estimate of θ_t is then given by the conditional expectation $\hat{\theta}_t = E[\theta_t | \mathscr{G}_t]$. It is called the *(nonlinear) filter* of the system θ_t based on the observed data $\mathscr{G}_t$.

We wish to obtain algorithms for computing the filter from the observed data $\{Y_s : s \leq t\}$, explicitly. For this purpose, we shall consider the conditional distribution of θ_t with respect to $\mathscr{G}_t$:

$$\pi_t(A) = P(\theta_t \in A | \mathscr{G}_t). \tag{4}$$

It is called the *filter of θ_t based on the observed data $\mathscr{G}_t$*. Now since θ_t has finite moments of any order, the conditional distribution π_t has finite moments of any order a.s. for every t.

We shall first obtain a Bayes formula for computing the conditional distribution π_t. Let P_X be the law of the C^k-valued Brownian motion $X(x, t)$ defined on $W_1 = C([0, T] : C^k)$ and let P_W be the law of a standard n-dimensional Brownian motion defined on $W_2 = C([0, T] : \mathbb{R}^n)$. The typical elements of W_1 and W_2 are denoted by $\tilde{X}$ and $\tilde{Y}$, respectively. Let $(\mathbb{R}^d, \mu)$ be the law of the initial value of the system process. The typical element of $\mathbb{R}^d$ is denoted by x. Consider a Stratonovich's stochastic differential equation on $P_X \times \mu \times P_W$:

$$\tilde{\theta}_t = x + \int_0^t \tilde{X}(\tilde{\theta}_s, \circ \mathrm{d}s) + \sum_{j=1}^n \int_0^t f_j(\tilde{\theta}_s, s)(\circ \mathrm{d}\tilde{Y}_s^j - h_j(\tilde{\theta}_s, s)\, \mathrm{d}s). \quad (5)$$

Then $\tilde{\theta}_t$ is a functional of $\tilde{X}$, x and $\tilde{Y}$. We denote it by $\tilde{\theta}_t = \tilde{\theta}_t(\tilde{X}, x, \tilde{Y})$. Define

$$\alpha_t(\tilde{X}, x, \tilde{Y}) = \exp\left\{\sum_{j=1}^n \int_0^t h_j(\tilde{\theta}_r, r)\mathrm{d}\tilde{Y}_r^j - \frac{1}{2}\sum_{j=1}^n \int_0^t h_j(\tilde{\theta}_r, r)^2\, \mathrm{d}r\right\}. \quad (6)$$

The following is a Bayes formula for computing π_t.

Theorem 6.3.1 *Set $Q = P_X \times \mu$. Then*

$$\pi_t(A)(\omega) = \frac{\displaystyle\int_{\tilde{\theta}_t(\tilde{X}, x, Y(\omega)) \in A} \alpha_t(\tilde{X}, x, Y(\omega))Q(\mathrm{d}\tilde{X}\, \mathrm{d}x)}{\displaystyle\int \alpha_t(\tilde{X}, x, Y(\omega))Q(\mathrm{d}\tilde{X}\, \mathrm{d}x)}. \quad (7)$$

Proof Let $\tilde{P}$ be the law of the triple $(X(\cdot, t), \theta_0, Y_t)$ on $W_1 \times \mathbb{R}^d \times W_2$. We first show that $\tilde{P}$ is absolutely continuous with respect to $Q \times P_W$. Set $\mathscr{F}_t = \sigma(X(\cdot, u): 0 \le u \le T) \times \sigma(Y_u : u \le t)$ and define

$$\beta_t = \exp\left\{-\sum_j \int_0^t h_j(\tilde{\theta}_s, s)(\mathrm{d}\tilde{Y}_s^j - h_j(\tilde{\theta}_s, s)\, \mathrm{d}s) - \frac{1}{2}\sum_j \int_0^t h_j(\tilde{\theta}_s, s)^2\, \mathrm{d}s\right\}.$$

Since $\tilde{Y}_t - \int_0^t h(\tilde{\theta}_s, s)\, \mathrm{d}s$ is a standard $(\mathscr{F}_t)$-Brownian motion with respect to $\tilde{P}$, β_t is a positive localmartingale with respect to $\tilde{P}$. Let $\{\tau_n\}$ be an increasing sequence of stopping times such that $\tilde{P}(\tau_n < T) \to 0$ as $n \to \infty$ and each stopped process $\beta_{t \wedge \tau_n}$ is a martingale. We define for each n a probability measure R_n by $R_n = \beta_{\tau_n}\tilde{P}$. Then by Girsanov's theorem (Theorem 2.3.14),

$$\tilde{Y}_t - \int_0^t h(\tilde{\theta}_s, s)\, \mathrm{d}s + \int_0^{t \wedge \tau_n} h(\tilde{\theta}_s, s)\, \mathrm{d}s$$

is an $(\mathscr{F}_t)$-Brownian motion with respect to R_n. Then $(\tilde{Y}_{t \wedge \tau_n}, R_n)$ is an $(\mathscr{F}_t)$-Brownian motion stopped at τ_n. Since $R_n = Q \times P_W$ on $\mathscr{F}_0$, we have the same equality on $\mathscr{F}_{\tau_n}$. Consequently the measure $Q \times P_W$ and $\tilde{P}$ are mutually absolutely continuous on the σ-field $\mathscr{F}_{\tau_n}$ for each n. Suppose now that B of $\mathscr{F}_T$ satisfies $Q \times P_W(B) = 0$. Since $B \cap \{\tau_n \ge T\} \in \mathscr{F}_{\tau_n}$, we have $\tilde{P}(B \cap \{\tau_n \ge T\}) = 0$. Let n tend to infinity. Then we get $\tilde{P}(B) = 0$. This proves that $\tilde{P}$ is absolutely continuous with respect to $Q \times P_W$.

Now let α_t be the Radon–Nikodym derivative of $\tilde{P}$ with respect to $Q \times P_W$. Then $\alpha_t = \beta_t^{-1}$ holds. Therefore α_t is represented by (6).

Set $\mathscr{G}_t = \sigma(\tilde{Y}(s) : s \le t)$. Then for any $A \in \mathscr{G}_t$ and $B \in \mathscr{G}_t$, we have

$$\begin{aligned}\tilde{P}(A \cap B) &= E_{Q \times P_w}[\alpha_t \chi(A)\chi(B)] = E_{Q \times P_w}[E_{Q \times P_w}[\alpha_t \chi(A)|\mathscr{G}_t]\chi(B)] \\ &= \tilde{E}[\alpha_t^{-1} E_{Q \times P_w}[\alpha_t \chi(A)|\mathscr{G}_t]\chi(B)] \\ &= \tilde{E}[\tilde{E}[\alpha_t^{-1}|\mathscr{G}_t] E_{Q \times P_w}[\alpha_t \chi(A)|\mathscr{G}_t]\chi(B)].\end{aligned}$$

Therefore,

$$\tilde{P}(A|\mathscr{G}_t) = \tilde{E}[\alpha_t^{-1}|\mathscr{G}_t] E_{Q \times P_w}[\alpha_t \chi(A)|\mathscr{G}_t]. \tag{8}$$

Setting $A = \Omega$ above, we get $\tilde{E}[\alpha_t^{-1}|\mathscr{G}_t] E_{Q \times P_w}[\alpha_t|\mathscr{G}_t] = 1$. Substitute this in (8). Then we obtain

$$\tilde{P}(A|\mathscr{G}_t) = \frac{E_{Q \times P_w}[\alpha_t \chi(A)|\mathscr{G}_t]}{E_{Q \times P_w}[\alpha_t|\mathscr{G}_t]} = \frac{\int \alpha_t \chi(A)\, dQ}{\int \alpha_t\, dQ}, \tag{9}$$

by Exercise 1.1.7. This proves the theorem. □

Remark In many applications, such as linear filters which we will study later, the functions $f_j(x, t), j = 1, \ldots, n$ in equation (1) are 0. Then the system process θ_t and the noise N_t are independent. In this case we can regard α_t as a functional of $\tilde{\theta}$ and $\tilde{Y}$. Then instead of formula (7) we have

$$\pi_t(A) = \frac{\int_{\tilde{\theta}_t \in A} \alpha_t(\tilde{\theta}, Y) P_\theta(d\tilde{\theta})}{\int \alpha_t(\tilde{\theta}, Y) P_\theta(d\tilde{\theta})} \tag{10}$$

where P_θ is the law of the system process. The above formula is called the *Kallianpur–Striebel formula.*

Stochastic partial differential equations for nonlinear filter

The Bayes formula or the Kallianpur–Striebel formula provides a method of computing the filter π_t through $\{Y_s : s \le t\}$. However when we want to compute π_t at every time t continuously the formula is not convenient, since it requires all the past data $\{Y_s : s \le t\}$ at every time t. It is desirable to get a recursive equation so that at every time $t + \Delta t$ the filter $\pi_{t+\Delta t}$ can be computed from π_t and the new data $\{Y_s : t \le s \le t + \Delta t\}$. For this purpose we shall obtain a stochastic partial differential equation governing the filter based on the observation process.

We shall first obtain a linear stochastic partial differential equation governing the so called 'unnormalized' conditional distribution $\rho_t(dx)$.

Let $(a(x, y, t), m(x, t), t)$ be the local characteristic of the C^k-valued Brownian motion $X(x, t)$. Define the first order differential operators $D_j(t)$ and $M_j(t), j = 1, 2, \ldots, n$ by

$$D_j(t)f = \sum_i f_j^i(x, t)\frac{\partial f}{\partial x^i}, \qquad M_j(t)f = D_j(t)f + h_j(x, t)f$$

and the second order one by

$$L_t f = \frac{1}{2}\sum_{i,j} a^{ij}(x, x, t)\frac{\partial^2 f}{\partial x^i \partial x^j} + \sum_i b^i(x, t)\frac{\partial f}{\partial x^i} - \sum_j h_j(x, t)D_j(t)f$$
$$- \frac{1}{2}\left\{\sum_j M_j(t)h_j(x, t)\right\} f, \tag{11}$$

where $b(x, t) = m(x, t) + c(x, t)$ and $c(x, t)$ is the correction term based on $a^{ij}(x, y, t)$. Consider a second order stochastic partial differential equation represented by the Stratonovich integrals

$$d\rho_t(f) = \rho_t(L_t f)\,dt + \sum_j \rho_t(M_j(t)f) \circ dY_t^j, \qquad \rho_0 = \mu_0. \tag{12}$$

Note that using the Itô integral, the above is equivalent to

$$d\rho_t(f) = \rho_t(A_t f)\,dt + \sum_j \rho_t(M_j(t)f)\,dY_t^j, \qquad \rho_0 = \mu_0 \tag{13}$$

where $A_t = L_t + \tilde{L}_t$ and $\tilde{L}_t$ is defined by (35) of Section 6.2. In the present case A_t is equal to the infinitesimal generator of the system process $\tilde{\theta}_t$ with respect to the probability $Q \times P_W$:

$$A_t f = \frac{1}{2}\sum_{i,j} a^{ij}(x, x, t)\frac{\partial^2 f}{\partial x^i \partial x^j} + \sum_i b^i(x, t)\frac{\partial f}{\partial x^i} + \frac{1}{2}\sum_j D_j(t)^2 f. \tag{14}$$

Theorem 6.3.2 *Equation* (12) *has a unique solution. It satisfies* $\rho_t(1) < \infty$ *a.s.* P *and*

$$\pi_t(f) = \frac{\rho_t(f)}{\rho_t(1)} \qquad a.s.\ P. \tag{15}$$

Proof Consider an equation on $Q \times P_W$ equivalent to (12):

$$d\tilde{\rho}_t(f) = \tilde{\rho}_t(L_t f)\,dt + \sum_j \tilde{\rho}_t(M_j(t)f) \circ d\tilde{Y}_t^j, \qquad \tilde{\rho}_0 = \mu_0 \tag{12'}$$

where $(\tilde{Y}_t^1, \ldots, \tilde{Y}_t^n)$ is a standard Brownian motion. It has a unique solution by Theorem 6.2.8. Since the coefficients of the 0-th order term of the operator $A_t \equiv \tilde{L}_t + L_t$ is identically 0, the solution is represented by (45) of Section 6.2 replacing $\tilde{\varphi}_{0,t}$ by $\tilde{\theta}_t$ and ρ_0 by μ_0, respectively, by Theorem

6.2.10. Here $\tilde{\theta}_t(x)$ is the stochastic flow determined by equation (5). In the present case $\tilde{\xi}_{0,t}(x)$ of (42) in the previous section is equal to the functional $\alpha_t(\tilde{X}, x, \tilde{Y})$ defined by (6). Then the solution of equation (12′) is represented by

$$\tilde{\rho}_t(f)(\tilde{Y}) = E_Q[f(\tilde{\theta}_t)\alpha_t(\tilde{X}, x, \tilde{Y})]. \tag{16}$$

Then Theorem 6.3.1 concludes the assertion of the theorem. □

The adjoint equation of equation (13) is

$$\mathrm{d}v(x,t) = A_t^* v(x,t)\,\mathrm{d}t + \sum_{j=1}^{n} M_j(t)^* v(x,t)\,\mathrm{d}Y_t^j. \tag{17}$$

It is called a *Zakai equation*. The Zakai equation has a unique solution if $h_j(x, t)$ are bounded by Corollary 6.2.9.

We will next obtain the stochastic partial differential equation for the conditional distribution π_t. The equation is not linear.

Theorem 6.3.3 *The filter π_t satisfies the following nonlinear stochastic partial differential equation*

$$\pi_t(f) = \pi_0(f) + \int_0^t \pi_s(A_s f)\,\mathrm{d}s$$
$$+ \sum_j \int_0^t \{\pi_s(fh_j(s)) - \pi_s(f)\pi_s(h_j(s)) + \pi_s(D_j(s)f)\}\{\mathrm{d}Y_s^j - \pi_s(h_j(s))\,\mathrm{d}s\}. \tag{18}$$

Proof Let ρ_t be the solution of equation (12). Since it satisfies (15), ρ_t has moments of any order a.s. Then equation (13) implies

$$\rho_t(1) = \rho_0(1) + \sum_j \int_0^t \rho_s(h_j(s))\,\mathrm{d}Y_s^j$$
$$= \rho_0(1) + \sum_j \int_0^t \pi_s(h_j(s))\rho_s(1)\,\mathrm{d}Y_s^j.$$

The solution is then represented by

$$\rho_t(1) = \rho_0(1)\exp\left\{\sum_j \int_0^t \pi_s(h_j(s))\,\mathrm{d}Y_s^j - \frac{1}{2}\sum_j \int_0^t \pi_s(h_j(s))^2\,\mathrm{d}s\right\}.$$

Consequently the inverse $\rho_t(1)^{-1}$ satisfies the equation

$$\mathrm{d}\rho_t(1)^{-1} = -\sum_j \pi_t(h_j(t))\rho_t(1)^{-1}\,\mathrm{d}Y_t^j + \sum_j \pi_t(h_j(t))^2 \rho_t(1)^{-1}\,\mathrm{d}t$$

by Itô's formula. We then have

$$\mathrm{d}(\rho_t(f)\rho_t(1)^{-1}) = \rho_t(f)\,\mathrm{d}\rho_t(1)^{-1} + \rho_t(1)^{-1}\,\mathrm{d}\rho_t(f) + d\langle \rho_t(1)^{-1}, \rho_t(f)\rangle,$$

where

$$\rho_t(f)\,d\rho_t(1)^{-1} = -\sum_j \pi_t(f)\pi_t(h_j(t))\,dY_t^j + \sum_j \pi_t(h_j(t))^2\pi_t(f)\,dt,$$

$$\rho_t(1)^{-1}\,d\rho_t(f) = \pi_t(A_tf)\,dt + \sum_j \pi_t(M_j(t)f)\,dY_t^j,$$

$$d\langle \rho_t(1)^{-1}, \rho_t(f)\rangle = -\sum_j \pi_t(h_j(t))\pi_t(M_j(t)f)\,dt.$$

Summing these terms, we get

$$d\pi_t(f) = \pi_t(A_tf)\,dt + \sum_j \{\pi_t(M_j(t)f) - \pi_t(f)\pi_t(h_j(t))\}\{dY_t^j - \pi_t(h_j(t))\,dt\}.$$

The above is equivalent to (18). The proof is complete. □

Theorem 6.3.4 *The probability valued process π_t satisfying* (18) *and having finite moments of any order is unique.*

Proof Let π_t be any solution of equation (20) having moments of any order. Set

$$\beta_t = \exp\left\{\sum_j \int_0^t \pi_s(h_j(s))\,dY_s^j - \frac{1}{2}\int_0^t \left(\sum_j \pi_s(h_j(s))^2\right) ds\right\} \tag{19}$$

and $\rho_t(f) = \beta_t\pi_t(f)$. Then ρ_t satisfies equation (13). Indeed, we have by Itô's formula

$$d\rho_t(f) = (d\beta_t)\pi_t(f) + \beta_t\,d\pi_t(f) + \langle \beta_t, \pi_t(f)\rangle. \tag{20}$$

Note that

$$(d\beta_t)\pi_t(f) = \sum_j \pi_t(h_j(t))\rho_t(f)\,dY_t^j,$$

$$\beta_t\,d\pi_t(f) = \rho_t(A_tf)\,dt$$

$$+ \sum_j \{\rho_t(h_j(t)f) - \pi_t(h_j(t))\rho_t(f) + \rho_t(D_j(t)f)\}\{dY_t^j - \pi_t(h_j(t))\,dt\},$$

$$d\langle \beta_t, \pi_t(f)\rangle = \sum_j \pi_t(h_j(t))\{\rho_t(h_j(t)f) - \pi_t(h_j(t))\rho_t(f) + \rho_t(D_j(t)f)\}\,dt.$$

Then we get equation (13) for ρ_t.

Now the solution ρ_t of equation (13) having moments of any order with the initial condition $\rho_0(f) = \mu_0(f)$ is unique. Since π_t is a probability, we have $\beta_t = \rho_t(1)$. Therefore we get $\pi_t(f) = \rho_t(f)/\rho_t(1)$, from which we conclude that π_t is unique. The proof is complete. □

Suppose that the initial distribution $\pi_0(dx)$ has a C^{k-1}-density function $p_0(x)$. Then $\pi_t(dx)$ has also a C^{k-1}-density $p(x, t)$, since $\rho_t(dx)$ has a C^{k-1}-

density. The density satisfies the following nonlinear stochastic partial differential equation

$$\begin{aligned} d_t p(x, t) &= A_t^* p(x, t)\, dt \\ &+ \sum_j \left\{ M_j(t)^* p(x, t) - p(x, t) \int_{\mathbf{R}^d} p(x, t) h_j(x, t)\, dx \right\} \\ &\times \left\{ dY_t^j - \left(\int_{\mathbf{R}^d} p(x, t) h_j(x, t)\, dx \right) dt \right\}, \end{aligned} \tag{21}$$

where A_t^* and $M_j(t)^*$ are the adjoint operators of A_t and $M_j(t)$, respectively. Equation (21) is called a *Kushner–Stratonovich equation.*

Equation (21) has a unique solution $p(x, t)$ which is rapidly decreasing if the initial function $p_0(x)$ is rapidly decreasing. Indeed, we have already seen the existence of the solution. Let $p(x, t)$ be an arbitrary rapidly decreasing solution. Set $\pi_t(f) = \int p(x, t) f(x)\, ds$. It is a solution of equation (19) having finite moments of any order. Therefore $p(x, t)$ is unique by Theorem 6.3.4.

Robustness of nonlinear filter

Let π_t be the filtering process. It is a functional of the observation process Y_s, $s \le t$. It is desirable that the filtering π_t is a continuous function of Y_s, $s \le t$ with respect to the uniform norm. The fact is easily verified if the operators $D_j(t)$, $j = 1, \ldots, n$ are identically 0, or equivalently the system process θ_t and the noise process N_t are independent. Indeed, in this case, we have by Itô's formula

$$\int_0^t h_j(\tilde{\theta}_s, s)\, dY_s^j = h_j(\tilde{\theta}_t, t) Y_t^j - \int_0^t Y_s^j d_s h_j(\tilde{\theta}_s, s). \tag{22}$$

Hence the functional $\alpha_t(\tilde{\theta}, Y)$ in the Kallianpur–Striebel formula (10) is a continuous functional of $Y_s : s \le t$. Consequently $\pi_t(A)$ given by the formula (10) is a continuous functional of $Y_s : s \le t$.

In the case where the $D_j(t)$, $j = 1, \ldots, n$ are not identically 0, $\pi_t(A)$ is not always a continuous functional, since $\tilde{\theta}_t$ depends on Y_t and may not be a continuous functional of $Y_s : s \le t$, in general. In the following, we show that $\tilde{\theta}_t$ is a continuous functional of $Y_s : s \le t$ provided that $f_j = (f_j^1 \cdots f_j^d)$, $j = 1, \ldots, n$ are time independent and commutative vector fields. Here we understand that $f = f_1$ is commutative in the case $n = 1$. Let $\psi_j(x, t)$ be the deterministic flow generated by the vector field $f_j(x)$. Then

$$\varphi_t(x) = \prod_{j=1}^n \psi_j(x, Y_t^j) \tag{23}$$

is the unique solution of the equation

$$d\varphi_t = \sum_{j=1}^{n} f_j(\varphi_t) \circ dY_t^j \tag{24}$$

starting at x at time 0 (see Example 4.9.8). Therefore φ_t is a continuous functional of $Y_s : s \leq t$. Consider now the stochastic differential equation

$$d\eta_t = (\varphi_t^{-1})_* X(\eta_t, \circ dt) - \sum_{j=1}^{n} h_j(\eta_t, t) f_j(\eta_t)\, dt, \tag{25}$$

where $(\varphi_t^{-1})_*$ is the differential of the map φ_t^{-1} (see Section 4.9). Let $\eta_t(x)$ be the stochastic flow generated by the above equation. Then $\theta_t(x) = \varphi_t(\eta_t(x))$ is a solution of the equation

$$d\theta_t = X(\theta_t, \circ dt) + \sum_{j=1}^{n} f_j(\theta_t)(\circ dY_t^j - h_j(\theta_t, t)\, dt) \tag{26}$$

by Theorem 4.9.5.

Now for each ω, we can regard (25) as an equation on $(W, \mathscr{B}, Q)$. Since the coefficients $(\varphi_t^{-1})_* X$ depend continuously on $Y_s : s \leq t$, the solution η_t depends continuously on $Y_s : s \leq t$ in the following sense.

$$Q(|\eta_t^{\omega} - \eta_t^{\omega'}| > \varepsilon) \to 0$$

holds as $Y_t(\omega) \to Y_t(\omega')$. Then the same property is valid for $\theta_t = \varphi_t(\eta_t)$. Therefore the functional α_t defined by (6) has the same property. Then the filter π_t represented by (7) is a continuous functional of $Y_s : s \leq t$.

Kalman filter

As an example we shall consider the case where the system process θ_t is governed by a linear stochastic differential equation on $\mathbb{R}^d$:

$$\theta_t = \theta_0 + \int_0^t F(s)\theta_s\, ds + \int_0^t G(s)\, dB_s, \tag{27}$$

where $F(s)$ and $G(s)$ are matrices, θ_0 is a Gaussian random variable and B_t is a standard Brownian motion independent of θ_0. The observation process is supposed to be given by

$$Y_t = \int_0^t H(s)\theta_s\, ds + N_t, \tag{28}$$

where $H(s)$ is a matrix and N_t is a standard Brownian motion independent of (θ_0, W_t).

The pair of processes θ_t and Y_t is Gaussian in this case. This fact makes the filtering problem much simpler. Indeed, the conditional expectation of θ_t with respect to $\sigma(Y_s : s \leq t)$ is equal to the orthogonal projection of θ_t to the L^2-subspace spanned by the linear sums of Gaussian random variables

$\{Y_s : s \le t\}$. Therefore the mean filter $\hat{\theta}_t \equiv \int x\pi_t(\mathrm{d}x)$ is represented as a linear functional of $\{Y_s : s \le t\}$. Because of this fact, the filter is called the *linear filter*. (Note that in general $\hat{\theta}_t$ is a nonlinear functional of $\{Y(s) : s \le t\}$. This is the reason why the filter introduced before is called nonlinear.)

Furthermore for almost all ω, π_t is a Gaussian distribution with mean $\hat{\theta}_t$ and covariance given by the matrix

$$P(t) = E[(\theta_t - \hat{\theta}_t)(\theta_t - \hat{\theta}_t)^t]$$

which represents the covariance of the filtering error. Indeed, since $\theta_t - \hat{\theta}_t$ is independent of $\mathscr{G}_t$, we have for any ξ of $\mathbb{R}^d$,

$$E[\exp\{i(\xi, \theta_t - \hat{\theta}_t)\}|\mathscr{G}_t] = E[\exp\{i(\xi, \theta_t - \hat{\theta}_t)\}] = \exp\{-(\xi, P(t)\xi)\},$$

so that

$$E[\exp\{i(\xi, \theta_t)\}|\mathscr{G}_t] = \exp\{i(\xi, \hat{\theta}_t) - (\xi, P(t)\xi)\},$$

proving that the conditional distribution $\pi_t \equiv P(\cdot|\mathscr{G}_t)$ is Gaussian with mean $\hat{\theta}_t$ and covariance $P(t)$ a.s. Since the Gaussian distribution π_t is completely characterized by its mean and covariance, it is now sufficient to obtain the mean filter $\hat{\theta}_t$ and its error covariance $P(t)$.

We shall first compute the mean filter $\hat{\theta}_t = \int x\pi_t(\mathrm{d}x)$ by applying Theorem 6.3.3. Note that the differential operator A_t of (14) is equal to the infinitesimal generator of θ_t:

$$A_t f = \frac{1}{2}\sum_{i,j}\left(\sum_k g_{ik}(t)g_{jk}(t)\right)\frac{\partial^2 f}{\partial x^i \partial x^j} + \sum_i\left(\sum_j f_{ij}(t)x_j\right)\frac{\partial f}{\partial x^i}, \tag{29}$$

where $G(t) = (g_{ij}(t))$ and $F(t) = (f_{ij}(t))$. Then setting $f(x) = x^i$ in (18), we get a stochastic differential equation for $\hat{\theta}_t = (\hat{\theta}_t^1, \ldots, \hat{\theta}_t^d)$ where $\hat{\theta}_t^i = \pi(x^i)$.

$$\begin{aligned}\hat{\theta}_t = \hat{\theta}_0 &+ \int_0^t F(s)\hat{\theta}_s\,\mathrm{d}s \\ &+ \int_s^t E[(\theta_s - \hat{\theta}_s)(\theta_s - \hat{\theta}_s)^t|\mathscr{G}_s]H(s)^t(\mathrm{d}Y_s - H(s)\hat{\theta}_s\,\mathrm{d}s).\end{aligned} \tag{30}$$

Now since (θ_t, Y_t) is Gaussian and $\theta_s - \hat{\theta}_s$ and $\mathscr{G}_s$ are independent, the conditional covariance matrix above does not depend on ω. It is equal to the matrix $P(s)$ of the filtering error. Then equation (30) can be written as

$$\hat{\theta}_t = \hat{\theta}_0 + \int_0^t (F(s) - P(s)H(s)^t H(s))\hat{\theta}_s\,\mathrm{d}s + \int_0^t P(s)H(s)^t\,\mathrm{d}Y_s. \tag{31}$$

We shall next obtain a differential equation governing $P(t)$. Set $f(x) = x^i x^j$ in equation (18). Then we obtain

$$
\begin{aligned}
E[\theta_t^i\theta_t^j|\mathscr{G}_t] \\
&= E[\theta_0^i\theta_0^j|\mathscr{G}_0] + \int_0^t \frac{1}{2}\left\{\sum_k g_{ik}(s)g_{jk}(s)\right\} ds \\
&\quad + \int_0^t \left\{\sum_k f_{ik}(s)E[\theta_s^k\theta_s^j|\mathscr{G}_s] + \sum_k f_{jk}(s)E[\theta_s^k\theta_s^i|\mathscr{G}_s]\right\} ds \\
&\quad + \sum_k \int_0^t \left\{\sum_l E[\theta_s^i\theta_s^j\theta_s^l|\mathscr{G}_s]h_{kl}(s) - \sum_l E[\theta_s^i\theta_s^j|\mathscr{G}_s]E[\theta_s^l|\mathscr{G}_s]h_{kl}(s)\right\} \\
&\qquad \times \left\{dY_s^k - \left(\sum_l h_{kl}(s)\hat{\theta}_s^l\right) ds\right\}. \qquad (32)
\end{aligned}
$$

Since π_s is a Gaussian distribution with mean $\hat{\theta}_s$ and covariance $P(s)$, we have

$$E[\theta_s^i\theta_s^j\theta_s^l|\mathscr{G}_s] = p_{ij}\hat{\theta}_s^l + p_{il}\hat{\theta}_s^j + p_{jl}\hat{\theta}_s^i + \hat{\theta}_s^i\hat{\theta}_s^j\hat{\theta}_s^l.$$

Therefore the above equation can be written in matrix notation as

$$
\begin{aligned}
E[\theta_t\theta_t^t|\mathscr{G}_t] &= E[\theta_0\theta_0^t|\mathscr{G}_0] + \frac{1}{2}\int_0^t G(s)G(s)^t\, ds \\
&\quad + \int_0^t \{E[\theta_s\theta_s^t|\mathscr{G}_s]F(s) + F(s)E[\theta_s\theta_s^t|\mathscr{G}_s]\}\, ds \\
&\quad + \int_0^t P(s)H(s)^t\, dI_s\hat{\theta}_s^t + \int_0^t \hat{\theta}_s(P(s)H(s)^t\, dI_s)^t, \qquad (33)
\end{aligned}
$$

where $I_t = Y_t - \int_0^t H(s)\hat{\theta}_s\, ds$. Further apply Itô's formula to (30). Then the matrix $(\hat{\theta}_t^i\hat{\theta}_t^j) = \hat{\theta}_t\hat{\theta}_t^t$ satisfies

$$
\begin{aligned}
\hat{\theta}_t\hat{\theta}_t^t &= \hat{\theta}_0\hat{\theta}_0^t + \int_0^t \{\hat{\theta}_s\hat{\theta}_s^t F(s) + F(s)\hat{\theta}_s\hat{\theta}_s^t\}\, ds + \int_0^t P(s)H(s)^t\, dI_s\hat{\theta}_s^t \\
&\quad + \int_0^t \hat{\theta}_s(P(s)H(s)^t\, dI_s)^t + \int_0^t P(s)H(s)^t H(s)P(s)\, ds. \qquad (34)
\end{aligned}
$$

Since $P(s) = E[\theta_s\theta_s^t|\mathscr{G}_s] - \hat{\theta}_s\hat{\theta}_s^t$, (33) and (34) imply

$$
\begin{aligned}
P(t) - P(0) &= \frac{1}{2}\int_0^t G(s)G(s)^t\, ds + \int_0^t \{P(s)F(s) + F(s)P(s)\}\, ds \\
&\quad - \int_0^t P(s)H(s)^t H(s)P(s)\, ds.
\end{aligned}
$$

Differentiating each term by t, we get a *matrix Riccati equation*

$$\frac{d}{dt}P(t) = P(t)F(t) + F(t)P(t) - P(t)H(t)^t H(t)P(t) + \frac{1}{2}G(t)G(t)^t. \tag{35}$$

We have thus proved the following Kalman filter.

Theorem 6.3.5 *Assume that the system process θ_t is given by* (27) *and the observation process by* (28). *Then the filter π_t is a Gaussian distribution valued process, whose mean $\hat{\theta}_t$ and covariance $P(t)$ can be computed using* (31) *and* (35), *respectively.* □

Exercise 6.3.6 (*cf.* Exercise 4.9.17) Assume that F, G and H in equations (27) and (28) are constant matrices. Assume further that the rank of the matrix $(F, FG, \ldots, F^{d-1}G)$ is d. Show that the error covariance $P(t)$ is nonsingular for any $t > 0$ for any initial $P(0)$ so that the linear filter π_t has a smooth density function.

6.4 Limit theorems for stochastic partial differential equations

Statement of theorems

Let us consider a family of stochastic partial differential equations with parameter $\varepsilon > 0$:

$$\begin{aligned}\frac{\partial u_\varepsilon}{\partial t} &= L_t u_\varepsilon + \sum_{i=1}^{d} f_\varepsilon^i(x, t, \omega)\frac{\partial u_\varepsilon}{\partial x^i} + f_\varepsilon^{d+1}(x, t, \omega)u_\varepsilon, \\ u_\varepsilon(x, t)|_{t=0} &= f(x),\end{aligned} \tag{1}$$

where L_t is the second order differential operator defined in Section 6.2. The objective of this section is to study the asymptotic behavior of the solutions $\{u_\varepsilon(x, t)\}$ as $\varepsilon \to 0$, when the coefficients $\{(f_\varepsilon^1, \ldots, f_\varepsilon^{d+1})\}$ converge to a white noise or the integrals $(F_\varepsilon^1, \ldots, F_\varepsilon^{d+1})$:

$$F_\varepsilon^i(x, t) = \int_0^t f_\varepsilon^i(x, r)\, dr \tag{2}$$

converge to a Brownian motion $(F^1(x, t), \ldots, F^{d+1}(x, t))$. We wish to show under certain conditions on $\{f_\varepsilon^i : \varepsilon > 0\}$ that $\{u_\varepsilon(x, t)\}$ and their derivatives $\{D_x^\alpha u_\varepsilon(x, t)\}$ converge uniformly on compact sets in L^p or in the sense of the law and the limit $u(x, t)$ satisfies the stochastic partial differential equation represented by the Stratonovich integrals:

$$\begin{aligned}du &= L_t u\, dt + \sum_{i=1}^{d} F^i(x, \circ dt)\frac{\partial u}{\partial x^i} + F^{d+1}(x, \circ dt)u, \\ u(x, t)|_{t=0} &= f(x).\end{aligned} \tag{3}$$

We begin by introducing assumptions on equations (1). We assume that the coefficients of the operator L_t belong to the class $C_b^{k,\delta}$ for some $k \geq 3$, $\delta > 0$ and there exists $a^{ij}(x, y, t)$ of $C_b^{k+1,\delta}$ non-negative definite and symmetric, such that $a^{ij}(x, t) = a^{ij}(x, x, t)$ holds. Further the last 0-th order coefficient $d(x, t)$ is bounded. We set

$$f_\varepsilon(x, t, \omega) = (f_\varepsilon^1(x, t, \omega), \ldots, f_\varepsilon^{d+1}(x, t, \omega))$$

and assume that for each ε it is bounded in (x, t) a.s. and belongs to the class $B_b^{k+1,\delta}$. Further both $D^\alpha f_\varepsilon$ and $D^\alpha F_\varepsilon$ are assumed to be integrable with respect to P for any x and t. The initial function f belongs to $C^{k,\delta}$ and is of polynomial growth. Then by Theorem 6.2.5, for each ε equation (1) has a unique solution with the slowly increasing property, which we shall denote by $u_\varepsilon(x, t)$. To ensure the convergence of solutions $\{u_\varepsilon : \varepsilon > 0\}$ we will introduce two Conditions $(\mathrm{F.1})_k$ and $(\mathrm{F.2})_k$. We set

$$\begin{aligned} \bar{F}_\varepsilon(x, t) &= E[F_\varepsilon(x, t)], & \bar{f}_\varepsilon(x, t) &= E[f_\varepsilon(x, t)], \\ \tilde{F}_\varepsilon(x, t) &= F_\varepsilon(x, t) - \bar{F}_\varepsilon(x, t), & \tilde{f}_\varepsilon(x, t) &= f_\varepsilon(x, t) - \bar{f}_\varepsilon(x, t) \end{aligned} \tag{4}$$

and

$$\mathscr{G}_{s,t}^\varepsilon = \sigma(f_\varepsilon(\cdot, u) : s \leq u \leq t). \tag{5}$$

Condition (F.1)$_k$

(i) *There exists a positive constant K not depending on ε such that $|D_x^\alpha \bar{f}_\varepsilon(x, t)| \leq K$ holds for any x, t and $|\alpha| \leq k$.*

(ii) *There exists a family of positive processes $K_\varepsilon(t)$, $\varepsilon > 0$ uniformly exponentially bounded in mean such that*

$$|E[D_x^\alpha \tilde{F}_\varepsilon(x, t) - D_x^\alpha \tilde{F}_\varepsilon(x, s) | \mathscr{G}_{t,T}^\varepsilon]|(1 + |D_y^\beta \tilde{f}_\varepsilon(y, t)| + |D^\beta \bar{f}_\varepsilon(y, t)|) \leq K_\varepsilon(t)$$

$$|E[D_x^\alpha \tilde{F}_\varepsilon(x, t) - D_x^\alpha \tilde{F}_\varepsilon(x, s) | \mathscr{G}_{0,s}^\varepsilon]|(1 + |D_y^\beta \tilde{f}_\varepsilon(y, s)| + |D^\beta \bar{f}_\varepsilon(y, s)|) \leq K_\varepsilon(s)$$

hold for any x, y, $s < t$ and α, β with $|\alpha| \leq k + 1$, $|\beta| \leq k$. □

Condition (F.2)$_k$

(i) *There exists a function $\bar{f}(x, t)$ belonging to the class $C_b^{k,\delta}$ where $\delta > 0$ such that for any $s < t$ the convergence*

$$\bar{F}_\varepsilon(x, t) - \bar{F}_\varepsilon(x, s) \xrightarrow[\varepsilon \to 0]{} \int_s^t \bar{f}(x, r)\, dr$$

holds uniformly on compact sets of spatial variables.

(ii) *For any $s < t$ and α with $|\alpha| \leq 2$ the convergence*

$$|E[D^\alpha \tilde{F}_\varepsilon(x, t) - D^\alpha \tilde{F}_\varepsilon(x, s) | \mathscr{G}_{t,T}^\varepsilon]|(1 + |\tilde{f}_\varepsilon(y, t)|) \to 0,$$

holds uniformly on compact sets of spatial variables.

(iii) *There exist bounded functions $A^{ij}(x, y, t), i, j = 1, \ldots, d+1$ belonging to the class C_b^{k+1}, symmetric and non-negative definite such that for any $s < t$ and α with $|\alpha| \le 1$ the convergence*

$$E\left[\int_s^t \{D_x^\alpha \tilde{F}_\varepsilon^i(x,t) - D_x^\alpha \tilde{F}_\varepsilon^i(x,r)\}\, dF_\varepsilon^j(y,r) \middle| \mathscr{G}_{t,T}^\varepsilon\right] \to \frac{1}{2}\int_s^t D_x^\alpha A^{ij}(x,y,r)\, dr$$

holds uniformly on compact sets of spatial variables in L^1. □

We will now give the definition of the weak convergence of the pairs of coefficients F_ε of equations (1) and their solutions u_ε. Let j be a non-negative integer. Let $C^{j,0}(\mathbb{R}^d \times [0, T] : \mathbb{R}^{d+1})$ be the linear space of all continuous maps F from $\mathbb{R}^d \times [0, T]$ into $\mathbb{R}^{d+1}$ such that $F(x, t)$ is j-times continuously differentiable with respect to x. The space $C^{j,0}(\mathbb{R}^d \times [0, T] : \mathbb{R}^1)$ is defined similarly. We denote its element by $u(x, t)$. Let W_j be the product space of $C^{j,0}(\mathbb{R}^d \times [0, T] : \mathbb{R}^{d+1})$ and $C^{j,0}(\mathbb{R}^d \times [0, T] : \mathbb{R}^1)$. Then for almost all ω, the pair $(F_\varepsilon, u_\varepsilon)(\omega)$ can be regarded as an element of W_j for $j \le k$. Hence its law is defined on the space $(W_j, \mathscr{B}(W_j))$ by

$$P_\varepsilon^j(A) = P\{\omega : (F_\varepsilon, u_\varepsilon)(\omega) \in A\}, \qquad A \in \mathscr{B}(W_j). \tag{6}$$

If the family of laws $\{P_\varepsilon : \varepsilon > 0\}$ converges weakly, the pair $(F_\varepsilon, u_\varepsilon)$ is said to *converge weakly as C^j-processes.*

Theorem 6.4.1. *Assume that the coefficients of equations* (1) *satisfy Conditions* (F.1)$_k$ *and* (F.2)$_k$ *for some $k \ge 2$. Then the family of pairs $\{(F_\varepsilon, u_\varepsilon)\}$ converges weakly as C^{k-2}-processes as $\varepsilon \to 0$. Further the following properties are valid with respect to the limit measure P_0^{k-2}.*

(i) *$F(x, t)$ is a C^k-Brownian motion with local characteristic $(A, \bar{f}, t)$.*
(ii) *$u(x, t)$ is a continuous C^k-semimartingale and satisfies equation* (3). □

We shall next consider the strong convergence.

Theorem 6.4.2 *Assume that $\{F_\varepsilon(x, t) : \varepsilon > 0\}$ are defined on the same probability space and converge to a random field $F(x, t)$ in probability for any x and t. Then under Conditions* (F.1)$_k$ *and* (F.2)$_k$ *where $k \ge 2$, the following convergences hold.*

(i) *$F(x, t)$ is a C^k-Brownian motion with local characteristic $(A, \bar{f})$. Furthermore,*

$$E\left[\sup_{|x| \le N, t} |D_x^\alpha F_\varepsilon(x,t) - D_x^\alpha F(x,t)|^p\right] \to 0 \tag{7}$$

holds for any α with $|\alpha| \le k-2$, $p > 1$ and $N > 0$.

(ii) *The solutions $u_\varepsilon(x, t)$ converge to $u(x, t)$ in the following sense:*

$$E\left[\sup_{|x|\le N, t} |D_x^\alpha u_\varepsilon(x, t) - D_x^\alpha u(x, t)|^p\right] \to 0 \tag{8}$$

holds for any α with $|\alpha| \le k - 2$, $p > 1$ and $N > 0$. Further, $u(x, t)$ is a continuous C^k-semimartingale and satisfies equation (3). □

Remark If the function $A^{ij}(x, y, t)$ in Condition $(F.2)_k$ is not symmetric, i.e. $A^{ij}(x, y, t) \ne A^{ji}(y, x, t)$, the limit $u(x, t)$ does not satisfy equation (3). A correction term is needed on the right hand side of (3). In fact, define the first order differential operator $H(t)$ by

$$H(t)u = \sum_{i=1}^{d} h^i(x, t)\frac{\partial u}{\partial x^i} + h^{d+1}(x, t)u, \tag{9}$$

where

$$h^i(x, t) = \frac{1}{2}\sum_{j=1}^{d} \frac{\partial}{\partial x^j}(A^{ij}(x, y, t) - A^{ji}(y, x, t))|_{y=x}, \qquad i = 1, \ldots, d + 1. \tag{10}$$

Then the limit $u(x, t)$ satisfies

$$du = L_t u\, dt + H(t)u\, dt + \sum_{i=1}^{d} F^i(\cdot, \circ dt)\frac{\partial u}{\partial x^i} + F^{d+1}(\cdot, \circ dt)u. \tag{11}$$

This will be shown after the proof of Theorem 6.4.1.

Proof of theorems

We may assume that the coefficients b^i, d, e of the operator L_t are identically 0. Let us first represent the solution of equation (1) using a backward stochastic flow. Let $X = (X^1, \ldots, X^d)$ be a C^k-Brownian motion with local characteristic $(a(x, y, t), -c(x, t), t)$ defined on another probability space $(W, \mathscr{B}, Q)$, where $c(x, t)$ is the correction term based on $a(x, y, t)$. On the product probability space $(\Omega \times W, \mathscr{F} \otimes \mathscr{B}, P \times Q)$, we consider the backward stochastic flow $((\varphi_\varepsilon)_{t,s}(x), (\xi_\varepsilon)_{t,s}(x, u))$, $x \in \mathbb{R}^d$, $u, v \in \mathbb{R}^1$, $0 \le s \le t \le T$ determined by the backward Stratonovich equation

$$(\varphi_\varepsilon^i)_{t,s}(x) = x + \int_s^t X^i((\varphi_\varepsilon)_{t,r}(x), \circ \hat{d}r) + \int_s^t f_\varepsilon^i((\varphi_\varepsilon)_{t,r}(x), r)\, dr, \tag{12}$$

where $\varphi_\varepsilon = (\varphi_\varepsilon^1, \ldots, \varphi_\varepsilon^d)$, and

$$(\xi_\varepsilon)_{t,s}(x, u) = u + \int_s^t f_\varepsilon^{d+1}((\varphi_\varepsilon)_{t,r}(x), r)(\xi_\varepsilon)_{t,r}(x, u)\, dr. \tag{13}$$

Then by Theorem 6.2.5, for each ε the solution of equation (1) is represented by

$$u_\varepsilon(x, t) = E_Q[(\xi_\varepsilon)_{t,0}(x, 1)f((\varphi_\varepsilon)_{t,0}(x))]. \tag{14}$$

Then the convergence of the solutions $\{u_\varepsilon(x, t)\}$ can be derived from the convergence of the backward flows $\{(\varphi_\varepsilon, \xi_\varepsilon)\}$.

Now observe that Conditions $(F.1)_k$ and $(F.2)_k$ are very close to Conditions $(C.1)_{k,\infty}$, $(C.2)_k$, $(C.3)\hat{}_{k,\infty}$ and $(C.4)_{k,\infty}$ in Chapter 5. We will show in the following that the flows $\{(\varphi_\varepsilon, \xi_\varepsilon)\}$ converge weakly as C^{k-1}-flows. We first prove the uniform L^p-estimates of the backward flows, which will imply the tightness of the laws by Theorem 1.4.7.

Lemma 6.4.3 *Assume Condition* $(F.1)_k$, $k \geq 2$. *Then for each* $p > 3$ *and* $N > 0$, *there exists a positive constant* C *such that for every* ε *and* $|\alpha| \leq k - 1$

$$E[|D_x^\alpha(\varphi_\varepsilon)_{t,s}(x)|^p] \leq C, \tag{15}$$

$$E[|D_x^\alpha(\varphi_\varepsilon)_{t,s}(x) - D_x^\alpha(\varphi_\varepsilon)_{t',s'}(x)|^p] \leq C(|s - s'|^{2-3/p} + |t - t'|^{2-3/p}) \tag{16}$$

hold for any $t > s$, $t' > s'$, $|x| \leq N$. *Further, similar estimates are valid for* $D_x^\alpha(\xi_\varepsilon)_{t,s}(x, 1)$.

Proof Set

$$\mathscr{G}_{s,t}^\varepsilon = \sigma(f_\varepsilon(\cdot, u), X(\cdot, u) - X(\cdot, v) : s \leq u, v \leq t).$$

Then $(X^1 + F_\varepsilon^1, \ldots, X^d + F_\varepsilon^d)$ is a forward–backward semimartingale adapted to $(\mathscr{G}_{s,t}^\varepsilon)$ with the same forward–backward local characteristic $(a^{ij}, -c^i + f_\varepsilon^i, t)$. The family $\{(a^{ij}, -c^i + f_\varepsilon^i, t)\}$ satisfy Conditions $(C.1)_{k,\infty}$, $(C.3)_{k,\infty}$ and $(C.3)\hat{}_{k,\infty}$ of Section 5.4 with respect to $\mathscr{G}_{s,t}^\varepsilon$, since the f_ε satisfy Condition $(F.1)_k$. Then for any $|x| \leq N$, $|\alpha| \leq k$ and ε, $D_x^\alpha(\varphi_\varepsilon)_{t,s}(x)$ satisfies (15) and (16) by Theorem 5.4.7. Next, apply Lemma 5.4.4 to the enlarged system (12) and (13). Then we find that $(\xi_\varepsilon)_{t,s}(x, 1)$ satisfy the estimates (15) and (16) with $t = t'$. Consider then the stochastic differential equation for $D_x^\alpha\xi_\varepsilon$. We can again apply Lemma 5.4.4 to prove that $D^\alpha(\xi_\varepsilon)_{t,s}(x, 1)$ satisfies the estimate of the form (15) and (16) with $t = t'$. Further inequality (16) is valid for $D^\alpha(\xi_\varepsilon)_{t,s}(x, 1)$ for any $t > s$ and $t' > s'$, which can be proved similarly to the proof of Theorem 5.4.6. □

We will now discuss the weak convergence of the backward flows $\{(\varphi_\varepsilon, \xi_\varepsilon) : \varepsilon > 0\}$. It is convenient to fix the time t. Then the law of $(X(t) - X(s), F_\varepsilon(t) - F_\varepsilon(s), (\varphi_\varepsilon)_{t,s}, (\xi_\varepsilon)_{t,s})$, $s \in [0, t]$ can be defined on the space

$$\hat{W}_j = C^{j,0}(\mathbb{R}^d \times [0, t] : \mathbb{R}^d) \times C^{j,0}(\mathbb{R}^d \times [0, t] : \mathbb{R}^{d+1}) \times C^{j,0}(\mathbb{R}^{d+1} \times [0, t] : \mathbb{R}^{d+1}),$$

where $j \leq k$. We denote the law by R_ε^j. The typical element of $\hat{W}_j$ is denoted by $(\hat{X}, \hat{F}, (\hat{\phi}, \hat{\xi}))$.

By the previous lemma, the laws $\{R_\varepsilon^j : \varepsilon > 0\}$ are tight if $j \leq k - 1$ (see the proof of Theorem 5.4.1). The local characteristics of $(X^1 + F_\varepsilon^1, \ldots, X^d + F_\varepsilon^d, F_\varepsilon^{d+1})$ satisfy Conditions $(C.2)_k$ and $(C.4)_k$ of Chapter 5 by Condition $(F.2)_k$. Therefore we have the following lemma by Theorem 5.4.1.

Lemma 6.4.4 *Assume Conditions* $(F.1)_k$, $(F.2)_k$, $k \geq 2$. *Then for each* $j \leq k - 1$, *the family of measures* $\{R_\varepsilon^j : \varepsilon > 0\}$ *converges weakly as* $\varepsilon \to 0$. *Concerning the limit measure* R_0^j, $(\hat{X}, \hat{F}, (\hat{\phi}, \hat{\xi}))$ *satisfies the following properties.*

(i) $\hat{X}(s)$ *is a* C^k*-Brownian motion with local characteristic* $(a, -c, t)$.

(ii) $\hat{F}(s)$ *is a* C^k*-Brownian motion with local characteristic* $(A, \bar{f}, t)$ *independent of* $\hat{X}$.

(iii) $(\hat{\phi}, \hat{\xi})$ *satisfies the backward Stratonovich stochastic differential equation*

$$\hat{\phi}_s^i(x) = x^i + \int_s^t \hat{X}^i(\hat{\phi}_r(x), \circ \hat{d}r) + \int_s^t \hat{F}^i(\hat{\phi}_r(x), \circ \hat{d}r), \qquad i = 1, \ldots, d, \tag{17}$$

$$\hat{\xi}_s(x, u) = u + \int_s^t \hat{\xi}_r(x, u) \hat{F}^{d+1}(\hat{\phi}_r(x), \circ \hat{d}r). \tag{18}$$

If the random fields $\{F_\varepsilon(x, t) : \varepsilon > 0\}$ converge to a C^k-valued Brownian motion strongly, then the associated backward stochastic flows converge strongly. In fact we have the following by Theorem 5.4.3.

Lemma 6.4.5 *Assume Conditions* $(F.1)_k$ *and* $(F.2)_k$, $k \geq 2$. *Assume further that* $\{F_\varepsilon(x, t) : \varepsilon > 0\}$ *are defined on the same probability space and* $F_\varepsilon(x, t)$ *converge to* $F(x, t)$ *a.s. for each* x. *Then we have the following for each* $p > 3$:

(i) $E[\sup_{|x| \leq N, t} |D^\alpha F_\varepsilon(x, t) - D^\alpha F(x, t)|^p] \to 0$ *if* $|\alpha| \leq k - 1$.

(ii) *let* $(\varphi_{t,s}, \xi_{t,s})$ *be the backward stochastic flow defined by the stochastic differential equations* (17) *and* (18) *replacing* $\hat{F}^i$ *by* F^i. *Then for each* $N > 0$ *and* $t \in [0, T]$,

$$E\left[\sup_{|x| \leq N, s \leq t} |D^\alpha(\varphi_\varepsilon)_{t,s}(x) - D^\alpha \varphi_{t,s}(x)|^p\right] \to 0, \tag{19}$$

$$E\left[\sup_{|x| \leq N, s \leq t} |D_x^\alpha(\xi_\varepsilon)_{t,s}(x, 1) - D_x^\alpha \xi_{t,s}(x, 1)|^p\right] \to 0, \tag{20}$$

hold if $|\alpha| \leq k - 1$. □

We now proceed to the proof of the convergence of the solutions $\{u_\varepsilon(x, t)\}$. It is necessary to show the tightness of $\{u_\varepsilon(x, t) : \varepsilon > 0\}$ in the space $C^{j,0}(\mathbb{R}^d \times [0, T] : \mathbb{R}^1)$ where $j \leq k - 2$. For this we need the uniform L^p-estimate of $u_\varepsilon(x, t)$.

Lemma 6.4.6 *For each $p > 3$ and positive N, there exists a positive constant K such that for every ε and $|\alpha| \leq k - 1$*

$$E[|D_x^\alpha u_\varepsilon(x, t)|^p] \leq K, \tag{21}$$

$$E[|D_x^\alpha u_\varepsilon(x, t) - D_x^\alpha u_\varepsilon(x, t')|^p] \leq K|t - t'|^{2-(3/p)} \tag{22}$$

hold for any $t, t', |x| \leq N$.

Proof Let us observe the representation of $u_\varepsilon(x, t)$ given by (14). We have for any α with $|\alpha| \leq k$

$$E_Q\left[\sup_{|x| \leq N} |D_x^\alpha\{(\xi_\varepsilon)_{t,0}(x, 1) f((\varphi_\varepsilon)_{t,0}(x))\}|^p\right] < \infty$$

almost surely by Lemma 6.4.3 and Kolmogorov's theorem (Theorem 1.4.1). Hence we can interchange the order of the derivation D_x^α and the integration E_Q in equality (14). Then we find that $u_\varepsilon(x, t)$ is a continuous C^k-process and satisfies

$$D_x^\alpha u_\varepsilon(x, t) = E_Q[D_x^\alpha\{(\xi_\varepsilon)_{t,0}(x, 1) f((\varphi_\varepsilon)_{t,0}(x))\}] \tag{23}$$

for any x, t if $|\alpha| \leq k$. By Lemma 6.4.3, the integrands of the above are uniformly L^p-bounded with respect to $P \times Q$ for any $p > 1$. Therefore $D_x^\alpha u_\varepsilon(x, t)$ is uniformly L^p-bounded with respect to P for any $p > 1$. This proves (21). Further the estimate (22) follows from Lemma 6.4.3 and equation (23) immediately. □

We are now able to prove Theorems 6.4.1 and 6.4.2. It is convenient to consider Theorem 6.4.2 first.

Proof of Theorem 6.4.2 The first assertion (7) has already been shown in Chapter 5. Let $(\varphi_{t,s}, \xi_{t,s})$ be the backward flow defined in Lemma 6.4.5. Set

$$u(x, t) = E_Q[\xi_{t,0}(x, 1) f(\varphi_{t,0}(x))]. \tag{24}$$

Then for each x and t, $u_\varepsilon(x, t)$ converges to $u(x, t)$ in L_p, since $((\varphi_\varepsilon)_{t,0}(x), (\xi_\varepsilon)_{t,0}(x, 1))$ are uniformly L^p-bounded and converge to $(\varphi_{t,0}(x), \xi_{t,0}(x, 1))$ in $L^p(P \times Q)$. Further since $\{u_\varepsilon(x, t) : \varepsilon > 0\}$ satisfy inequalities (21) and (22) we can show that the family of laws $\{P_\varepsilon^{k-2}\}$ of $\{(F_\varepsilon, u_\varepsilon)\}$ defined on W_{k-2} is tight, similarly to the proof of Theorem 5.4.1. Then we get the strong convergence of $u_\varepsilon(x, t)$ similarly to the proof of Theorem 5.2.8. The last assertion of the theorem is obvious since $u(x, t)$ of (24) satisfies the stochastic partial differential equation (3). □

Proof of Theorem 6.4.1 We can regard F_ε as a $W_{k-2} = C^{k-2,0}(\mathbb{R}^d \times [0, T] : \mathbb{R}^{d+1})$-valued random variable. Since they converge weakly, we can define W_{k-2}-valued random variables $\{F'_\varepsilon : \varepsilon > 0\}$ on a suitable probability $(\Omega', \mathscr{B}', P')$ such that for each ε F_ε and F'_ε have the same law and $\{F'_\varepsilon : \varepsilon > 0\}$ converges to F' a.s. P' (see Theorem 1.1.5). Let $u'_\varepsilon(x, t)$ be the solution of equation (1) associated with F'_ε and $u'(x, t)$ be the solution of equation (3) associated with $F'(x, t)$. Then $\{(F'_\varepsilon, u'_\varepsilon)\}$ converges to (F', u') strongly in the sense of Theorem 6.4.2. Since for each ε $(F_\varepsilon, u_\varepsilon)$ and $(F'_\varepsilon, u'_\varepsilon)$ have the same law, $\{(F_\varepsilon, u_\varepsilon)\}$ converges weakly and the limit law satisfies properties (i) and (ii) of Theorem 6.4.1 by Theorem 6.4.2. The proof is complete. □

Finally let us consider the case where $A^{ij}(x, y, t)$ in Condition $(F.2)_k$ is not symmetric. Then the assertions of Lemma 6.4.4 are valid if we make the following correction in equations (17) and (18).

$$\phi_s^i(x) = x + \int_s^t \hat{X}^i(\phi_r(x), \circ \hat{d}r) + \int_s^t \hat{F}^i(\phi_r(x), \circ \hat{d}r) + \int_s^t h^i(\phi_r(x), r)\, dr, \tag{25}$$

$$\xi_s(x, u) = u + \int_s^t \hat{F}^{d+1}(\phi_r(x), \circ \hat{d}r)\xi_r(x, u) + \int_s^t h^{d+1}(\phi_r(x), r)\xi_r(x, u)\, dr. \tag{26}$$

Since the limit $u(x, t)$ is represented by (24), it satisfies the stochastic partial differential equation (11).

Frequently used notation and assumptions

(1) *Seminorms* Let $\alpha = (\alpha_1, \ldots, \alpha_d)$ be a multi index of non-negative integers and $|\alpha| = \alpha_1 + \cdots + \alpha_d$. Set

$$D^\alpha f = D_x^\alpha f = \frac{\partial^{|\alpha|} f}{(\partial x^1)^{\alpha_1} \ldots (x^d)^{\alpha_d}}.$$

Let m, N be non-negative integers, $0 < \delta \le 1$ and $p > 1$. Let K be a compact subset of a domain $\mathbb{D}$ of $\mathbb{R}^d$ and $B_N = \{x \in \mathbb{R}^d; |x| \le N\}$.

$$\|f\|_{m;K} = \sup_{x \in K} \frac{|f(x)|}{(1 + |x|)} + \sum_{1 \le |\alpha| \le m} \sup_{x \in K} |D^\alpha f(x)|.$$

$$\|f\|_{m;N} = \|f\|_{m;B_N}, \qquad \|f\|_m = \|f\|_{m;\mathbb{D}}.$$

$$\|f\|_{m,p;N} = \left(\sum_{|\alpha| \le m} \int_{B_N} |D^\alpha f(x)|^p \, dx \right)^{1/p}.$$

$$\|f\|_{m+\delta;K} = \|f\|_{m;K} + \sum_{|\alpha| = m} \sup_{\substack{x, y \in K \\ x \ne y}} \frac{|D^\alpha f(x) - D^\alpha f(y)|}{|x - y|^\delta}.$$

$$\|f\|_{m+\delta} = \|f\|_{m+\delta;\mathbb{D}}.$$

$$\|g\|_{m;K}^\sim = \sup_{x, y \in K} \frac{|g(x, y)|}{(1 + |x|)(1 + |y|)} + \sum_{1 \le |\alpha| \le m} \sup_{x, y \in K} |D_x^\alpha D_y^\alpha g(x, y)|.$$

$$\|g\|_{\delta;K}^\sim = \sup_{\substack{x, y, x', y' \in K \\ x \ne x', y \ne y'}} \frac{|g(x, y) - g(x', y) - g(x, y') + g(x', y')|}{|x - x'|^\delta |y - y'|^\delta}.$$

$$\|g\|_{m+\delta;K}^\sim = \|_{m;K}^\sim + \sum_{|\alpha| = m} \|D_x^\alpha D_y^\alpha g\|_{\delta;K}^\sim.$$

$$\|F\|_{m;N}^\sim = \|F\|_{m,\infty;N}^\sim = \sup_{t \in [0,T]} \|F(t)\|_{m;N}.$$

$$\|F\|_{m,p;N}^\sim = \sup_{t \in [0,T]} \|F(t)\|_{m,p;N}.$$

(2) *Function spaces*

$C^m = C^{m,0} = \{f(x), x \in \mathbb{D}; m\text{-times continuously differentiable}\}$.

$C^{m,\delta} = \{f \in C^m; \|f\|_{m+\delta;K} < \infty$ for any compact subset K of $\mathbb{D}\}$.

$C_b^m = C_b^{m,0} = \{f \in C^m; \|f\|_m < \infty\}, \qquad C_b^{m,\delta} = \{f \in C^{m,\delta}; \|f\|_{m+\delta} < \infty\}$.

$\tilde{C}^m = \{g(x, y), x, y \in \mathbb{D}; m\text{-times continuously differentiable with respect to each } x \text{ and } y\}$.

$\tilde{C}^{m,\delta} = \{g \in \tilde{C}^m; \|g\|_{m+\delta;K}^\sim < \infty$ for any compact subset K of $\mathbb{D}\}$.

$\tilde{C}_b^m = \tilde{C}_b^{m,0} = \{g \in \tilde{C}^m; \|g\|_m^\sim < \infty\}, \qquad \tilde{C}_b^{m,\delta} = \{g \in \tilde{C}^{m,\delta}; \|g\|_{m+\delta}^\sim < \infty\}$.

$W_m = W_{m,\infty} = C([0, T]; C^m)$.

$\hat{W}_m = C([0, T]; G^m)$ where G^m is the group of C^m-diffeomorphisms.

$W_{m,p} = C([0, T]; H^{\mathrm{loc}}_{m,p})$ where $H^{\mathrm{loc}}_{m,p} = \{f : \mathbb{R}^d \to \mathbb{R}^d;\ \|f\|_{m,p;N} < \infty$ for all $N\}$.

class $C^{m,\delta}$: the set of all continuous functions $f(x, t)$ on $\mathbb{D} \times [0, T]$ such that $f(t) \equiv f(\cdot, t)$ belongs to $C^{m,\delta}$ for every t and satisfies $\int_0^T \|f(t)\|_{m+\delta;K}\,dt < \infty$ for any compact subset K of $\mathbb{D}$.

class $C_b^{m,\delta} = \{f \in$ class $C^{m,\delta};\ \int_0^T \|f(t)\|_{m+\delta}\,dt < \infty\}$.

class $C_{ub}^{m,\delta} = \{f \in$ class $C_b^{m,\delta};\ \|f(t)\|_{m+\delta}$ is bounded$\}$.

class $\tilde{C}^{m,\delta}$: the set of all continuous functions $g(x, y, t)$ on $\mathbb{D} \times \mathbb{D} \times [0, T]$ such that $g(t) \equiv g(\cdot, \cdot, t)$ belongs to $\tilde{C}^{k,\delta}$ for every t and satisfies $\int_0^T \|f(t)\|_{k+\delta;K}\,dt < \infty$.

The classes $\tilde{C}_b^{k,\delta}$ and $\tilde{C}_{ub}^{k,\delta}$ are defined similarly.

(3) *Local characteristic of a semimartingale* $(a(x, y, t), b(x, t), A_t)$.

class $B^{m,\delta}$: $\int_0^T \|a(t)\|^{\sim}_{m+\delta;K}\,dA_t < \infty$ and $\int_0^T \|b(t)\|_{m+\delta;K}\,dA_t < \infty$ a.s. for any compact subset K of $\mathbb{D}$.

class $B_b^{m,\delta}$: $\int_0^T \|a(t)\|^{\sim}_{m+\delta}\,dA_t < \infty$ and $\int_0^T \|b(t)\|_{m+\delta}\,dA_t < \infty$ a.s.

class $B_{ub}^{m,\delta}$: $\|a(t)\|^{\sim}_{m+\delta}$ and $\|b(t)\|_{m+\delta}$ are bounded.

(4) *Conditions*

Remarks on references

Chapter 1

Sections 1.1 and 1.2 are of standard character. The martingale theory stated without proof in Section 1.2 is due to Doob [26]. Also see Ikeda–Watanabe [49], Liptser–Shiryaev [91], Meyer [99] and others. More information on the Feller Markov process is found in Blumenthal–Getoor [13].

There are extensive studies on ergodic theory and recurrent potential theory for Markov processes, specially for those of discrete time parameter. Results in Section 1.3 are adapted from Revuz [112]. The recurrent process in the sense of Harris was introduced by Harris [43] for the Markov process with discrete parameter. Theorems 1.3.8 and 1.3.9 are known as Orey's theorems, see Orey [106]. Our discussion on recurrent potential theory is adapted from Neveu [104]. Theorem 1.3.5 is shown in H. Watanabe [124] under a slightly different assumption in the case where the state space is compact.

Kolmogorov's regularity and tightness criterion for stochastic processes stated in Section 1.4 is found in Prohorov [111]. The extension to random fields is due to Totoki [122]. We refer to Kono [71] for the proof of Theorem 1.4.1. Theorem 1.4.4 is developed from Čencov [18].

Chapter 2

The useful notion of the localmartingale given in Section 2.1 is due to Itô–Watanabe [62]. The quadratic variation of martingales was studied by Meyer [100]. Most of the material in Section 2.2 in adapted from Kunita [77].

The stochastic integral was defined by Itô [52] and was extended to what in this book we call the Itô integral by Kunita–Watanabe [81], Meyer [100] and Motoo–Watanabe [102]. Another stochastic integral was introduced by Stratonovich [116]. The relation between these two integrals (Theorem 2.3.5) is discussed in Itô [60].

The orthogonal basis of martingales (Theorems 2.3.8 and 2.3.9), Itô's formula (Theorem 2.3.11) and its application (Theorem 2.3.13) are due to

Kunita–Watanabe [81]. The proof of Burkholder's inequality (Theorem 2.3.12) is due to Getoor–Sharpe [37].

Chapter 3

The material in this chapter is an expansion of two sets of lecture notes, Kunita [77] and [79]. The regularity theorems for semimartingales with spatial parameter (Theorems 3.1.1–3.1.4) are adapted from the former.

The stochastic integral based on a *C*-valued Brownian motion (Section 3.2) and the stochastic differential equation based on such a Brownian motion (Section 3.4) were introduced by Le Jan [86]. Borkar [14] discusses other related stochastic integrals. Generalized Itô's formulas (Theorems 3.3.3 and 3.3.4) have been introduced by many authors in various cases. Ventzel [123] seems to be the first contribution to this problem. The formula is called Itô–Ventzel's formula in some texts. Much later Bismut [9] and the author [77] tackled this problem again. See also Sznitman [120]. The present form of the formula is due to Kunita [79]. The formulas for the differential calculus for stochastic integrals (Theorems 3.3.3 and 3.3.4) are in exactly the same form as the well known classical formula but here the rigorous proof is given first.

Chapter 4

Apart from the stochastic differential equation, the Brownian flow itself was first studied by Harris [44]. He introduced infinitesimal means and infinitesimal covariances in certain special Brownian flows called homogeneous and isotropic flows. The problem of finding random infinitesimal generators of Brownian flows was studied by Baxendale [5], Fujiwara–Kunita [35], Le Jan [86] and Le Jan–Watanabe [90]. In the second reference, stochastic flows with jumps called *Lévy flows* are studied. The results of Section 4.3 are closely related to the works by Baxendale [6], Darling [23] and Le Jan [88], [89].

The fact that the solutions of a stochastic differential equation define a stochastic flow of homeomorphisms or diffeomorphisms has been proved by various methods. One is to regard the equation as a stochastic differential equation defined on a Hilbert manifold consisting of diffeomorphisms and solve it directly to get the flow of diffeomorphisms. See Baxendale [4] and Elworthy [29], [31]. Another method is to approximate the equation by a sequence of stochastic ordinary equations and show that the associated sequence of the flows converges to the desired flow. See Bismut [10],

Ikeda–Watanabe [49] and Malliavin [93]. A further method which we adopt in this book appeared in Kunita [74] with the aid of Varadhan. See also Kunita [77], [79]. If the coefficients of a stochastic differential equation are not Lipschitz continuous, the solution of the equation may or may not define a flow of homeomorphisms. See Harris [45] and Yamada–Ogura [129].

The continuity and differentiability with respect to the parameter of solutions of stochastic differential equations was studied by Blagovescěnskii–Freidlin [12]. The existence of the weak derivatives is discussed in Gihman–Skorohod [40], [41]. The proof given in Section 4.6 is a revision of Kunita [77].

Most of the material in Section 4.7 is adapted from Kunita [77]. Theorem 4.7.8 was also proved by Elworthy [30] using a different method. Recently Taniguchi [121] showed that the diffeomorphic property of solutions of an equation defined on open subset of Euclidean space. Sections 4.8–4.9 are adapted from Kunita [77]. Theorem 4.9.7 was announced by S. Watanabe [126]. A formula similar to Corollary 4.9.8 is found in Bismut [10]. Theorem 4.9.10 in the case of the nilpotent Lie algebra is due to Yamato [130].

Chapter 5

Most of the material in Chapter 5 is adapted from Kunita [79]. Lemma 5.2.5 is an extension of Khasminskii [67]. The use of Sobolev's space for proving the tightness and regularity of infinite dimensional stochastic processes was proposed by Kushner [82]. The method using truncation is due to Kesten–Papanicolaou [66]. There are many works on the central limit theorems concerning stochastic ordinary differential equation. Borodin [15], Khasminskii [68], Papanicolaou–Kohler [108] and others. Theorem 5.6.1 is a refinement of these works in the sense that they converge as stochastic flows. The two mixing lemmas (Lemmas 5.6.2 and 5.6.3) are taken from Ibragimov–Linnik [46] and Kesten–Papanicolaou [66] respectively. Theorem 5.6.4 is related to the work by Papanicolaou–Stroock–Varadhan [110]. Theorem 5.7.1 includes the approximating theorems for stochastic differential equations studied by Bismut [10] and Ikeda–Watanabe [49]. A support theorem of stochastic flow similar to Theorem 5.7.6 is stated in Ikeda–Watanabe [49].

Chapter 6

The first order nonlinear stochastic partial differential equation was studied by Kunita [75]. The linear first order equation was studied by Ogawa [105] and Funaki [36] before the above paper. The second order equation was studied by many authors. The works by Krylov–Rozovsky [73] and Pardoux [110] use the method of partial differential equations. The more probabilistic method using stochastic flows adopted in this book is due to Kunita [76]. See also Davis [24].

There are many studies on nonlinear filtering problems. Theorem 6.3.1 is due to Kallianpur–Striebel [64]. Theorem 6.3.2 is due to Zakai [131]. Theorem 6.3.3 is due to Fujisaki–Kallianpur–Kunita [34]. The nonlinear stochastic partial differential equation of Theorem 6.3.3 is called the FKK equation in some texts. The robustness of the nonlinear filters is discussed by Clark [21] and Davis [24]. The related problem is discussed in Chaleyat-Maurel and Michel [19]. Theorem 6.3.5 can be found in Kalman–Bucy [65].

A somewhat different approach to limit theorems for stochastic partial differential equations can be found in Bouc–Pardoux [16], Kushner–Huang [83] and H. Watanabe [125].

References

[1] Adams, R. A. 1975, *Sobolev spaces*, Academic Press, New York.

[2] Arnold, L. 1974, *Stochastic differential equations: Theory and Applications*, John Wiley–Interscience, London.

[3] Arnold, L. & Wihstutz, V. 1986, Lyapunov exponents: a survey, Lyapunov exponents, *Lect. Notes Math.* **1186**, 1–26.

[4] Baxendale, P. 1980, Wiener processes on manifolds of maps, *Proc. Royal Soc. Edinburgh* **87A**, 127–52.

[5] Baxendale, P. 1984, Brownian motions in the diffeomorphism group *I*, *Compositio Mathematica* **53**, 19–50.

[6] Baxendale, P. 1986, The Lyapunov spectrum of a stochastic flow of diffeomorphisms, Lyapunov exponents, *Lect. Notes in Math.* **1186**, 322–37.

[7] Baxendale, P. & Harris, T. E. 1986, Isotropic stochastic flows, *Ann. Prob.* **14**, 1155–79.

[8] Billingsley, P. 1968, *Convergence of probability measures*, John Wiley, New York.

[9] Bismut, J. M. 1981, A generalized formula of Ito and some other properties of stochastic flows, *Z. Wahrscheinlichkeitstheorie. Verw. Geb.* **55**, 331–50.

[10] Bismut, J. M. 1981, Mécanique aléatoire, *Lect. Notes Math.* **866**.

[11] Bismut, J. M. & Michel, D. 1981, Diffusions conditionnelles, I. Hypoellipticité partielle, *J. Functional Analysis* **44**, 174–211.

[12] Blagovescĕnskii, Yu. N. & Freidlin, M. I. 1961, Certain properties of diffusion processes depending on a parameter, *Soviet Math. Dokl.* **2**, 633–6.

[13] Blumenthal, R. M. & Getoor, R. K. 1968, *Markov processes and potential theory*, Academic Press, New York.

[14] Borkar, V. S. 1984, Evolution of interacting particles in a Brownian medium, *Stochastics* **14**, 33–79.

[15] Borodin, A. N. 1977, A limit theorem for solutions of differential equations with random right-hand side, *Theor. Probab. Appl.* **22**, 482–97.

[16] Bouc, R. & Pardoux, E. 1984, Asymptotic analysis of PDE's with wide-band noise disturbances and expansion of the moments, *Stochastic Analysis and Applications* **2**, 369–422.

[17] Carverhill, A. 1985, Flows of stochastic dynamical systems; ergodic theory, *Stochastics* **14**, 273–317.

[18] Čencov, N. N. 1958, Limit theorems for some classes of random functions, *Proc. All-Union Conf. Theory Prob. and Math. Statist. Elevan*, (Selected *Transl. Math. Statist. and Prob.* **9**(1970), 37–42).

[19] Chaleyat–Maurel, M. & Michel, D. 1986, Une propriété de continuité en filtrage nonlinéaire, *Stochastics* **19**, 11–40.

[20] Chen, K. T. 1962, Decomposition of differential equations, *Math. Ann.* **146**, 263–78.

[21] Clark, J. M. C. 1978, The design of robust approximations to the stochastic differential equations of nonlinear filtering, Communication Systems and Random Process Theory, *NATO Advanced Study Institute Series*, Sijthoff and Noordhoff, Alphen aan den Rijn.

[22] Courant, R. & Hilbert, D. 1962, *Methods of mathematical physics*, II, Interscience, New York.

[23] Darling, R. W. R. 1988, Ergodicity of a measure-valued Markov chain induced by random transformations, *Probab. Th. Rel. Fields*, **77**, 211–29.

[24] Davis, M. H. A. 1986, Nonlinear filtering and stochastic flows, *Proc. Intern. Congr. Math.* Berkeley, 1000–10.

[25] Dellacherie, C. 1974, Intégrals stochastiques par rapport aux processus de Wiener ou de Poisson, Séminaire de Probab. VIII, *Lect. Notes Math.* **381**, 25–6.

[26] Doob, J. L. 1967, *Stochastic processes*, John Wiley, New York.

[27] Dowell, R. M. 1980, Differentiable approximations to Brownian motion on manifolds, unpublished Ph.D. thesis, University of Warwick, Coventry.

[28] Dynkin, E. B. 1965, *Markov processes*, I, II, Springer Verlag, Berlin.

[29] Elworthy, K. D. 1978, 'Stochastic dynamical systems and their flows', in A. Friedman and M. Pinsky, (eds.) *Stochastic Analysis*, 79–95, Academic Press, New York.

[30] Elworthy, K. D. 1982, Stochastic flows and the C_0-diffusion property, *Stochastics* **6**, 233–8.

[31] Elworthy, K. D. 1982, Stochastic differential equations on manifolds, *Lond. Math. Soc. Lect. Note Ser.* 70, Cambridge University Press.

[32] Friedman, A. 1964, *Partial differential equations of parabolic type*, Prentice–Hall, Englewood, New Jersey.

[33] Friedman, A. 1975, 1976, *Stochastic differential equation and applications*, I, II, Academic Press, New York.

[34] Fujisaki, M., Kallianpur, G. & Kunita, H. 1972, Stochastic differential equations for the nonlinear filtering problem, *Osaka J. Math.* **9**, 19–40.

[35] Fujiwara, T. & Kunita, H. 1985, Stochastic differential equations of jump type and Lévy processes in diffeomorphisms group, *J. Math. Kyoto Univ.* **25**, 71–106.

[36] Funaki, T. 1979, Construction of a solution of random transport equation with boundary condition, *J. Math. Soc. Japan* **31**, 719–44.

[37] Getoor, R. K. & Sharpe, M. J. 1972, Conformal martingales, *Invent. Math.* **16**, 271–308.

[38] Gihman, I. I. 1947, A method of constructing random processes, *Dokl. Akad. Nauk. SSSR*, **58**, 961–4, (in Russian).

[39] Gihman, I. I. 1950, Certain differential equations with random functions, *Ukr. Mat. Zh.* **2**, No. 3, 45–69, (in Russian).

[40] Gihman, I. I. & Skorohod, A. V. 1972, *Stochastic differential equations*, Springer Verlag, Berlin.

[41] Gihman, I. I. & Skorohod, A. V. 1979, *The theory of stochastic processes* III, Springer Verlag, Berlin.

[42] Girsanov, I. V. 1960, On transforming a certain class of stochastic processes by absolutely continuous substitution of measures, *Theory Prob. Appl.* **5**, 285–301.

[43] Harris, T. E. 1956, The existence of stationary measures for certain Markov processes, *Proc. 3rd Berkeley Symp. on Mathematical Statistics and Probability*, **Vol 2**, 113–24.

[44] Harris, T. E. 1981, Brownian motions on the homeomorphisms of the plane, *Ann. Probab.* **9**, 232–54.

[45] Harris, T. E. 1984, Coalescing and noncoalescing stochastic flows in R_1, *Stochastic Proc. Appl.* **17**, 187–210.

[46] Ibragimov, I. A. & Linnik, Yu. V. 1971, *Independent and stationary sequences of random variables*, Wolters–Noordhoff, Groningen.

[47] Ichihara, K. & Kunita, H. 1974, A classification of the second order degenerate elliptic operators and its probabilistic characterization, *Z. Wahrscheinlichkeitstheorie Verw. Geb.* **30**, 235–54.

[48] Ikeda, N., Nakao, S. & Yamato, Y. 1977, A class of approximations of Brownian motion, *RIMS, Kyoto Univ.* **13**, 285–300.

[49] Ikeda, N. & Watanabe, S. 1981, *Stochastic differential equations and diffusion processes*, North Holland-Kodansha, Tokyo.

[50] Ikeda, N. & Watanabe, S. 1984, Stochastic flow of diffeomorphisms, in M. Pinsky & Marcel Dekker (eds.) *Stochastic Analysis and Applications*, 179–98.

[51] Itô, K. 1942, *Differential equations determining Markov processes*, Zenkoku Shijo Danwakai. (English translation, *Kiyoshi Itô, Selected papers*, Springer Verlag, Berlin, 1987.)
[52] Itô, K. 1944, Stochastic integral, *Proc. Japan. Acad. Tokyo*, **20**, 519–24.
[53] Itô, K. 1946, On a stochastic integral equation, *Proc. Japan Acad. Tokyo*, **22**, 32–5.
[54] Itô, K. 1950, Stochastic differential equations in a differentiable manifold, *Nagoya Math. J.* **1**, 35–47.
[55] Itô, K. 1951, On stochastic differential equations, *Memoirs of American Math. Soc.*, **4**.
[56] Itô, K. 1951, On a formula concerning stochastic differentials, *Nagoya Math. J.* **3**, 55–65.
[57] Itô, K. 1953, Stochastic differential equations in a differential manifold (2), *Mem. Coll. Sci. Univ. Kyoto Math.* **28**, 81–5.
[58] Itô, K. 1961, *Lectures on stochastic processes*, Tata Institute of Fundamental Research, Bombay.
[59] Itô, K. 1963, The Brownian motion and tensor fields on Riemannian manifold, *Proc. Intern. Congr. Math. Stockholm*, 536–9.
[60] Itô, K. 1965, Stochastic differentials, *Appl. Math. Opt.* **1**, 374–81.
[61] Itô, K. & McKean, H. P. Jr. 1965, *Diffusion processes and their sample paths*, Springer Verlag, Berlin.
[62] Itô, K. & Watanabe, S. 1965, Transformation of Markov processes by multiplicative functionals, *Ann. Inst. Fourier, Grenoble* 15, **1**, 13–30.
[63] Kallianpur, G. 1980, *Stochastic filtering theory*, Springer Verlag, New York.
[64] Kallianpur, G. & Striebel, C. 1968, Estimation of stochastic system: Arbitrary system process with additive white noise observation errors, *Ann. Math. Statist.* **39**, 785–801.
[65] Kalman, R. E. & Bucy, R. S. 1961, New results in linear filtering and prediction theory, *Trans. ASME Ser. D, J. Basic Eng.* **83**, 95–108.
[66] Kesten, H. & Papanicolaou, G. C. 1979, A limit theorem for turbulent diffusion, *Commun. math. Phys.* **65**, 97–128.
[67] Khasminskii, R. Z. 1963, Principle of averaging for parabolic and elliptic differential equations and for Markov processes with small diffusion, *Theory Probab. Appl.* **8**, 1–21.
[68] Khasminskii, R. Z. 1966, A limit theorem for the solutions of differential equations with random right-hand sides, *Theory Probab. Appl.* **11**, 390–406.
[69] Kifer, Y. 1986, *Ergodic theory of random transformations*, Birkhauser, Boston.
[70] Kobayashi, S. & Nomizu, K. 1963, *Foundations of differential geometry* I, John Wiley, New York.
[71] Kono, N. 1979, Real variable lemmas and their applications to sample properties of stochastic processes, *J. Math. Kyoto Univ.* **19**, 413–33.
[72] Krylov, N. V. & Rozovsky, B. L. 1977, On the Cauchy problem for linear stochastic partial differential equations, *Izv. Akad. Nauk. SSSR*, **41**, 1329–47. (English translation *Math. USSR Izvestija* **11**(1977), 1267–84.)
[73] Krylov, N. V. & Rozovsky, B. L. 1982, Stochastic partial differential equations and diffusion processes, *Usp. Mat. Nauk* **37**:**6**, 75–95. (English translation *Russian Math. Survey* **37**:**6**(1982), 81–105.)
[74] Kunita, H. 1981, On the decomposition of solutions of stochastic differential equations, Stochastic Integrals, *Lect. Notes Math.* **851**, 213–55.
[75] Kunita, H. 1982, Stochastic partial differential equations connected with non-linear filtering, Nonlinear filtering and stochastic control, *Lect. Notes Math.* **972**, 100–69.
[76] Kunita, H. 1984, First order stochastic partial differential equations, in K. Itô (ed.), *Stochastic Analysis*, Kinokuniya, Tokyo, 249–69.
[77] Kunita, H. 1984, Stochastic differential equations and stochastic flows of diffeomorphisms, Ecole d'été de Probabilités de Saint-Flour 12, 1982 *Lect. Notes Math.* **1097**, 143–303.
[78] Kunita, H. 1986, Convergence of stochastic flows connected with stochastic ordinary differential equations, *Stochastics* **17**, 215–51.

[79] Kunita, H. 1986, *Lectures on stochastic flows and applications*, Tata Institute of Fundamental Research, Bombay, Springer Verlag, Berlin.

[80] Kunita, H. 1986, Stochastic flows and stochastic partial differential equations, *Proc. Intern. Congr. Math. Berkeley*, 1021–31.

[81] Kunita, H. & Watanabe, S. 1967, On square integrable martingales, *Nagoya Math. J.* **30**, 209–45.

[82] Kushner, H. 1969, An application of the Sobolev imbedding theorems to criteria for the continuity of the processes with a vector parameter, *Ann. of Math. Statist.* **40**, 517–26.

[83] Kushner, H. J. & Huang, H. 1985, Limits for parabolic partial differential equations with wide band stochastic coefficients and an application to filtering theory, *Stochastics* **14**, 115–48.

[84] Kushner, H. J. & Huang, H. 1985, Weak convergence and approximations for partial differential equations with stochastic coefficients, *Stochastics* **15**, 209–45.

[85] Kusuoka, S. & Stroock, D. 1984, The partial Malliavin calculus and its application to non-linear filtering, *Stochastics* **12**, 83–142.

[86] Le Jan, Y. 1982, Flots de diffusions dans $\mathbb{R}^d$, *C. R. Acad. Sci., Paris, Ser.* I **294**, 697–9.

[87] Le Jan, Y. 1984, Equilibre et exposants de Lyapounov de certains flots browniens, *C. R. Acad. Sci. Paris, Ser.* I **298**, 361–4.

[88] Le Jan, Y. 1985, On isotropic Brownian motions, *Z. Wahrscheinlichkeitstheorie Verw. Geb.* **70**, 609–20.

[89] Le Jan, Y. 1987, Equilibre statistique pour les produits de difféomorphismes aléatoires indépendants, *Ann. Inst. Henri Poincaré, Probabilités et Statistiques*, **23**, 111–20.

[90] Le Jan, Y. & Watanabe, S. 1984, Stochastic flows of diffeomorphisms, in K. Itô (ed.) *Stochastic analysis*, Kinokuniya, Tokyo, 307–32.

[91] Liptser, R. S. & Shiryaev, A. N. 1977, *Statistics of Random Processes* I, Springer Verlag, New York.

[92] Malliavin, P. 1977, Un principe de transfert et son application au calcul des variations, *C. R. Acad. Sci. Paris, Ser. A* **284**, 187–9.

[93] Malliavin, P. 1978, Stochastic calculus of variation and hypoelliptic operators, in K. Itô (ed.) *Proc. Int. Symp. SDE Kyoto* 1976 Kinokuniya, Tokyo, 195–263.

[94] Malliavin, P. 1978, *Géométrie différentielle stochastique*, Les Presses de l'Université de Montréal, Montréal.

[95] Matsumoto, H. 1986, Convergence of driven flows of diffeomorphisms, *Stochastics* **18**, 343–55.

[96] Matsushima, Y. 1972, *Differentiable manifolds*, Marcel Dekker, New York.

[97] McKean, H. P. Jr. 1969, *Stochastic integrals*, Academic Press, New York.

[98] Métivier, M. 1982, *Semimartingales: a course on stochastic processes*, Walter de Gruyter, Berlin.

[99] Meyer, P. A. 1966, *Probability and potentials*, Blaisdell, Waltham.

[100] Meyer, P. A. 1967, Integrales stochastiques I–IV, Séminaire de Probab. I, *Lect. Notes Math.* **39**, 72–162.

[101] Meyer, P. A. 1981, Flot d'une equation différentielle stochastique, Séminaire de Probab. XV, *Lect. Notes. Math.* **850**, 103–17.

[102] Motoo, M. & Watanabe, S. 1965, On a class of additive functionals of Markov processes, *J. Math. Kyoto Univ.* **4**, 429–69.

[103] Neveu, J. 1972, *Martingales à temps discret*, Masson et Cie, Paris.

[104] Neveu, J. 1972, Potential markovien récurrent des chaines de Harris, *Ann. Inst. Fourier, Grenoble* **22,2**, 85–130.

[105] Ogawa, S. 1973, A partial differential equation with the white noise as a coefficient, *Z. Wahrscheinlichkeitstheorie Verw. Geb.* **28**, 53–71.

[106] Orey, S. 1971, *Lecture notes on limit theorems for Markov chain transition probabilities*, Van Nostrand, London.

[107] Palais, R. S. 1957, A global formulation of the Lie theory of transformation groups, *Memoirs of Amer. Math. Soc.* **22**.

[108] Papanicolaou, G. C. & Kohler, W. 1974, Asymptotic theory of mixing stochastic ordinary differential equations, Commun. *Pure Appl. Math.* **27**, 614–68.

[109] Papanicolaou, G. C., Stroock, D. W. & Varadhan, S. R. S. 1977, Martingale approach to some limit theorems, 1976 Duke Turbulence Conf. *Duke Univ. Math. Ser.* III, Chapter VI, 1–120.

[110] Pardoux, E. 1979, Stochastic partial differential equations and filtering of diffusion processes, *Stochastics* **3**, 127–67.

[111] Prohorov, Yu. V. 1956, Convergence of random processes and limit theorems in probability theory, *Theor. Prob. Appl.* **1**, 157–214.

[112] Revuz, D. 1975, *Markov chains*, North-Holland, Amsterdam.

[113] Rogers, L. C. G. & Williams, D. 1987, *Diffusions, Markov Processes and Martingales, vol. 2: Itô Calculus*, Wiley, New York.

[114] Rozovsky, B. L. 1973, On the Itô–Ventzel formula, *Vestnik of Moscow Univ.* **1**, 26–32 (in Russian).

[115] Shu, J. G. 1982, On the mollifier approximation for solutions of stochastic differential equations, *J. Math. Kyoto Univ.* **22**, 243–54.

[116] Stratonovich, R. L. 1968, *Conditional Markov processes and their application to the theory of optimal control*, Amer. Elsevier, New York.

[117] Stroock, D. W. 1982, *Lectures on topics in stochastic differential equations*, Tata Institute of Fundamental Research, Bombay, Springer Verlag, Berlin.

[118] Stroock, D. W. & Varadhan, S. R. S. 1972, On the support of diffusion processes with applications to the strong maximum principle, *Proc. 6th Berkeley Symp. Math. Statist. Probab.* III, 333–59, Univ. California Press, Berkeley.

[119] Stroock, D. W. & Varadhan, S. R. S. 1979, *Multidimensional diffusion processes*, Springer Verlag, Berlin.

[120] Sznitman, A. S. 1982, Martingales dépendent d'un paramétre: une formule d'Ito, *Z. Wahrscheinlichkeitstheorie Verw. Geb.* **60**, 41–70.

[121] Taniguchi, S. 1989, Stochastic flows of diffeomorphisms on an open set in R^n, *Stochastics and Stochastic Reports* **28**, 301–15.

[122] Totoki, H. 1961, A method of construction of measures on function spaces and its applications to stochastic processes, *Mem. Fac. Sci. Kyushu Univ. Ser. A. Math.* **15**, 178–90.

[123] Ventzel', A. D. 1965, On equations of the theory of conditional Markov processes, *Theory Probab. Appl.* **10**, 357–61.

[124] Watanabe, H. 1964, Potential operator of a recurrent strong Feller process in the strict sense and boundary value problem, *J. Math. Soc. Japan* **16**, 83–95.

[125] Watanabe, H. 1988, Averaging and fluctuations for parabolic equations with rapidly oscillating random coefficients, *Probab. Th. Rel. Fields* **77**, 359–78.

[126] Watanabe, S. 1980, Flow of diffeomorphisms defined by stochastic differential equations on manifolds and their differentials and variations, *Suriken Kokyuroku* **391**, 1–23, (in Japanese).

[127] Watanabe, S. 1983, 'Stochastic flows of diffeomorphisms', Proc. Fourth Japan–USSR Symp. Probab. Theory, *Lect. Notes Math.* **1021**, 699–708.

[128] Wong, E. & Zakai, M. 1965, On the relation between ordinary and stochastic differential equations, *Int. J. Eng. Sci.* **3**, 213–29.

[129] Yamada, T. & Ogura, Y. 1981, On the strong comparison theorems for solutions of stochastic differential equations, *Z. Wahrscheinlichkeitstheorie Verw. Geb.* **56**, 3–19.

[130] Yamato, Y. 1979, Stochastic differential equations and nilpotent Lie algebras, *Z. Wahrscheinlichkeitstheorie. Verw. Geb.* **47**, 213–29.

[131] Zakai, M. 1969, On the optimal filtering of diffusion processes, *Z. Wahrscheinlichkeitstheorie Verw. Geb.* **11**, 230–43.

Index

Printed in the United States
By Bookmasters